MICROSYSTEM DESIGN

ABOUT THE COVER

The author gratefully acknowledges the cover photograph by Felice Frankel, Artist in Residence at the Massachusetts Institute of Technology and coauthor of *On the Surface of Things: Images of the Extraordinary in Science.*

This particular image, taken with Nomarski optics, presents a wafer-bonded piezoresistive pressure sensor. It is fabricated in the sealed-cavity process developed by Professor Martin Schmidt of the Massachusetts Institute of Technology with his graduate students, Lalitha Parameswaran and Charles Hsu. The piezoresistors are clearly visible, and the slight contrast across the central diaphragm region shows that the diaphragm is actually slightly bent by the pressure difference between the ambient and the sealed cavity beneath.

MICROSYSTEM DESIGN

Stephen D. Senturia
Massachusetts Institute of Technology

KLUWER ACADEMIC PUBLISHERS
Boston / Dordrecht / London

Distributors for North, Central and South America:
Kluwer Academic Publishers
101 Philip Drive
Assinippi Park
Norwell, Massachusetts 02061 USA
Telephone (781) 871-6600
Fax (781) 871-6528
E-Mail <kluwer@wkap.com>

Distributors for all other countries:
Kluwer Academic Publishers Group
Distribution Centre
Post Office Box 322
3300 AH Dordrecht, THE NETHERLANDS
Telephone 31 78 6392 392
Fax 31 78 6546 474
E-Mail <orderdept@wkap.nl>
          Electronic Services <http://www.wkap.nl>

 Library of Congress Cataloging-in-Publication Data

Senturia, Stephen D., 1940-
   Microsystem design / Stephen D. Senturia
      p. cm.
   Includes bibliographical references and index.
   ISBN 0-7923-7246-8 (alk. paper)
   1. Microelectromechanical systems--Design and construction. 2. System design.
   3. Microelectronics. I. Title.

TK7875.S46 2000
621.381--dc21   00-048768

Printed on acid-free paper.
Printed in the United States of America

To Bart Weller,
for his generosity, and
To Angelika Weller,
for her indomitable spirit

Contents

Foreword

It is a real pleasure to write the Foreword for this book, both because I have known and respected its author for many years and because I expect this book's publication will mark an important milestone in the continuing worldwide development of microsystems. By bringing together all aspects of microsystem design, it can be expected to facilitate the training of not only a new generation of engineers, but perhaps a whole new type of engineer – one capable of addressing the complex range of problems involved in reducing entire systems to the micro- and nano-domains. This book breaks down disciplinary barriers to set the stage for systems we do not even dream of today.

Microsystems have a long history, dating back to the earliest days of microelectronics. While integrated circuits developed in the early 1960s, a number of laboratories worked to use the same technology base to form integrated sensors. The idea was to reduce cost and perhaps put the sensors and circuits together on the same chip. By the late-60s, integrated MOS-photodiode arrays had been developed for visible imaging, and silicon etching was being used to create thin diaphragms that could convert pressure into an electrical signal. By 1970, selective anisotropic etching was being used for diaphragm formation, retaining a thick silicon rim to absorb package-induced stresses. Impurity- and electrochemically-based etch-stops soon emerged, and "bulk micromachining" came into its own. Wafer bonding (especially the electrostatic silicon-glass bond) added additional capability and was applied to many structures, including efforts to integrate an entire gas chromatography system on a single wafer. Many of these activities took place in university laboratories, where sensor research could make important contributions that complemented industry. The work was carried out primarily by electrical engineers trained in microelectronics, who often struggled to understand the mechanical aspects of their devices. There were no textbooks to lead the sensor designer through all the relevant areas, information on which was widely scattered.

The demand for improved automotive fuel economy and reduced emissions in the late 70s took integrated pressure sensors into high-volume production. By the early 1980s, pressure sensors with on-chip readout electronics were also in production and bulk micromachining was being applied to flowmeters, accelerometers, inkjet print heads, and other devices. At this point, the field of "integrated sensors" began to organize itself, establishing independent meetings to complement special sessions at microelectronics conferences.

Surface micromachining came on the scene in the mid-80s and quickly led to applications in accelerometers, pressure sensors, and other electromechanical structures. Microactuators became the focus for considerable work, and the notion of putting entire closed-loop systems on a chip became a real goal. The field needed an acronym, and "MEMS" (MicroElectroMechanical Systems) was gradually adopted, in spite of the fact that many of the devices were not really mechanical. The term "microsystems" also became increasingly common in referring to the integration of sensors, actuators, and signal-processing electronics on a common (but not necessarily monolithic) substrate. The field at this point began to see the long-needed entry of mechanical engineers, but it was still centered in academia. And there were still few, if any, courses in sensors or MEMS. It was a research focus for people trained mostly in electrical engineering and physics, and the mechanics, chemistry, or materials information needed as an adjunct to microelectronics had to be dug out by people not primarily trained in those fields. There were still no textbooks on microsystems.

Beginning in the late 80s, MEMS received increasing emphasis worldwide. In the US, the Emerging Technologies Program of the National Science Foundation selected it as a focal point, and in 1992 the Defense Advanced Research Projects Agency (DARPA) did as well; suddenly funding went up dramatically and so did the number of players. Similar investments were being made in Europe and Asia, so that after more than 25 years the field finally reached critical mass. The 1990s saw MEMS-based inkjet print heads, pressure sensors, flowmeters, accelerometers, gyros, uncooled infrared imagers, and optical projection displays all enter production. Emphasis on full microsystems increased. More advanced devices are now being developed, including DNA analyzers, integrated gas chromatography systems, and miniature mass spectrometers.

Microsystem design now cuts across most disciplines in engineering and is the focus for courses, and degree programs, at many major universities. These courses, which must be open to individuals with a wide range of backgrounds, need material covering an equally wide range of subjects in a coherent, unified way. This book will be a major help in meeting these challenges. The author has been a principal figure guiding the development of microsystems for more than two decades, through both his research contributions and his leadership of the conferences and journals that have defined the field. The book itself is comprehensive, a friendly tutor for those just entering the field and a resource for long-time practitioners. It is all here — the technology, the modeling, the analysis methods, and the structures. The important principles of materials, mechanics, and fluidics are here so that the designer can understand and predict these aspects of advanced structures. And the electronics is here so that he or she can also understand the signal readout and processing challenges and the use of on-chip feedback control. Noise is covered so that the basic limitations to accuracy and resolution can likewise be anticipated. Finally, case studies tie everything together, highlighting important devices of current interest.

Because this text brings together all of the topics required for microsystem design, it will both accelerate development of the field and give rise to a new type of engineer, the microsystems engineer, who can combine knowledge from many disciplines to solve problems at the micro- and nano-levels. Microsystems are now much more than a specialized sub-area of microelectronics; in substantial measure, they are the key to its future, forming the front-ends of global information technology networks and bridges from microelectronics to biotechnology and the cellular world.

Finding solutions to many of today's problems will require microsystems engineers to develop devices that we cannot yet imagine at a scale that we cannot see. Certainly, many challenges lie ahead, but the basic principles presented here for meeting them will remain valid. And as the physicist Richard Feynman said in proposing microsystems over forty years ago, there really *is* plenty of room at the bottom!

Kensall D. Wise
The University of Michigan Ann Arbor, MI

Preface

The goal of this book is to bring together into one accessible text the fundamentals of the many disciplines needed by today's microsystems engineer. The subject matter is wide ranging: microfabrication, mechanics, heat flow, electronics, noise, and dynamics of systems, with and without feedback. And because it is very difficult to enunciate principles of "good design" in the abstract, the book is organized around a set of Case Studies that are based on real products, or, where appropriately well-documented products could not be found, on thoroughly published prototype work.

This book had its roots in a graduate course on "Design and Fabrication of MEMS Devices" which my colleague Prof. Martin Schmidt and I co-taught for the first time in the Fall of 1997. I then offered it as a solo flight in the Spring of 1999. Our goal was to exploit our highly interdisciplinary student mix, with students from electrical, mechanical, aeronautical, and chemical engineering. We used design projects carried out by teams of four students as the focus of the semester, and with this mix of students, we could assign to each team someone experienced in microfabrication, another who really understood system dynamics, another with background in electronics, and so on. Lectures for the first two-thirds of the semester presented the material that, now in much expanded form, comprises the first sixteen chapters of this book. Then, while the teams of students were hard at work on their own design problems (more on this below), we presented a series of lectures on various case studies from current MEMS practice.

In creating this "written-down" version of our course, I had to make a number of changes. First and foremost, I greatly expanded both the depth and breadth of the coverage of fundamental material. In fact, I expanded it to such an extent that it is now unlikely I can cover it all within a one-semester course. Therefore, I expect that teachers will have to make selections of certain topics to be emphasized and others that must be skipped or left to the students to read on their own.

A second change, and perhaps a more important one for teachers using this book, is that I took our collection of homework problems compiled from the 1997 and 1999 versions of the course and used them as the worked-out examples in the text chapters. While this had the effect of greatly enriching the content of the book, it created a temporary deficiency in homework problems that has only been partially repaired by the rather modest set of new homework problems that I created for the printed book. Thanks to the world-wide web, though, we now have an efficient mechanism for distributing additional homework problems as they become available. (See "Note to Teachers" below.)

I am hoping to provide, both by example and by presentation of the underlying fundamentals, an approach to design and modeling that any engineering student can learn to use. The emphasis is on lumped-element models using either a network representation or a block-diagram representation. Critical to the success of such an approach is the development of methods for creating the model elements. Thus, there is a chapter on the use of energy methods and variational methods to form approximate analytical solutions to problems in which energy is conserved, and a chapter on two different approaches to creating lumped models for dissipative systems that obey the heat flow equation or its steady-state relative, the Laplace equation. The approach to modeling is based, first, on the use of analytical methods and, second, on numerical simulations using MATLAB, SIMULINK, and occasionally MAPLE. Every numerical example that appears in this book was done using the Student Edition of Version 5 of MATLAB and SIMULINK. In a few places, I provide comparisons to the results of meshed numerical simulations using finite-element methods, but that is not the main purpose of the book.

The Case Studies that form Part V of the book were selected to sample a multidimensional space: different manufacturing and fabrication methods, different device applications, different physical effects used for transduction. In making the selection, I had the invaluable assistance of extended discussions with Dr. Stephen Bart and Dr. Bart Romanowicz of Microcosm Technologies, as well as creative suggestions from Prof. Martin Schmidt. It is well known in the MEMS world that I emphasize the importance of packaging. It will therefore come as no surprise that the first chapter in Part V is a Case Study on packaging, using an automotive pressure sensor as an example. The remaining Case Studies deal with a piezoresistively-sensed bulk-micromachined silicon pressure sensor, a capacitively-sensed surface-micromachined polysilicon accelerometer, a piezoelectrically excited and sensed bulk-micromachined resonant quartz rate gyro, two types of electrostatically actuated optical projection display, two approaches to the construction of single-chip systems for amplification of DNA, and a surface micromachined catalytic sensor for combustible gases.

As stated earlier, design problems are an important part of our teaching of this material. A good design problem is one that requires the student team to confront the problem's specifications and constraints, develop a system architecture for approaching the problem, work through the definition of a microfabrication process, assign device dimensions, and model the device and system behavior so as to meet the specifications. Examples have included the design of a flow controller, a particular type of pressure sensor or accelerometer, a resonant strain sensor (including the circuit that drives it), a system for polymerase chain reaction (PCR), a temperature-controlled hot stage for catalytic chemical sensors, and many more. Sample design problems will be posted on the web site. We like problems that involve feedback, because this forces the creation of a decent open-loop dynamic device model for insertion into the loop. Our experience has been that it is primarily through the design problems that students recognize how truly empowering the mastery of the fundamental subject matter is. They also learn a lot about teamwork, successful partitioning of a problem, intra-team communication, and very important realities about design tradeoffs. Only when the team realizes that by making one part of the job easier they may be making a different part harder does the true merit of the design problem emerge. Our students have benefitted from the volunteer effort of several of our graduate students and post-docs who have served as mentors to these design teams during the semester, and we recommend that teachers arrange such mentorship when using design problems in their own courses.

If Prof. Schmidt and I had to pick one subject that has been the hardest to teach in the context of our course, it would be process integration: creation of realistic fabrication sequences that produce the desired result with no evil side effects (such as putting on gold at a point in the process where high-temperature steps remain). Students who have already had a laboratory class in microfabrication technology have a huge advantage in this regard, and we encourage students to take such a class before embarking on our design course.

The opportunity to write this book was provided by a sabbatical leave from my regular duties at M.I.T., somewhat stretched to cover a total of thirteen months, from August 1999 through August 2000. Because of the tight time schedule, a few short-cuts had to be made: First, I did not attempt to provide comprehensive references on the various topics in this book. I have included basic recommended reading at the end of each chapter and also citations to key published work where appropriate. But citations to the work of many people are omitted, not because of any lack of respect for the work, but out of the need to get the job done in a finite time and to have it fit in a finite space. Others, such as Kovacs and Madou, have written books with extensive references to the published literature, and I urge readers of this book to have these books at hand as supplementary references. Second, I did not create a CD-ROM with the MATLAB, SIMULINK, and MAPLE models on it. My reasoning was that the world-wide web now offers a much more efficient way to distribute such material and that, by using the web, I would have some flexibility in revising, correcting, and improving the models over time. Third, I did not include very much data on different material properties and process steps. My hope is that there is just enough data to permit some useful examples to be worked out. And, thankfully, additional data is now much easier to find, thanks to the compilation of various on-line databases, such as at the mems.isi.edu web site.

A Note to Teachers

As mentioned above, the supply of homework problems in this book is not what I would like it to be, in terms of either quantity or intellectual depth. Rather than delay publication of the book solely for the purpose of creating new homework problems, I decided to take advantage of the world-wide web to create a depository for supplementary material. The URL for the web site for this book is:

http://web.mit.edu/microsystem-design/www

Posted on this web site will be the inevitable errata that I fail to catch before publication, additional homework and design problems, and MATLAB, SIMULINK and MAPLE models that support both the examples in the various chapters and the new homework problems.

Anyone using this book is invited to submit materials for posting on this web site, using the on-line submission form. This is particularly important if you find an error or have an idea about how the index can be improved, but new homework and design problems and numerical models are also welcome. They will be posted with full attribution of the source.

As this manuscript leaves my hands for the publisher, I still have not decided how to handle solutions to homework problems. By the time the book appears in print, there will be a tentative policy. Teachers using the book are invited to send me comments at sds@mit.edu on what solution-handling method would be most helpful.

Finally, when you submit something to the web site, the web-site submission form will request that you grant me permission to publish the material, both on the web site and in future editions of this book. This will greatly strengthen the body of material that all future microsystem-design students can access.

Stephen D. Senturia
Brookline, MA

Acknowledgments

A book of this type relies on inputs from many sources. I already mentioned in the Preface the contributions of three key individuals: Martin Schmidt, with whom I developed the original version of the course on which this book is based; and Stephen Bart and Bart Romanowicz of Microcosm Technologies, with whom I spent many stimulating hours throughout much of the year discussing how to present this complex field in a coherent and meaningful way. Steve and Bart contributed significantly to the process used to select the Case Studies for this book and provided many helpful and insightful comments on a variety of details.

The Case Studies of Part V each had a champion who provided materials and commentary. All of the following individuals were of immense help in enabling what could be presented here: David Monk, William Newton, and Andrew McNeil of the Motorola Sensor Products Division (Chapters 17 and 18), Michael Judy of Analog Devices and Tom Kenny of Stanford (Chapter 19), Josef Berger, David Amm, and Chris Gudeman of Silicon Light Machines (Chapter 20), Brad Sage of Systron Donner (Chapter 21), Andreas Manz of Imperial College and David Moore of Cambridge University (Chapter 22), and Ronald Manginell of Sandia National Laboratories (Chapter 23).

Students and various staff members at MIT have also played a role in the creation of this book. I particularly want to thank Mathew Varghese, who worked with me to develop the models used in Chapter 22, and Erik Deutsch, who committed precious time to collecting material data and associated references. Dr. Arturo Ayon was helpful in providing data and documentation for several examples in the book. Postdocs Mark Sheplak (now at the University of Florida), Reza Ghodssi (now at the University of Maryland) and Carol Livermore served as design-team mentors and helped sharpen our insights into this aspect of teaching. And there have been about fifty students who have now completed our MIT course. Their excitement about the field has provided significant motivation for getting this book written.

It is highly desirable to "pre-test" textbook material before committing it to press. My first opportunity for such a pre-test was a two-week short course on Microsystem Design for engineering students at the Institute of Microtechnology (IMT) and staff members at the Swiss Center of Electronics and Microtechnologies (CSEM) in Neuchâtel, Switzerland, held during June 1998. I am indebted to Prof. Nico de Rooij of IMT and Phillipe Fischer of the Swiss Institute for Research in Microtechnology (FSRM) for arranging the course, and to Mark Grétillat for his demonstrations of the MEMCAD system as part of the course. This activity, which took place using terse hand-written notes, helped me learn what would make an effective text that could serve the needs of both students and active professionals.

A more extensive pre-test occurred during the Spring of 2000. Professors Mark Sheplak and Toshi Nishida of the University of Florida used the first three parts of an early draft as the basis for a graduate course on MEMS technology and devices. Their feedback, and the feedback from their students, has led to a number of improvements in scope and presentation, and their very careful reading of the text identified many small things that needed correction.

I have benefited from the efforts of many other colleagues who have read sections of the draft and provided comments. I particularly want to mention Prof. Olav Solgaard of Stanford, Prof. G. K. Ananthasuresh of the University of Pennsylvania, Dr. Srikar Vengallatore of MIT, and MIT graduate students Joel Voldman, Mathew Varghese, and Erik Deutsch.

Perhaps the most significant reader has been my wife, Peg Senturia, who, as a non-engineering writer, is able to see issues about presentation and subject order that we more technical types overlook. In particular, her suggestions on chapter order in the first part of the book and her editorial efforts on the introductory material have greatly improved the clarity and accessibility of the design approach presented here. Beyond that, words fail to express the gratitude I feel, first, for her putting up with my preoccupation during what has been an intense thirteen months and, second, for doing so in a shared at-home workspace. Our bond of love has deepened through this project.

Finally, I want to thank three of my faculty colleagues at MIT for the classes I took during the Fall semester of 1993. My department has an Adler Scholarship which allows a faculty member a term off from teaching provided that he or she takes a course for credit and gets a grade! As an Adler Scholar during the Fall of 1993, I enrolled in Prof. Jacob White's course on numerical methods, did every homework problem and got a decent grade. During that same semester, I sat in on two mechanical engineering classes on modeling of dynamic systems, one taught by Prof. Kamal Youcef-Toumi, the other by Prof. Neville Hogan. The experience of that semester of total immersion into modeling and numerics planted the seeds from which this book has grown.

Stephen D. Senturia

I
GETTING STARTED

Chapter 1

INTRODUCTION

What's in a name? ...A rose by any other name would smell as sweet.
—W. Shakespeare in *Romeo and Juliet*

1.1 Microsystems vs. MEMS

In Europe, they are called "Microsystems." In the United States, and increasingly elsewhere, they are called "Microelectromechanical Systems," or "MEMS." While the European name is more general, more inclusive, and in many ways more descriptive, the MEMS acronym is catchy, unique, and is taking hold worldwide.[1] Regardless of what we call them, the questions are: What are they? How are they made? What are they made of? And how do we design them?

1.1.1 What are they?

"Microsystems," literally, are "very small systems" or "systems made of very small components." They do something interesting and/or useful. The name implies no specific way of building them and no requirement that they contain any particular type of functionality. "Microelectromehcanical systems (MEMS)," on the other hand, takes a position: *Micro* establishes a dimensional scale, *electro* suggests either electricity or electronics (or both), and *mechanical* suggests moving parts of some kind. But the MEMS concept has grown to encompass many other types of small things, including thermal, magnetic, fluidic, and optical devices and systems, with or without moving parts. So let's

---

[1] Two recent data points: Australia and Israel have, apparently, adopted the MEMS usage in recent national status reports.

not argue about it. Let us admit that both terms cover the domain of interest and get on with it.

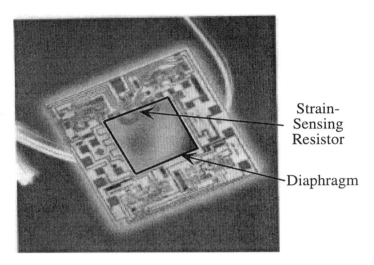

Strain-
Sensing
Resistor

Diaphragm

Figure 1.1. A bulk-micromachined silicon integrated pressure sensor, the subject of the Case Study in Chapters 17 and 18. (Copyright of Motorola; used by permission.)

In practice, Microsystems/MEMS share several common features: First, MEMS involve both electronic and non-electronic elements, and perform functions that can include signal acquisition (sensing), signal processing, actuation, display, and control. They can also serve as vehicles for performing chemical and biochemical reactions and assays.

Second, MEMS are "systems" in the true sense, which means that important system issues such as packaging, system partitioning into components, calibration, signal-to-noise ratio, stability, and reliability must be confronted.

Third, the most successful MEMS have been those which involve paradigm shifts from the "macro" way of doing things, more than simply reducing the size scale. Examples include the ink-jet print head, which enables high-quality color printing at very low cost, and thin-film magnetic disk heads, which continue to enable startling improvements in the mass-memory industry. Other examples include silicon pressure sensors and silicon and quartz sensors for the measurement of acceleration and rotation. Microfluidic devices are beginning to enable astonishing improvements in the speed of biochemical analysis.

Finally, some MEMS involve large arrays of microfabricated elements. Examples include uncooled infrared imaging devices and both reflective and refractive projection displays.

This book will address a broad spectrum of MEMS examples, drawn from a wide range of disciplines and application areas.

Figure 1.2.   A surface-micromachined integrated silicon accelerometer, the subject of the Case Study of Chapter 19. (Source: Analog Devices, Inc., reprinted with permission.)

1.1.2   How are they made?

Most MEMS devices and systems involve some form of lithography-based microfabrication, borrowed from the microelectronics industry and enhanced with specialized techniques generally called "micromachining." The batch fabrication that is characteristic of the microelectronics industry offers the potential for great cost reduction when manufacturing in high volume.

Lithographic techniques generally require the use of flat substrates. Silicon is often used even when there are no electronic components in the device because the tools and instruments needed for microfabrication are designed to match the characteristics of silicon wafers. Lithography offers in-plane sub-micron precision on dimensional scales from micron to millimeter. Thin-film deposition and etching techniques in combination with wafer-bonding techniques allow patterning of the third dimension, making possible the creation of movable parts. The combination of lithography with thin-film methods tends to result in structures characterized by extrusion of two-dimensional features into the third dimension. New fabrication methods provide additional freedom

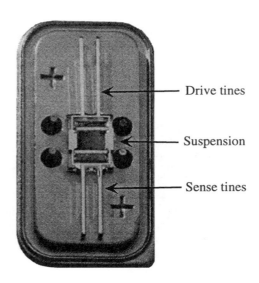

Figure 1.3. A quartz rate gyroscope, the subject of the Case Study of Chapter 21. (Source: Systron Donner, reprinted with permission.)

to sculpt more general three-dimensional structures, but these have not yet entered high-volume manufacturing.

Because MEMS fabrication usually involves some steps shared with conventional microelectronics, there is an almost reflexive urge to make fully integrated microsystems, *i.e.*, integrated circuits that include mechanical or other non-electronic elements on the silicon chip along with the electronic part of the system. Some manufacturers are aggressively pursuing this path. However, an alternate strategy is to partition the microsystem into subsystems that are fabricated separately, then assembled into a compact system during the packaging operation. The debate over the "right" approach is a continuing one, with no clear answer. However, individual MEMS practitioners (including this author) have strong opinions. It is clear that the system architecture and its partitioning into components have an enormous impact on the details of how the system is built. Further, the encapsulation and packaging of the components into a system can greatly impact the design.

1.1.3   What are they made of?

The choice of materials in a microsystem is determined by microfabrication constraints. Integrated circuits are formed with various conductors and insulators that can be deposited and patterned with high precision. Most of these are inorganic materials (silicon, silicon dioxide, silicon nitride, aluminum, and tungsten), although certain polymers are used as well. Microfabrication that extends beyond conventional microelectronics opens up a much broader range of materials and a corresponding set of additional techniques such as electroplating of metals, and molding and embossing of plastics.

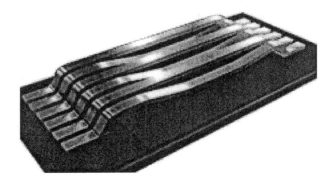

Figure 1.4. Schematic view of one pixel of an electrostatically actuated diffractive optical display, the subject of the Case Study of Chapter 20. (Source: Silicon Light Machines, reprinted with permission.)

The range of materials for microsystems has now become very broad. This offers many choices. One role of the design process is to evaluate the relative merits of different ways of building a proposed device.

Since the performance of MEMS devices depends on the constitutive properties of the materials from which they are made, the increased diversity of material choices carries with it a requirement for measurement and documentation of their properties. Many of these materials are used in thin-film form, and it is well known that thin-film properties can differ from bulk properties. Hence, certain properties that are critical in device performance, for example, the elastic modulus or residual stress of a suspended beam, must be monitored in manufacturing to ensure repeatability from device to device. This demands new methods of material property measurement, a subject of increasing importance in the microsystems field.

1.1.4 How are they designed?

As we shall discuss at length in Chapter 2, the design of microsystem requires several different levels of description and detail. On one level, the designer must document the need and specifications for a proposed microsystem, evaluate different methods by which it might be fabricated, and, if the device is to become a commercial product, further evaluate the anticipated manufactured cost. At another level, for each proposed approach, one must deal with details of partitioning the system into components, materials selection and the corresponding fabrication sequence for each component, methods for packaging and assembly, and means to assure adequate calibration and device uniformity during manufacture. Clearly, this can be a daunting job.

Quantitative models play a key role in the design process by permitting prediction of performance prior to building a device, supporting the troubleshooting of device designs during development, and enabling critical evaluations of failure mechanisms after a device has entered manufacturing. Models can be developed by hand in analytical form, or they can take the form of numerical simulations carried out on high-speed workstations. Experience suggests that there is a natural progression from approximate analytical models early in the design cycle to more detailed and comprehensive numerical simulations later in the design cycle, continuing into device development and manufacture.

1.2   Markets for Microsystems and MEMS

The humorous proverb, occasionally found in Chinese fortune cookies, states: "Forecasting is difficult, especially the future." Various organizations from time to time carry out market surveys in MEMS, hoping to be able to forecast the future. While fraught with difficulty, these studies do serve a useful purpose in identifying application areas, and attempting to project how each application area will grow. The numbers in such studies are best-guess estimates, often predicting overly quick growth, but many organizations use them as one input to strategic decisions on where to invest research and development resources.

Table 1.1.   System Planning Corporation Market Survey (1999) with an estimate of 1996 product volume and a forecast of 2003 sales (in millions of US$) [1].

MEMS Device and Applications	1996	2003
Inertial Measurement: Accelerometers and rate gyros	350-540	700-1400
Microfluidics: Ink-jet printers, mass-flow sensors, biolab chips	400-500	3000-4450
Optics: Optical switches, displays	25-40	440-950
Pressure Measurement: Automotive, medical, industrial	390-760	1100-2150
RF Devices: Cell phone components, devices for radar	none	40-120
Other Devices: Microrelays, sensors, disk heads	510-1050	1230-2470

A market study by the System Planning Corporation [1] provides a particularly comprehensive view of the MEMS/Microsystems field, and, with large error bars, projects growth of individual markets. The market estimates in Table 1.1 are taken from this study.

The table shows estimates of the worldwide markets in 1996 and projected corresponding markets in 2003. The large uncertainties, even in the 1996 estimates, are due to the difficulty of assessing the MEMS component of a market. For example, ink-jet print heads are a MEMS component that enables the manufacture and sale of a printer. How much of the value of the printer belongs in the MEMS market? If only the ink-jet print head itself, then the market is judged as small; if the full printer assembly, then the market is huge. In each case, the surveyors had to make judgments about how to slice between MEMS as a component and MEMS as a full system. It is also difficult to obtain accurate figures even when the slicing decision is clear. Nevertheless, the survey is interesting, and shows how broad the reach of MEMS is expected to become. Clearly, the growth of MEMS will impact many disciplines.

1.3   Case Studies

The field of MEMS/Microsystems is so rich and varied, that it is a challenge to select a small set of devices that illustrate the field. Although there are a number of examples sprinkled throughout the book, the most detailed ones are the Case Studies contained in Chapters 17 through 23.

Table 1.2 (following page) summarizes the six devices used for the Case Studies. Four are based on commercial products either already in production or close to it. The other two are based on prototype development. The examples are drawn from a broad range of MEMS applications: pressure measurement, inertial sensing (acceleration and rotation), optical applications, microfluidic chemical systems, and thermally based chemical sensing. The pressure example also serves as the Case Study for Chapter 17 on packaging. And while these examples have broad coverage, it is clear from the previous section that many wonderful and interesting examples cannot be presented at this same level of detail. Choices must be made, and in this case, the choices have been made to encounter as many of the fundamental issues of MEMS design as possible with a limited set of examples.[2]

Figure 1.1 (page 4) shows an integrated pressure sensor. The silicon die is approximately 3×3 mm. A central portion, outlined for clarity, is where the silicon has been thinned from the back using *bulk micromachining* to create a diaphragm about 20 μm thick. Differential pressure applied on the two sides of the diaphragm causes it to bend. The bending produces elastic strain. A strain-sensitive *piezoresistor* is located near the edge of the diaphragm, where the strain is large. As the diaphragm bends, this resistor changes its value, producing a detectable voltage in the readout circuit. This particular

---

[2]An *unlimited* set of examples can be found in the list of Suggested Reading at the end of this chapter.

Case Study	Partitioning	Technology	Transduction	Packaging
Pressure sensor	Monolithic	Bulk micromachining with bipolar circuitry plus glass frit wafer bonding	Piezoresistive sensing of diaphragm deflection	Plastic
Accelerometer	Monolithic	Surface micromachining with CMOS circuitry	Capacitive detection of proof-mass motion	Metal can
Resonant rate gyroscope	Hybrid	Bulk micromachined quartz	Piezoelectric sensing of rotation-induced excitation of resonant mode	Metal can
Electrostatically driven display	Hybrid	Surface micromachining using XeF_2 release	Electrostatic actuation of suspended tensile ribbons	Bonded glass device cap plus direct wire bond to ASIC
DNA amplification with PCR	Hybrid	Bonded etched glass	Pressure-driven flow across temperature-controlled zones	Microcapillaries attached with adhesive
Catalytic combustible gas sensor	Hybrid	Surface micromachined with selective deposition of catalyst	Resistance change due to heat of reaction of combustible gas	Custom mounting for research use

Table 1.2.   Summary of the Case Studies examined in this book.

product has the readout circuit integrated onto the same chip as the mechanical diaphragm and, hence, is an *integrated*, or *monolithic*, pressure sensor.

Figure 1.2 (page 5) shows a monolithic accelerometer mounted on and wire-bonded to a standard integrated circuit header. The accelerometer is fabricated with a technique called *surface micromachining*. The structure is built as a unit, after which sacrificial layers are removed by selective etching to free the moving parts. The proof mass of the accelerometer is suspended above the surface of the wafer by a 2 μm gap, and is retained with beams that serve as elastic springs. A series of fingers attached to the proof mass are located between pairs of fixed fingers attached to the substrate. When the device experiences an in-plane acceleration in a direction perpendicular to the fingers, the differential capacitance between the fixed and movable fingers changes. Devices of this type are used to trigger the airbags in automobiles.

Figure 1.3 (page 6) shows a quartz rate gyroscope that is manufactured with bulk micromachining. The device exploits the piezoelectric property of quartz, which couples electric fields to elastic deformation. The drive tines are set into vibration in the plane of the device. If the device is then rotated about the tine axis, out-of-plane bending of the drive tines results. The central suspension couples this out-of-plane motion to the sense tines, which bend out of plane in response. The out-of-plane bending of the sense tines is detected by electrodes evaporated onto the edges of the tines.

Figure 1.4 (page 7) shows in schematic form a portion of an electrostatically actuated diffractive optical display. It is fabricated with surface micromachining. A long array of thin ribbons is organized into pixels of 6 ribbons made of silicon nitride coated with aluminum. Within each pixel, alternate ribbons can be actuated to move them by one-half wavelength of light, producing a strong diffraction pattern. This diffracted light is collected by a lens, and, in combination with a mechanically scanned mirror, projected onto a screen to create a high-resolution display. The electronic drive circuitry is built on separate custom-designed chips that are bonded onto a common substrate with the optical device, and wire-bonded together.

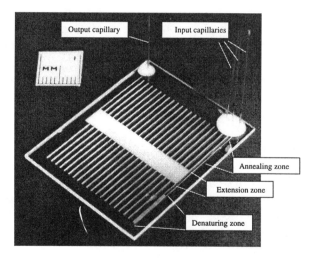

Figure 1.5.   A continuous-flow chamber for polymerase chain reaction (PCR) amplification of deoxyribonucleic acid (DNA), the subject of the Case Study of Chapter 22. (Source: Andreas Manz, Imperial College; reprinted with permission.)

Figure 1.5 shows a polymerase-chain-reaction (PCR) device made of two bonded glass wafers, each of which has patterns etched into it prior to bonding. The result is a tortuous flow channel which terminates in wells to which input and output capillaries are bonded. Flow is driven with pressure. The three labeled zones are held at different temperatures, so that a liquid sample flowing through the device experiences a cyclic variation in temperature. This kind of thermal cycle is precisely what is need to perform amplification of genetic material, deoxyribonucleic acid (DNA).

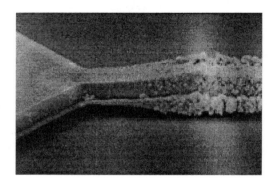

Figure 1.6.   A catalyzed combustible-gas sensors, the subject of the Case Study of Chapter 23. (Source: Sandia National Laboratories, reprinted with permission.)

Figure 1.6 shows a magnified view of a combustible gas sensor. A thin filament of polycrystalline silicon is suspended above a substrate. The filament has a granular platinum coating, which serves as a catalyst. The temperature of the filament is raised by passing current through it. The presence of combustible gases that oxidize on the catalyst incrementally increases the temperature, which shifts the resistance. The drive circuit, which is built separately, detects a reduction in the power needed to maintain the filament at its set-point temperature.

1.4   Looking Ahead

Following an overview of the design process in Chapter 2, which presents an explanation of the different types of models that are required in the design process, we briefly survey microfabrication methods to establish a context for thinking about microsystems. Part II, on Lumped Models, explores a specific modeling choice (lumped-element equivalent-circuit models), with appropriate links to system-level (input-output) and physical modeling and simulation. Then, in Part III, we examine the different energy domains that appear in MEMS (elasticity, electromagnetism, fluids, and so on), and develop modeling methods appropriate to each. System issues are addressed in Part IV, specifically the electronic components of a microsystem, feedback, and random noise. Finally, in Part V, the Case Studies that illustrate the various aspects of modeling and design are examined. Where possible, comparisons between the relatively elementary models developed here and more sophisticated numerical simulations are presented. However, a complete introduction to numerical methods is well beyond what can be covered in a text of this type.

The tools we shall use in developing our models are, first, pencil and paper, second, algebraic support from the symbolic capabilities of MATLAB and Maple V, and, finally, selected numerical simulations using MATLAB and SIMULINK.[3] An industrial designer or researcher will, in addition, use microelectronic computer-aided-design systems to perform layout of photomasks, and CAD-for-MEMS systems, now commercially available, to perform needed numerical simulations of physical device behavior and system-level simulations of circuit and overall system behavior.

Related Reading

J. Brysek, K. Petersen, J. R. Mallon, L. Christel, and F. Pourahmadi, *Silicon Sensors and Microstructures*. Freemont, CA, USA: NovaSensor, 1990.

J. W. Gardner, *Microsensors*. Chichester, UK: Wiley, 1994.

A. D. Khazan, *Transducers and Their Elements*. Englewood Cliffs, NJ, USA: PTR Prentice Hall, 1994.

G. T. A. Kovacs, *Micromachined Transducers Sourcebook*. New York: WCB McGraw-Hill, 1998.

M. Lambrechts and W. Sansen, *Biosensors: Microelectrochemical devices*. Bristol, UK: Institute of Physics, 1992.

M. Madou, *Fundamentals of Microfabrication*, pp. 35–37. New York: CRC Press, 1997.

N. Maluf, *An Introduction to Microelectromechanical Systems Engineering*. Boston: Artech House, 2000.

R. S. Muller, R. T. Howe, S. D. Senturia, R. L. Smith, and R. M. White, eds., *Microsensors*. New York: IEEE Press, 1991.

NovaSensor, *Silicon Sensors and Microstructures*, Fremont CA: NovaSensor, 1990.

L. Ristic, ed., *Sensor Technology and Devices*. Boston: Artech House, 1994.

---

[3] MATLAB and SIMULINK are registered trademarks of The Mathworks, Inc. Maple and Maple V are registered trademarks of Waterloo Maple Inc. All examples presented in this book were done using Maple V and/or Version 5 of the Student Editions of MATLAB and SIMULINK.

H. J. De Los Santos, *Introduction to Microelectromechanical (MEM) Microwave Systems*. Boston: Artech House, 1999.

R. G. Seippel, *Transducers, Sensors & Detectors*. Reston, VA, USA: Reston Publishing Company, 1983.

M. Tabib-Azar, *Microactuators: Electrical, Magnetic, Thermal, Optical, Mechanical, Chemical and Smart Structures*. Boston: Kluwer Academic Press, 1998.

W. S. Trimmer, ed., *Micromechanics and MEMS: Classic and Seminal Papers to 1990*. New York: IEEE Press, 1997.

S. J. Walker and D. J. Nagel, "Optics and MEMS," Technical Report NRL/MR/6336–99-7975, Naval Res. Lab., Washington DC, 1999. PDF version available at http://mstd.nrl.navy.mil/6330/6336/moems.htm.

K. S. Wise, ed., *Proc. IEEE Special Issue: Integrated Sensors, Microactuators, & Microsystems (MEMS)*, vol. 86, no. 8, August, 1988.

Problems

1.1 A Library Treasure Hunt: By consulting recent issues of at least two of the following MEMS-oriented journals, locate articles that illustrate (a) mechanical sensing, (b) use of MEMS for optics, (c) a chemical or biological system, and (d) an actuator. The journals are: *IEEE/ASME Journal of Microelectromechanical Systems, Journal of Micromechanics and Microengineering, Sensors and Actuators, Sensors and Materials, Analytical Chemistry,* and *Biomedical Microdevices*.

1.2 An On-Line Treasure Hunt: By consulting the events calendar at the web site http://mems.isi.edu (or some other MEMS-oriented web site), identify conferences scheduled for the coming year that cover some combination of (a) mechanical sensing, (b) use of MEMS for optics, (c) chemical or biological systems, and (d) actuators. On the basis of material published at the individual conference web sites and of an examination of available previously published programs and/or proceedings from these conferences (in your library or on the web), try to select which conference would be most appropriate for each of the four topic areas. (And if one is taking place in your vicinity, try to attend. Most meetings welcome the assistance of student volunteers.)

Chapter 2

AN APPROACH TO MEMS DESIGN

If technology-driven design is a hammer looking for a nail, then market-driven design is a nail looking for a hammer. A good designer will match hammers to nails, regardless of which comes first.

—Anonymous

2.1 Design: The Big Picture

We now begin our discussion of microsystem design. As a first step, we examine the big picture, including difficult subjects such as creativity and invention, market opportunities, and choices of technologies, system architectures, and manufacturing methods. We shall find that *modeling*, the principal subject of this book, plays a critical role in analyzing the choices that must be made.

2.1.1 Device Categories

The design of a complete microsystem or a MEMS component depends on the type of microsystem or component it is. We distinguish between three major categories:

Technology Demonstrations: Components or systems intended to drive development activity, either to test a device concept or push the capabilities and limits of a particular fabrication technology. Small numbers of working devices are needed.

Research Tools: Components, and occasionally full systems, having the purpose of enabling research or performing a highly specialized task, such as measuring the value of selected material properties. The numbers of working devices varies by application; quantitative accuracy is required.

Commercial Products: Components and full systems intended for commercial manufacture and sale. The numbers of working devices and the degree of precision and accuracy needed depend on the market.

Each category has different requirements for the end product. Technology Demonstrations might require only a handful of working devices. Device-to-device consistency is not usually of primary importance. Commercial Products, on the other hand, might require millions of devices per year. Device-to-device consistency and high manufacturing yield are likely to be of paramount importance. Research Tools, intended, for example, to perform a measurement of a quantity of interest, must typically be calibrated, and should have repeatable behavior, device to device and measurement to measurement.

The boundaries between these categories are not firm. A device could start out as a Technology Demonstration and, eventually, end up as a highly successful Commercial Product, serving as a Research Tool at an intermediate development step. But by categorizing microsystems this way, we can gain important insights into the design process.

2.1.2   High-Level Design Issues

There are five high-level issues that affect the design and development of new devices and products (while we recognize that there are three categories of devices to consider, the term "product" is used below to represent all three):

Market: Is there a need for this product? How large is the market? How fast will it develop? We have already seen that estimates exist to help answer these questions.

Impact: Does this product represent a paradigm shift? That is, is it a new way of accomplishing something, or does it enable a new kind of system? Most truly successful MEMS devices have been paradigm shifts.

Competition: Are there other ways to make an equivalent product? Are there other organizations building a similar product? Both types of competition are important.

Technology Is the technology to make and package the product available? Is it in-house? Or must it be acquired through vendors?

Manufacturing: Can the product be manufactured at an acceptable cost in the volumes required?

The relative importance of these five issues depends on the category of microsystem or MEMS component under consideration (see Table 2.1). What we observe in the table is that Technology is critical to all three categories, everything is critical for Commercial Products, and, depending on end use, there is a moderate need to attend to all issues for the Research Tools. For example, device-to-device repeatability is required for Research Tools, and that begins to sound like a manufacturing issue. In addition, if there is no market need for the Research Tool and it won't have any impact, why build it at all? On the other hand, one could embark on a Technology Demonstration project to try out a new device concept with only a vague conception of end use in a real market. If the new device works and truly has impact (in the paradigm-shifting sense), then one presumes a market can be developed.

Table 2.1.   Relative Importance of High-Level Design Issues

Category	Markets	Impact	Competition	Technology	Manufacturing
Technology Demonstration		++		+++	
Research Tools	++	++	+	+++	++
Commercial Products	+++	+++	+++	+++	+++

2.1.3   The Design Process

The quotation at the beginning of this chapter refers to "market-driven" and "technology-driven" designs. A market-driven design is one where the need for a specific capability is well understood, and a search is made for device concepts and corresponding fabrication methods that can realize the desired capability at an acceptable cost. A technology-driven design is one where a specific set of technological capabilities have been developed, and a search is made for device concepts and markets that can use this technology. Perhaps the best kind of design is an "opportunistic" design which matches a market opportunity with a new technology, achieving a paradigm-shifting way of meeting that market need, and doing so cost-effectively. One of the pioneers of MEMS, Dr. Kurt Petersen, has introduced the concept of a *starving market* to capture these design opportunities, markets that are desperate for the capability that can be provided by a suitable MEMS device or system [2].

The fact is that in a fast-developing technology like MEMS, market identification and development often lags technology development. For example, it is first necessary to prove that one can build a small mirror that can be tipped with electrostatic forces before one attempts to build an array of a million such mirrors to create a projection display.[1] But once one understands what kinds of device performance is possible with new technologies, market-driven design enters the picture. Here are additional examples:

The world was not sitting around with a specification for ink-jet printing technology waiting for someone to develop it. Instead, what the world wanted was faster high-resolution color printers, regardless of technology. Innovative individuals explored the ink-jet technology as a way of meeting this market need. Technology Demonstrations were required to see what was possible, out of which grew a capability that has had a revolutionary effect on the printing industry.

Similarly, in the automotive market, the safety mandate for airbags in cars created a significant driving force for accelerometer development.[2] The demonstration that one could use microfabrication technology to make freely moving capacitor plates that would respond to inertial forces predated the conception and design of a silicon automotive accelerometer. And while there has been significant competition from other methods for making this acceleration measurement, at this point the silicon accelerometer has basically displaced all other technological approaches for controlling airbag deployment.

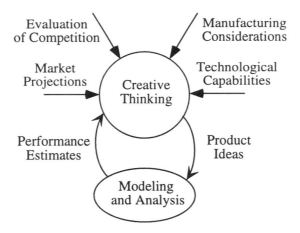

Figure 2.1.   Illustrating high-level design issues and their relation to modeling and analysis. The modeling and analysis block is expanded in Fig. 2.3.

One can think of the design process as a highly iterative flow-chart, as in Figure 2.1. Whether market-driven or technology-driven, the creative designer conceives of a type of device and a possible technology for building it. The

---

[1] See Chapter 20 for a discussion of displays.
[2] See Chapter 19 for a discussion of accelerometers.

designer must then analyze whether the device conception, as implemented in that technology, will actually perform as desired. This analysis step, supported by the construction of *models*, is in the critical path for all decisions affecting design. Modeling and analysis do not replace the creative thinking needed to conceive of product ideas that will have impact, but they support the entire design process.

2.2   Modeling Levels

Modeling and analysis of devices and systems is a complex subject. Modeling occurs at many levels and uses a variety of modeling paradigms, each one selected to be appropriate for that level. Figure 2.2 illustrates a highly simplified overview of these modeling levels.

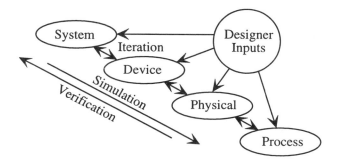

Figure 2.2.   Different modeling levels for microsystems.

We identify four modeling levels – System, Device, Physical, and Process – each linked to the next by double-headed arrows symbolizing iterative exchange of information between levels. At the top is the *system level*. This is the home of block-diagram descriptions and lumped-element circuit models, either of which can be used, and both of which lead to a coupled set of *ordinary differential equations (ODE's)* to describe the dynamical behavior of the system. Often these equations are written in a specific form as a coupled set of first-order ordinary differential equations for the *state variables* of the system. In this form, the equations are referred to as the *state equations* for the system.

At the bottom is the *process level*. This is where the process sequence and photomask designs for device manufacture are created. Process modeling at this level is a highly sophisticated numerical activity for which a number of commercial CAD tools have been developed, referred to generically as *technology CAD* or *TCAD*. For the MEMS designer, the importance of TCAD is that it can predict device geometry from the masks and process sequence. Also, because the material properties can depend on the detailed process sequence, the designer must know the proposed process in order to assign the correct material properties when modeling the device.

The *physical level* addresses the behavior of real devices in the three-dimensional continuum. The governing equations are typically *partial differential equations (PDE's)*. Various analytical methods can be used to find closed-form solutions in ideal geometries, but the modeling of realistic devices usually requires either approximate analytical solutions to the PDE's or highly meshed numerical solutions. A variety of numerical modeling tools using either finite-element, boundary-element, or finite-difference methods are available for simulation at the physical level.

While meshed representations of the PDE's of continuum physics are useful in physical simulation, such representations are typically too cumbersome when dealing with entire devices and their associated circuitry. Instead, we go to the *device level* and create what are called *macro-models* or *reduced-order models* in a form that captures the essential physical behavior of a component of the system, and simultaneously is directly compatible with a system-level description.

An ideal macro-model is analytic, rather than numerical. Designers can think more readily with analytic expressions than with tables of numeric data. The macro-model should capture all the essential device behavior in a form that permits rapid calculation and direct insertion into a system-level simulator. The macro-model should be energetically correct, conserving energy when it should, and dissipating energy when it should. It should have the correct dependence on material properties and device geometry. It should represent both static and dynamic device behavior, both for small-amplitude (linear) excitation, and large-amplitude (presumably nonlinear) excitation. Finally, the macro-model should agree with the results of 3D simulation at the physical level, and with the results of experiments on suitable test structures. This is a tall order! Not every model we shall encounter will reach this ideal. But the requirements are clear.

An important feature of Fig. 2.2 is the presence of the various Designer Inputs. The designer can create models directly at the system level, or directly at any of the lower levels. For example, one can specify a physical device description along with all its dimensions and material properties, then use physical simulation tools to calculate device behavior, capture this behavior in a reduced-order model, and insert it into a system-level block diagram. Or, if one is early in the design process, one can simply use a parameterized reduced-order model to represent a particular device and defer until later the specification of device dimensions to achieve the desired performance.

Generally, when one moves to a lower level to get information needed at a higher level, we use the term *simulation*, although the term *modeling* would do as well. And when one starts at a lower level and performs analysis to move to the next higher level, we use the term *verification*, although, again, the term *modeling* would serve equally well. The word "verification" is puzzling to some. A way to think about it is as the testing of a hypothesis. Suppose at the device level a designer thinks that a particular shape of device is capable of producing a particular capacitance change when a force is applied to it. The designer can either test this hypothesis by building the device and performing experiments, or he[3] can move down one level, to the physical level, and use the laws of elasticity and electricity to predict the change in capacitance in response to the applied force. The results of that modeling activity, whether done analytically or numerically, are returned to the device level and can be used to test, *i.e.* verify, the designer's hypothesis.

Experiment, of course, is the ultimate verification, but when designing devices and systems, one hopes to learn a great deal about the system through modeling before the relatively costly step of building experimental prototypes.[4] Just prior to committing to prototypes, one would like to be able to take the proposed prototype design, described at the process level with masks and a specific process sequence, and go up each step of the verification ladder using the actual detailed design.

2.2.1   Analytical or Numerical?

An important strategic issue is what mix of analytical and numerical tools to use in design. In this book, we shall make substantial use of SIMULINK for system-level modeling using block diagrams, and MATLAB for selected numerical device-level and physical-level simulation. However, intelligent use of such numerical tools ultimately depends on a solid understanding of the underlying principles of the devices and their operation, and this understanding is best achieved by studying analytical (algebraic) models in detail. Furthermore, the design insights provided by analytic models are invaluable, especially the insights into the effects of varying either device dimensions or material properties. Thus, for the purposes of this book, the answer is: "analytical first, numerical second."

In the commercial MEMS world, especially as devices enter high-volume manufacturing, there is increased emphasis on numerical tools at all levels. Since the mid-1990's, a merchant CAD industry targeted at the MEMS and Microsystems fields has developed. The advantage of using numerical CAD tools, especially at the physical level, is that subtle second-order effects can be accurately captured without requiring detailed analytical model development. For some of the examples in this book, comparisons will be made between the

---

[3]We acknowledge that good designers come in both genders. With apologies in advance, we use "he" in place of the more accurate, less-offensive, but more tedious "he or she."
[4]Modeling is often called *numerical prototyping*.

approximate analytical models developed here and the corresponding, more accurate results of three-dimensional physical simulation, in order to give a feeling for the accuracy and limitations of the analytical models.

Before exploring an example that illustrates the various modeling levels, we will further expand this picture to illustrate both the richness and the complexity of the modeling challenge.

2.2.2   A Closer Look

Figure 2.3 shows a much expanded view of the "Modeling and Analysis" block of Fig. 2.1. The unshaded blocks capture the basic modeling levels illustrated in Fig. 2.2. The shaded blocks are new, and reflect some of the added complexity of designing real microsystems.

At the top of Fig. 2.3, a product idea is converted into a system architecture. This high-level architectural step requires immediate attention to a very detailed design question: whether or not to merge the system electronics with the non-electronic devices as a single monolithic device. This decision about *physical system partitioning* affects everything in the design, especially the device *packaging*.

While numerical prototyping may assist in making this very important decision, business judgment may play an even bigger role. As an example, one manufacturer of automotive accelerometers committed from the start to a fully integrated monolithic design, while several others adopted the so-called "two-chip" design, in which a custom integrated circuit and a mechanical MEMS device are fabricated separately, then packaged together. Obviously, the design details are very different for these two approaches. The relative success in the marketplace depends on device performance and manufacturing cost. Because the full manufacturing cost (including device yield) cannot be predicted accurately prior to the implementation of the complete product technology, there is significant business risk associated with this kind of high-level architectural decision.

In addition to the physical partitioning of the system into components, the system is usually also partitioned into *subsystems* for analysis purposes. What is shown in the unshaded blocks of Fig. 2.3 is a representation of the various modeling levels and their interrelationships for one of the subsystems. Working down, one finds lumped-element subsystem modeling, and below that, the construction of individual device models, both for the electronic and non-electronic components of the subsystem. Interactions with the package must be included at this level. For both kinds of components, individual device models are needed, and these depend on detailed device dimensions and on the underlying physical laws governing device behavior. The device dimensions and material properties are directly related to the the photomasks and process steps.

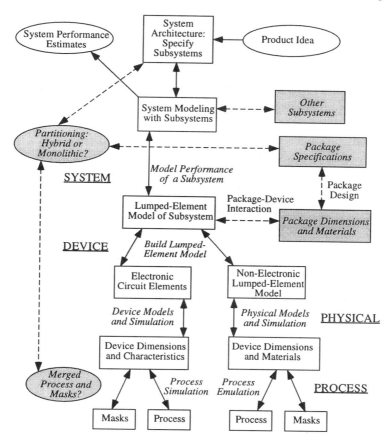

Figure 2.3. An expanded view of the "Modeling and Analysis" block of Fig. 2.1. The various modeling levels of Fig. 2.2 are indicated, and correspond to the entries in italics between the unshaded blocks. The shaded blocks are additional aspects of the design process not captured in Fig. 2.2.

The various blocks in this diagram can be traversed in whatever order best suits the designer, depending on the level of maturity of the design. When testing a new concept, one might work exclusively at the circuit and system level until suitable specifications for components are developed, then move down in the chart to worry about detailed design of device geometry and material selection, as well as a specific circuit to interact with the non-electronic components. Or, one could start with a proposed device shape, and iterate through a range of device dimensions, examining how the dimensional variations affect performance of the entire system. By the time a device is in development, and/or ready for manufacturing, detailed modeling from the bottom up is often performed for the purpose of design verification, using meshed simulations at the physical level and extensive circuit (or block-diagram) simulation incorporating reduced-order device models extracted from the physical simulation results.

2.3   Example: A Position-Control System

This section presents an example of a feedback position-control system to illustrate the modeling at various levels. No attempt is made at this point to do detailed analysis, or even to explain exactly why the models look the way they do. Those details constitute the rest of the book.

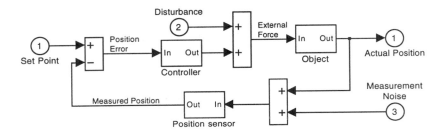

Figure 2.4.   A position-control system.

Fig. 2.4 shows a position-control system in rather abstract block-diagram form.[5] The goal of this system is to apply the correct external force to an object so that its actual position agrees with the desired position. The desired position is provided by the value of Input 1, labeled *set point.* In order to achieve this goal, the actual position is measured with a position sensor. However, because all measurements involve noise, we show a noise input (Input 3) being added to the actual position at the sensor input. The sensor output provides a measured object position, possibly compromised by noise, and also by calibration errors in the sensor. The measured position is subtracted from the set point, and their difference, called the *position error* (or simply, the *error*), is used as input to a *controller*, which amplifies the error and converts it to the correct time-varying external force that will result in moving the object to the desired position, thereby shrinking the error to zero. This entire feedback process of sensing and control takes place in the presence of unwanted sources of motion, called *disturbances.* An example of a disturbance would be the force on the object due to noise-induced acoustic vibration of its supports.

This system, which is a classic feedback system, has already been partitioned into subsystems for analysis purposes: the object, the controller, and the position sensor. As an illustration of subsystem modeling, we will consider

---

[5]This figure, along with many others later in this book, is taken directly from a SIMULINK model window.

just the object itself, and treat it as an inertial mass, supported on a spring, with its motion damped by a linear dashpot. The corresponding block diagram for this case is shown in Fig. 2.5.

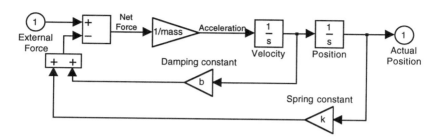

Figure 2.5. The "object" of Fig. 2.4 represented as a spring-mass-dashpot subsystem in block-diagram form . The input is the external force and the output is the actual position.

The input to this subsystem is the external force on the inertial mass. The restoring force from the spring, represented as the triangular gain block with value *k*, is found by multiplying the position by the spring constant *k*. The damping force is found by multiplying the velocity of the mass by the damping constant *b*. These two forces oppose the motion of the inertial mass, and, hence, are subtracted from the external force to yield the net force on the inertial mass. The triangular gain block with value *1/mass* converts this force to the acceleration of the mass, which is then integrated twice to yield position (the $1/s$ blocks are integrators, the notation being based on the Laplace transform of the integration operation). If a time-varying external force is applied at the input, the time-varying position can be readily calculated.

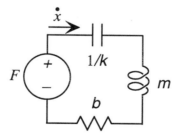

Figure 2.6. The "object" of Fig. 2.4 represented as a spring-mass-dashpot subsystem in equivalent-circuit form . The voltage source is the external force and the "current" is the velocity of the mass.

Alternatively, one could represent the spring-mass-dashpot subsystem with the equivalent circuit of Fig. 2.6, in which the force *F* is analogous to voltage, the velocity of the object $\dot{x}$ analogous to current, with the mass represented by an electrical inductor of value m, the spring by an electrical capacitor with capacitance $1/k$, and the dashpot with a resistor of value b.

What is the difference between these representations? At the important level, namely, in predicting performance, there is no difference. Both models give the same dynamic behavior for the spring-mass-dashpot system. So why choose one or the other? The answer to this question depends on the environment being used for simulation. In SIMULINK, the block diagram with integrators is easily constructe, and is easy to use as a macro-element in building more complex systems. But when we wish to connect a mechanical element to electrical circuits, the equivalent circuit representation has advantages because the same modeling environment used for circuits, such as the widely used program SPICE, can then also be used for the mechanical elements.

It is even possible to create hybrid representations, in which equivalent circuit models are used for some subsystems, and block diagram representations are used for other susbsystems. This can be done by creating new types of circuit blocks that do not correspond to actual physical circuits, but have the correct underlying input-output behavior and can be connected together in a circuit-like topology. The languages used to express these types of models are called *hardware description languages*, or *HDL*, for short. We shall not use HDL models in this book.

The next level of modeling, the device level, identifies specific values of the mass, spring constant, and dashpot constant based on the details of the device geometry and material properties. We shall defer this part of the example until much later, when we have a richer array of tools with which to address these issues.

2.4   Going Forward From Here

So, considering both the high-level design issues, involving assessments of markets, impact, manufacturing, technology, and competition, and the more detailed modeling and design issues, what will we cover in this book? We will assume that the basic evaluation of *what to build* has been made and that candidate technologies for *how to build it* have been identified. Our focus will be on developing the analysis tools to assist in (1) the assessment of different device designs and their relative performance and (2) the prediction of performance (in great detail) for the design ultimately selected. Along the way, we shall have to consider the following issues:

System Architecture: What does the whole system look like, including microfabricated components, electronics, and the package?

System Partitioning: What options are there for changing how the total system is partitioned among components? How does it affect the packaging?

Transduction Methods: What transduction methods will be used? Are there variations that would improve performance, or reduce cost?

Fabrication Technologies With which technology will the microfabricated parts be built? Does the choice of fabrication technologies differ if the system is partitioned differently? Which approach will yield the most robust design in terms of repeatable device dimensions, material properties, and performance?

Domain-Specific Knowledge: Devices typically involve at least two energy domains. Are we correctly applying the laws of physics and chemistry in the prediction of device behavior? Is our level of approximation acceptable?

Electronics: What is the interface between the physical device and the electronic part of the system, and how do the details of the electronics, including device noise, affect overall system performance?

These issues intersect many different design and modeling levels. We choose to begin at the process level. In the following two chapters, we shall examine microfabrication technologies, which define the overall design space within which which microsystem design must occur. Then we will move to the device level, intentionally skipping over the physical level until after we have introduced the language of lumped-element modeling. Once this is done, we return to a systematic exploration of the physical level, one energy domain at a time. We then look at the electronic components, and system-level issues of feedback and noise. With all this preparation, we are finally ready to consider the Case Studies of Part V, which address the full set of issues for each of the case-study examples.

Related Reading

A. Nathan and H. Baltes, *Microtransducer CAD: Physical and Computational Aspects*, Vienna: Springer, 1999.

B. Romanowicz, *Methodology for the Modeling and Simulation of Microsystems*, Boston: Kluwer Academic Press, 1998.

S. D. Senturia, CAD Challenges for Microsensor, Microactuators, and Microsystems, *Proc. IEEE*, vol. 86, no. 8, pp. 1611-1626, 1998.

Problems

2.1 A Library Treasure Hunt: Consult the references given thus far (under "Related Reading" in Chapters 1 and 2, and in Problem 1.1) and find an example of a microsystem or MEMS device that has been modeled using block-diagram representations, and an example of a microsystem that has been modeled with equivalent-circuit lumped-element models. Explain how the elements of the model correspond to the elements of the device in question.

2.2 An On-Line Treasure Hunt: Search the web to find at least one company that specializes in CAD for MEMS. Determine from the company web site whether they provide tools for dealing with all four design levels, and with packaging interactions.

2.3 Another Treasure Hunt: Search the web or the published literature to find data on a MEMS accelerometer. Determine, to the extent possible from the published material, the physical system partitioning, and whether the product uses position-control feedback as part of its acceleration measurement.

Chapter 3

MICROFABRICATION

In small proportions we just beauties see;
And in short measures, life may perfect be.

—Ben Jonson

3.1   Overview

The microsystem examples of Chapter 1 included MEMS devices fabricated with a variety of methods. In this and the next chapter, we survey microfabrication methods and provide some guidelines for the design of complete fabrication sequences. The presentation is necessarily brief. Fortunately, there now exist excellent references on microfabrication technologies, both for conventional microelectronics [3, 4, 5, 6, 7] and for the specialized MEMS techniques called *micromachining* [8, 9].

Microfabrication, as practiced in the microelectronics and MEMS fields, is based on *planar* technologies : constructing the electronic devices and MEMS components on (or, in some cases, in) substrates that are in the form of initially flat wafers. Because the microelectronics industry has made huge investments to develop wafer-based process technologies, there is a correspondingly huge advantage for MEMS designers to exploit these same process steps, or variants based on those steps. We will present the key microfabrication methods in two main sections: wafer-level processes (including wafer bonding), and pattern transfer (including isotropic and anisotropic etching). Along the way, we will describe the various types of micromachining process steps. Common practice is to identify *bulk micromachining* with processes that etch deeply into the substrate, and *surface micromachining* with processes that remove sacrificial layers from beneath thin-film structures, leaving free-standing mechanical structures. This distinction is descriptive of complete process sequences, but is not particularly helpful when discussing the individual process steps since many of the steps are used in process sequences of either type. What matters, ultimately, is to find a process sequence that creates the desired structure in an accurate and manufacturable manner and at an acceptable cost. Because process design involves a subtle combination of technical factors, manufacturing experience, and business judgment, it is most effective to learn by example using the illustrative processes presented here and in the Case Studies.

3.2   Wafer-Level Processes

3.2.1   Substrates

Planar substrates of choice include single-crystal silicon, single-crystal quartz, glass, and fused (amorphous) quartz. Some attention is also now being given to gallium arsenide, since a variety of optoelectronic devices can be fabricated with that material. All are available in wafer form in sizes that are compatible with standard microelectronics processing equipment. As the microelectronics industry moves toward larger and larger wafer sizes [200 mm (8") diameter wafers are now standard, with 300 mm (12") diameter coming next], there is pressure on MEMS fabricators to shift to increasing wafer sizes to maintain compatibility with production equipment. However, MEMS fabrication obeys different economics than standard microelectronics. It is not uncommon for a few 25-wafer runs of 100 mm (4") wafers to supply a full year's production of a single product. Therefore, compared to conventional microelectronics, there is less pressure to go to larger wafer sizes. In fact, careful planning and a well-designed product mix are required to make cost-effective use of the production capacity of MEMS fabrication facilities, even at the smaller wafer sizes.

3.2.1.1   Silicon Wafers

Single-crystal-silicon wafers are classified by the orientation of the surface relative to the crystalline axes. The nomenclature is based on the *Miller indices*, which are described in detail in the references [3, 7, 8]. Silicon is a cubic material, constructed from two interpenetrating face-centered cubic lattices of atoms. Figure 3.1 illustrates the simpler case of a *unit cell* of a simple cubic lattice having identical atoms at the corners of a cube. This unit cell forms the repeat unit for a complete lattice. Relative to the axes formed by the cube edges, a particular direction is denoted with square brackets, such as [100], and the corresponding plane perpendicular to this direction is denoted with parentheses as (100). Because of the cubic symmetry, there are six directions that are symmetrically equivalent to [100]. The complete set of equivalent directions is written with angle brackets as <100>, and the corresponding set of six symmetrically equivalent planes normal to these six directions is written with curly brackets as {100}.

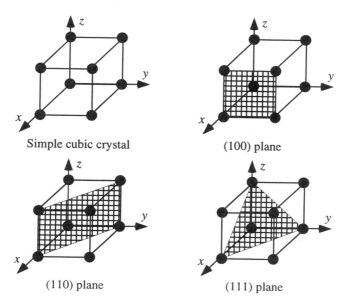

Simple cubic crystal   (100) plane

(110) plane   (111) plane

Figure 3.1.   Illustrating the different major crystal planes for a simple cubic lattice of atoms.

Crystals that are oriented with one of the {100} planes as its surface are called (100) wafers. If a face diagonal of the unit cell is normal to the wafer surface, it is called a (110) wafer. And if a cube diagonal is normal to the wafer surface, it is called a (111) wafer. MOS technologies conventionally use (100) wafers because of the low defect density that can be achieved at the interface between silicon and silicon dioxide. Bipolar-transistor technologies historically used (111) wafers, but are now also fabricated with (100) wafers. (110) wafers are a specialty orientation used for some selective-etching applications.

The complete silicon unit cell is shown in Fig. 3.2. Each atom is identical, but shading is used to clarify their positions. Every atom is tetrahedrally bonded to four neighbors. The illustration on the left shows an interior atom bound to one corner atom and three face-center atoms, but in fact, since every atom is identical, every atom has the exact same bonding structure and local environment. While it is difficult to see, even with a full three-dimensional crystal model in hand, a (111) oriented surface has the highest density of atoms per unit area. Further, each atom in a (111) surface is tetrahedrally bonded to three atoms beneath the surface, leaving only one bond potentially "dangling" at the free surface. In contrast, atoms on (110) or (100) surfaces are tetrahedrally bonded to only two atoms beneath the surface, and have two potential "dangling" bonds. The fact that silicon can be etched anisotropically

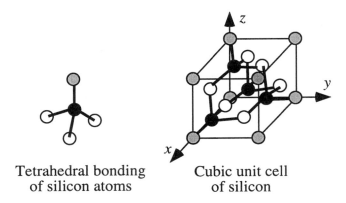

Tetrahedral bonding   Cubic unit cell
of silicon atoms   of silicon

Figure 3.2.   The tetrahedral bonding of each atom in the silicon unit cell. For clarity, atoms at the corners of the cubic unit cell are shaded, those in the center of the faces are white, and those interior atoms that are displaced by 1/4 of the body diagonal from either a face or corner atom are shaded black.

by certain etchants is attributed, in part, to this difference in bonding of the atoms on the different crystal surfaces.

Most silicon crystals are grown from a highly purified melt using the Czochralski method, abbreviated CZ (see [3, 7] for details). A small seed crystal with a preselected orientation is inserted into a heated crucible containing a highly purified melt. The seed is gradually pulled out of the melt while the crucible containing the melt is rotated. The melt temperature and pulling speed are controlled to balance crystal growth rate with pulling rate. An alternative method is called float zone, abbreviated FZ. Starting with a polysilicon rod, a radio-frequency heater creates a local melted zone that is dragged from one end of the rod to the other. To start the growth, a seed crystal can be used at one end of the rod assembly.

Table 3.1.   Specifications for high-quality Czochralski silicon wafers, adapted from [7].

Property	Value
Diameter	up to 300 mm
Thickness	500μm for 100mm, 675μm for 150 mm
Wafer bow	$\leq 25\mu$m
Wafer taper	$\leq 15\mu$m
Oxygen	5-25 ppm
Carbon	1-5 ppm
Heavy metals	≤ 1 ppb

The quality of silicon crystals is specified in terms of their chemical impurities and structural imperfections such as point defects (atoms missing or out of place) and dislocations (places where the crystal planes don't fit perfectly together because of either extra planes of atoms or imperfect stacking of the planes into a screw-like assembly). CZ wafers typically have higher amounts of residual chemical impurities, such as carbon, oxygen and heavy metals, compared to FZ wafers because the molten zone tends to carry impurities with it as it sweeps from one end of the rod to the other. Oxygen as an impurity in CZ wafers has some benefit in certain microelectronic device fabrication sequences, as the presence of oxygen promotes the migration of point defects during high-temperature process steps away from the surfaces where devices are fabricated. On the other hand, the higher purity FZ wafers may be required in devices where heavy-metal contamination must be minimized. Dislocation density is also affected by high-temperature processing. However, crystals can now be grown virtually free of the most disruptive forms of dislocations. Some specifications for a high-quality silicon wafer are shown in Table 3.1. Wafer bow refers to the amount of out-of-plane bending of the wafer, while wafer taper refers to thickness variation across a wafer diameter. Wafers are supplied in standard diameters and thicknesses, and are polished to mirror smoothness on either one or both sides. Nonstandard thicknesses can also be purchased, although at premium prices.

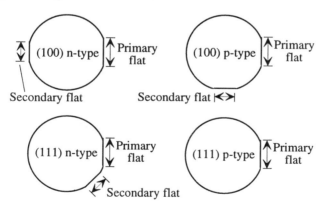

Figure 3.3. Primary and secondary wafer flats are used to identify orientation and type.

Wafers are also characterized by their *doping level*, n-type or p-type. Doping, discussed further in Section 3.2.5, refers to the process by which impurities are intentionally added to modify the electrical conductivity and conductivity type. In the case of CZ crystal growth, dopants are added to the starting melt, and become incorporated uniformly as the crystal grows. The implications of doping on electrical conductivity and electronic device behavior are discussed in Chapter 14. For the moment, it is sufficient to note that introduction of

Group III atoms such as boron produces p-type material, while introduction of Group V atoms such as phosphorus and arsenic produce n-type material.

A pattern of flat edges (or notches for 200 mm wafers and above) is ground into the wafer that encodes the wafer orientation and type, as shown in Fig. 3.3. The relative orientation of the secondary wafer flat to the larger primary wafer flat provides the identification.

3.2.1.2   Quartz Wafers

Quartz is a hexagonal material, with the z-axis conventionally identified as the hexagonal axis. Because of its piezoelectric properties (to be discussed in detail in Chapter 21), single-crystal quartz plays an important role in MEMS. Large quartz crystals are grown from seeds and then sliced into flat wafers. The standard nomenclature for single-crystal quartz substrates is not based on Miller indices. Some of the basic orientations (or "cuts"), such as X-cut and Z-cut quartz, refer to the crystalline axes normal to the plane of the wafer, but others, such as AT-cut quartz, refer to off-axis orientations that are selected for specific temperature insensitivities of their piezoelectric or mechanical properties. Details on the different cuts of quartz can be found in any standard reference on piezoelectrics, such as Mason [10] or Ikeda [11].

We shall not discuss quartz microfabrication in detail, except to comment on the use of anisotropic wet etchants to create micromechanical structures (see Section 3.3.4.2).

3.2.2   Wafer Cleaning

Before a wafer can be subjected to microelectronic processes that involve high temperatures, it must be cleaned. The standard set of wafer cleaning steps is called the *RCA cleans* (see [7]). The first step is removal of all organic coatings in a strong oxidant, such as a 7:3 mixture of concentrated sulfuric acid and hydrogen peroxide ("pirhana"). Then organic residues are removed in a 5:1:1 mixture of water, hydrogen peroxide, and ammonium hydroxide. Because this step can grow a thin oxide on silicon, it is necessary to insert a dilute hydroflouric acid etch remove this oxide when cleaning a bare silicon. The HF dip is omitted when cleaning wafers that have intentional oxide on them. Finally, ionic contaminants are removed with a 6:1:1 mixture of water, hydrochloric acid, and hydrogen peroxide. Note that the cleaning solutions do not have metallic cations. These RCA cleans must be performed before every high-temperature step (oxidation, diffusion, or chemical vapor deposition).

3.2.3   Oxidation of Silicon

One of the great virtues of silicon as a semiconductor material is that a high-quality oxide can be thermally grown on its surface. The chemical reaction is straightforward: a molecule of oxygen reacts directly with the silicon, forming silicon dioxide. This is accomplished in tube furnaces, operated at temperatures from about $850°$ C to $1150°$ C. In *dry oxidation*, pure oxygen is used as the oxidant, flowed through the oxidation furnace with a background flow of nitrogen as a diluent. The oxidation rate depends on the arrival of oxygen at the silicon-oxide interface. The oxygen must diffuse through the oxide to reach this interface, so as the oxide gets thicker, this arrival rate decreases. As a result, a bare silicon wafer grows oxide relatively quickly, but an already-oxidized wafer, subjected to the same conditions, adds relatively little additional oxide.

This dependence of oxidation rate on oxide thickness has been captured in the *Deal-Grove* model of oxidation kinetics (discussed in detail in [3, 7]). If x_i is the initial oxide thickness present on the wafer when the oxidation begins, then the final oxide thickness x_f is given by

$$x_f = 0.5 A_{DG} \left[\sqrt{1 + \frac{4B_{DG}}{A_{DG}^2}(t + \tau_{DG})} - 1 \right] \qquad (3.1)$$

where A_{DG} and B_{DG} are temperature-dependent constants (see below), t is the oxidation time, and τ_{DG} is given by

$$\tau_{DG} = \frac{x_i^2}{B_{DG}} + \frac{x_i}{(B_{DG}/A_{DG})} \qquad (3.2)$$

There are two asymptotic regions. At short times, the square root can be expanded to yield the *linear rate model*:

$$x_f = \frac{B_{DG}}{A_{DG}}(t + \tau_{DG}) \qquad (3.3)$$

At long times, we obtain the *parabolic rate model*:

$$x_f = \sqrt{B_{DG}t} \qquad (3.4)$$

Table 3.2.   Deal-Grove rate constants for the dry oxidation of silicon [7]. The value of τ_{DG} is the recommended value when starting with a bare wafer.

Temperature (°C)	A_{DG} (μm)	B_{DG} (μm^2/hr)	B_{DG}/A_{DG} (μm/hr)	τ_{DG} (hr)
920	0.235	0.0049	0.0208	1.4
1000	0.165	0.0117	0.071	0.37
1100	0.090	0.027	0.30	0.067

Table 3.2 gives the rate constants for the oxidation of silicon under dry conditions. The value of τ_{DG} in the Table is what should be used for a bare

wafer and corresponds to an initial oxide thickness of about 27 nm. Hence, the Deal-Grove model is accurate only for oxides thicker than about 30 nm. Most oxides used in MEMS devices meet this criterion. The gate oxides of modern MOS transistors are thinner, on the order of 10 nm. The kinetics of oxidation for such thin oxides is a separate subject [7].

Table 3.3.   Deal-Grove rate constants for the wet oxidation of silicon [7]. The value of τ_{DG} is the recommended value when starting with a bare wafer.

Temperature (°C)	A_{DG} (μm)	B_{DG} (μm^2/hr)	B_{DG}/A_{DG} (μm/hr)	τ_{DG} (hr)
920	0.05	0.203	.406	0
1000	0.226	0.287	1.27	0
1100	0.11	0.510	4.64	0

The diffusion rate of oxygen through oxide can be significantly enhanced if there is water vapor present. Water breaks a silicon-oxygen-silicon bond, forming two OH groups. This broken-bond structure is relatively more mobile than molecular oxygen; hence, the oxidation rate is faster. The water vapor can be provided by oxidizing hydrogen to steam in the furnace. The Deal-Grove parameters for wet oxidation are given in Table 3.3.

Dry oxidation is typically used when the highest-quality oxides are required. An example is the thin gate oxides of MOS transistors, which are on the order of 10 nm thick. Wet oxidation is used to make thicker oxides, from several hundred nm up to about 1.5 μm. When still thicker oxides are required, high-pressure steam oxidation or chemical vapor deposition methods[1] are used.

Example 2.1

Determine how long it takes to grow 1 μm of oxide on a bare silicon wafer at 1000 °C under both wet and dry conditions.

SOLUTION: This is a reasonably thick oxide, so we try the parabolic rate law. The oxidation time is

$$t = \frac{x_f^2}{B_{DG}}$$

The results are 3.48 hr (209 min) under wet conditions and 85.5 hr under dry conditions.

---

[1]Chemical vapor deposition is discussed in Section 3.2.6.2.

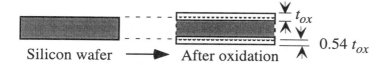

Figure 3.4. Thermal oxidation consumes some of the wafer thickness. Only 54% of the final oxide thickness appears as a net increase in wafer thickness. The remaining 46% appears as a conversion of silicon to oxide within the original wafer.

Because the density of silicon atoms in silicon dioxide is lower than in crystalline silicon, the conversion of silicon to silicon dioxide makes the overall wafer thickness increase, as illustrated in Fig. 3.4. About 54% of the total oxide thickness appears as added thickness. The remaining 46% is conversion of substrate material to oxide.

3.2.4   Local Oxidation

When a portion of a silicon wafer is covered with an oxygen diffusion barrier, such as silicon nitride, oxidation cannot occur. As a result, protected regions of a wafer remain at their original heights, while unprotected regions are converted to oxide. This is illustrated schematically in Fig. 3.5.

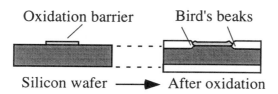

Figure 3.5. Illustrating selective oxidation of silicon with a silicon nitride thin film serving as an oxidation barrier. In practice, a very thin strain-relief oxide, of order 5 nm thick, is placed beneath the nitride to relieve strains generated during the oxidation process. This oxide is not shown in the diagram.

During the oxidation processes, stresses are generated that slightly lift the protective nitride at its edges, creating a tapered oxide called a *bird's beak*. In so-called *LOCOS* MOS processes, named for the use of *Local Oxidation of Silicon*, the semi-recessed thick oxide regions simplify the process of isolating individual transistors and interconnecting them without creating parasitic transistors beneath the interconnections. The transistors themselves are formed in the protected regions, after the nitride and thin strain-relief oxide have been removed.

3.2.5   Doping

Doping is the process whereby minute quantities of impurities are added to semiconductor crystals at substitutional sites, thereby modifying the electrical characteristics of the material. Silicon is a group-IV semiconductor. Dopant atoms from group III of the periodic table, such as boron, create mobile charge carriers in the silicon that behave like positively charged species. They are called *holes*. Materials in which holes are the predominant charge carriers are called *p-type*. Dopant atoms from group V of the periodic table, such as phosphorus, arsenic, and antimony, create mobile charge carriers that behave like negatively charged species, called *"electrons"*, where the quotation marks are added to remind us that these are special *conduction electrons*, as differentiated from the normal valence electrons on the silicon atoms. Materials in which electrons are the predominant charge carriers are called *n-type*.

When a region contains dopants of both types, it is the *net dopant concentration* in a region that determines its conductivity type. Thus an initially p-type region can be converted to an n-type region by adding more n-type dopant than the amount of p-type dopant originally present. This process is called *counterdoping*, and is critical for the manufacture of both microelectronic devices and MEMS.

The process of doping a wafer consists of two steps: *deposition*, which is the delivery of the correct amount of dopant atom to a region near the wafer surface, usually by ion implantation, and *drive-in*, which is the redistribution of the dopant atoms by diffusion.

3.2.5.1   Ion Implantation

Ion implantation is a process in which a particle accelerator shoots a beam of dopant atoms directly into the wafer. The beam is scanned across the wafer surface as the wafer is rotated so as to achieve uniform deposition. By monitoring the ion current, the total *ion dose* Q_I, in atoms/cm^2, can be accurately controlled.

This simplest way to think about ion implantation is that it introduces a very thin sheet of a known number of dopant atoms at a particular depth into the substrate. The depth of this sheet, called the *projected range*, depends on the ion energy, the ion species, and the material into which the implantation is performed. Figure 3.6 presents selected data on projected ranges in silicon.

The fact that ion implantation has a finite range means that surface layers can be used to mask regions of the wafer, preventing the dopant ion from reaching the wafer (see Fig. 3.7). Photoresist[2] or dielectric films such as silicon dioxide or silicon nitride are commonly used as implant masks, permitting selective

---

[2]Photoresist is discussed in Section 3.3.

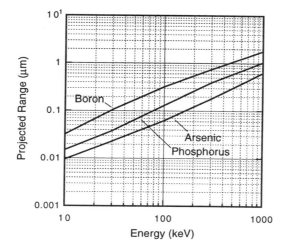

Figure 3.6. Projected ranges of ions implanted into silicon (redrawn from [7]).

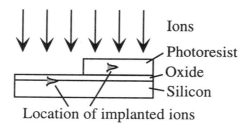

Figure 3.7. Illustrating the use of a photoresist mask to keep the implant from reaching the silicon in selected regions.

doping of different regions of the wafer surface. As a first approximation, one can use the silicon range data for silicon dioxide and other inorganic materials. For the heavier ions (phosphorus and arsenic), the projected range in oxide is about 20% less than in silicon. The projected range in photoresist is about 2-3 times greater than in silicon, so if a photoresist mask is used to protect a region of the wafer from receiving an implant, its thickness needs to be about 3 times the projected range in silicon. Of course, these estimates are much too crude for detailed device design, but are very useful at the initial "crayon engineering" phase of conceptual device and process prototyping. Computer simulations using process-modeling programs such as SUPREM[3] are used in

---

[3] SUPREM stands for Stanford University Process Engineering Module, developed by the group of Prof. Robert Dutton at Stanford. Commercial equivalents of this popular process-modeling software are available from a variety of microelectronic CAD vendors.

final device design to set the exact energy of an implant to achieve a desired dopant location.

If we look more closely at the dopant distribution following ion implantation, the ion profile can typically be described by a Gaussian of the form:

$$N_I(x) = N_{I,p} \exp \left[-\frac{(x - R_p)^2}{2(\Delta R_p)^2} \right]$$ (3.5)

where $N_I(x)$ is the dopant concentration, N_p is the peak dopant concentration, given by

$$N_{I,p} = \frac{Q_I}{\sqrt{2\pi} \Delta R_P}$$ (3.6)

and where R_p is the *projected range*. ΔR_p is the *projected range standard deviation*, and is a measure of the statistical spread of implanted ions (the width of the implanted sheet). ΔR_p is typically on the order of 25-30% of R_p. Following ion implantation, an anneal at about 900°C is required to assist the implanted atoms to reach their required substitutional positions in the crystalline lattice and to reduce lattice defects (atoms out of place) caused by the ion bombardment. If it is also desired to shift the atom positions by driving them deeper into the substrate, drive-in diffusion anneals at higher temperatures are required, as discussed in Section 3.2.5.2 below.

Because direct ion bombardment into a surface can knock surface atoms loose (a process called sputtering), it is normal to do ion implantation through a thin protective layer, such as a few tens of nm of silicon dioxide. This oxide also tends to scatter the ions slightly as they pass through, providing a range of impact angles into the silicon. This reduces an effect called *channeling*, which is the deep penetration of an atom that is moving exactly along a high symmetry direction and happens to strike the crystal between atoms. The presence of the oxide layer affects the penetration depth into the substrate, and must be taken into account, for example, with the aid of a process simulation program such as SUPREM. By proper choice of surface layer thickness and implant energy, it is possible to arrange that the peak of the implant concentration is located at, or only slightly beneath, the silicon surface. This yields an as-implanted dopant profile that decreases with depth into the wafer. Alternatively, if a higher implant energy is used, regions beneath the surface can be doped more heavily than the surface itself. The final dopant distribution, however, can be modified by the diffusion that takes place during any high-temperature process step that occurs after the implant. Dopant diffusion is discussed below.

3.2.5.2   Drive-In Diffusion

Ion implantation provides a certain dose of dopants in a layer near the silicon surface. To redistribute these dopants (as well as to remove any residual defects produced by the implantation process), high-temperature anneals in a suitable atmosphere are used. Usually, one chooses to do such *drive-in* anneals beneath a protective oxide layer to prevent dopant evaporation.

The process for redistribution is called *diffusion*, a flux of dopants from regions of high concentration toward regions of lower concentration.[4] Mathematically, diffusion is represented as having the flux proportional to the the concentration gradient of the dopant. The proportionality constant is called the *diffusion constant*. Diffusion constants for the various dopant atoms are thermally activated, having the typical form

$$D = D_0 e^{-E_A/k_B T} \tag{3.7}$$

where k_B is Boltzmann's constant (1.38×10^{-23} J/K), T is temperature in Kelvins, D_0 is a constant and E_A is the activation energy, typically in the range of 0.5 to 1.5 eV, depending on the diffusion mechanism.

Table 3.4. Parameters governing the temperature dependence of diffusion in silicon [7].

Ion	D_0 (cm^2/sec)	E_A (eV)
Boron	0.72	3.46
Phosphorus	3.85	3.66

At 1100°C, the dopant diffusivities for boron and phosphorus are both about 1.4×10^{-13} cm^{2}sec^{-1}, but they have different temperature dependences, as illustrated in Table 3.4. Drive-in diffusions require high temperatures, typically in the 1000° to 1150°C range. The net effect of a drive-in diffusion is to take the original dose produced during the deposition step and spread it out into a Gaussian profile with a width proportional to $\sqrt{Dt}$, where D is the diffusion constant for the species at the anneal temperature, and t is the drive-in time. For example, if an ion implant step is used to introduce a narrow layer of dopant atoms very near the surface with total dose Q_I atoms per square cm, the dopant profile as a function of time during a drive-in anneal takes the form

$$N(x,t) = \frac{Q_I}{\sqrt{\pi D t}} \exp\left[-\left(\frac{x^2}{4Dt}\right)\right] \tag{3.8}$$

If a complete process sequence involves several high-temperature steps, each one enables some diffusive motion of dopants. Thus, the accumulated Dt product for a complete process sequence is a rough measure of how much overall dopant motion to expect. When building MOS transistors, too much

---

[4]The physics of diffusion and of the related subject of heat flow are discussed in Chapter 11.

dopant motion can destroy device performance. Therefore, process sequences in which transistors are being fabricated have strict limits on the total Dt product.

When a p-dopant is diffused into an n-type substrate, or vice versa, a *pn junction* is created. The electrical implications of such a junction are discussed in Chapter 14. What is physically important is the *junction depth*, the distance beneath the silicon surface at which the concentration of introduced dopant of one type equals the background concentration of the wafer. Using Eq. 3.8 as the dopant distribution, if the background concentration of the wafer is N_D, then the resulting junction depth x_j is

$$x_j = \sqrt{(4Dt) \ln \left(\frac{Q}{N_D \sqrt{\pi Dt}} \right)} \qquad (3.9)$$

Example 2.2 illustrates a junction-depth calculation.

There are many second-order effects that affect diffusion profiles. Boron can dissolve slowly into overlying oxide, slightly depleting the surface concentration. Diffusion rates are also affected by the presence of point defects and, to some degree, by other dopants and by whether or not oxidation is taking place during the drive-in. Detailed modeling of these effects requires numerical simulation of the process steps using a program such as SUPREM.

3.2.6   Thin-Film Deposition

Many microelectronic process steps involve the deposition and subsequent patterning of a thin film. There is a wide variety of methods for performing such depositions, which are generally referred to as *additive processes.*]

3.2.6.1   Physical Vapor Deposition

Physical vapor deposition, abbreviated PVD, covers two major methods: *evaporation*, and *sputtering*. Evaporation is used primarily for metals. The surface of a metal sample held in a crucible is heated with an incident electron beam. The flux of vapor atoms from the crucible is allowed to reach the wafer. Such evaporation must be done under high-vacuum conditions. Evaporation with an e-beam is quite directional, allowing interesting shadowing effects to be used. For example, a tilted substrate allows one side of raised features to be coated, while the other side is shadowed. Such a method is used to deposit electrodes on the sides of etched quartz microstructures used in accelerometers and rate gyroscopes.

Sputtering is a process in which chemically inert atoms, such as argon, are ionized in a glow discharge (also called a *plasma*). The ions are accelerated into a target by the electric field in the so-called *dark space* at the boundary of the plasma. Atoms from the target are knocked out (the sputtering process),

Example 2.2

An n-type substrate with a background doping $N_D = 10^{15}$ cm^{-3} is doped by ion implantation with a dose of boron atoms Q of 10^{15} cm^{-2}, located very close to the surface of the silicon. The wafer is then annealed at 1100°C. How long should the drive-in anneal be to achieve a junction depth of 2 μm? What is the surface concentration that results?

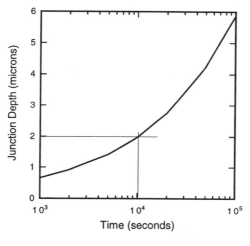

Example 2.2

Solution: Since the equation relating junction depth to diffusion time is transcendental and cannot be algebraically inverted, one can either plot x_j *vs.* t and read off the required time, or use an iterative numerical solution of the transcendental equation. The graph is shown above, with the result that the anneal time is 10^4 seconds, or 2.8 hours. The surface concentration is found by substituting this value for t into Eq. 3.8, and evaluating $N(0, 2.8$ hr$)$. The result is 1.8×10^{19} cm^{-3}.

and these atoms are allowed to reach the substrate. Sputtering takes place in a low-pressure gas environment. It is less directional than e-beam evaporation and typically can achieve much higher deposition rates. As a result, it is the metallization method of choice in most microelectronic manufacturing. Sputtering can also be used with non-metallic targets. Dielectric films, such as silicon dioxide, can be sputtered, although chemical vapor deposition methods, described later, are preferred for these films. However, some specialty materials, such as the piezoelectric films zinc oxide and aluminum nitride, are well-suited to sputtering.

3.2.6.2   Chemical Vapor Deposition

Chemical vapor deposition, abbreviated CVD, refers to a class of deposition methods in which precursor materials are introduced into a heated furnace. A chemical reaction occurs on the surface of the wafer, resulting in deposition. A wide variety of materials can be deposited by CVD methods: silicon films are formed from the decomposition of silane (SiH_4), with hydrogen as a gaseous byproduct; silicon nitride is formed by the reaction of dichlorosilane (SiH_2Cl_2) and ammonia (NH_3), with hydrogen and HCl vapor as gaseous byproducts; silicon dioxide is formed by the reaction of silane or a silane derivative with a suitable oxidizing species, such as oxygen or nitrous oxide, or from the decomposition of an organosiloxane. CVD depositions are typically performed under low-pressure conditions (less than 1 Torr) and usually involve an inert diluent gas, such as nitrogen, in addition to the reacting species. The abbreviation LPCVD is used to identify *low-pressure* CVD processes, although, in practice, most commercially important CVD process are also low-pressure processes. With the exception of silicon epitaxy (see below), CVD processes typically involve temperatures in the range of 500° to 850° C. These temperatures are too high to perform on wafers that have been metallized with either aluminum or gold, which form eutectics with silicon at 577 °C and 380 °C, respectively. Thus, in designing process flows, it is necessary to perform all CVD processes before any depositions with these metals. Tungsten, on the other hand, is able to withstand CVD temperatures.

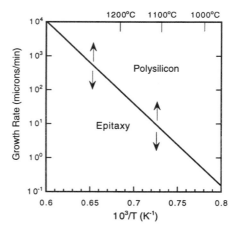

Figure 3.8.   The relation between deposition rate, temperature, and the morphology of the deposited CVD silicon film (redrawn after Fig. 7-8 of [7]).

If CVD deposition occurs on a single-crystal substrate, it is possible for the underlying crystal to serve as a template for the deposited material to develop as an extension of the single crystal. This process, called *epitaxy* is widely used for silicon and also for various III-V compounds, such as GaAs. Taking the case of silicon as an example, there is a relation between substrate temperature, deposition rate, and crystal growth. If at a given temperature, the deposition rate is slow enough to provide ample time for atoms to migrate along the surface and find equilibrium lattice sites, then epitaxy is possible. At higher deposition rates, the deposition occurs as polycrystalline material called *polysilicon*. Figure 3.8 illustrates this relation. It is seen that the maximum deposition rates for epitaxy drop dramatically with temperature. One interesting feature of silicon epitaxy is that if a portion of the single-crystal wafer is exposed, while the rest is covered with silicon dioxide, epitaxial growth can occur over the single-crystal regions while polysilicon is being deposited over the oxide.

Deposition rates can be enhanced if the deposition occurs in a glow-discharge plasma. The abbreviation PEVCD is used to identify such *plasma-enhanced* CVD processes. PECVDis used to deposit dielectric films at lower temperature (below 400 °C) than could be achieved without the plasma assist. This is extremely important in constructing insulation layers for use with metallization.

With the exception of the special conditions needed for epitaxy, films deposited by LPCVD are either amorphous, in the case of the inorganic dielectrics, or polycrystalline, in the case of polysilicon or metals. The material properties can depend *in detail* on the exact process, including such issues as temperature, gas flow rates, pressures, and the chemistry or morphology of the layer on which the film is deposited. The mass density of a deposited film, for example, depends on how many voids are present (regions where atoms could fit, but, because of the relatively low temperature of deposition, do not get filled up). Thermally grown silicon dioxide has fewer voids than LPCVD silicon dioxide, which, in turn, has fewer voids than lower-temperature PECVD silicon dioxide. In addition, the CVD oxides have some residual hydrogen. Post-deposition anneals are used to *densify* the films.

3.2.6.3   Electrodeposition

Electrodeposition, or electroplating, is an electrochemical process in which metal ions in solution are deposited onto a substrate. Metals that are well suited for plating are gold, copper, chromium, nickel, and magnetic iron-nickel alloys (permalloy). Most plating involves control with an applied electric current. Plating uniformity depends on maintaining a uniform current density everywhere the plating is done. Features of different areas, and regions at the corners of features, may plate at different rates.[5] Plated metals often exhibit rougher surfaces than evaporated or sputtered films.

Plating is used in microelectronics to deposit copper interconnect on silicon integrated circuits, and it is widely used in creating copper and magnetic-material microstructures in magnetic MEMS sensors and actuators. Plating is also used in making high-aspect-ratio microstructures, generically called HARM when using molds made with standard optical lithography [12], and called LIGA when the molds are made using X-ray lithography with highly collimated synchrotron sources [13].

3.2.6.4   Spin Casting

Thin films can be deposited from solution by a technique called spin casting. The material to be deposited, typically a polymer or chemical precursor to a polymer, is dissolved in a suitable solvent. Solution is applied to the wafer, and the wafer is spun at high speed. Centrifugal forces, in combination with the surface tension of the solution and the viscosity of the solution, spread the film to a uniform thickness. The spinning also allows for some of the solvent to evaporate, which increases the film viscosity. After spinning, baking in an oven is used to remove the remaining solvent and, depending on the material, to perform further chemical reactions.

Spinning is the standard method for depositing photoresist, a photosensitive polymer layer used in photolithography (discussed below). It can also be used to deposit polyimide films, which are insulating polymers that can withstand temperatures to about 400°C. The deposited polymer, in this case, is a polyamic acid. The post-spinning bake removes solvent and also converts the amic acid to an imide, with the loss of a water molecule. Some polyimides are also photosensitive, and can be patterned lithographically and then used as molds for plating of HARM structures.

3.2.6.5   Sol-Gel Deposition

Some materials, especially various oxides, can be formed from the sintering of a deposited gel layer that contains a suitable precursor. An example is a partially polymerized gel of organosiloxane compounds that can be dissolved in a solvent and spin-cast onto the wafer. Subsequent baking crosslinks the siloxane with the removal of the volatile organic moiety, and further densifies the siloxane into silicon dioxide. *Spin-on glasses* are examples of silicon dioxide deposited in this manner. It is also possible to deposit various piezoelectric materials, such as lead zirconate titanate (abbreviated PZT), by this method.

---

[5] A method that does not depend on current density is called *electroless plating*. It involves pre-treating the substrate with an activation solution and then using a plating solution that contains the necessary reactants. Electroless copper, for example, is used for the traces on high-performance circuit boards.

3.2.6.6   Thin Film Stress

A major problem with deposited thin films is the control of the state of *mechanical stress*. The mechanical behavior of stressed films is discussed in Chapters 8 and 9. For the moment, it is sufficient to state that virtually all deposited thin films have some degree of in-plane stress and that the stress distribution is sensitive to processing conditions, and to subsequent thermal history such as post-deposition anneals. Thermally grown silicon dioxide also is stressed. Thin-film stresses can lead to curvature of the wafer itself and to warpage and bending of suspended structures.

3.2.7   Wafer Bonding

Wafer bonding [14] is a method for firmly joining two wafers to create a stacked wafer layer. Wafer bonding is used both in MEMS device fabrication and, especially, in device packaging. There are three main types of wafer-bonding processes: direct bonding, field-assisted bonding, and bonding with an intermediate layer.

3.2.7.1   Direct Wafer Bonding of Silicon

The direct bonding of silicon wafers to one another requires high temperatures, on the order of 1000°C.[6] The process is illustrated in Fig. 3.9.

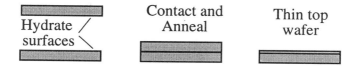

Figure 3.9.   Illustrating the direct fusion bonding of two silicon wafers.

The first step is the cleaning and hydration of the surfaces to be bonded. They must be smooth and completely particle-free. Contaminant particles create gaps which cause the bonding to fail locally. Hydration typically occurs during the wafer-cleaning operation. The surfaces to be bonded are then contacted and pressed together, using hydrogen bonding of the hydrated surfaces to provide a modest degree of adhesion. The contacted pair is placed in a high-temperature furnace to fuse the two wafers together. The resulting bond is as strong as the silicon itself. After bonding, the top wafer can be thinned by mechanical grinding and polishing, or by wet etching (see Section 3.3.4.1).

A variation on silicon-to-silicon bonding is illustrated in Fig. 3.10. The bottom wafer, in this case, has a thermal oxide on it. The surfaces are cleaned

---

[6]Bonding can also be achieved at lower temperatures, as low as 800°C, but more aggressive surface cleaning is required, and the resulting bond strength is reduced.

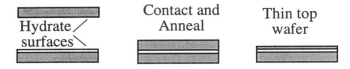

Hydrate / surfaces \   Contact and Anneal   Thin top wafer

Figure 3.10.   When one of the wafers has an oxide on its surface, the resulting structure is a thin film of silicon on insulating oxide. This type of structure is called silicon-on-insulator, or SOI.

and hydrated, then contacted and annealed, and the top wafer is thinned. The result is a layer of silicon on an insulating oxide, referred to generically as *silicon-on-insulator*, or SOI. Because this SOI structure is formed by bonding followed by etchback of the top layer, it is referred to as BESOI.[7]

Bonding can also be done on patterned wafers. Figure 3.11 illustrates the steps. One of the two wafers has a cavity etched into it. After that, the surfaces are cleaned and hydrated, contacted, and annealed. The upper wafer is then thinned, leaving a thin silicon diaphragm over the cavity. Piezoresistors can then be implanted into the diaphragm, making a pressure sensor.[8]

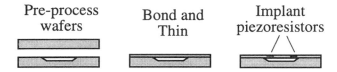

Pre-process wafers   Bond and Thin   Implant piezoresistors

Figure 3.11.   Illustrating the fabrication of a piezoresistive pressure sensor using wafer bonding.

This sealed cavity example has another use. After bonding and thinning, the exterior of the wafer is virtually identical to a blank wafer. Therefore, it can be used in a standard microelectronics fabrication line to build transistors on the regions next to the cavity. After the transistors are built, subsequent etch steps in the diaphragm region can be used to create moveable microstructures on the same wafer as the circuitry [15].

Depending on the ambient atmosphere in which the wafers are sealed, the sealed cavity may have residual gas in it. Oxygen in the cavity reacts with the walls during the high-temperature anneal, forming a thin oxide on the cavity walls, but nitrogen or other inert species remain. After the device is thinned, the diaphragm is potentially vulnerable to damage when heated, because the pressure increase of the residual gas in the cavity causes the thinned diaphragm

---

[7] An alternate method for making very thin SOI layers is to implant a heavy dose of oxygen into a silicon wafer at high energy, so that the peak oxygen concentration is well below the silicon surface. Annealing then converts the heavily implanted region to oxide beneath a surface layer that remains as single-crystal silicon.
[8] Piezoresistors are discussed in Chapter 18.

to bend, and possibly break. By controlling the ambient gas composition during bonding, cavity pressures after bonding can be reduced to a safe zone [15].

It is also possible, using commercially available equipment, to align already-patterned wafers to one another prior to bonding, with accuracies of a few microns. This makes possible microstructures that cannot be fabricated using a single wafer.

3.2.7.2   Anodic Bonding

In the previous section, hydrogen bonding was used to hold the wafers in intimate contact, and a high-temperature anneal was used to fuse the two wafers chemically. An alternative method, which is restricted to certain glasses bonded to conductors, is called *field-assisted bonding*, or *anodic bonding*. The apparatus is illustrated in Fig. 3.12.

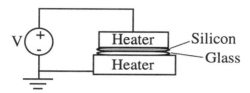

Figure 3.12.   Apparatus for anodic bonding of silicon to glass.

The mechanism responsible for anodic bonding is the mobility of sodium ions in the glass. Pyrex 7740 glass, which has a thermal expansion close to that of silicon, is a good candidate. When the silicon wafer is placed on the glass, and the two are heated to temperatures on the order of 500°C, a positive voltage (300 - 700 V) applied to the silicon repels sodium ions from the glass surface, creating a net negative charge at the glass surface. The force of attraction between the positively charged silicon wafer and the negatively charged glass surface brings the two surfaces into intimate contact. At the elevated temperature, they can fuse together. Contact is typically initiated at a single point by applying a load, and, as contact is established, it spreads out to cover the rest of the wafer. Bonding time ranges from seconds to minutes and can be monitored by measuring the current in the circuit. When bonding is completed, this current drops to zero. As with silicon-to-silicon bonding, this bonding method is very susceptible to interference from particulate contaminants.

3.2.7.3   Bonding with an Intermediate Layer

We are all familiar with the use of adhesives or solders to bond and laminate structural elements together. In the case of microelectronics, candidates for the role of the adhesive must meet the thermal and cleanliness requirements of microfabrication. Glass frits, which are powdered slurries of relatively low-melting glasses, can be applied selectively to parts of the surface of one of the wafers using techniques such as screen printing. After a low-temperature bake to reduce the fluid content of the frit layer, the wafers are contacted and annealed, flowing the glass into a continuous layer that bonds the wafers together.

An alternative "adhesive" method is to use gold layers and apply moderate pressure to the two wafers at a temperature of about 300°C. This is called *thermocompression bonding*. Or conventional solders can be used, provided the mating surfaces have been pre-coated with a material which the solder can easily wet.

Finally, one can use polymeric adhesives, such as polyimides, silicones, or epoxy resins, provided that the cleanliness and thermal requirements of the device process are not compromised.

3.3   Pattern Transfer

Integrated circuits and microfabricated MEMS devices are formed by defining patterns in the various layers created by wafer-level process steps. Pattern transfer consists of two parts: a photo-process, whereby the desired pattern is photographically transferred from an optical plate to a photosensitive film coating the wafer, and a chemical or physical process of either removing or adding materials to create the pattern. Most processes are subtractive, removing material by etching unwanted material away chemically. A few processes are additive, such as doping or plating. Another additive process is called lift-off, discussed later in this section.

3.3.1   Optical Lithography

Optical lithography is very much like the photographic process of producing a print from a negative. Contact prints are made when the negative is placed directly onto a sheet of photosensitive paper. When light is shined through the negative onto the surface of the paper, the varied light intensity reaching the paper exposes the image, which is then made visible through a chemical development process. Enlargements or reductions are made by interposing suitable lenses between the negative and the photosensitive paper. Both methods are used in optical lithography.

The enabling materials of optical lithography are *photoresists*, polymeric optically-sensitive materials that are deposited onto the wafer surface by spin casting. Following spinning, the resists are prebaked at low temperature to remove solvent, but are not fully hardened. Completion of the hardening process occurs after optical exposure.

Figure 3.13 illustrates the lithographic process analogous to contact printing. A *photomask* contains the pattern to be transferred as a set of opaque and

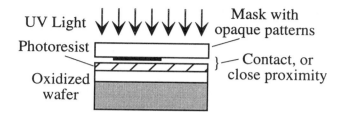

Figure 3.13. Illustrating contact or proximity photolithography.

transparent regions. It is brought into contact with an oxidized silicon wafer coated with photoresist. Ultraviolet light is directed through the mask onto the wafer, exposing the unprotected portions of the resist, which change their chemical properties as a result of the light exposure. Unlike regular photography, optical lithography as practiced by the integrated circuits industry does not use any gray scale. The photochemical processes in the photoresist are relatively high in contrast, and develop sharp boundaries between exposed and protected regions. Contact lithography is one of the standard processes used in MEMS manufacture.

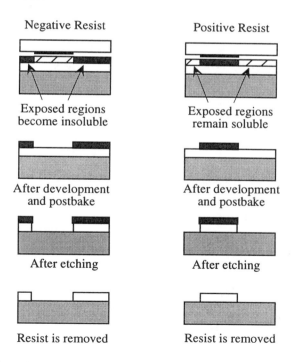

Figure 3.14. Illustrating how negative and positive photoresists result in different patterns from the same mask.

What happens after exposure depends on the specific chemistry of the photoresist. There are two types, illustrated in Fig. 3.14. *Negative photoresist* functions much like the photographic printing process. Regions that are clear in the negative become dark in the transferred image. Specifically, the regions of the photoresist that are exposed to the ultraviolet light become cross-linked and insoluble in the developer, while the protected regions remain soluble. After immersion in the developer or exposure to a continuous spray of developer, the soluble portions are removed. The net result is a transfer of pattern into the photoresist so that after etching, the opaque regions of the mask become regions cleared of photoresist. To transfer the pattern into the oxide, the resist must first be hardened by baking to make it more chemically inert. After this postbake, the silicon dioxide can be removed by an etching process, to be discussed later. Following the etching, the photoresist is removed, leaving the mask pattern transferred into the oxide layer.

Positive photoresist works oppositely to negative photoresist. The chemistry of the photoresist is different. Regions exposed to the UV light become more soluble in the developer than the protected regions. After development and postbake, the protected regions of resist remain on the wafer so that after etching, the opaque regions of the mask remain as oxide and the clear regions are removed. A good way to remember what each resist does is the mnemonic: "Positive resist protects."

Because direct contact between the wafer and the mask can eventually cause damage to the mask, a variant of the contact lithography is to leave a small air gap between the mask and the photoresist-covered wafer. This is called *proximity lithography*. The achievable resolution is somewhat less than with contact lithography, because diffraction can occur at the edges of the opaque regions. This process was used in the early days of integrated circuit technology, but has now been largely replaced by step-and-repeat projection lithography, described later.

When using contact lithography, the mask must be the same size as the wafer, and every feature to be transferred must be placed on the mask at its exact final size. An alternative method, which has become the standard in the integrated-circuit industry, is to use *projection step-and-repeat lithography* to expose the photoresist. This method is illustrated schematically in Fig. 3.15. The projection optics operate as in a photographic enlarger, except the mask image is *reduced* by a factor of 5 or 10 when projected onto the wafer. This has several advantages. First, the mask can be drawn with larger feature sizes: a 0.3 μm feature on a wafer requires a relatively easy-to-create 3 μm feature on the mask. Second, because of the reduction, only a portion of the wafer is exposed at one time. The exposed region is typically one "chip," a complete integrated circuit. Chip sizes of 1 to 1.5 cm can be made in this way. In order to expose the entire wafer, the wafer is moved after each exposure to

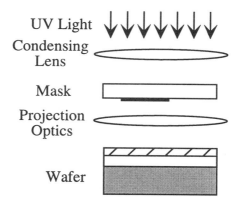

Figure 3.15. Illustrating projection lithography. The projection optics typically reduce the mask features by a factor of 5 or 10 when projected onto the wafer.

bring the next chip under the projection optics. This is called *step-and-repeat* lithography. While it requires rather expensive equipment to do the projection and precision positioning needed for the step-and-repeat operation, it has the distinct advantage that alignment of successive patterns is done locally, one chip at a time. It is not necessary to maintain rigid positional accuracy of successive layers at the 0.1 μm level across the full diameter of the wafer, a task that becomes impossibly difficult as wafer sizes increase.

Another alternative to contact lithography is the scanning 1:1 projection aligner, which illuminates an arc-shaped portion of the mask, and has an optical system that accurately projects this region onto the wafer. Then, both the mask and the wafer are scanned synchronously, projecting the scanned mask image onto the scanned wafer.

Because the "stepper" has a field of view on the wafer on the order of 1 cm, many MEMS devices cannot be fabricated with stepping lithography, but must use contact or scanning projection lithography. But where a stepper can be used for MEMS fabrication, the increased accuracy of the patterns can be an advantage. A rule of thumb is that features down to 1-2 μm can be successfully patterned with contact or scanning projection lithography on wafers that already have some structure present, with alignment accuracies on the order of a few tenths of microns. If finer features are needed, the stepper becomes a necessity. This may require device redesign to fit into the size of one chip. Alternatively, it is possible (but difficult) to stitch patterns together so that devices that cover more than one chip in area can be built with a stepper.

Regardless of whether contact or projection lithography is used, there can be systematic errors in the pattern-transfer process. The regions of the photoresist that remain on the wafer may not match exactly the corresponding mask regions, because the optical exposure and development step may result in pattern expansion or reduction. Further, the etching step, to be discussed more below, can undercut the resist slightly (this is indicated schematically in Fig. 3.14, and in more detail in Fig. 3.17 (page 58). The net effect of the optical and etching steps is to introduce a systematic error between the drawn mask dimension and the final dimension of the patterned feature on the wafer. This systematic error is called a *process bias*. First-order compensation of process bias can be achieved by determining through experiment what the bias is, and then adjusting the dimensions of the mask features so that the final fabricated dimension is correct.

There are also random errors in pattern transfer. These can arise from errors in registering one pattern to what is already present on the wafer, or from slight variations in resist exposure, or etching times. As we shall discover later, the behavior of MEMS devices can depend strongly on device dimensions. Therefore, both process biases and random patterning errors can result in variations in device performance that must be corrected with suitable calibration procedures.

3.3.2   Design Rules

When designing a complete process, each layer must be aligned to the previously patterned features. Because there can be small misalignments at each step, robust process design calls for the use of minimum mandatory offsets between mask features in successive layers. An example is shown in Fig. 3.16, in which the upper layer is required to extend beyond the boundaries of the via by a minimum distance.

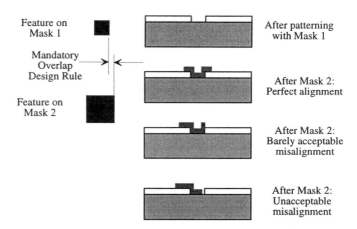

Figure 3.16.   Illustrating design rules that permit successful fabrication of devices in the presence of misalignment errors.

Figure 3.16 shows what can happen during misalignment. The feature in the upper layer patterned with Mask 1 can be incorrectly placed with respect to the lower feature patterned with Mask 2. At the bottom is shown an unacceptable misalignment. Design rules specify a minimum overlap between such mask features so that even in the worst expected misalignment, the fabricated structure is acceptable. The rules are developed from statistical information on alignment errors obtained from actual device processing, and are structured so that, with high probability, every process step produces acceptable devices.

Integrated-circuit processes have many design rules governing every step of the process sequence, and every mask layer. In MEMS devices, other than prudent use of minimum overlaps and consistency with the minimum feature sizes that can be produced by the lithographic process, there are no standard design rules. The device designer must assess the impact of possible alignment errors at every step, and adjust mask features accordingly.

3.3.3 Mask Making

Since photomasks are the basis for lithography, it is reasonable to ask where the masks themselves come from. Production of masks is a specialized process that can be purchased from vendors. There are two basic methods. Both start with fused silica (amorphous quartz) plates (5" square plates are typical) that have been coated with a thin chromium layer, and then with a photoresist. The type of resist depends on the method to be used to create the mask.

An *optical pattern generator* requires an optical photoresist. It has sets of shutters that permit the exposure of rectangular-shaped regions. The mask fabrication is much like a step-and-repeat lithography process: a rectangle is exposed, the wafer is repositioned, and another rectangle is exposed. This process continues until the entire mask pattern is created in the photoresist. The resist is then developed and baked, and the chromium etched into the correct pattern. The resist is removed, leaving a chrome-on-quartz mask. A variant of this procedure is to use a photographic emulsion on the quartz plate that becomes optically opaque when exposed and developed. Emulsion masks are cheaper to make, but are more susceptible to scratching and surface damage than chrome masks.

A design for use with an optical pattern generator is, at its most basic, a list of rectangles specifying their size, orientation (within limits imposed by the pattern generator), and their location on the wafer. The pattern-generator code is created by the CAD tools used for *mask layout*, converting the drawings of the designer into a transfer format suitable for pattern-generator use.

A second method for making masks is to use an electron-beam lithography machine to write the required pattern into an *electron-beam resist*, a polymer that is transparent to light, but can be depolymerized by the penetration of an electron beam, rendering the material locally soluble. The electron-beam machine is much like an electron microscope, with an electron gun, lateral controls that direct the beam to different regions of the sample, and blanking controls that turn the beam on and off as needed. A typical electron beam resist is poly-(methylmethacrylate) (abbreviated PMMA) . It behaves much like a positive photoresist, in that it is removed where the e-beam exposure occurs. Following exposure, the resist is developed, the chromium is etched, and the resist is removed.

A major advantage of e-beam masks is that there are few restrictions on the shapes of features. Many MEMS devices employ curved structures. To form these with optical pattern generators requires superposition of many rectangles, forming a somewhat jagged edge that washes out into a smooth curve because of the finite spatial resolution of the mask-making and photolithography processes. With an e-beam, however, the curved feature can be written directly.

If the feature sizes of interest are relatively large, greater than about 50 μm, an alternate low-cost mask-making method can be used. The artwork is drawn using any standard personal-computer software, and then printed onto transparencies using a high-resolution laser printer. Depending on the wafer size, one can fit several mask layers onto a single transparency. Since commercial quartz masks can cost $1000 per layer and may require several weeks to obtain from a vendor, the single-transparency approach has obvious appeal when it can be used.

3.3.3.1   Double-Sided Lithography

Many MEMS devices require patterning on both sides of a wafer. To achieve the required positional alignment between front and back side, special tooling is required. One first aligns the back-side mask to fiducial marks on the tool, then places the wafer over this mask, and aligns the wafer to the tool using alignment features on the front. Within alignment tolerances, the wafer and back-side mask are then aligned to each other. The back side of the wafer is then exposed from the back.

3.3.3.2   Soft Lithography

So-called *soft lithography* is a relatively recent development for transferring patterns [16]. Unlike conventional optical lithography, it uses a molded polymeric body to accomplish physical pattern transfer, just like a rubber stamp used to press onto an ink pad then onto paper. The mold is formed by casting a silicone rubber, poly(dimethylsiloxane) (PDMS), onto a master that contains the desired relief pattern. The master can be formed by conventional lithography and etching. The molded parts are stripped from the master, and are then used as flexible printing devices, especially for non-planar substrates. When coated with the material to be transferred (typically a thin organic film, such as a self-assembled monolayer (SAM)), then pressed onto the desired surface, patterned material transfer can be achieved. The production of the patterns and molds can be done in a few steps, permitting rapid prototyping to explore new ideas. While this new approach is not yet a standard manufacturing method, it appears to hold significant promise for the future of MEMS.

3.3.4   Wet Etching

We have already referred several times to the process of removing material through the openings in a suitable masking layer, such as patterned photoresist. We now briefly survey the various etching methods used in microelectronic fabrication.

3.3.4.1   Isotropic Wet Chemical Etching

The oldest form of etching is immersion of the patterned substrate in a suitable liquid chemical that attacks the exposed region of the substrate, and leaves the protected regions alone. Acid etching of metal plates (engraving) has been used by artists for many centuries to make master plates for printing artworks.[9]

The rate of etching and the shape of the resulting etched feature depend on many things: the type of substrate, the specific chemistry of the etchant, the choice of masking layer (and the tightness of its adhesion to the substrate), the temperature (which controls reaction rates), and whether or not the solution is well stirred (which affects the rate of arrival of fresh reactants at the surface). Depending on the temperature and mixing conditions, the etching reaction can be either *reaction-rate controlled*, dominated by temperature, or *mass-transfer limited*, determined by the supply of reactants or the rate of removal of reaction products.[10]

Most wet etching is *isotropic*, in that the rate of material removal does not depend on the orientation of the substrate. However, when etching single-crystal substrates with certain etchants, orientation-dependent etching can occur, as discussed in the next section.

Figure 3.17 illustrates the use of isotropic wet etching through the openings in a masking layer to pattern a thin film on a substrate. The assumption is that the etchant does not attack either the masking layer or the substrate. Note the undercut of the mask feature. This is because of the isotropic etch behavior; provided there is a sufficient supply of reactants, lateral etching occurs at about the same rate as vertical etching. Adhesion of the mask to the thin film is also

---

[9]And also for printing money. The *U.S. Bureau of Printing and Engraving* is where United States currency is manufactured.

[10]Chemical reactions are, to some degree, reversible. The rate of the reverse reaction depends on the concentration of reaction products. If these are not removed from the reaction site, the net etching rate can decrease. Stirring is a good way to remove reaction products.

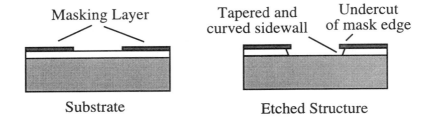

Figure 3.17. Illustrating pattern transfer by isotropic wet etching through the openings in a masking layer. Because the wet-etching is isotropic, the mask is undercut, and the sidewall is typically tapered and curved (the curve detail is not shown in the figure).

important. If this adhesion is weak, enhanced etching can actually occur at the film-mask interface, exaggerating the sloping of the sidewall. Not shown in the figure is the fact that the sidewall, in addition to having a slope, is actually somewhat curved. Such tapered sidewalls can be an advantage when attempting to cover the etched feature with an additional film. Perfectly vertical steps are harder to cover.

Table 3.5. Selected wet etchants.

Material	Etchant
Thermal or CVD silicon dioxide	Buffered hydrofluoric acid (5:1 NH$_4$F:conc HF)
Silicon nitride	Hot phosphoric acid
Polysilicon	KOH or ethylene diamine/pyrochatecol (EDP)
Aluminum	PAN (phosphoric, acetic, nitric acids)
Copper	Ferric chloride
Gold	Ammonium iodide/iodine alcohol

Table 3.5 illustrates the variety of materials that can be etched with wet etchants. A much more complete database of etchants, including data on etch rates and selectivity, can be found in [17] and [18]. The *selectivity* of the etchant for the desired material, compared to other materials that may be present, is important. For example, HF etches silicon dioxide, but also etches silicon nitride slowly. If nitride is the masking material, one must be concerned with how long it must remain exposed to the etchant. This is not a problem when simply patterning a thin film beneath the mask, but in sacrificial processes, collectively referred to as *surface micromachining*, in which long etch times may be required to remove all of the oxide beneath structural elements, the etch selectivity of protective layers is important.

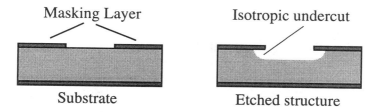

Figure 3.18. Isotropic wet etching deeply into a substrate. The masking material is required on both sides of the wafer in this case. If the wafer edges are not protected, they will be etched as well. This example illustrates a single chip, and does not show the wafer edge.

Wet etching can also be used to etch deeply into the substrate. Whenever the substrate is heavily etched, it is common to refer to the process as *bulk micromachining* [19]. Figure 3.18 illustrates the kind of etched features that result when using isotropic wet etchants. An example is the use of nitric-acetic-hydroflouric acid mixtures to etch silicon, or the use of hot buffered HF to etch deeply into glass. One method of making the microflow channels used in microfluidic biochemical analysis systems is to perform this type of wet etching in a glass substrate.

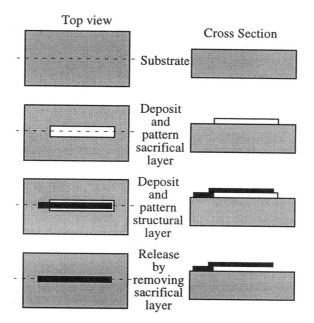

Figure 3.19. Illustrating surface micromachining: the use of an isotropic wet etchant to remove a sacrificial layer beneath a cantilever beam.

Surface micromachining [20] is illustrated in Fig. 3.19. A sacrificial layer is deposited and patterned on a substrate. Then, a structural material is deposited and patterned so that part of it extends over the sacrificial layer and part of it provides an anchor to the substrate. Finally, the sacrificial layer is removed, leaving a cantilever beam of the structural-layer material attached at one end to the substrate.

Table 3.6.   Surface micromachining material systems.

Structural	Sacrificial	Release Etch	Isolation	Ref.
Polysilicon	SiO_2	Buffered HF	$Si_3N_4 + SiO_2$	[20]
Polyimide	Aluminum	PAN etch	SiO_2	[21]
LPCVD Si_3N_4 + Al	Polysilicon	XeF_2	SiO_2	[22]
Aluminum	Photoresist	Oxygen plasma	SiO_2	[23]

There is enormous flexibility in the design of surface micromachining processes. One needs three, or possibly four, different materials: a substrate (or a suitable thin-film coating over a substrate to provide the anchoring surface), a sacrificial material, a structural material, and, possibly, an electrical insulation material to isolate the structural elements from the substrate. This last role can be played by a coating over the substrate. The etchant that is used to release the structure must etch the sacrificial layer quickly, and the remaining layers very slowly, if at all. For example, 5:1 buffered HF etches thermal oxide at about 100 nm/min, but etches silicon-rich silicon nitride at a rate of only 0.04 nm/min [18]. Table 3.6 lists various combinations of materials and etchants that have been successfully used in surface micromachining.

During the release etch, when it is typically necessary to exploit the undercutting of the structural elements to remove thin films over distances large compared to the isolation-layer thickness, the etch selectivity becomes critical. In the case of silicon dioxide as a sacrificial material with silicon nitride as the isolation material, if the stoichiometry is adjusted to make the nitride silicon rich, not only does the etch rate in HF decrease significantly, but its residual stress decreases, allowing thicker layers to be deposited without cracking or other deleterious mechanical effects.

One significant technology problem in surface micromachining is the unintended adhesion of released mechanical elements to the substrate. When using a wet release etch, the surface tension during drying can pull compliant beams into contact with the substrate, and during the final drying, they can adhere firmly together. This phenomenon is generically called *stiction*. Methods of avoiding stiction include (1) the use of self-assembled molecular monolayers

(SAM's) to coat the surfaces during the final rinse with a thin hydrophobic layer, reducing the attractive force, (2) the use of vapor or dry-etching release methods, such as XeF_2, (3) various drying methods (freeze drying and drying with supercritical CO_2) that remove the liquid without permitting surface tension to act, and (4) temporary mechanical support of the moveable structure during release using posts of photoresist or some other easily removed material.

Surface micromachining is proving to be commercially important. By far the most widespread material system used to date is silicon-rich silicon nitride as a coating on an oxidized silicon substrate to provide electrical isolation and an anchor, silicon dioxide as the sacrificial layer, polysilicon as the structural material, and buffered HF as the etchant. This combination is used, for example, in Analog Devices surface micromachined accelerometers, as discussed in Chapter 19. The use of LPCVD Si_3N_4 with aluminum electrodes as the structural material, polysilicon as the sacrificial layer, and the vapor etchant XeF_2 as the release etch is used in the Silicon Light Machines optical projection display, as discussed in Chapter 20. Texas Instruments [23] uses aluminum as the structural material in their projection display, with photoresist as the sacrificial material removed by plasma etching.[11]

3.3.4.2   Anisotropic Wet Etching

When etching a single crystal, certain etchants exhibit orientation-dependent etch rates. Specifically, strong bases, such as potassium hydroxide (KOH), tetramethyl ammonium hydroxide (TmAH), and ethylene diamine pyrochatecol (EDP), exhibit highly orientation-dependent etch characteristics in silicon [19]. Hydroflouric acid exhibits similar orientation-dependent effects when etching single-crystal quartz [24].

It has been suggested [25, 26] that the mechanism responsible for orientation dependent etching in silicon is the detailed bond structure of the atoms that are revealed in different surface planes. For example, {100} and {110} planes have atoms with two bond directed back into the crystal, and two bonds that must somehow "dangle," with unfulfilled valence. Atoms in {111} planes, however, have three bonds directed back into the crystal and only one bond that dangles. Hence atoms in {111} planes, to first order, are "more tightly bound" to the rest of the crystal. What is observed experimentally is that the {111} planes etch significantly less quickly than {100} or {110} planes in strong bases, and this permits the creation of amazing structures.

The etching reaction is the breaking of a silicon-silicon bond with the insertion of OH groups, and hence is enhanced in strong bases, which have an abundance of OH^- ions. Ultimately, the etching forms $Si(OH)_4$, which is solu-

---

[11]Plasma etching is discussed in Section 3.3.5.

ble in strong bases, along with the consumption of four molecules of water and the release of two molecules of hydrogen gas. More details on the reactions and proposed mechanisms can be found in the references.

Suitable masks for these strong-base etchants are silicon dioxide, which does etch, but rather slowly, and silicon nitride, which is nearly untouched. If a wafer with a (100) orientation and a suitably patterned masking layer is put into KOH (or some other strong-base etchant), the etch proceeds rapidly until all of the exposed crystal faces have {111} normals.

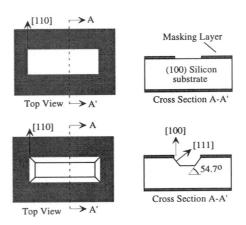

Figure 3.20. When a (100) oriented wafer with mask features aligned to <110> directions is placed in an anisotropic etchant, exposed {100} planes etch rapidly, but {111} planes etch slowly.

Figure 3.20 shows a rectangular opening in a masking layer on a (100) silicon wafer. The edges of the rectangle are perfectly aligned with the symmetrically equivalent <110> directions in the plane of the wafer. As etching proceeds, the exposed {100} planes etch rapidly while the {111} planes etch slowly. Every {111} plane intersects a (100) surface along one of the <110> directions. Hence, starting from every straight mask edge that is aligned with one of the <110> directions, a slow-etching {111} plane becomes revealed by the etch. It makes a 54.7° angle with the plane of the wafer, the characteristic feature of anisotropically etched bulk-micromachined structures. Because the etch rate of the {111} planes is not zero, there is some small mask undercut, not shown in the Figure.

Allowing the structure of Fig. 3.20 to remain in the etch eventually results in the meeting of {111} planes initiated at opposite edges of the mask. The result is a self-limiting V-groove, as illustrated in Fig. 3.21. Again, there is a small amount of mask undercut, not shown in the Figure, because the etch

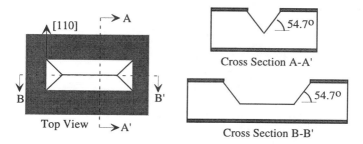

Figure 3.21.   If the structure of Fig. 3.20 is allowed to continue, a self-terminating V-groove is formed, with all surfaces bounded by {111} planes.

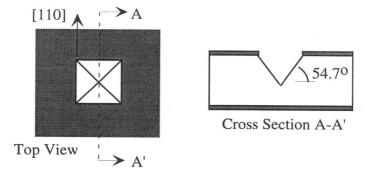

Figure 3.22.   A square <110>-oriented mask feature results in a pyramidal pit.

rate of the {111} planes is not strictly zero. Figure 3.22 shows that a square <110>-oriented mask feature results in a pyramidal pit.

Figure 3.23 shows an extremely important feature of anisotropic etching. When the opening in the mask layer has a convex corner, the {111} planes that initiate from the edges making up that corner are rapidly etched away, undercutting the convex-corner portion of the mask. A plausible explanation for this behavior is that if the reason for the slow etching of {111} planes is that three of the surface-atom bonds are directed into the silicon and only one is dangling, then at a convex corner where two {111} planes meet, this stabilizing feature is no longer present. Atoms exactly at such a boundary must have two dangling bonds, and thus are etched away, exposing other fast-etching planes. Regardless of whether this simple explanation is exactly correct, such corners are found experimentally to undercut rapidly. If the etch is allowed to go to completion, the protruding part of the mask layer is completely undercut, leaving a cantilever beam hanging over a V-groove like a diving board over a swimming pool. At the A-A' cross section, the mask-layer structure is completely suspended.

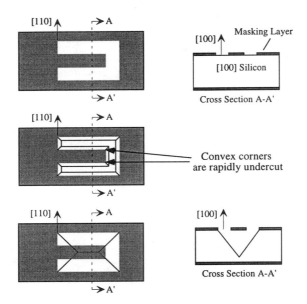

Figure 3.23. Convex corners where {111} planes meet are not stable. They are rapidly undercut. This permits creation of suspended structures.

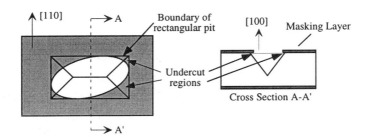

Figure 3.24. Any mask-layer feature, if etched long enough, will result in a rectangular V-groove pit beneath a rectangular that is tangent to the mask features, with edges oriented along <110> directions.

This undercutting phenomenon extends to other shapes. Figure 3.24 shows an elliptical feature. Anisotropic etching of this feature results in a rectangular V-groove pit with its edges parallel to <110> directions. Large areas of the mask layer can be undercut in this manner.

Undercutting has important implications for alignment accuracy. Suppose it is desired to align a rectangular feature so that its edges are perfectly parallel to <110> directions but, because of a lithographic error, the pattern is actually rotated a few degrees off target. Figure 3.25 illustrates a drastic case, with a 5° misalignment. Once again, the bounding rectangular etch pit is what

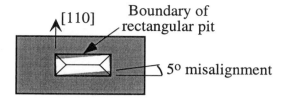

Figure 3.25. The effect of misalignment is to enlarge the etched region. This figure shows the effect of a 5° misalignment for a rectangular feature.

determines the final etched shape. In practice, wafer flats identify a [110] direction to better than one degree. However, if extremely accurate alignment is needed, then it is necessary to pre-etch an alignment feature into the wafer using the anisotropic etchant so that the exact <110> in-plane directions are delineated. An example of such a feature would be a circle. In an anisotropic etchant, the circular mask feature tends to form a square with edges tangent to the circle along <110> directions.

The etch rate in anisotropic etchants decreases rapidly when etching heavily boron-doped material, with concentrations in excess of about 5×10^{19} cm^{-3}. This means that heavily doped p-type silicon, denoted p+ silicon, can be used as a masking material.

Because anisotropic etches require long exposure to the etchant, the etch-rate ratio of the fast-etching planes to the {111} planes is extremely important. In 20 weight-percent solution of KOH at 85 °C, the etch rate of the {100} planes is 1.4 μm/min, and the selectivity to the {111} planes is 400:1. The etch rate of the masking layer is also important. Possible masking materials for KOH are silicon nitride, which hardly etches at all, SiO_2, which etches at about 1.4 nm/min, and p+ silicon, which offers between a 10:1 and 100:1 reduction in etch rate over lightly doped silicon, depending on the etchant and the etch temperature.

As an example, to etch all the way through a 500 μm thick wafer, which requires about 6 hours in 20% KOH at 85°C, would require at least a 504 nm of silicon dioxide as a masking layer. Alternatively, as little as 50 nm of silicon nitride would suffice. Processes that include these deep etches must be designed with both absolute etch rate and mask etch rate in mind. Additional data on the temperature- and doping-dependences of the etch rates for silicon and for various masking materials can be found in references [8, 19, 25, 26].

3.3.4.3 Etch Stops for Wet Etching

While it is always possible to use a measured etch rate and a specified time to determine the depth of etched features, it is highly desirable to be able to use a well-defined structural feature to stop an etch. There is a rich array of possible etch-stop structures in use. This section introduces a few of them.

Electrochemical Etch Stop

When a silicon wafer is biased with a sufficiently large anodic potential relative to the etchant, it tends to oxidize. This process is called *electrochemical passivation*, because if there is oxide on the surface, the oxide masks the etch.[12] This passivation step is a redox reaction and requires current. If the supply of that current is blocked with a reverse biased pn junction, the passivation cannot occur and the etching can proceed.

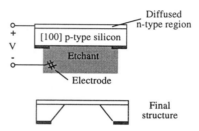

Figure 3.26. Illustrating the electrochemical etch stop. The reverse biased pn junction prevents the required passivation current from flowing. As a result, the p-silicon remains unpassivated and is etched. When the junction is reached, the passivation current can flow, so the n-silicon becomes passivated and etching ceases.

The use of this electrochemical property to control the thickness of an etched diaphragm is illustrated in Fig.3.26. A p-type wafer has a diffused n-region at its upper surface. The depth of the junction can be accurately controlled by the combination of ion implantation and drive in. The backside of the wafer is covered with a masking layer that contains an opening. When this wafer is placed into the anisotropic etchant with an anodic bias applied as shown, the reverse biased p-n junction prevents the passivation current from flowing. As a result, the p-silicon remains unpassivated and hence etches. However, the n-silicon becomes passivated and is not etched. Furthermore, as anisotropic etching proceeds, when the junction is reached there is no longer anything blocking the flow of passivation current. The n-silicon quickly passivates, terminating the etch. The resulting diaphragm thickness is equal to the depth of the original junction, at least ideally.[13] In practice, it is necessary to control the junction leakage and current supply paths to be certain, first, that there is no

---

[12]It is also called *anodization*, although that usage refers to treatments of aluminum rather than to silicon etching.

[13]The resulting structure can be used to make a pressure sensor, as discussed in Section 4.3.1 and in Chapter 18.

pathway that can provide enough current to passivate the p-region, and, second, that once the junction is reached, all features have enough current supply so that they passivate at the same point.

Figure 3.26 is only schematic, in that it does not show the container for the etchant and the fixturing that is typically used. In practice, the wafer is immersed into the etchant with only an edge available for contact. The top surface n-type surface would typically be protected with additional mask-layer material. In addition, contact to the p-type layer is often made to assure that any leaking current through the pn junction can be captured before it causes passivation.

p+ Etch Stop

If instead of using an n-type diffusion into a p-wafer, a heavily boron doped p+ layer is formed by ion implantation and diffusion, the anisotropic etch will terminate without requiring application of anodic bias. This greatly simplifies the tooling required to perform the etch. However, the etch-rate selectivity is not quite as high as for passivating oxides. Furthermore, the p+ diaphragm will have residual tensile stress that might affect the performance of any device, such as a pressure sensor, that depends on the mechanical properties of structure. Finally, it is not possible to diffuse piezoresistors into already-heavily-doped p+ silicon. So while the p+ etch stop offers some advantages and can be used with great effect in some processes, it also has some important limitations.

Dielectric Etch Stop

If the n-layer of Fig. 3.26 were replaced by a material that is not etched, for example, silicon nitride, then the result of the anisotropic etch is a silicon nitride diaphragm suspended over a hole through the silicon. LPCVD silicon nitride is very brittle, and has a very high residual tensile stress, making it prone to fracture. However, if the ratio of dichlorosilane to ammonia is increased during silicon nitride deposition, a silicon-rich nitride layer is deposited which has lower stress. Membranes of silicon rich nitride can be used as the basis for mechanical devices. Masks for X-ray lithography can also be made in this fashion, depositing gold on the membrane to create the masking regions.

No Etch Stop: Making Through Holes

If no etch stop is used, a hole is produced all the way through the wafer. The sloped sidewalls of the holes also offer an opportunity for making interconnects between the back and the front of a wafer [27].

3.3.5   Dry Etching

Chemically reactive vapors and the reactive species in glow-discharge plasmas are highly effective etchants. These are classified into two groups: vapor etching, and plasma-assisted etching. Within the plasma-assisted etching class are three variants that have very different characteristics.

3.3.5.1   Vapor Etching

There is one vapor etchant that has become commercially important in micromachining processes. The gas xenon diflouride, XeF_2, is a highly selective vapor etchant for silicon, with virtually no attack of metals, silicon dioxide, or other materials [22]. As a result, it is ideal for the dry release of surface micromachined structures in which polysilicon is used as the sacrificial layer. This process is used in the manufacture of the electrostatically actuated projection display chip discussed in Chapter 20.

Vapors are also used in some packaging operations, for cleaning and degreasing components prior to packaging.

3.3.5.2   Plasma-Assisted Etching

A low-pressure glow-discharge plasma produces ionized species in abundance. When directed to the surface of the wafer, these ions can produce both sputtering effects, removing material by direct ion-beam bombardment, and chemical-reaction effects, converting surface atoms to volatile species that can be removed by the vacuum pump.

The least selective use of ions produced in a plasma is for the sputtering away of material. This can be done with a blanket exposure to a plasma, for example an argon plasma, removing material from all parts of the wafer. Selectivity is achieved by the relative rate of sputtering of different materials. This process is called *ion milling*. A variant of this process is *focused ion beam milling*, using an argon ion source in combination with focusing electrodes so that ions from the plasma only strike the surface within a small region. Direct-write patterning can be accomplished by scanning the ion-beam location across the wafer surface. Masking of ion-beam removal is difficult because the physical process of sputtering has no chemical specificity.

When the ions have some chemical reactivity, such as the ions produced in an oxygen plasma , the result of reaction of the ionic species with the surface of the wafer can be a volatile species that is removed by the vacuum pump. An example is the use of blanket oxygen-plasma exposure to remove photoresist from wafers after completion of lithography processes, converting the polymer to carbon dioxide and water. This process is referred to as *ashing*.

There is a rich array of plasma chemistries that can be used for etching all of the microelectronic thin films: oxides, nitrides, metals, and silicon. A small selection is illustrated in Table 3.7. A much more complete database of etchants, including data on etch rates and selectivity, can be found in [18]. Most etch chemistries involve either flourinated or chlorinated species. Recipes are developed to achieve the desired chemical selectivity, for example, etching oxide and

Table 3.7. Examples of etch gases for plasma etching of selected materials [3, 7, 8]. Many other choices are available. The specific mixture and etch conditions must be tailored to achieve required selectivity and etch profile.

Material to be etched	Etch gas
Silicon or polysilicon	CF_4, SF_6
Silicon dioxide	CF_4/H_2
Silicon nitride	CF_4/O_2
Organics	O_2, O_2/CF_4, O_2/SF_6
Aluminum	BCl_3

stopping on nitride, or etching silicon and stopping on oxide. Generally, with a relatively small number of exceptions, the selectivity that can be achieved with plasma etching is not as great as with wet etching. As a result, the setting of the plasma parameters (radio-frequency power levels, gas mixtures and flow rates, and pressure) can be critical to achieve the desired result.

The shape of plasma-etched feature is a strong function of the etching conditions. The higher the base pressure in the plasma, the more isotropic the etch profile. As the pressure is reduced in a plasma etcher, the etch rate generally reduces, and the etch becomes more directional because the ions that are accelerated through the dark space at the edge of the plasma strike the surface with a definite orientation. At higher pressure, the trajectories of these atoms are randomized by collisions, producing a less directional ion flux at the wafer surface.

At the limit of low pressures and a correspondingly high degree of directionality, the process is called *reactive ion etching*, essentially directing a flux of reactive ions normal to the surface. Sidewalls are not significantly etched because the ions do not strike them. As a result, nearly vertical sidewall features can be produced.

Deep Reactive Ion Etching (DRIE)

A dramatic, and commercially highly important variant of reactive ion etching is a relatively new process called *deep reactive ion etching*, or DRIE. This process takes advantage of a side-effect of a glow discharge, the tendency to create polymeric species by chemical crosslinking. In fact, most plasma processes are a critical race between deposition of polymeric material from the plasma and the removal of material from the surface. In well-designed plasma chemistries, removal dominates. However, in a new process developed by workers at Bosch [28], the deposition of polymer from the etchant is used to great advantage.

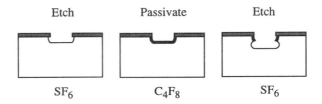

Figure 3.27. DRIE of silicon with the Bosch process exploits alternating etching and deposition steps, producing an extremely vertical sidewall etched into silicon. There is a small amount of scalloping on the sidewall, highly exaggerated in the figure.

Figure 3.27 illustrates the DRIE process. A photoresist mask can be used. For very deep etches, a combination of photoresist and oxide may be required. The etch proceeds in alternating steps of reactive-ion etching in an SF_6 plasma and polymer deposition from a C_4F_8 plasma. During the etch process, the polymer is rapidly removed from the bottom of the feature but lingers on the sidewall, protecting it from the SF_6 etchant. As a result, the silicon beneath the first cut is removed during the second etch cycle, but the top of the feature does not become wider. Eventually the polymer protecting on the sidewalls is eroded. At that point, another polymer deposition step is used, and the cycle is repeated. The etch profiles achieved with this process are simply astonishing.

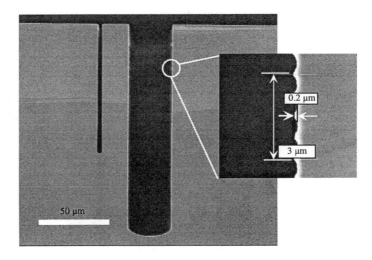

Figure 3.28. Photo of features etched in silicon with DRIE. Note that the slender trench etches at a slower rate than the wider trench. Typical scalloping of a sidewall is illustrated with the inset. (Source: Arturo Ayon, MIT; reprinted with permission.)

Figure 3.28 shows a photograph of deep features etched into silicon with the Bosch process. At low magnification, the structures appear perfectly smooth and the sidewalls are perfectly vertical. At higher magnification, a slight scalloping of the walls corresponding to the alternation between etching and passivation is observed.

The Bosch process is revolutionizing the development of micromachining processes. It is now possible to cut features all the way through the wafer thickness with a precision of a few microns. Figure 3.29 shows a silicon microturbine fabricated with DRIE. The rotor in the center is free to rotate.

Journal
bearing
deep etch

2 mm

Figure 3.29. A microfabricated silicon turbine produced with DRIE. Note the slender journal bearing cut, which makes the central portion free to rotate. (Source: C.-C. Lin, MIT; reprinted with permission.)

The combination of DRIE with wafer bonding has enabled remarkable three-dimensional structures to be microfabricated. When a rotor wafer is combined with additional wafers that provide flow paths and necessary access ports, a complete air-bearing-supported spinning rotor can be created, as shown in Figs. 3.30 and 3.31. The rotor is spun with air directed at the turbine blades. Speeds in excess of 1 million rpm have been achieved [29].

3.3.6 Additive Processes: Lift-Off

Most additive processes were discussed in the section on wafer-level processes, including the patterning of metals by electroplating through a patterned masking layer. There is one more additive process that is important: *lift-off*. Lift-off is used with metals that are difficult to etch with plasmas. Figure 3.32 illustrates the technique.

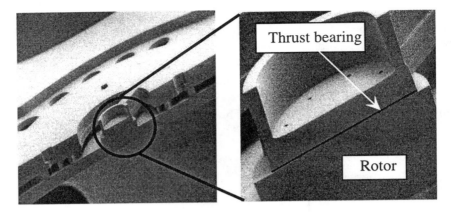

Figure 3.30.   Cross-sections of microbearing rig structures fabricated with the combination of DRIE and silicon wafer bonding. The thrust-bearing gap has been highlighted for visibility. (Source: Reza Ghodssi, MIT; reprinted with permission.)

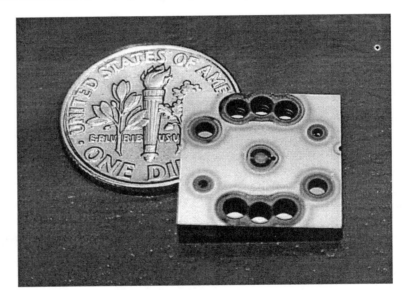

Figure 3.31.   A fully assembled five-wafer lamination of silicon wafers to create a complete microbearing rig. (Source: Reza Ghodssi and Luc Frechette; reprinted with permission.)

A wafer is coated with a resist, and is exposed and developed so as to create a slightly re-entrant resist profile. This same effect can be created with a two-part masking layer in which the upper layer is patterned, and the second layer beneath is slightly undercut. When metal is evaporated from a directional source, such as an e-beam heated crucible, the resist profile shadows the side walls. Provided that the metal thickness is only a fraction of the resist thickness,

Re-entrant resist profile

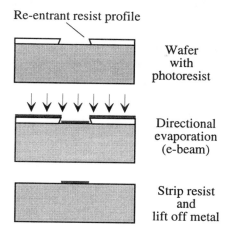

Wafer
with
photoresist

Directional
evaporation
(e-beam)

Strip resist
and
lift off metal

Figure 3.32. Illustrating the lift-off method for patterning evaporated metals.

as shown in the Figure, discontinuous metal is deposited. When the resist is stripped, the metal on top of the resist is "lifted-off," while the metal deposited directly into the opening of the resist remains.

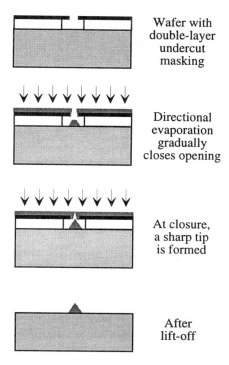

Wafer with
double-layer
undercut
masking

Directional
evaporation
gradually
closes opening

At closure,
a sharp tip
is formed

After
lift-off

Figure 3.33. The use of a modified lift-off process to create sharp tips.

A variant of this process can be used to make metal tips. Called the *Spindt process* for its inventor [30], it exploits the deposition of the evaporated metal onto the edges of the masking layer, gradually closing off the opening (see Fig. 3.33). The evaporated layer at the substrate becomes conical in shape, finally reaching a sharp point when the masking layer closes off completely. This process is used to create very sharp tips for field-emission displays and vacuum-microelectronic devices.

3.3.7   Planarization

During the process of creating microfabricated structures, the initially flat wafer surface develops a topography. While topographical features may be essential to creating the desired structure, they can also interfere with successful process flow. There are three primary difficulties with non-planar surfaces: etching processes can leave behind unetched "stringers;" non-planar wafers cannot be successfully bonded; and, lithography is compromised, both by the difficulty of obtaining uniform resist deposition over non-planar surfaces and by the limited depth of focus of high-resolution lithographic exposure tools.

3.3.7.1   Stringers

Stringers are an important detail that can affect an entire process flow. Figure 3.34 illustrates the problem. A thin film is deposited over a topographic feature, such as a portion of a sacrificial layer. When this film is patterned, it can be difficult to remove the material on the sidewall of the lower layer. This remaining thin slice is called a stringer. Considerable care is required to design processes in such a way that stringers are not produced at some point in the process sequence.

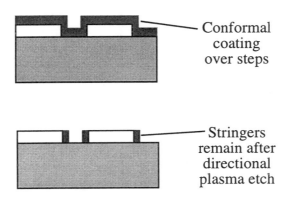

Figure 3.34. Illustrating the formation of stringers during the plasma etching of a thin film deposited over a step.

A particularly difficult place for stringer formation occurs when a sacrificial layer is deposited over topographical features with a thickness comparable to the sacrificial-layer thickness, as illustrated in Fig. 3.35.

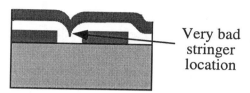

Very bad
stringer
location

Figure 3.35.   When a sacrificial layer is deposited over topographical feature, a cusp is formed. If another structural layer is deposited over this cusp, the cusp region is extremely difficult to clear in a subsequent etch, so a stringer remains after etching.

Because it is difficult to remove stringers, it is best to design the topography of the device so that they are prevented. One good method for doing this is to planarize the surface before depositing structural layers. There are two widely used planarization methods: chemical mechanical polishing, and resist etchback. Resist etchback depends, in turn, on the planarization that results from spin-casting of polymer films.

3.3.7.2   Chemical Mechanical Polishing

Wafer polishing is a process in which the wafer is pressed against a rotating platen on which an abrasive slurry is provided. When this slurry includes chemicals that help dissolve the removed material, the process is referred to as *chemical mechanical polishing*, or CMP. CMP has become a critical process in the microelectronics industry for creating multi-level interconnect in integrated circuits. Successive dielectric layers are planarized before performing the patterning step of opening vias to permit connection to lower layers. The vias are then filled with metal, after which, the next layer of conductor is deposited and patterned. By using planarized layers, extremely accurate lithography can be performed over complex subsurface topography.

In MEMS, CMP is playing an increasingly important role. One important use is to prevent stringer formation during surface micromachining processes. Referring again to Fig. 3.35, if the intermediate sacrificial layer is made thick enough, it can be polished back to a flat surface before depositing the upper layer, removing the cusp that causes the stringer. A second use is to create flat surfaces prior to wafer bonding. Because wafer bonding requires intimate surface contact, remnants of surface topography can prevent good bond formation. A third use, like that in integrated circuits, is to provide a good planar surface for lithographic definition of features.

CMP is not a perfect process. It is primarily a *mechanical* polishing. There-fore, the rate of material removal depends on the local force applied to the surface. This force, and how it is distributed, depends in detail on the local topography. A flat smooth surface will polish much slower than a surface with a few peaks on it. This means that polishing rates exhibit a strong *pattern dependence*. As a result, the so-called "field regions" of a wafer, with no features, will polish at a different rate than regions that have local topography. Dummy structures can be placed in the field region to achieve uniform polishing rates across the wafer.

Different materials polish at different rates, even with the same applied force. It is sometimes necessary to add thin polish-stop layers (for example, silicon nitride) over features that need to be revealed by a polishing step, but which should not themselves become thinner during polishing.

3.3.7.3   Planarization with Polymers

Spin-cast polymers tend to planarize substrate topographies. There is an analogy to varnishing a floor. During application, the surface tension creates a flat surface. As solvent is lost, the film becomes thinner. When the viscosity becomes high enough that the polymer can no longer flow in response to surface forces, the drying film develops topography that mirrors the underlying topography, but is less pronounced. To get a more planar surface, one applies more than one coating of polymer just as one applies multiple coats of varnish to get a perfectly smooth floor.

Polyimides are often used for planarization. Because these materials can function at temperatures above 300°C, they can remain as part of the structure. But if the structure has mechanical loads applied to it, the response of the polyimide may vary with ambient humidity because, like all polymers, it tends to absorb moisture. When it does, the residual stress in the polymer changes, and this can affect the mechanical behavior of the structure.

Another polymer that can be used for planarization is the SU-8 resist [31], which permits formation of extremely thick layers that can planarize large topographical features in a single application. The resist is based on epoxy-resin chemistry, and once hardened at 200 °C, is difficult to etch or remove. It can be used as a structural component of devices intended for low-temperature use (below the hardening temperature).

3.3.7.4   Resist Etchback

Planarization by a polymer can be exploited to planarize an oxide layer over which the polymer is deposited. Photoresist is deposited over oxide topography so as to planarize that topography. Wafer is then plasma etched with an etch recipe that etches the photoresist at about the same rate as the underlying oxide. This results in a much more planar oxide surface.

3.4 Conclusion

This whirlwind tour through microfabrication process steps is not sufficiently detailed to produce experts. However, it does provide an overview of many of the steps used in modern MEMS device process sequences and permits discussion of some of the trade-offs between process and device performance that will arise throughout the rest of this book. In the following chapter, we discuss how combinations of these process steps are combined to make complete process flows.

Related Reading

S. A. Campbell, *The Science and Engineering of Microelectronic Fabrication*. New York: Oxford University Press, 1996.

R. A. Colclaser, *Microelectronics: Processing and Device Design*. New York: Wiley, 1980.

Fairchild Corporation, *Semiconductor & Integrated Circuit Fabrication Techniques*. Reston, VA: Reston, 1979.

G. T. A. Kovacs, *Micromachined Transducers Sourcebook*. New York: WCB McGraw-Hill, 1998.

M. Madou, *Fundamentals of Microfabrication*, pp. 35–37. New York: CRC Press, 1997.

S. M. Sze, *Semiconductor Devices: Physics and Technology*. New York: Wiley, 1985.

J. L. Vossen and W. Kern, *Thin Film Processes*, New York: Academic Press, 1978.

S. Wolf and R. N. Tauber, *Silicon Processing for the VLSI Era*, vol. 1: Process Technology. Sunset Beach, CA, USA: Lattice Press, second ed., 2000.

G. T. A. Kovacs, N. I. Maluf, and K. E. Petersen, "Bulk micromachining of silicon," *Proc. IEEE*, vol. 86, pp. 1536–1551, 1998.

J. M. Bustillo, R. T. Howe, and R. S. Muller, "Surface micromachining for microelectromechanical systems," *Proc. IEEE*, vol. 86, pp. 1552–1574, 1998.

M. A. Schmidt, "Wafer-to-wafer bonding for microstructure formation," *Proc. IEEE*, vol. 86, pp. 1575–1585, 1998.

H. Guckel, "High-aspect-ratio micromachining via deep X-ray lithography," *Proc. IEEE*, vol. 86, pp. 1586–1593, 1998.

Problems

3.1   Using the Deal-Grove model, determine dry oxidation times required to produce 100 nm of oxide on a bare silicon wafer at temperatures of 920, 1000, and 1100°C.

3.2   Using the Deal-Grove model, determine the wet and dry oxidation times needed to grow 0.5 μm of oxide on a bare silicon wafer at 1000°C.

3.3   It is desired to implant phosphorus into silicon through 0.1 μm of silicon dioxide. Select an implant energy to place the peak implanted concentration at the oxide-silicon interface.

3.4   What mask feature size is required to produce a 400 μm diaphragm of thickness 20 μm on a silicon wafer that is 500 μm thick? What is the edge-length variation for the diaphragm if the actual wafer thickness for different wafers varies between 490 and 510 μm? Assuming that the sensitivity of a pressure sensor varies as the inverse fourth power of the diaphragm edge length, what percentage variation in sensitivity can be attributed to variations in wafer thickness?

Chapter 4

PROCESS INTEGRATION

Castles in the air – they're so easy to take refuge in. So easy to build, too.
—Henrik Ibsen, in *The Wild Duck*

4.1 Developing a Process

In the previous chapter, we examined a variety of wafer-level and patterning technologies. The question now becomes, "How do we design a process sequence so that the device we imagine (our castle in the air) can actually be built?"

There are two ways to approach this question. One is to elucidate some general principles of process design, the main subject of this chapter. A second is to study examples of successful process sequences, and learn by assimilation.

The most important principle in process design is to understand who is in charge. Technologists like to think that they are in charge, and in some sense, they are. But what drives process design is a set of three severe taskmasters: device performance, manufacturability, and cost. When process sequences can be invented that simultaneously serve all three of these ends, a major paradigm shift in the way devices are made can result. On the other hand, technological capabilities that fail to meet these criteria are eventually discarded.

4.1.1 A Simple Process Flow

We begin by illustrating a process flow for making a junction diode [32]. The process consists of two parts, a list of individual process steps, shown in Table 4.1, and a set of photomasks used for lithographic definition of the patterned features, shown in Fig. 4.1. As is the custom in making sketches of microelectronic process flows, the horizontal scales and vertical scales in the device cross-sections of Fig. 4.1 are not commensurate. Lateral device dimen-

Table 4.1. Process steps for fabricating a diode with the masks of Fig. 4.1. Specific conditions for oxide growth, implant energy and dose, and drive-in diffusion must be determined to complete the process specification.

Step	Description
Starting material	(100) p-type silicon, 1×10^{15} cm^{-3} boron
1. Clean	Standard RCA cleans with HF dip
2. Oxide	Grow 0.1 μm SiO$_2$
3. Protect front	Spin cast photoresist on front and prebake
4. Backside implant	Implant boron to achieve 10^{19} cm^{-3} at surface after all anneals
5. Strip	Strip photoresist
6. Photolithography	Spin cast resist, prebake, expose Mask 1 (n-region), develop, postbake
7. Implant	Ion implantation of phosphorus to achieve 1×10^{19} cm^{-3} surface concentration after drive-in
8. Strip	Strip photoresist
9. Clean	RCA clean without HF dip
10. Drive-in	Drive in diffusion to achieve 1 μm junction depth
11. Photolithography	Mask 2 (contacts)
12. Etch oxide	Etch oxide with buffered HF (BHF) to open contacts
13. Strip	Strip photoresist
14. Clean	Pre-metal clean (RCA without HF dip)
15. Metal	Evaporate 1 μm aluminum
16. Photolithography	Mask 3 (aluminum)
17. Etch aluminum	PAN etch (phosphorus-acetic-nitric acids)
18. Strip	Strip photoresist
19. Backside metal	Evaporate aluminum on backside
20. Sinter	Anneal contacts at 425 °C, 30 min

sions can be ten's or many hundreds or thousands of μm, while dimensions in the thickness dimension might be only a few μm. Nor are the individual layers drawn to strict thickness scale, since very thin layers would then become impossible to see.

This process illustrates many of the issues that arise in process design. The first issue is the selection of starting material. The type, orientation, and doping of the wafer must be specified. In this case, a lightly doped p-type (100) wafer is chosen. The light doping makes it relatively easy to build diodes with moderately deep (1 μm) junctions. The proposed packaging scheme for these diodes is to use the back side as one contact through the *die-attach* of the die to its package, and to use a *wire bond* for the second contact. Therefore, attention must be given to providing a bond pad on the front side (see Mask 3) and to making good electrical contact to the back side of the wafer. Low-resistance non-rectifying contacts (so-called *Ohmic* contacts) to lightly doped

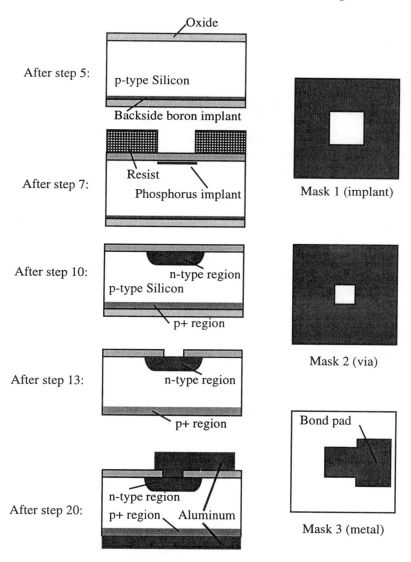

Figure 4.1. Mask set and selected cross-sections for the diode process.

silicon are difficult to achieve. Therefore, the back side is implanted with boron to increase the doping near the surface. However, the drive-in of this dopant is deferred until after the n-type dopant (phosphorus) has been introduced. Both drive-ins can be done at once. Prior to any implants, a 0.1 μm thick oxide is grown (for example, at 1100 °C for 1 hour in dry oxygen). Then, during the backside implant, the front side of the wafer is protected with a coating of photoresist to prevent scratching or accidental damage during handling. This resist is removed after the implant, reaching the top figure in Fig. 4.1.

The first photolithography step involves coating the front side with photoresist, prebaking it to remove solvent, exposing it with Mask 1, developing the resist, then postbaking the resist to harden it. The mask set for this process is shown at the right of Fig. 4.1. The *mask polarity* has been chosen for use with positive photoresist. If negative photoresist were being used, the dark and clear regions of the masks would be reversed. Mask 1 is used at this point to pattern the photoresist, which then serves as an implant mask. The phosphorus implant energy is chosen to be 180 keV based on the range data of Fig. 3.6, with a projected range of about 0.2 μm in silicon. Because the first 0.1 μm of material is oxide, the actual projected range is slightly deeper, with the net result that the peak phosphorus concentration will be slightly more than 0.1 μm beneath the silicon surface. The resist thickness must be at least three times the projected range in silicon, or 0.6 μm. In practice, a 1 μm thick resist would be used to assure no implantation in the projected regions. This resist, which is shown in the second cross-section of Fig. 4.1, is removed before the drive-in diffusion.

The phosphorus dose Q_I must be selected to yield the desired junction depth and surface concentration after the drive-in anneal. The constraints are that the junction depth is to be 1 μm and the surface concentration after the drive-in is to be 1×10^{19} cm^{-3}. Therefore, we require:

$$N_I(0, t) = \frac{Q_I}{\sqrt{\pi D t}} = 10^{19} \text{ cm}^{-3} \tag{4.1}$$

and

$$N_I(1\mu\text{m}, t) = N_I(0, t) \exp\left[-\frac{(1\mu\text{m})^2}{4Dt}\right] = 10^{15} \text{ cm}^{-3} \tag{4.2}$$

Because there are three variables (Q_I, t, and the drive-in temperature) and only two equations, there is no unique solution. The conditions for this drive-in should be matched to the backside boron implant so as to assure a satisfactory surface concentration of boron. A possible choice is 1100 °C, where boron and phosphorus have comparable diffusion constants. At lower temperatures, the boron would diffuse more than the phosphorus, requiring a higher backside initial dose to achieve the same final surface concentration[1]. Assuming 1100 °C for the drive-in, the diffusion constant is 1.4×10^{-13} cm^2/sec. Therefore, the diffusion time is 1940 sec, or 32 min. The corresponding phosphorus dose

---

[1]However, other process considerations might dictate the use of a lower temperature for this drive-in. An increase in backside boron implant dose to compensate for the more rapid diffusion is readily accomplished.

requirement is 2.9×10^{14} ions/cm^2. Because of similar diffusion characteristics, this same dose for the boron backside implant will achieve the required backside surface boron concentration of 10^{19} cm^{-3}. We have now reached the middle figure of Fig. 4.1. The backside region with higher boron concentration is denoted p+ to distinguish it from the more lightly doped p-type substrate.

The next step is the etching of *vias* (also called *contact openings* or, simply, *contacts*) using photolithography and Mask 2. The process of etching the contacts on the front side of the wafer also removes the oxide from the back side of the wafer, preparing both the front and back for metallization. Following a wafer clean, aluminum is evaporated onto the front of the wafer, and is then patterned with Mask 3. After stripping the photoresist, blanket aluminum is deposited on the back. The final step is a *metal sinter* step, an anneal at about 425 °C in an inert atmosphere such as nitrogen, to improve the silicon-aluminum contact. There are several issues here: the exposed silicon surfaces tend to regrow thin oxides during cleaning steps, and since these wafers have oxide on them, the HF dip is typically omitted. Aluminum, however, is a good reducing agent, and can reduce and dissolve a small amount of oxide. The purpose of the metal sinter is to promote this oxide dissolution, and to permit small amounts of silicon-aluminum interdiffusion. It is important to limit the length of this sinter, because if too much interdiffusion is permitted to occur, pit-shaped voids can develop at the silicon surface when silicon diffuses into the aluminum. These voids become filled with aluminum, and if the junctions are shallow, spikes of metal can actually create contacts all the way through the junction to the p-type substrate, ruining the diode. In our example, with a 1 μm junction depth, this is not a severe problem. To reduce the amount of interdiffusion, an aluminum:silicon alloy is often used for integrated-circuit metallization instead of pure aluminum. Or barrier materials that restrict interdiffusion, such as TiW or TiN, are deposited beneath the metallization.

4.1.2   The Self-Aligned Gate: A Paradigm-Shifting Process

We present a second microelectronic process example, this one illustrating how a device-performance constraint led to the development of a wholly new process sequence, with enormous impact.

The earliest metal-oxide-semiconductor field-effect transistors (MOSFETs) used metal gates, as illustrated in Fig. 4.2. The operating principles and electrical performance of these devices are discussed in Chapter 14. For now, consider the *drain* and *source* diffusions to be n+ regions diffused into a p-type substrate. A thin oxide (the *gate oxide*) is required between the central portion of the surface between source and drain (called the *channel*) and the gate electrode. For the purpose of the present discussion, we can treat the processing of the source, drain, and gate electrode as a geometry problem.

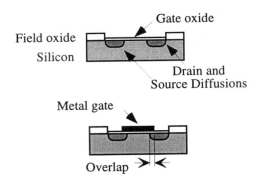

Figure 4.2. The gate electrode overlaps the underlying source and drain diffusions. The resulting capacitance limits the speed of the device.

In the process shown in Fig. 4.2, the source and drain regions are diffused into the wafer before the gate electrode is deposited and patterned. To prevent photolithographic misalignment of the gate relative to the edges of the source and drain, a minimum overlap is required. This overlap produces drain-to-gate capacitance that limits the switching speed of the transistor.

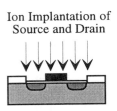

Figure 4.3. The use of a self-aligned gate as the mask for the source/drain implant reduces the gate-to-drain overlap.

Figure 4.3 illustrates an alternate way of patterning the source and drain. If the gate material is deposited first and then patterned, it can serve as a mask for the ion implantation of the source and drain. As a result, the edges of the source and drain regions are perfectly aligned with the edge of the gate electrode. This is called a *self-aligned* process step.

The self-aligned process only works when ion implantation is used as the method for introducing the dopants into the substrate. However, the use of a metal for the gate is not compatible with the requirement to perform a high-temperature source/drain drive-in diffusion after the implant. Therefore, in spite of the fact that polysilicon is a much worse conductor than a metal, the industry shifted to a polysilicon gate, which can easily survive the source/drain drive-in. During the drive-in, there is a small amount of lateral diffusion so that the edges of the source and drain end up slightly overlapping the gate, but the resulting drain-to-gate capacitance is much smaller than in the original process. Greatly increased transistor speed is the result. Furthermore, there is some economy of process steps, since the same implant that dopes the substrate to form the source and drain can be used to dope the polysilicon.

The introduction of the self-aligned polysilicon gate was a significant paradigm shift in the manufacture of integrated circuits. There are many other examples. They are all, ultimately, driven either by improved device performance, improved manufacturability, or lower cost. Technologists take note!

4.2 Basic Principles of Process Design

Good process design is a creative art, supported by careful engineering analysis and experiment. Table 4.2 lists a number of critical process-design issues that should be considered. The various items in the table are discussed in the remainder of this chapter, with an emphasis on MEMS aspects of process design rather than on strictly micro*electronic* process design.

Table 4.2. Process design issues.

Device geometry
Backside processing
Institutional constraints
System partitioning
Packaging
Process partitioning
Cleaning requirements
Cross-contamination constraints
Thermal constraints
Material property control
Mechanical stability
Process accuracy
Alignment features
Wafer architecture
Die separation

4.2.1 From Shape to Process and Back Again

In designing process sequences for MEMS devices, there is a complex iteration that reasons from the desired shape to a plausible process and back again. At the highest level, one asks whether it is *possible* to build a desired device structure with a given process sequence. For example, an accelerometer consists of an elastically supported proof mass in combination with some means of detecting proof-mass motion. It is definitely *possible* to build accelerometers with bulk micromachining, surface micromachining, a silicon-on-glass process, or DRIE together with wafer bonding. How does one select?

Sometimes the decision is based on the economics of an already-committed capital investment.[2]   That is, a company that already owns an integrated-circuit fabrication facility with some unused capacity might be driven toward surface-micromachined technologies using polysilicon as a structural layer, since the integrated-circuit process already has a well-integrated polysilicon process step. On the other hand, it is difficult to make a large proof mass with surface micromachining, and this translates into a corresponding difficulty in making highly sensitive accelerometers. Each organization must struggle with this decision, because it has huge implications. Included in this decision is the physical partitioning question discussed in Chapter 2: whether to integrate the electronics with the MEMS device. There is, at present, no "right" answer to this question. Each organization makes its own assessment. Ultimately, the market will decide.

Modeling of expected device performance plays a key role in this decision process. Very often, the manufactured cost of a part produced in volume is dominated by the achievable yield, and this, in turn, can be dominated by some minute detail of a particular process step. Unwise choices early in the decision process can lead to economic failure later. The more thoroughly a proposed device can be understood prior to committing major capital investment to its manufacture, the better the chance of producing a manufacturable part at an acceptable cost.

4.2.2   Process Design Issues

Having now painted a specter of economic disaster resulting from bad choices in process design, we now discuss the various issues listed in Table 4.2 in the hope of increasing the chance for good process design. What is offered below should be read as guidelines, not guarantees.

4.2.2.1   Device Geometry

It can be very difficult to visualize the actual device geometry that will result from a given process sequence and mask set. Advanced CAD-for-MEMS systems, such as MEMCAD,[3] provide a solid-modeling capability that constructs and visualizes a three-dimensional rendering of the device starting from a process description and mask set. The alternative to such automatic model construction is to draw cross-sections through every conceivable plane of interest. Among the things that are often revealed by judicious cross-section

---

[2] An example of an institutional constraint.
[3] Microcosm Technologies, Cambridge, MA.

studies are bad stringer locations, unwanted structural features at the edges of devices, or failure to create and uncover suitable bonding pads prior to final packaging. Process flows must often be iterated many times until the device is satisfactory when viewed from all possible directions.

4.2.2.2   Backside Processing

Wafers that go into furnaces for high-temperature oxidation or CVD processes are affected on the back as well as the front of the wafer. It is necessary to track all changes to the back when evaluating a proposed process, even if the device itself is built entirely on one surface of a wafer. Residual films on the back of the wafer can interfere with electrical contact, or can produce wafer bow because of uncompensated thin-film stress.

4.2.2.3   System Partitioning and Packaging

The first step in a good process design is to understand the *system* into which the MEMS device and its associated circuit must fit. The question of whether or not to integrate some of the electronics with the MEMS device colors the entire process design. As a default, this author recommends firmly *against* integrating any electronics with the MEMS device. The argument *for* integration must be made on the basis of a plausible performance benefit, a manufacturing benefit, or a cost benefit. When that argument can be made, then integration of the electronics may be appropriate. At present, there are a few companies that manufacture MEMS devices that combine electronics with a mechanical part; most do not. The Case Studies of Part V include examples of both monolithic and hybrid microsystems.

The second step is to decide how the part is to be packaged. Does the device require vacuum packaging? Does it require environmental access? Can some of the packaging be done at the wafer level using bonding operations? These questions only get answered iteratively as the process is defined, but they need to be addressed from the very beginning.

If the decision is made to combine electronics monolithically with the MEMS device, the wafer-fabrication process is basically a standard integrated-circuit process with selected enhancements needed to build the MEMS part of the device. This is the case for the Analog Devices family of accelerometer products, one of which serves as a Case Study in Chapter 19.

On the other hand, if the decision is to separate the electronic circuitry from the MEMS device, there is considerable freedom in choosing the process for the MEMS device. Choices of substrates, processes steps, and device dimensions can be made independent of the constraints of an integrated-circuit process. When the electronics part of the system is built on a separate custom-designed integrated circuit (an *Application-Specific Integrated Circuit*, or "ASIC"), provision is needed for economical packaging of the MEMS part with the ASIC.

Many manufacturers are using a "two-chip" package, in which two chips are mounted together onto a lead frame and then molded into a single plastic package. Packaging issues are discussed in more detail in Chapter 17.

4.2.2.4   Process Partitioning and Contamination Constraints

It is common practice in the integrated-circuits industry to *partition* a process into a *front-end process*, which, among other things, includes all of the high-temperature steps, and a *back-end process*, which typically involves interconnect and packaging. In the diode example presented earlier in this chapter, the steps up to the contact opening would be considered the front end, and the rest the back end. The front-end process is particularly sensitive to chemical contaminants, and both front- and back-end processes are sensitive to particulate contaminants.

Many MEMS devices incorporate materials that are not permitted in IC manufacturing facilities. Gold and copper, for example, are electrically active impurities in silicon, and are rapid diffusers. Therefore, front-end silicon processing facilities will not allow these materials to enter. If the back-end process shares the same equipment set, then the back-end process may be restricted as well. Further, packaging requirements that are unique to MEMS devices may call for adhesives or low-temperature polymeric materials that do not satisfy the cleanliness requirements of IC facilities. Finally, as a precaution against chemical or particulate contamination, some wafer-fabrication facilities will not permit wafers to be removed from the process line and then returned at a later point in the processing. For all these reasons, the designer of a MEMS process must understand the restrictions on process flow imposed by the various facilities in which the fabrication is to occur, and must design the process flow so that potential IC contaminants are never brought into facilities where IC processing takes place. When using a set of standard IC steps as part of a MEMS process flow, it is wise to partition the MEMS process sequence so that the standard steps all appear as a single sub-sequence, with the special micromachining steps placed either before or after this sub-sequence. The sealed-cavity wafer-bonded sensor process described in Chapter 3 is an example of effective process partitioning.

4.2.2.5   Thermal Constraints

Attention must be paid to the thermal requirements of individual process steps and to the merging of these steps into a process flow that does not compromise any material. As a trivial example, photoresist and other polymeric materials must always be removed before a wafer goes into any high-temperature processing step, such as oxidation, diffusion, or LPCVD deposition.

The diode process presented earlier in this chapter showed an example of using a single high-temperature step to drive in both the phosphorus and boron implants. This is an example of economical use of high-temperature steps. More generally, one must recognize that *every high temperature step has an effect*: dopants move by diffusion, morphologies of polycrystalline materials can change, and residual stresses can change (discussed below). In the IC industry, where diffusion is a major concern, process design involves limiting the total "$\sqrt{Dt}$" of a process, which is shorthand code for tracking the total impact of thermal history on dopant diffusion.

Metals also create thermal constraints. Metals cannot be deposited onto a substrate which must subsequently be exposed to a high-temperature environment. Therefore, dielectric films that will be used to insulate metals must be deposited by PECVD or some other low-temperature process such as polyimide deposition and cure. PECVD films are not as robust as their LPCVD counterparts. They have higher void content, are more easily etched, and can contain residual hydrogen that can be released during anneals at elevated temperatures. And polymer films have strict temperature limits that must be observed to avoid degradation.

Packaging also imposes thermal constraints on the device design. Depending on how the device is to be attached to the package and how the package is to be sealed, there may be requirements for the thermal survivability of the materials within the device.

4.2.2.6 Material Property Control

Considerable attention must be given to control of the material properties that form the final device. All thin film materials can have residual stresses, and these stresses can have profound effects on the mechanical behavior of devices. Furthermore, the stresses can be very process-dependent. For example, the final stress in LPCVD polysilicon, after deposition and subsequent high-temperature anneal, can vary from highly compressive to moderately tensile, depending on the deposition temperature, the doping, and the annealing temperature and time. Oxides have compressive stresses that change during high-temperature anneals, nitrides have tensile stresses, and metals can be either tensile or compressive depending on deposition conditions and thermal history.

Control of thin-film stress, or rather, the lack of it, can cripple what is otherwise a perfectly plausible process sequence. Thus, process design where thin films are involved must be done hand-in-hand with experimental determination of the thin-film stresses and their variability and with device modeling to determine the effects of these stresses and their variation. This author has witnessed examples of brilliant device concepts crashing against insurmountable barriers of stress control. In virtually every unsuccessful case within the author's experience, insufficient attention was given to stress control *early in the process development*.

4.2.2.7   Mechanical Stability of Interim Structures

Micromachined structures that must be released in some fashion can go through an intermediate state in which they are prone to fracture. For example, silicon dioxide typically has a compressive residual stress while silicon nitride has a high tensile stress. Suspended structures that include oxide have a tendency to buckle and fracture. Therefore, process sequences that involve etching away of silicon dioxide as a sacrificial layer must be checked to assure that at no point in the etching process is an intermediate structure created that can buckle and break because of the stress in the not-yet-etched oxide.

A similar problem comes up when using the sealed-cavity wafer-bonded process described in Chapter 3. When making accelerometers with this process, the sealed cavity is cut open by a plasma etch that patterns the proof mass over the cavity. Because there is residual gas in the cavity and the plasma etch takes place at low pressures, it is possible to shatter the structure just before the release is complete. In this case, the solution is to create a pressure-relief pathway in a non-critical part of the device structure [33]. This relief path is opened first, after which the rest of the release etch can proceed.

4.2.2.8   Process Accuracy

Real process steps do not make exact replicas of the features on a mask. The photoresist exposure process can produce process biases, and there can be random variations as well. Thicknesses of deposited films can also vary across a wafer and from wafer to wafer. While it is possible to correct for an average process bias by changing the dimensions of a feature in a mask, the process variation at every step is an intrinsic part of life in the microfabrication world. Device designs that are intended for manufacture must incorporate a sensitivity analysis based on expected processing variations in order that the calibration and trim procedure can be designed to cover the expected distribution in device performance.

4.2.2.9   Alignment Features

It is necessary that every photolithography step be accurately aligned to the structure already present on the wafer. This requires not only the inclusion of alignment marks on the mask, but an investment of thought as to which previous layer will hold the alignment targets. For example, ion implantation leaves no visible features on the wafer. If, early in a process flow, one wishes to implant a portion of the wafer and not other portions, an alignment target must first be established by some process step that leaves visible features in the wafer. Then subsequent process steps can construct features that are aligned to the invisible implants.

4.2.2.10   Wafer Architecture

Not all locations on a wafer are equivalent. Dry etching rates, in particular, can be somewhat non-uniform across a wafer. This can lead to some variations in feature sizes from device to device because some devices must be over-etched to provide enough time for the rest of the devices to be correctly etched. As a result, some attention should be given to placement of devices on the wafer. Devices too close to the edge may fail to yield.

A related issue is the processing that occurs at the edge of a wafer. This involves both photoresist processing and thin-film deposition. Wafer edges are tapered slightly during manufacture to minimize chipping and handling accidents. When photoresist is spun onto a wafer, it can form a slightly thicker coating at the wafer edge (an *edge bead*). Many spin-coaters have provision for removing this edge bead, but when that occurs, the edge of the wafer may see a different effective process flow from the flat central region of the wafer. CVD processes tend to coat the edges of wafers, except possibly where they are supported by the wafer holder in the furnace. Since wafer-bonding operations, especially, depend on the characteristics of the entire wafer, some care and though about what is happening to the edge of the wafer is extremely important.

4.2.2.11   Die Separation

Attention must be given to the method of die separation, and how it interacts with any release etch that must occur. The typical die-sawing operation is wet and dirty. Individual chips must be carefully cleaned after sawing. However, if these chips contain exposed delicate mechanical structures, contamination is sure to occur. And small gaps provide places for liquid to be held by surface-tension forces, ultimately leading to sticking down of moveable parts.

It is highly desirable to encapsulate moving parts at the wafer level prior to die separation. This may be accomplished, for example, by bonding silicon to suitably patterned glass wafers. Absent such a wafer-level packaging scheme, the die separation step becomes a high-risk step, with the potential for costly yield losses unless device handling procedures are robust.

4.3   Sample Process Flows

This section contains elementary process sequences that illustrate how to build different types of structures. These examples offer the opportunity to discuss a variety of device- and process-design issues and how they interact with each other.

4.3.1   A Bulk-Micromachined Diaphragm Pressure Sensor

We present here a relatively simple process for building a bulk-micromachined piezoresistively sensed diaphragm pressure sensor, a first step in thinking about the pressure-sensor Case Study device of Chapters 17 and 18. We will presume that KOH is to be used as the anisotropic etchant, and that an electrochemical etch stop will be used to control diaphragm thickness, which is designed for 10 μm. With this diaphragm thickness, the diaphragm size needed for the particular sensitivity range is presumed to be 500 μm. We will do a particularly simple design, one with a so-called *half bridge*: one piezoresistor located on the diaphragm near the edge to sense bending, and an identical resistor located away from the diaphragm. The two resistors are connected in series, requiring three contact pads: power, ground, and the contact point between the two resistors.

The resistors are to be covered with a thin oxide so that they are protected from inadvertent ionic contamination. This oxide could, perhaps, be the same oxide that is used as an etch mask for the anisotropic etch. However, in that case, the oxide would have to be quite thick, at least 500 nm, and the stress in the oxide might affect the diaphragm bending stiffness. An alternative approach, the one used here for illustration, is to use a much thinner oxide for protection and to overcoat this oxide with LPCVD nitride for the KOH etch mask. By adjusting the relative thicknesses of the compressively stressed oxide and the tensile-stressed nitride, approximate stress balance of the dielectric coatings can be achieved along with the considerably more robust protection of the piezoresistors provided by the nitride-over-oxide combination.

The sequence of steps is also important. Do we etch the diaphragm first, or implant the piezoresistors? The choice made here is to implant the piezoresistors first because then the wafer requires no special fixturing during the implant and the implant can be done by a standard commercial vendor. It is therefore necessary to prepattern the wafer with alignment marks because the implant leaves no visible features on the wafer. With these considerations, we come up with the five-mask process flow of Table 4.3. Note that there are two photolithography steps after etching the diaphragm. This means that the vacuum chucks needed to hold the wafer during photoresist spinning must be designed to match the wafer layout so that vacuum is not applied beneath any of the diaphragms but, instead, is applied to regions where the wafer has its full thickness.

The starting material is selected to be a lightly-doped p-type wafer with a 10 μm epitaxial layer that is moderately-doped n-type. This type of wafer can be purchased from vendors specializing in epitaxy. As an alternative, it would be possible to implant and diffuse a 10 μm deep junction, but this would require a very heavy implant and a very long diffusion. Note that the wafer diameter is specified, since wafer thicknesses vary with wafer diameter.

Table 4.3. Process steps for fabricating a bulk-micromachined pressure sensor with the masks of Fig. 4.5.

Step	Description
Starting material	100 mm (100) p-type silicon, 1×10^{15} cm^{-3} boron, with 10 μm n-type epi-layer, 5×10^{16} cm^{-3} phosphorus
1. Clean	Standard RCA cleans with HF dip
2. Oxide	Grow 0.17 μm SiO$_2$
3. Photolithography	Mask 1 (alignment)
4. Etch	Etch alignment marks into SiO$_2$
5. Strip	Strip Photoresist
6. Photolithography	Mask 2 (piezoresistors)
7. Implant	Ion implantation of boron to achieve 1×10^{19} cm^{-3} at surface after drive-in
8. Strip	Strip photoresist
9. Clean	RCA clean, no HF dip
10. Drive-in	Drive in diffusion to achieve 0.2 μm junction depth
11. Clean	RCA clean, no HF dip
12. Nitride	Deposit 50 nm LPCVD silicon nitride
13. Backside photolithography	Mask 3 (diaphragm)
14. Etch	Remove nitride and oxide from back of wafer
15. KOH Etch	Etch back side with KOH using electrochemical etch stop
16. Photolithography	Mask 4 (vias)
17. Etch	Plasma etch nitride and oxide from vias
18. Strip	Strip photoresist
19. Clean	Pre-metal clean (RCA without HF dip)
20. Metal	Evaporate 1 μm aluminum
21. Photolithography	Mask 5 (aluminum)
22. Etch aluminum	PAN etch (phosphorus-acetic-nitric acids)
23. Strip	Strip photoresist
24. Sinter	Anneal contacts at 425 °C, 30 min
25. Test	Test for piezoresistor value at wafer level
26. Die separation	Separate devices with die saw
27. Mount	Mount in package (failed die not used)
28. Wire bond	Form wire bonds to package
29. Seal	Close and seal package
30. Test	Final test and calibration

Alternatively, one could specify a wafer thickness, for example, of 500 $\pm$ 5 μm. Following cleaning, a thermal oxide of thickness 0.17 μm is grown. This choice of thickness is dictated by our goal of creating stress balance in the dielectric layers. As-deposited LPCVD silicon nitride has about 1 giga-Pascal of tensile stress[4] whereas thermal oxide grown at temperatures above 950°C has a compressive stress of about 300 mega-Pascal. To achieve stress balance, the thickness of the oxide times its stress should equal the thickness of the nitride times its stress. We plan to use 50 nm of nitride as an etch mask for the KOH etch, which thus suggests we should use 167 nm, or 0.17 μm, for the oxide thickness, provided it is grown above 950°C. The oxidation conditions to achieve this oxide thickness are the subject of Problem 4.1.

Mask 1 contains the alignment marks needed to locate the otherwise invisible implanted piezoresistors with respect to the rest of the device features. Figure 4.4 illustrates the principle. In order to align a mask to a wafer, there must be a visible feature already present on the wafer, called an *alignment target*. Alignment targets are typically produced during an etch step. Subsequent masks then have matching features on them which, when aligned to the target, produce perfect alignment of all features on the mask to the features already on the wafer. The figure shows a commonly used form, a box within another box.

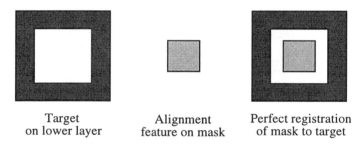

| Target
on lower layer | Alignment
feature on mask | Perfect registration
of mask to target |

Figure 4.4.   Illustrating alignment targets and features.

Patterning of alignment targets is part of the process design. The appearance of the target depends on whether one is using positive or negative resist, and creating the target by etching a depression into the substrate or a film or by leaving raised topography in a layer that is mostly etched away. It is also important to keep track of the visibility of the alignment targets in multi-mask processes, because some intermediate steps might obscure the visibility of targets in lower layers. In the example here, separate alignment targets for each mask layer will be etched into the thermal oxide using Mask 1. There are no visibility problems. The most critical alignment, the positioning of the backside etch relative to the piezoresistor locations, comes early in the process with only 50 nm of nitride as an added layer. The only deposited layer that potentially obscures visibility is the aluminum. In this case, the topography of the alignment target initially etched into the oxide should provide a corresponding topographic feature on

---

[4]Stress and its unit, the Pascal, are thoroughly explained in Chapter 8.

the aluminum surface to permit alignment of the metal mask. Alternatively, it would be possible to add an alignment target to the via mask, and use that target for alignment of the metal mask.

To get precise registration of one mask level to another, it is desirable that the alignment targets be small, with overlaps between mask feature and target close to the resolution of the patterning process. But if the alignment feature size is much smaller than typical device features, as often happens in MEMS devices, special care is need to set etch recipes so that the alignment features are faithfully patterned. This can result in over- or under-etching of device features. A good middle ground is to make the alignment marks large enough so that their patterning is well within process tolerances, but not so large that resulting alignment uncertainties might produce unacceptable device structures.

The number of alignment targets and their placement on the wafer depend on the lithography system being used. A common practice is to include alignment targets on every chip. While not every target will be used during the process of physically aligning the mask to the wafer, inspection of the alignment of targets across the wafer provides a rapid confirmation of overall alignment accuracy.

Once the alignment targets have been patterned into the oxide and the photoresist has been stripped, the lithography for implanting of the piezoresistors can be done. The ion energy, dose, and drive-in conditions needed to meet the specification in Table 4.3 are the subject of Problem 4.2.

The mask for the piezoresistors is shown in Fig. 4.5 together with the remaining three masks and a not-to-scale device cross-section. Note that one of the piezoresistors is located near what will become the edge of the diaphragm, and one is located over unetched substrate. The wide regions on the implant mask are to provide low-resistance interconnect between the three bond pads on the left and the two piezoresistors.[5] This mask layout is for illustration only, and has not been optimized for chip area. Nor has it been designed to provide exactly equivalent interconnect resistance to each piezoresistor. These are issues that would have to be addressed if this device were being commercialized.

After implant of the piezoresistors, LPCVD nitride is deposited. The backside lithography is then done to open the etch window for the KOH etch. The nitride is removed, then the oxide. Whether plasma or wet etching is used here depends on the tolerances of the device. Wet etching of the oxide will slightly undercut the nitride mask, and because the masking oxide does etch slowly in KOH, further undercutting the nitride, the final diaphragm boundary will be slightly larger than that which would result from the mask dimension. Determination of the correct backside mask feature size (Problem 4.3) must

---

[5] An alternative design would be to locate the vias that connect metal to implant close to the diaphragm edge, and use metal interconnect to connect to the bond pads. This has the advantage of separating the wire-bond area from the via, but leaves more exposed metal on the chip surface that might become subject to corrosion.

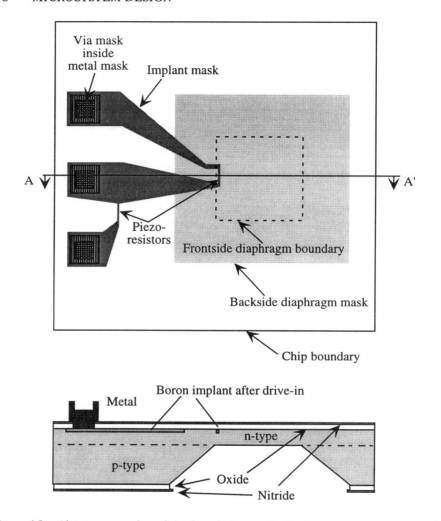

Figure 4.5. Above: an overlay of the four device masks for the pressure sensor example. Below: a not-to-scale cross-section through A-A'.

include three things: (1) the size of the final diaphragm, (2) the extra area required for the sloped sidewalls produced by the KOH etch, (3) correction for mask undercut during etching. This undercut is shown greatly exaggerated in the cross-section of Fig. 4.5.

If the backside etch mask is not accurately aligned with the <110> crystalline directions, the diaphragm size will increase, as discussed in Figure 3.25 of the previous chapter. This addressed in Problem 4.4.

Once the diaphragm is etched, the rest of the process is just like the diode example at the beginning of this chapter. Vias are etched and metal is evaporated and patterned. Then, prior to die separation, each piezoresistor circuit is tested electrically, and chips with bad circuits are marked. Following die separation with a die saw, individual good devices are placed into their packages and wire-bonded. Final test and calibration can then take place.

4.3.2   A Surface-Micromachined Suspended Filament

This example concerns a very simple two-mask process for creating suspended polysilicon filaments using surface micromachining It is the starting point for thinking about the more complex suspended-filament process discussed in the Case Study of Chapter 23.

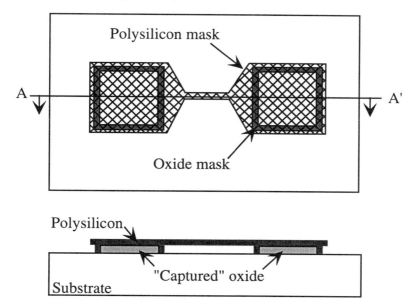

Figure 4.6.   Masks and not-to-scale cross-section through A-A' for the suspended filament process. The mask polarities assume negative resist for the oxide mask and positive resist for the polysilicon mask.

A silicon substrate is used in this process, as silicon is not attacked by the release etch. A 2 μm oxide is deposited by CVD. This oxide is patterned with the oxide mask, creating two square trenches. The width of these trenches is designed to be about twice the thickness of the polysilicon so that when polysilicon is deposited, they fill completely. There follows a 2 μm deposition of polysilicon which is then patterned with the polysilicon mask. The final step is the release etch, immersion in buffered HF to remove all the oxide except that which is surrounded by the polysilicon-filled trench. The filament is thus suspended above the substrate by the thickness of the sacrificial oxide, but is strongly supported by the blocks of "captured" oxide within the polysilicon trench.   The captured oxide also provides supported pedestals onto which bond-pad metal could be deposited.

4.4   Moving On

The discussion and examples presented here may manage to scratch the surface of the very large subject of process design, but the field is endless, truly without limit. This is an area where imagination and creativity can accomplish wonders. Trusting that the reader is well endowed with both, we now move on to Part II, and to the specifics of device and system modeling.

Problems

4.1   Determine the oxidation conditions that will grow the 0.17 μm of thermal oxide in the pressure-sensor process of Table 4.3.

4.2   Determine the ion energy, dose, and drive-in conditions that will meet the piezoresistor specification in the pressure-sensor process of Table 4.3.

4.3   Referring to the pressure sensor of Section 4.3.1 assume that the KOH solution etches the (100) silicon at 1.4 μm/min and also etches SiO_2 at 7.5 nm/min. Determine the size of the backside etch mask feature that will result in a 500 μm diaphragm edge length on a wafer that is exactly 500 μm thick prior to epitaxy. How much variation in edge length is caused by a 1% variation in wafer thickness?

4.4   Calculate the effect on diaphragm size of a 1° misalignment of the backside etch mask to the <110> direction, using the assumptions of Problem 4.3. If the sensitivity of the pressure sensor varies as the inverse fourth power of the diaphragm size, what percent variation in sensitivity is caused by this 1° misalignment? Based on this result and on Problem 4.3, which produces more variation in sensitivity, a 1° alignment error or a 1% wafer-thickness variation?

4.5   Create a process and mask set that will form oval-cross-section flow channels in a fused quartz substrate by etching identical semi-oval channels into two substrates, and then performing an aligned wafer bond. A possible masking material for wet etching into the quartz is thin polysilicon coated with chromium.

4.6 Using combinations of deposited oxide and polysilicon films, create
a mask set and process flow that will produce a wheel that can rotate
freely about a hub that extends over the top of the wheel. A rough
sketch is shown below. Both features are made of polysilicon. Shading
is used to distinguish the hub from the rotor, which is free to rotate. It
is also free to fall down to touch the substrate, but is shown lifted to
emphasize that it is not connected to the substrate.

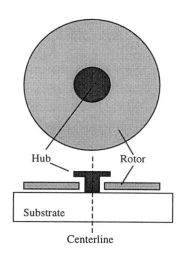

II
MODELING STRATEGIES

Chapter 5

LUMPED MODELING
WITH CIRCUIT ELEMENTS

5.1   Introduction

In this Chapter, we begin our discussion of device and system modeling. We choose to use small sets of electric circuit elements to represent the behavior of devices. Actual devices, of course, exist in a three-dimensional physical continuum and their behavior is governed by the laws of physics, chemistry, and biology. But through analysis, we can extract simplified device representations that are readily expressible with equivalent electric circuits. In doing so, we gain access to an immensely powerful set of intellectual tools that have been developed for understanding electrical circuits. Circuit analogies also permit efficient modeling of the interaction between the electronic and the non-electronic components of a microsystem.

A further advantage of circuit models is that they are intrinsically correct from an energy point of view. In contrast, the block-diagram models illustrated with the SIMULINK examples in Chapter 2 trace signal flow only, and do not have an explicit representation for power and energy. Nor do they capture the reciprocity of energy exchange that is characteristic of transducer behavior. For all these reasons, we begin with circuit-element modeling. Block-diagram modeling returns in Chapter 7 when we consider the dynamic behavior of systems.

Unlike 3D physical objects, which are bounded by surfaces, circuit elements are abstractions that have two or more discrete *terminals* to which potential difference (voltages) can be applied and into which electric currents can flow. Kirchhoff's Laws (see Section 5.4.1) govern the relationships among the voltages and currents that must be satisfied when circuit elements are connected into complete circuits.

Before introducing the circuit elements themselves, we present general-izations of the electrical voltage and current variables, with emphasis on the exchange of energy between circuit elements.

5.2   Conjugate Power Variables

We begin with the concept of a *lumped element*, a discrete object that can exchange energy with other discrete objects. Specifically, we focus on two components, A and B, that can exchange energy as illustrated schematically in Fig. 5.1.[1]

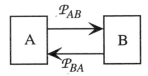

Figure 5.1.   Two lumped elements exchanging energy.

We define the *power flow* (*i.e.*, the energy flow per unit time) from A to B as $\mathcal{P}_{AB}$ and the reverse power flow from B to A as $\mathcal{P}_{BA}$. Since both $\mathcal{P}_{AB}$ and $\mathcal{P}_{BA}$ must be greater than or equal to zero, we can write each of them as the square of a real number, r_1 or r_2:

$$\mathcal{P}_{AB} = r_1^2 \tag{5.1}$$

$$\mathcal{P}_{BA} = r_2^2 \tag{5.2}$$

The net power flow from A to B is the difference

$$\mathcal{P}_{Net} = \mathcal{P}_{AB} - \mathcal{P}_{BA} \tag{5.3}$$

Inserting the appropriate squared term, we find

$$\mathcal{P}_{Net} = r_1^2 - r_2^2 \tag{5.4}$$

which can be factored to yield

$$\mathcal{P}_{Net} = (r_1 + r_2)(r_1 - r_2) \tag{5.5}$$

In other words, the net power flow from A to B can always be written as a product of two real numbers. This leads us to the concept of *conjugate power*

---

[1]This discussion is based on Neville Hogan's notes [34].

variables, a pair of quantities whose product yields the net power flow between two elements.

In the most general case, we define the two (time-dependent) conjugate power variables as an *effort* $e(t)$ and a *flow* $f(t)$. Associated with the flow is a time-dependent *generalized displacement* $q(t)$, given by

$$q(t) = \int_{t_0}^{t} f(t)\, dt + q(t_0) \tag{5.6}$$

The dimension of the product $e \cdot f$ is power. Therefore, the dimension of the product $e \cdot q$ is energy.

We can also define a time-dependent *generalized momentum* $p(t)$, given by

$$p(t) = \int_{t_0}^{t} e(t)\, dt + p(t_0) \tag{5.7}$$

The dimension of the product $p \cdot f$ is also energy.

Table 5.1. Examples of conjugate power variables.

Energy Domain	Effort	Flow	Momentum	Displacement
Mechanical translation	Force F	Velocity $\dot{x}, v$	Momentum p	Position x
Fixed-axis rotation	Torque τ	Angular velocity ω	Angular momentum J	Angle θ
Electric circuits	Voltage V, v	Current I, i	...	Charge Q
Magnetic circuits	Magnetomotive force MMF	Flux rate $\dot{\phi}$	...	Flux ϕ
Incompressible fluid flow	Pressure P	Volumetric flow Q	Pressure momentum Γ	Volume V
Thermal	Temperature T	Entropy flow rate $\dot{S}$	...	Entropy S

There are many energy domains,[2] and for each one, it is possible to define a set of conjugate power variables that can serve as the basis of lumped modeling with equivalent circuit elements. Table 5.1 identifies some of them.[3]

5.3   One-Port Elements
5.3.1   Ports

A *port* is a pair of terminals on a circuit element that must carry the same current *through* the element, entering the element at one of the terminals and leaving the element at the other. In anticipation of generalizing this concept to other energy domains, we call the current the *through variable*. The voltage is applied between the two terminals. We call the voltage difference between the terminals the *across variable*. The appropriately signed (see below) product of current and voltage is the power entering the circuit element at that port.

Circuit elements with only two terminals always have one port. There are five types of ideal one-port elements. There are also circuit elements with multiple terminals. Their behavior can be represented with more complicated multi-port elements. In this chapter, we introduce all of the one-port and several of the standard two-port elements.

5.3.2   The Variable-Assignment Conventions

In electric circuits, the effort variable is the across variable (voltage), while the flow variable is the through variable (current). We will call this particular assignment the $e \rightarrow V$ convention because effort and voltage are linked. A lumped element labeled with the effort and flow variables is shown in Fig. 5.2.

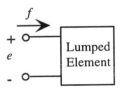

Figure 5.2.   Sign conventions for one-port elements in the $e \rightarrow V$ convention.

---

[2] The individual energy domains are the subject of Part III of this book.

[3] We observe in Table 5.1 that the symbols used for the variables in different energy domains overlap, which has the potential for confusion. This is a very serious problem, because each discipline has its own "standard" symbol usage. While some attempt has been made in this book to differentiate similar symbols by choice of type face, it is impossible to create a unique non-overlapping set of symbols that can be readily understood by everyone and easily related to other basic texts in the various disciplines. Be careful, therefore, to check the context when encountering formulae, either in this text, or elsewhere. Fortunately, overlapping symbols don't usually appear together in the same expression. And where similar standard symbols must appear together, they will be differentiated with subscripts or other suitable identifiers so they can be kept straight.

Standard sign conventions are useful to ensure that the algebraic sign of the power is calculated correctly. Both effort and flow are algebraic quantities and can be of either sign. However, once we assign a reference direction for positive flow into one of the terminals, we choose to use that same terminal as the positive terminal for defining effort. When we follow this convention, the product of e and f is the power *entering* the element.

We now wish to make an analogy to other energy domains and, therefore, must decide which variable gets assigned as the through variable, and which as the across variable. While it is logical to assume that, by analogy to the electric-circuit case, one should always assign the effort as the across variable and the flow as the through variable, it turns out that there are actually four conventions, each of which is useful for selected cases. The conventions are summarized in Table 5.2, and are discussed below.

Table 5.2. Different conventions for assigning circuit variables.

Convention	Across Variable	Through Variable	Product	Principal Use
$e \rightarrow V$	e	f	power	electric circuit elements
$f \rightarrow V$	f	e	power	mechanical circuit elements
Thermal	T	$\dot{Q}$	Watt-Kelvin	thermal circuits
HDL	q	e	energy	HDL circuit representation of mechanical elements

As we will see, the $e \rightarrow V$ convention has the advantage that potential energy is *always* associated with energy storage in capacitors. The other conventions lack this consistency. In fluidic systems, the $e \rightarrow V$ convention assigns pressure (effort) as the across variable, analogous to voltage, and the volume flow rate (flow) as the through variable, analogous to current. In the mechanical energy domain, the $e \rightarrow V$ convention assigns force to voltage and velocity to current, making displacement analogous to electric charge.[4] Many authors, however, choose the opposite assignment for mechanical and acoustic elements, namely, assigning velocity (flow) as the across variable, and force (effort) as the through variable. We call this the $f \rightarrow V$ convention.[5] While the $f \rightarrow V$ convention seems at first illogical, it turns out that the transcription from mechanical

---

[4]Tilmans has published two comprehensive papers on the use of the $e \rightarrow V$ convention for modeling linear mechanical and electromechanical devices [35, 36].

[5]A good discussion of these two conventions (but with different names) can be found in Beranek [37]. See also Romanowicz [38] for use of the $f \rightarrow V$ convention.

drawings to equivalent circuits is somewhat easier using the $f \rightarrow V$ convention than with the $e \rightarrow V$ convention.

In the thermal energy domain, while temperature T is a perfectly good across variable, the entropy flow rate turns out not to be a good choice for the through variable (this is discussed in Chapter 11). Instead, we use the heat current $\dot{Q}$, which has units of Watts. Note that the product between the across and through variables is not power, but a hybrid unit of Watt-Kelvins. When we use this *thermal convention*, we must use the heat energy Q, which is analogous to charge in the electric domain, to track energy flow. Of course, since charge is a conserved quantity in electrical systems, there is no risk of failing to conserve heat energy when using the thermal convention.

The fourth convention, called "HDL," is used when formulating abstract circuit elements in hardware-description languages.[6] In HDL usage, as in the thermal domain, it is not necessary to use conjugate power variables as long as one is careful to write the underlying constitutive properties of the element correctly and to keep explicit track of the energy. The designer can select pairs of variables that best help him think about the design. In mechanical systems, this leads to the displacement being assigned as the across variable and the force as the through variable. The product of these variables is energy, not power.

Because there are multiple conventions, it is important to be clear about which one is being used. This book uses the $e \rightarrow V$ convention for everything except the thermal domain. When examining papers by other authors, be sure to check the assignment convention being used. Any of the conventions can be used to reach the same conclusions about device behavior, but the intermediate vocabulary is convention-dependent.

5.3.3 One-Port Source Elements

Figure 5.3 shows two independent source elements, a *flow source* and an *effort source* (analogous to current and voltage sources in electrical circuits). The definition of the effort source is that the effort e equals the source value $e_o(t)$ for any flow f. The flow source (assuming it is connected to a network that provides a path for the flow) is defined as having a flow f equal to the source value $f_o(t)$ for any value of the effort e. These are clearly *active* elements, in that they supply power to other elements whenever the product of e and f is negative.

---

[6]Hardware-description languages are discussed by Romanowicz [38].

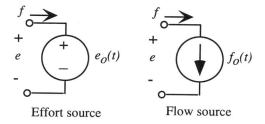

Effort source   Flow source

Figure 5.3.   Independent source elements in the $e \to V$ convention.

5.3.4   One-Port Circuit Elements

There are three basic one-port circuit elements: the generalized resistor, which is a dissipative element, and two energy-storage elements: the generalized capacitor and the generalized inductor, also called an *inertance*. These elements are shown in Figure 5.4.

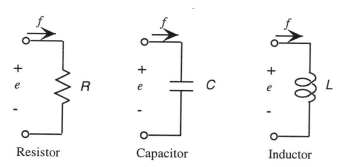

Resistor   Capacitor   Inductor

Figure 5.4.   Basic one-port elements in the $e \to V$ convention.

5.3.4.1   The Generalized Resistor

The characteristic of the generalized resistor is defined directly in terms of e and f. That is, we write either e as a function of f [$e = e(f)$], or f as a function of e [$f = f(e)$]. An important characteristic of these functions is that they go through the origin (if not, the element can be redefined as containing one of the source elements in combination with a resistor whose characteristic does go through the origin). A second important characteristic is stated in terms of the quadrants of the $e - f$ plane the functions occupy. If the functions fall entirely in the first and third quadrants, so that the product ef is always positive, the element is a purely *dissipative* element (also called a *passive* element), absorbing power for any allowed effort or flow condition. If the graph enters either the second or fourth quadrant at any point, the element can, over some portion of its characteristic, deliver net power to other elements. In this case, the element is considered an *active* element.

A *linear resistor* is characterized by a single scale factor between effort and flow (analogous to Ohm's law for resistors):

$$e = Rf \tag{5.8}$$

where R is the *resistance* of the resistor. The Ohm's Law version, of course, for standard electrical variables, is

$$V = RI \tag{5.9}$$

5.3.4.2   The Generalized Capacitor

The generalized capacitor is defined in terms of a relation between effort and displacement, that is, $e = \Phi(q)$, where Φ is a well-behaved function that goes through the origin of the $e - q$ plane. We associate capacitors with *potential energy*. Whenever a capacitor has a non-zero effort (hence also a non-zero displacement), there is stored energy in the capacitor.

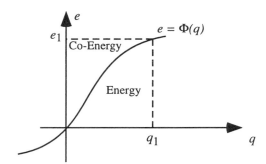

Figure 5.5.   Illustrating energy and co-energy for a generalized capacitor.

Referring to Figure 5.5, we can write

$$\mathcal{W}(q_1) = \int_0^{q_1} e \, dq = \int_0^{q_1} \Phi(q) \, dq \tag{5.10}$$

where $\mathcal{W}(q_1)$ is the *stored potential energy* in the capacitor defined by $\Phi(q)$ when it has displacement q_1. In Figure 5.5, $\mathcal{W}(q_1)$ is the area beneath the curve, labeled as "Energy."

We will also use Figure 5.5 to define another energy quantity, the *co-energy* $\mathcal{W}^*(e)$. The first definition we shall use is

$$\mathcal{W}^*(e) = eq - \mathcal{W}(q) \tag{5.11}$$

Thus, the co-energy is clearly the area between the e axis and the curve as labeled in Figure 5.5. Based on this visualization, we can also write an integral definition:

$$W^*(e_1) = \int_0^{e_1} q \, de = \int_0^{e_1} \Phi^{-1}(e) \, de \qquad (5.12)$$

where we have used the inverse of $\Phi(q)$ to represent the relation $q = \Phi^{-1}(e)$.

An electrical example of a generalized capacitor is a parallel-plate capacitor, having area A and a medium of dielectric permittivity ε between the plates (see Fig. 5.6).[7]

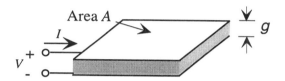

Figure 5.6. A parallel-plate capacitor; voltage is the across variable and current is the through variable.

In general, for a linear capacitor, we can write

$$Q = CV \qquad (5.13)$$

where Q is the charge on the capacitor, V is the voltage across the capacitor, and C is the capacitance of the capacitor. As derived in Appendix B, the charge density on the plate (neglecting fringing fields) has magnitude $\varepsilon V/g$, where g is the gap between the plates. The total charge on the plate is, therefore, $\varepsilon A V/g$, and the capacitance is given by

$$C = \frac{\varepsilon A}{g} \qquad (5.14)$$

The energy stored in the capacitor is

$$W(Q) = \frac{Q^2}{2C} \qquad (5.15)$$

and the co-energy $W^*(V)$ is

$$W^*(V) = \frac{CV^2}{2} \qquad (5.16)$$

---

[7]The definition of dielectric permittivity is reviewed in Appendix B.

For a linear capacitor, $\mathcal{W}(Q)$ and $\mathcal{W}^*(V)$ are numerically equal to one another; for a nonlinear capacitor (of which we will see many examples throughout the book), they differ, and the difference is important.

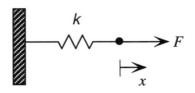

Figure 5.7.   Hooke's Law spring.

We now examine a mechanical element, continuing to use the $e \rightarrow V$ convention. Figure 5.7 shows the schematic for a Hooke's Law spring, for which the force F is related to the displacement x by the linear spring constant k:

$$F = kx \tag{5.17}$$

The stored energy in the spring when displaced by x_1 is

$$\mathcal{W}(x_1) = \int_0^{x_1} F(x)dx = \frac{1}{2}kx_1^2 \tag{5.18}$$

If we assume that the two generalized displacements, Q for the capacitor and x for the spring should lead to the same form of stored potential energy function, we are led to the conclusion that the circuit analog for a spring is a capacitor, with

$$C_{\text{spring}} = \frac{1}{k} \tag{5.19}$$

That is, when we build equivalent circuits for mechanical assemblies in the $e \rightarrow V$ convention, each spring gets represented by a capacitor equal to the compliance (rather than the stiffness) of the spring. The bigger the compliance, the bigger the displacement for a given force, just as for an electrical capacitor: the larger the capacitance, the bigger the charge for a given voltage.

5.3.4.3   The Generalized Inertance

The generalized inertance (or generalized inductor), which is represented by the inductance symbol in Figure 5.4, is defined by a functional relation between the flow f and the momentum p of the form $f = \Psi(p)$. This function must go through the origin of the $f - p$ plane. The correspondence with the definitions used for the generalized capacitance is quite close, as illustrated in Figure 5.8. In particular, the energy and co-energy are defined as

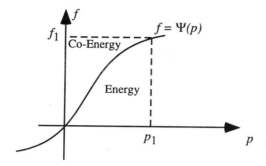

Figure 5.8. The generalized inertance.

$$W(p_1) = \int_0^{p_1} \Psi(p)\,dp \tag{5.20}$$

and, in the general definition,

$$W^*(f) = pf - W(p) \tag{5.21}$$

leading to the integral definition for the co-energy

$$W^*(f_1) = \int_0^{f_1} \Psi^{-1}(f)\,df \tag{5.22}$$

In this equation, the inverse function $\Psi^{-1}(f)$ is used to define p in terms of f.

In the $e \rightarrow V$ convention inertances are used for *stored kinetic energy*. Since not every row in Table 5.1 has a generalized momentum, inertances are encountered only in some energy domains. The most obvious example is an inertial mass, defined by the relation

$$p = mv \tag{5.23}$$

where p is the usual linear momentum, m is the mass, and v is the velocity. For this element, the function $\Psi(p)$ would be written

$$\Psi(p) = \frac{p}{m} \tag{5.24}$$

The stored energy at a particular momentum p_1 would be

$$W(p_1) = \frac{p_1^2}{2m} \tag{5.25}$$

while the co-energy $W^*(v_1)$ would be

$$W^*(v_1) = \frac{mv_1^2}{2} \tag{5.26}$$

Similar definitions apply to rotational motion in terms of angular velocity and angular momentum.

5.4   Circuit Connections in the $e \rightarrow V$ Convention

Perhaps the most difficult aspect of using lumped circuit elements to build device macro-models is to determine how to connect them together. We are all familiar with series and parallel connections for electrical elements. But how do these apply to mechanical or other elements? There are two basic concepts that hold for the $e \rightarrow V$ convention:[8]

Shared Flow and Displacement: Elements that share a common flow, and hence a common variation of displacement, are connected *in series*.

Shared Effort: Elements that share a common effort are connected *in parallel*.

Figure 5.9 shows a spring-mass-dashpot assembly, like that used to illustrate the object in the position-control system of Chapter 2. The mass is connected via a spring to a fixed support, being pulled by a force F. Also shown is a *dashpot*, a mechanical damping element analogous to an electrical resistor. All three elements share the same displacement. That is, if the mass moves to the right, the spring must stretch, and the dashpot end is displaced. Using the reasoning above, all three elements should be connected in series. The correct equivalent circuit is shown on the right.[9] We recognize this as the same circuit introduced in Chapter 2.

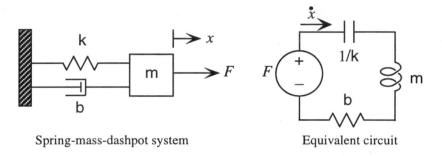

Spring-mass-dashpot system   Equivalent circuit

Figure 5.9.   Translating mechanical to electrical representations.

5.4.1   Kirchhoff's Laws

The various circuit elements are connected at *nodes*, and are governed by generalizations of Kirchhoff's Laws. These are:

---

[8]When the $f \rightarrow V$ convention is used, the role of series and parallel connections is reversed.

[9]Here we see another overlap of notation – the jagged line that is used in mechanical drawings to represent springs and in electrical drawings to represent resistors. The resistor is the electrical equivalent of the dashpot while the capacitor is the electrical equivalent of the spring.

Kirchhoff's Current Law (KCL): The sum of all currents (flows) entering a node is zero.

Kirchhoff's Voltage Law (KVL): The oriented sum of all voltages (efforts) around any closed path is zero.

The first of these is readily understood. The algebraic sign assigned to a flow at a node depends on the reference direction for that flow. The second refers to an *oriented sum*. This means that a consistent set of algebraic signs must be used for the efforts when traversing a closed path. A convenient mnemonic is to use the sign first encountered (of the + and - pair).

For the circuit example of Figure 5.9, since all elements share the same current, the application of KCL is trivial. To use KVL, it is necessary to label the effort across each element, as in Figure 5.10. One can use any internally consistent sign assignment for each element. In this figure, the effort variable for each of the mechanical circuit elements is assigned the + sign at the terminal where the flow $\dot{x}$ enters the element. The KVL equation for this circuit then becomes:

$$-F + e_k + e_m + e_b = 0 \qquad (5.27)$$

The physical interpretation of this equation is that the total applied force F has three components: the part of the force that is accelerating the mass (e_m), the part that is required to stretch the spring (e_k), and the part that is appears as a damping force in the dashpot (e_b).

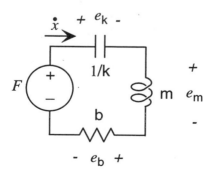

Figure 5.10. Fully labeled equivalent circuit.

5.5   Formulation of Dynamic Equations

Once the circuit model is correctly made, any method of formulating dynamical equations is permitted. The realm of electrical circuit analysis is rich in methods, from the use (for strictly linear circuits) of Laplace transform and impedance methods to the more general method of determining state variables and writing state equations. We will illustrate both for the circuit of Figure 5.10.

5.5.1   Complex Impedances

If complex impedances are used (the s-plane of the Laplace transform), each linear resistor is represented by itself, each linear capacitor by an impedance[10]

$$Z_C(s) = \frac{1}{sC} \tag{5.28}$$

and each linear inductor by an impedance

$$Z_L(s) = sL \tag{5.29}$$

For each impedance, the relation between effort and flow is the same:

$$e(s) = Z(s)f(s) \tag{5.30}$$

Thus, if $F(s)$ is the Laplace transform of the applied force, the KVL equation is written

$$F(s) = \left(sm + b + \frac{k}{s}\right)\dot{x}(s) \tag{5.31}$$

where $\dot{x}(s)$ is the Laplace transform of the velocity of the mass. This leads to the transfer function

$$\frac{\dot{x}(s)}{F(s)} = \frac{1}{sm + b + \frac{k}{s}} \tag{5.32}$$

When rationalized, this becomes

$$\frac{\dot{x}(s)}{F(s)} = \frac{s}{s^2 m + sb + k} \tag{5.33}$$

This transfer function has a single zero[11] at $s = 0$, set up by the requirement that under DC steady state conditions, the spring stretches to a fixed length, and the velocity of the mass goes to zero. It also has a pair of poles at

---

[10]Use of the s-plane and complex impedances is thoroughly discussed in [39].

[11]A *zero* a root of the numerator polynomial in s of a transfer function, while a *pole* is a root of the denominator polynomial.

$$s_1, s_2 = -\frac{b}{2m} \pm \sqrt{\left(\frac{b}{2m}\right)^2 - \frac{k}{m}} \tag{5.34}$$

We recognize this as the usual damped resonant behavior of a second-order system. All linear-system methods based on the Laplace transform, Bode plots, or other impedance- and pole-zero-based analysis can then be used. The dynamic behavior of this example system is analyzed in Chapter 7.

5.5.2 State Equations

An alternate approach, one particularly suited for the nonlinear examples we shall encounter, is based on *state equations*. In this approach, the state variables are identified, and a set of coupled first-order differential equations are written for the system. Depending on the level of complexity, either analytical or numerical methods may be appropriate for exploring the behavior of the solutions.

A *state-determined system* is a system in which the time derivatives of a set of variables describing the state, called the *state variables*, are determined by the present state and the instantaneous values of any inputs applied to the system. That is, a system with n state variables is described by n first-order differential equations whose right-hand side contains the state variables and inputs, but not their time-derivatives.

The simplest way to understand state variables in this context is to note that that every energy-storage element (capacitor or inductor) that can have its initial condition *independently* specified contributes one state variable to the total dynamical picture. For example, two capacitors in parallel with each other (or two inductors in series) would contribute only one state variable because it would not be possible to specify different initial voltages on the two capacitors (or independent initial currents in the two inductors).

For the example of Figure 5.10, two possible independent state variables are the displacement x associated with the capacitor (*i e.*, the charge) and the momentum p associated with the inertance. Harking back to the definitions of the generalized displacement and momentum, we see that for the p state variable,

$$\frac{dp}{dt} = e_m = F - kx - b\frac{p}{m} \tag{5.35}$$

(where we have inserted the characteristic of a linear dashpot, $e_b = bv$ and replaced v by p/m). The corresponding state equation for the x state variable is

$$\frac{dx}{dt} = \frac{p}{m} \tag{5.36}$$

State variables are not unique. Any pair of linearly independent quantities that can be derived from the original pair (x and p) can serve as state variables for the system. Thus, another perfectly acceptable set of state variables would be:

$$x_1 = x \tag{5.37}$$
$$x_2 = \dot{x} = v \tag{5.38}$$

With these definitions, the pair of state equations becomes

$$\dot{x}_1 = x_2 \tag{5.39}$$
$$\dot{x}_2 = -\frac{k}{m}x_1 - \frac{b}{m}x_2 + \frac{F}{m} \tag{5.40}$$

The state-determined nature of the system is clear, with the time derivative of the pair of state variables depending only on the present state (x_1 and x_2) and the applied input (F). We will return to the issue of dynamical equations in Chapter 7.

5.6   Transformers and Gyrators

Some elements have more than one port. There are two important two-port elements that are used to aid in translating variables from one energy domain to another. These are the *transformer* and the *gyrator*. They both have the characteristic that they neither store nor dissipate energy. In technical parlance, they are *lossless* and *memoryless* (*i.e.*, unlike capacitors and inductors, they contribute no state variables to the system).

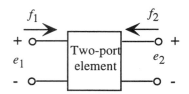

Figure 5.11.   General two-port element.

Figure 5.11 shows a schematic of a two-port element. For a lossless two-port, the total power entering the element through both ports must be zero. Therefore, the element must satisfy the following equation:

$$e_1 f_1 + e_2 f_2 = 0 \tag{5.41}$$

There are two ways of constructing linear elements that satisfy this constraint. One is the *transformer*; the second is the *gyrator*. The definitions are given below, along with circuit symbols in Fig. 5.12:

TRANSFORMER:

$$\begin{pmatrix} e_2 \\ f_2 \end{pmatrix} = \begin{pmatrix} n & 0 \\ 0 & -\frac{1}{n} \end{pmatrix} \begin{pmatrix} e_1 \\ f_1 \end{pmatrix} \tag{5.42}$$

GYRATOR:

$$\begin{pmatrix} e_2 \\ f_2 \end{pmatrix} = \begin{pmatrix} 0 & n \\ -\frac{1}{n} & 0 \end{pmatrix} \begin{pmatrix} e_1 \\ f_1 \end{pmatrix} \tag{5.43}$$

In both definitions, the quantity n is a scale factor that might have dimensions, depending on the two energy domains on either side of the element. The scale factor for a transformer has a special name, the *turns ratio*, from its physical origin in the ratio of the number of turns of primary and secondary coils wound on a common ferromagnetic core.

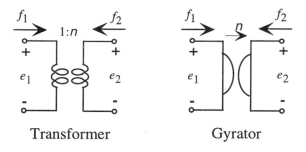

Figure 5.12.   Circuit symbols for the transformer and gyrator.

5.6.1   Impedance Transformations

One effect of transformers and gyrators is that they modify the effects of other impedances within the circuit. Figure 5.13 shows an impedance $Z(s)$ connected to the output of a transformer with turns ratio n.

It can be readily shown from the transformer equation that the impedance measured at the input terminals $Z_{in}(s)$, which is equal to $e_1(s)/f_1(s)$, is equal to $Z(s)/n^2$.

Similarly, referring to Fig. 5.14, it is readily shown that when a gyrator has an impedance $Z(s)$ connected at its output, the equivalent impedance at its input is given by

$$Z_{in}(s) = \frac{n^2}{Z(s)} \tag{5.44}$$

Thus, if a capacitor is connected at the output of a gyrator, with impedance $1/sC$, the equivalent impedance at the input is an inductor, with impedance

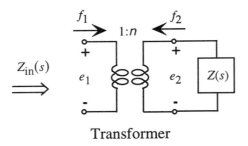

Transformer

Figure 5.13.   A transformer with an impedance connected at its output.

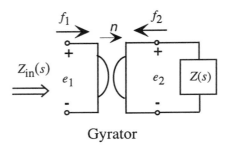

Gyrator

Figure 5.14.   A gyrator with an impedance connected at its output.

sn^2C. We use this fact in the next section to create a model of an electrical inductor.

5.6.2   The Electrical Inductor

A schematic view of an electrical inductor is shown in Fig. 5.15. A coil of n turns is wound on a ferromagnetic core that has an air gap of area A and height g, with g presumed to be much smaller than the circumferential length of the core L_m.

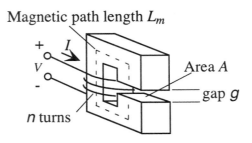

Figure 5.15.   An electrical inductor built on a magnetically permeable core.

Current in the coil establishes a magnetic field[12] which, because of the high permeability of the magnetic core, establishes flux ϕ in a *magnetic circuit* consisting of the core plus the air gap. The line integral of the magnetic field $\mathcal{H}$ around the magnetic circuit is the *magnetomotive force*, abbreviated MMF, and denoted by F_{MM}.

$$F_{MM} = \mathcal{H}_\mu L_m + \mathcal{H}_g g \tag{5.45}$$

where $\mathcal{H}_\mu$ is the $\mathcal{H}$-field in the permeable ring and $\mathcal{H}_g$ is the $\mathcal{H}$-field in the gap. In the permeable material, $\mathcal{B}_\mu = \mu \mathcal{H}_\mu$ and in the air gap $\mathcal{B}_g = \mu_0 \mathcal{H}_g$, where μ_0 is the permeability of free space. The flux through any cross-sectional surface S in the magnetic circuit is defined as

$$\phi = \int_S \boldsymbol{B} \cdot d\mathbf{A} \tag{5.46}$$

For convenience in this example, we assume that the cross-sectional area is constant around the magnetic circuit. Since the flux is continuous around the circuit, we conclude that $\mathcal{B}_\mu = \mathcal{B}_g$, or

$$\mathcal{H}_\mu = \frac{\mu_0}{\mu} \mathcal{H}_g \tag{5.47}$$

Thus we can now solve for the flux density in the gap:

$$\mathcal{B}_g = \mu_0 \mathcal{H}_g = \left(\frac{\mu_0}{g + \frac{\mu_0}{\mu} L_m} \right) F_{MM} \tag{5.48}$$

from which we obtain the flux in the gap ϕ.

$$\phi = \left(\frac{\mu_0 A}{g + \frac{\mu_0}{\mu} L_m} \right) F_{MM} \tag{5.49}$$

We recognize this as a relation between applied force F_{MM} and a generalized displacement, the flux ϕ. Thus, this element looks like a spring. It stores magnetic potential energy by establishing flux in response to an applied MMF.

The *reluctance* of a magnetic circuit element is defined as the ratio of MMF to flux. Thus, it is directly analogous to the stiffness of a spring. The higher the reluctance, the less flux for a given MMF, just as with a spring, the stiffer the spring, the less displacement for a given force. In equation form

$$F_{MM} = \mathcal{R}\phi \tag{5.50}$$

where $\mathcal{R}$ is the reluctance, or, to put it in the form analogous to the capacitor

---

[12]Magnetic fields are reviewed in Appendix B.

$$\phi = \frac{1}{\mathcal{R}} F_{MM} \tag{5.51}$$

$\mathcal{R}$ is directly analogous to the spring constant of a spring, and should be represented in the magnetic energy domain by a capacitor of value $1/\mathcal{R}$, where

$$\frac{1}{\mathcal{R}} = \frac{\mu_0 A}{g + \frac{\mu_0}{\mu} L_m} \tag{5.52}$$

We note that in the limit $\mu \gg \mu_0$, this reduces to an expression that looks very much like a parallel-plate capacitor. That is, in this limit, the equivalent capacitor in the magnetic energy domain has value

$$C_{\mathrm{mag}} = \frac{1}{\mathcal{R}} = \frac{\mu_0 A}{g} \tag{5.53}$$

So, *where is the inductor?* We thought this element was an inductor, not a capacitor. Yet it stores potential energy, like a capacitor. The answer lies in the choice of energy domain. We chose to model this inductor in the magnetic energy domain. But in a circuit, it must interact with real currents and voltages. The connection between the circuit domain and the magnetic domain is through a gyrator, as shown on the left-hand side of Fig. 5.16.

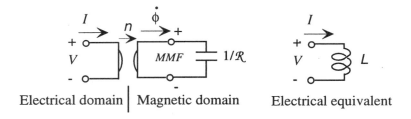

Electrical domain | Magnetic domain   Electrical equivalent

Figure 5.16.   A lumped equivalent circuit for the electrical inductor.

The governing equations for the gyrator, taking note of the reference direction for $\dot{\phi}$ in the figure, are written

$$\dot{\phi} = \frac{1}{n} V \tag{5.54}$$

$$F_{MM} = nI \tag{5.55}$$

where the gyrator parameter n is, in this case, the number of turns in the coil. The first gyrator equation is Faraday's Law of Induction, which says that the induced EMF in the coil is the time rate of change of the linked flux, while the second gyrator equation is just the definition of the MMF in terms of current derived in Appendix B. It is readily shown from these two equations that

$$V = s \left(\frac{n^2}{\mathcal{R}} \right) I \tag{5.56}$$

which is our familiar electrical inductor, with inductance L given by

$$L = \frac{n^2}{\mathcal{R}} \tag{5.57}$$

Thus, a completely equivalent representation is the simple electrical inductor shown on the right-hand side of Fig. 5.16

The stored energy in the inductor when carrying current I can be found by considering the stored potential energy in the capacitor. This energy is

$$\mathcal{W}_L = \frac{\phi^2}{2C} = \frac{\phi^2 \mathcal{R}}{2} \tag{5.58}$$

The corresponding co-energy is

$$\mathcal{W}_L^* = \frac{F_{MM}^2}{2\mathcal{R}} \tag{5.59}$$

If we use the gyrator relation between F_{MM} and I, we find that the co-energy is the familiar result:

$$\mathcal{W}_L^* = \frac{1}{2} L I^2 \tag{5.60}$$

At this point, the reader may be wondering why we go through this complex example, if the final result is to be a simple inductor. The answer is found in the next chapter on transducers. If there are moveable parts associated with the inductor, then the connection between the electrical domain and the mechanical domain is best explained with the gyrator-plus-capacitor model, which shows the magnetic domain explicitly as an intermediate stage between electrical and mechanical domains.

Related Reading

L. Beranek, *Acoustics*, New York: McGraw-Hill, 1954, Chapter 3.

A. G. Bose and K. N. Stevens, *Introductory Network Theory*, New York: Harper & Row, 1965.

B. Romanowicz, *Methodology for the Modeling and Simulation of Microsystems*, Boston: Kluwer Academic Press, 1998.

S. D. Senturia and B. D. Wedlock, *Electronic Circuits and Applications*, New York: Wiley, 1975; now available from Krieger reprints.

H. Tilmans, Equivalent circuit representation of electromechanical trans-ducers: I. Lumped-parameter systems, *J. Micromech. Microeng.*, vol. 6 , pp. 157-176, 1996; erratum published in vol. 6, p. 359, 1996.

H. Tilmans, Equivalent circuit representation of electromechanical trans-ducers: II. Distributed-parameter systems, *J. Micromech. Microeng.*, vol. 7 , pp. 285-309, 1997.

Problems

5.1   Confirm the impedance-transforming properties of the transformer and gyrator with a detailed calculation.

5.2   Make an equivalent circuit for the mechanical system in the figure below.

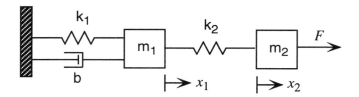

5.3   For the circuit below, use complex impedances to find the transfer function $I_1(s)/V(s)$.

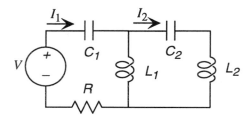

5.4   Develop a set of state equations for the circuit of Problem 5.3.

5.5   A water reservoir is filled by a pump that delivers a steady flow rate, 24 hours a day. The level in the reservoir varies throughout the day, depending on water usage in the community. Create a circuit model that captures this behavior (only the form of the circuit is requested – no calculations). This will require decisions on what type of element to use to model the input, the reservoir itself, and the variable water usage.

Chapter 6

ENERGY-CONSERVING TRANSDUCERS

6.1   Introduction

By definition, *transducers* always involve at least two energy domains. They are energy-conversion devices. This means that we can expect to find that multi-port elements are required for their proper modeling. In this chapter, we shall examine *energy-conserving* transducers in great detail. In later chapters, we shall also examine transducers that involve energy dissipation, such as transducers based on thermal effects or on electrical resistance. For energy-conserving transducers, a correct treatment of the stored energy leads naturally to a self-consistent and extremely compact representation. We shall start with a simple example, then add additional lumped elements to represent inertial masses and extrinsic energy-loss mechanisms, thereby creating lumped-element models that have at least first-order representations of all the important features of many types of sensors and actuators.

6.2   The Parallel-Plate Capacitor

We shall begin by modeling a parallel-plate capacitor with one moveable plate, as in Fig. 6.1. We define this capacitor as having plate area A, a gap g, and, when charged, a charge $+Q$ on one plate and $-Q$ on the other (of course, Q is a signed quantity, so either sign of charge is representable this way).

Because of the opposite charges on the two plates, there is a force of attraction between the plates. In fixed-plate electrical capacitors, we do not ever think about this force, but it is *always* present whenever the capacitor is charged. When one plate is moveable, the inclusion of this mechanical force becomes essential. Without some mechanical support providing an equal and opposite force on the upper plate, the two plates would move together. Thus, we are led immediately to the idea that this element has two ports: an electrical

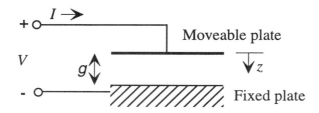

Figure 6.1.   A parallel-plate capacitor with a moveable upper plate.

port, with voltage and current (the time derivative of charge) as the conjugate variables, and a mechanical port with force and velocity (the time derivative of displacement) as the conjugate variables.

We will now compare two ways of reaching a final charged state for this capacitor (Fig. 6.2). The first involves charging the capacitor to charge Q at fixed gap g, assuming that some friendly magician is supplying the opposing external force to keep the plates from crashing together. We can calculate the electrical stored energy for this step. There is no mechanical work done by the external force, since the gap is fixed. The second method involves charging the capacitor to charge Q at zero gap (or as close to zero as possible), then lifting the upper plate to create the gap g. Because we charge the capacitor at zero gap, the voltage across the capacitor is virtually zero (effectively, the capacitance is infinite at zero gap; one can transfer unlimited amounts of charge without requiring finite voltage). Hence, there is no electrical stored energy at zero gap. However, mechanical work must be done by the external force to lift the plate to gap g. We expect these two methods should lead to the same total stored energy in the final state.

6.2.1   Charging the Capacitor at Fixed Gap

The capacitance of this moveable-plate parallel-plate capacitor, with fringing fields neglected, is given by

$$C = \frac{\varepsilon A}{g} \tag{6.1}$$

where ε is the permittivity of the material between the gap, presumed to be air; hence $\varepsilon \approx \varepsilon_0$, the permittivity of free space.

The stored energy, in general, for a lumped capacitive element with effort e and displacement q is

$$\mathcal{W}(q) = \int_0^q e(q)dq \tag{6.2}$$

For the electrical capacitor in the $e \to V$ convention,

$$e \to V \quad \text{and} \quad q \to Q$$

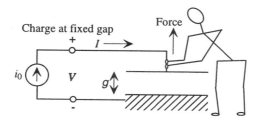

Charge at fixed gap

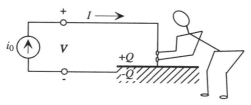

Charge at zero gap, then...

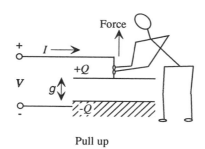

Pull up

Figure 6.2. · Two ways of charging a capacitor.

and since

$$Q = CV \qquad (6.3)$$

the expression for the stored energy becomes

$$W(Q) = \int_0^Q \frac{Q}{C} dQ \qquad (6.4)$$

which leads to

$$W(Q) = \frac{Q^2}{2C} = \frac{Q^2 g}{2\varepsilon A} \qquad (6.5)$$

6.2.2 Charging the Capacitor at Zero Gap, then Lifting

Let us imagine that we have the two plates in direct contact (zero gap), and we now have our friendly magician transfer a charge $+Q$ from the bottom plate to the top plate while they are still in contact, leaving the bottom plate with a charge $-Q$. (These magicians are quite powerful!) At zero gap, the electrical stored energy is zero.

The force between the plates set up by the charges $+Q$ and $-Q$ on opposite plates depends on the electric field set up by the charges. This field is

$$\mathcal{E} = \frac{Q}{\varepsilon A} \tag{6.6}$$

and the corresponding force is

$$F = \left(\frac{Q}{2}\right)\mathcal{E} = \frac{Q^2}{2\varepsilon A} \tag{6.7}$$

where the factor of 2 beneath the Q is to avoid double-counting of charge.[1] We note that this force is *independent of g*!

If we now pull on the upper plate and raise it by a distance g, the work done (all of which is stored potential energy) is

$$W(g) = Fg = \frac{Q^2 g}{2\varepsilon A} \tag{6.8}$$

which is the *same* stored energy as we found when calculating the electrostatic stored energy $W(Q)$ at fixed gap.

We conclude that we can put stored energy into this system in two ways: either fix the gap and increase the charge, or keep the charge fixed and increase the gap. This means that the stored energy is actually a function of two variables, $W(Q, g)$. Further, we can construct a set of differentials to describe how the stored energy changes if either quantity changes, using the basic definitions of stored energy in each energy domain:

$$dW = F \, dg + V \, dQ \tag{6.9}$$

Using this formulation, we can write

$$F = \left. \frac{\partial W(Q, g)}{\partial g} \right|_Q \tag{6.10}$$

[1]The origin of the factor of 2 in the denominator of Eq. 6.7 is interesting, and illustrates the importance of being careful with singular functions, such as "infinitessimally thin" sheets of charge. If we think of the charge on the surface of a conductor as a sheet of zero thickness, then the electric field rises instantaneously from zero inside the conductor to its full value between the plates. Naively, computing the force by taking the charge-field product involves taking a product of an impulse (the sheet charge) and a step function (the electric field). This is an ill-defined mathematical concept, since the impulse only takes on meaning inside of integrals. But it is readily shown that if the sheet of charge is assumed to have a small but nonzero thickness, thereby smoothing the transition from zero to full field, the factor of 1/2 in the force emerges naturally. This result is independent of the detailed shape of the charge distribution.

leading to

$$F = \frac{Q^2}{2\varepsilon A} \tag{6.11}$$

and

$$V = \frac{\partial \mathcal{W}(Q,g)}{\partial Q}\bigg|_g \tag{6.12}$$

leading to

$$V = \frac{Qg}{\varepsilon A} = \frac{Q}{C} \tag{6.13}$$

As a final result, we note that

$$\frac{\partial F}{\partial Q}\bigg|_g = \frac{\partial V}{\partial g}\bigg|_Q = \frac{Q}{\varepsilon A} \tag{6.14}$$

This is simply a way of stating that the second derivative of the stored energy does not depend on the order of differentiation:

$$\frac{\partial^2 \mathcal{W}(Q,g)}{\partial Q \partial g} = \frac{\partial^2 \mathcal{W}(Q,g)}{\partial g \partial Q} \tag{6.15}$$

This is not at all surprising for a scalar potential energy function. The efforts F and V, considered as elements of a vector, constitute the gradient of a scalar. But the gradient of a scalar has no curl. Therefore Eq. 6.15 must apply.

6.3 The Two-Port Capacitor

We have now demonstrated that the stored energy in this parallel-plate capacitor is a function of an electrical variable (charge) and a mechanical variable (displacement, or gap). This leads us to construct a *two-port capacitor* to describe this element.

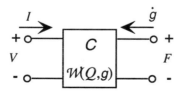

Figure 6.3. A two-port capacitive transducer

The two-port capacitor is a *single* capacitor, with a *single* stored energy, but *two* ports. The stored energy is written, as above,

$$\mathcal{W}(Q,g) = \frac{Q^2 g}{2\varepsilon A} \tag{6.16}$$

and the effort variables are the appropriate partial derivatives of W

$$F = \left. \frac{\partial W(Q,g)}{\partial g} \right|_Q = \frac{Q^2}{2\varepsilon A} \qquad (6.17)$$

$$V = \left. \frac{\partial W(Q,g)}{\partial Q} \right|_g = \frac{Qg}{\varepsilon A} \qquad (6.18)$$

The circuit element is written as a single box with two ports. At the electrical port, the effort variable is V, and the flow variable is the current $I = \dot{Q}$; at the mechanical port, the effort variable is the force F separating the plates, and the flow variable is the velocity of the moveable plate $\dot{g}$.

6.4   Electrostatic Actuator

Let us now apply these ideas to the basic electrostatic actuator by adding a spring between the moveable plate and a fixed support (Fig. 6.4). Structurally, this means that the spring is replacing the friendly magician on the mechanical side, creating an upward force on the moveable plate if the plate moves down from its rest position. We assume that the rest-position gap (with zero spring force and zero capacitor charge) is g_0. Figure 6.4 illustrates the assembly in the upper part and an equivalent circuit in the lower part.

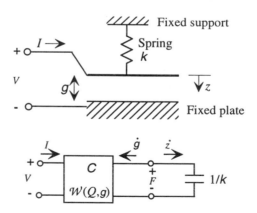

Figure 6.4.   The basic electrostatic actuator: a moveable capacitor plate attached to a spring.

Some care is needed in handling the spring, because its standard flow variable $\dot{z}$ is in a direction opposite to $\dot{g}$. That is, since the end of the spring and the moveable plate share the same displacement, their flow variables are in series. However, a decrease in gap is an increase in the length of the spring – hence, the opposite orientation of reference directions.

6.4.1 Charge Control

We now consider operating this actuator with an ideal current source connected to the electrical port (Fig. 6.5), thereby controlling directly the addition or removal of charge from the capacitor.

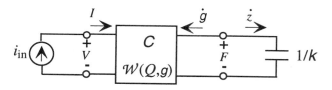

Figure 6.5. Charge control of an electrostatic actuator.

To analyze this system, we do what mechanical engineers call "following the *causal path*." In this case, assuming we started from an initially uncharged state at $t = 0$, we can easily determine the charge:

$$Q = \int_0^t i_{in}(t)dt \qquad (6.19)$$

The force is determined by this value of Q:

$$F = \frac{Q^2}{2\varepsilon A} \qquad (6.20)$$

The characteristic of the spring requires that the displacement of the end of the spring be given by

$$z = \frac{F}{k} \qquad (6.21)$$

and the gap is obtained from

$$g = g_0 + \int_0^t \dot{g}\,dt \qquad (6.22)$$

where g_0 is the gap in the absence of any voltage or spring force. Since

$$\dot{g} = -\dot{z} \qquad (6.23)$$

the integral for $\dot{g}$ must simply yield $-z$. Therefore,

$$g = g_0 - z \qquad (6.24)$$

or,

$$g = g_0 - \frac{Q^2}{2\varepsilon Ak} \qquad (6.25)$$

Finally, the voltage across the capacitor at the electrical port is

$$V = \frac{Qg}{\varepsilon A} = \frac{Q\left(g_0 - \dfrac{Q^2}{2\varepsilon A k}\right)}{\varepsilon A} \tag{6.26}$$

We see that as Q increases, the force of attraction between the plates increases as Q^2. At equilibrium, this force of attraction must be balanced by the force from the spring, and this requires the spring to be stretched from its rest position. The greater the charge, the more the spring must be stretched. The gap reduces to zero at a charge $\hat{Q}$, determined by

$$\frac{\hat{Q}^2}{2\varepsilon A} = kg_0 \tag{6.27}$$

The voltage V also goes to zero when the gap goes to zero.

6.4.2   Voltage Control

What happens if we use a voltage source instead of a current source at the electrical port (Fig. 6.6) so that the voltage V is specified by the source value V_{in}? First of all, we expect that controlling the voltage will have the effect of controlling the charge, since the element is still a parallel plate capacitor. However, as we shall see, this moveable-plate capacitor is a nonlinear element. Therefore the energy and co-energy are not the same.[2] Since we are controlling voltage rather than charge, we *must* use the co-energy of the system rather than the energy.

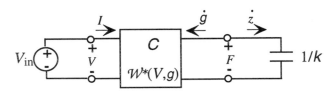

Figure 6.6.   Voltage control of an electrostatic transducer.

We recall that the purely electrical co-energy is defined as

$$\mathcal{W}^*(V) = QV - \mathcal{W}(Q) \tag{6.28}$$

We now have a situation where there can be both mechanical and electrical stored potential energy. But the co-energy definition for the electrical domain

---

[2]See Section 5.3.4.2.

remains the same as long as we continue to use the gap g as the mechanical-domain independent variable. That is,

$$\mathcal{W}^*(V,g) = QV - \mathcal{W}(Q,g) \tag{6.29}$$

Differentially, this becomes

$$d\mathcal{W}^*(V,g) = (QdV + VdQ) - d\mathcal{W}(Q,g) \tag{6.30}$$

But

$$d\mathcal{W}(Q,g) = Fdg + VdQ \tag{6.31}$$

Therefore, we find, in differential form

$$d\mathcal{W}^*(V,g) = QdV - Fdg \tag{6.32}$$

From this expression, we obtain the equations

$$Q = \left.\frac{\partial \mathcal{W}^*(V,g)}{\partial V}\right|_g \tag{6.33}$$

$$F = -\left.\frac{\partial \mathcal{W}^*(V,g)}{\partial g}\right|_V \tag{6.34}$$

In this case, we have a mixed set of variables. The generalized displacement Q is defined in terms of a derivative with respect to a generalized effort V, while the generalized effort F is defined in terms of a derivative with respect to a generalized displacement g. These mixed-variable representations are reminiscent of the various thermodynamic functions.[3]

We can apply these results, using the co-energy function calculated at fixed gap, to obtain

$$\mathcal{W}^*(V,g) = \int_0^V QdV = \int_0^V \frac{\varepsilon A}{g}VdV \tag{6.35}$$

from which we find

$$\mathcal{W}^*(V,g) = \frac{\varepsilon AV^2}{2g} \tag{6.36}$$

Applying the differential relations, we obtain

$$Q = \frac{\varepsilon A}{g}V \tag{6.37}$$

---

[3] The enthalpy, Helmholtz free energy, and Gibbs free energy are all co-energies that use various combinations of the primary thermodynamic quantities: pressure (an effort), volume (a displacement), temperature (an effort), and entropy (a displacement). The enthalpy is the co-energy with respect to the pressure-volume product, the Helmholtz free energy is the co-energy with respect to the temperature-entropy product, and the Gibbs free energy is the co-energy with respect to both products.

which agrees with the basic expression $Q = CV$. For the force, we obtain

$$F = -\left.\frac{\partial W^*(V,g)}{\partial g}\right|_V = -\left[-\frac{\varepsilon A V^2}{2g^2}\right] = \frac{\varepsilon A V^2}{2g^2} \qquad (6.38)$$

This agrees with our previous result $Q^2/2\varepsilon A$.

So, either way is correct. We can use a charge-displacement representation along with the energy, or a voltage-displacement representation with the co-energy. When using the co-energy, we must be careful to observe the minus sign on the equation defining the force (many red faces have been caused by inadvertent neglect of this sign).

If we now examine the complete circuit, the causal path has been changed. The voltage determines the force, which stretches the spring, thus determining the change in gap, from which, in combination with the voltage, we can obtain the charge.

$$F = \frac{\varepsilon A V_{in}^2}{2g^2} \qquad (6.39)$$

and

$$g = g_0 - z \qquad (6.40)$$

as before, with

$$z = \frac{F}{k} \qquad (6.41)$$

Thus

$$g = g_0 - \frac{\varepsilon A V_{in}^2}{2kg^2} \qquad (6.42)$$

Note that this is now a *cubic* equation for the gap. Further, we note that there is an implicit feedback in this equation. As we increase the voltage, the gap decreases, with the amount of decrease growing as the gap gets smaller. Thus, there is positive feedback in this system, and at some critical voltage, the system goes unstable, and the gap collapses to zero. This phenomenon is called pull-in (or "snap-down"). The details of the gap-voltage relationship are examined in the next section. For the moment, we note that once the stable gap g is identified, the charge is found from:

$$Q = \frac{\varepsilon A}{g} V_{in} \qquad (6.43)$$

6.4.3   Pull-In

The voltage-controlled parallel-plate electrostatic actuator exhibits an important behavior called *pull-in*. In order to explain it, we consider the *stability* of the equilibrium that must exist between the electrostatic force pulling the plate down and the spring force pulling the spring up. Stability analysis involves perturbing the position slightly and asking whether or not the net force tends to return to the equilibrium position.

The net force on the upper plate at voltage V and gap g, using a sign convention that assigns a positive sign for forces that increase the gap, is

$$F_{\text{net}} = \frac{-\varepsilon A V^2}{2g^2} + k(g_0 - g) \tag{6.44}$$

where g_0 is the gap at zero volts and zero spring extension. At a point of equilibrium, F_{net} is zero. If we now ask how F_{net} varies with a small perturbation of the gap to $g + \delta g$, we can write

$$\delta F_{\text{net}} = \left. \frac{\partial F_{\text{net}}}{\partial g} \right|_V \delta g \tag{6.45}$$

If δF_{net} is positive for positive δg, then g is an *unstable* equilibrium point, because a small increase δg creates a force tending to increase it further. If δF_{net} is negative, then g is a stable equilibrium point.

We can evaluate δF_{net} using Eq. 6.44.

$$\delta F_{\text{net}} = \left(\frac{\varepsilon A V^2}{g^3} - k \right) \delta g \tag{6.46}$$

In order for g to be a stable equilibrium, the expression in parenthesis must be negative, which means that

$$k > \frac{\varepsilon A V^2}{g^3} \tag{6.47}$$

Clearly, since the equilibrium gap decreases with increasing voltage, there will be a specific voltage at which the stability of the equilibrium is lost. This is called the *pull-in voltage*, denoted V_{PI}. At pull-in, there are two equations that must be satisfied: the original requirement that $F_{\text{net}} = 0$, and the new requirement that

$$k = \frac{\varepsilon A V_{\text{PI}}^2}{g_{\text{PI}}^3} \tag{6.48}$$

It is readily shown that the pull-in occurs at

$$g_{\text{PI}} = \frac{2}{3} g_0 \tag{6.49}$$

and, at this value of gap, the equilibrium voltage is

$$V_{\text{PI}} = \sqrt{\frac{8kg_0^3}{27\varepsilon A}} \tag{6.50}$$

We can gain additional insight into the pull-in phenomenon by examining the two components of F_{net} graphically. It is helpful to use normalized variables. Specifically, we normalize the voltage to the pull-in voltage, and we then examine the normalized *displacement* of the plate from its equilibrium position, using g_0 as the normalization constant. Thus

$$v = \frac{V}{V_{PI}} \tag{6.51}$$

$$\zeta = 1 - \frac{g}{g_0} \tag{6.52}$$

Using these variables, the condition for equilibrium is that

$$\frac{4}{27} \frac{v^2}{(1 - \zeta)^2} = \zeta \tag{6.53}$$

Figure 6.7 shows the two sides of this equation plotted simultaneously for the specific value $v = 0.8$. There are two intersections, meaning two equilibrium points. However, only one of them is stable.

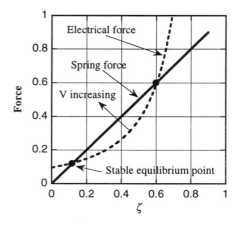

Figure 6.7.   Electrical and spring forces for the voltage-controlled parallel-plate electrostatic actuator, plotted for $V/V_{PI} = 0.8$.

As the voltage is increased, the dashed electrical force moves in the direction of the arrow, and the two equilibrium points move toward each other, exactly merging when $\zeta = 1/3$ and $v = 1$. For normalized voltages greater than 1, the curves never intersect, and there is no stable equilibrium.

Another way to explore pull-in is to make a graph of the normalized equilibrium gap (g/g_0) as a function of normalized voltage. This is shown in Fig. 6.8, with stable displacements occurring for normalized gaps greater than 2/3, and unstable collapse to zero gap occurring once the pull-in voltage is reached.

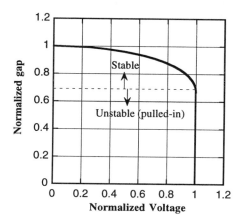

Figure 6.8. Normalized gap as a function of normalized voltage for the electrostatic actuator.

6.4.4 Adding Dynamics to the Actuator Model

So far, we have dealt exclusively with either charge-controlled or voltage-controlled equilibrium behavior of the parallel-plate electrostatic actuator. We can now add elements that give the system interesting dynamical behavior. Specifically, we can include a mass to represent the mechanical inertia of the moving plate, a dashpot to capture the mechanical damping forces that arise from the viscosity of the air that must be squeezed out when the plates move together and drawn in when the plates move apart,[4] and a source resistor for the voltage source that drives the transducer. This enhanced model is shown in Fig. 6.9.

It is now clear from the equivalent circuit that the two-port capacitor is the essential element of this actuator, with one port in the electrical domain and one port in the mechanical domain. To create the governing equations for this actuator, we simply write KVL equations in the two domains, and manipulate the equations into state form. Because this is a nonlinear system, we cannot use Laplace transforms or complex impedances, but must instead work with the full set of nonlinear governing equations. For the electrical domain:

$$\dot{Q} = I = \frac{1}{R}\left(V_{\text{in}} - \frac{Qg}{\varepsilon A}\right) \qquad (6.54)$$

---

[4]This effect is called *squeezed-film damping*. See Chapter 13.

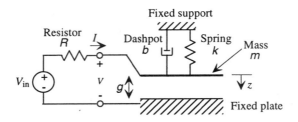

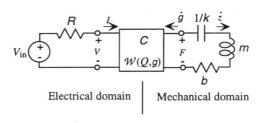

Electrical domain | Mechanical domain

Figure 6.9.   A complete electrostatic actuator, with added elements representing the inertia of the moveable element, mechanical damping, and the source resistance of the electrical network.

and for the mechanical domain, using the force as

$$F = \frac{Q^2}{2\varepsilon A} \tag{6.55}$$

we obtain

$$\frac{Q^2}{2\varepsilon A} + b\dot{g} + m\ddot{g} + k(g - g_0) = 0 \tag{6.56}$$

To get this into state form, we must identify three state variables. If we select as state variables

$$x_1 = Q \tag{6.57}$$
$$x_2 = g \tag{6.58}$$
$$x_3 = \dot{g} \tag{6.59}$$

the state equations become:

$$\dot{x}_1 = \frac{1}{R}\left(V_{\text{in}} - \frac{x_1 x_2}{\varepsilon A}\right) \tag{6.60}$$

$$\dot{x}_2 = x_3 \tag{6.61}$$

$$\dot{x}_3 = -\frac{1}{m}\left(\frac{x_1^2}{2\varepsilon A} + k(x_2 - g_0) + bx_3\right) \tag{6.62}$$

The solution of these equations, and the dynamic behavior of these kinds of systems generally, is the subject of the next chapter.

6.5   The Magnetic Actuator

Figure 6.10 shows a schematic view of a magnetic actuator, in which a slab of permeable material, called the armature, is able to move laterally into the gap of a magnetic-core inductor and is supported by a spring.

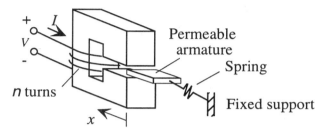

Figure 6.10.   A magnetic actuator.

The first question is how to determine the force that the actuator creates. Our procedure is to find an expression for the reluctance of the inductor as a function of x, to use this in the magnetic co-energy, and then to take the negative gradient with respect to x to find the force.

A complete solution of this problem is quite difficult but, if we make a few simplifying assumptions, we can make the problem tractable. The approach is based on the analysis of the inductor in Section 5.6.2. We recall that in the absence of the moveable armature, the flux density B_g in the gap g was

$$B_g = \left(\frac{\mu_0}{g + \frac{\mu_0}{\mu} L_m} \right) F_{MM} \tag{6.63}$$

where μ is the permeability of the core and the armature, μ_0 is the permeability of free space, F_{MM} is the magnetomotive force (equal to nI), and L_m is the magnetic path length through the core. Now we have a moving armature, which introduces two different gaps into the problem, as illustrated in the cross-section of Fig. 6.11.

If we assume that when the armature moves into the gap by an amount x, that fraction of the area has a much smaller gap g_μ, then the flux density in that portion of the structure is increased to

$$B_{g_\mu} = \left(\frac{\mu_0}{g_\mu + \frac{\mu_0}{\mu} L_m} \right) \mathcal{M} \tag{6.64}$$

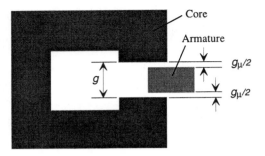

Figure 6.11.   Cross-section of the magnetic actuator of Fig. 6.10.

This is equivalent to a complete neglect of any field-spreading effects between the two parts of the gap and is somewhat analogous to the neglect of fringing fields in parallel-plate capacitors. With this assumption, the total flux ϕ is

$$\phi = \left[\left(\frac{\mu_0 A}{g + \frac{\mu_0}{\mu} L_m}\right)\left(1 - \frac{x}{x_0}\right) + \left(\frac{\mu_0 A}{g_\mu + \frac{\mu_0}{\mu} L_m}\right)\left(\frac{x}{x_0}\right)\right] F_{MM} \quad (6.65)$$

where x_0 is a scale factor to indicate the fraction of the gap area affected by motion of the armature. The quantity in square braces is simply $1/\mathcal{R}$ for the structure, and since it is $1/\mathcal{R}$ that enters the magnetic co-energy, we can leave the result in that form.

The magnetic co-energy becomes

$$\begin{aligned} W^*(F_{MM}, x) &= \frac{F_{MM}^2}{2}\left[\left(\frac{\mu_0 A}{g + \frac{\mu_0}{\mu} L_m}\right)\left(1 - \frac{x}{x_0}\right)\right. \\ &\quad \left. + \left(\frac{\mu_0 A}{g_\mu + \frac{\mu_0}{\mu} L_m}\right)\left(\frac{x}{x_0}\right)\right] \end{aligned} \quad (6.66)$$

The mechanical force is

$$F = -\left.\frac{\partial W^*(F_{MM}, x)}{\partial x}\right|_{F_{MM}} \quad (6.67)$$

which yields

$$F = \frac{F_{MM}^2}{2x_0} \left[\frac{\mu_0 A (g - g_\mu)}{g g_\mu + \left(\frac{\mu_0 L_m}{\mu} \right)(g + g_\mu) + \left(\frac{\mu_0 L_m}{\mu} \right)^2} \right] \qquad (6.68)$$

We see that the force varies as the square of F_{MM} and is independent of the displacement x, but it is still very difficult to visualize the contributions of the various terms. However, if we assume that

$$g_\mu \ll g \qquad (6.69)$$

and further assume that

$$\frac{\mu_0 L_m}{\mu} \ll g \qquad (6.70)$$

both of which are very reasonable assumptions, then the force reduces to

$$F = \frac{F_{MM}^2}{2x_0} \left(\frac{\mu_0 A}{g_\mu + \frac{\mu_0 L_m}{\mu}} \right) \qquad (6.71)$$

Essentially, the force arises from the increase in magnetic stored energy when, due to lateral motion of the armature within the gap, air with permeability μ_0 is replaced by armature material with the larger permeability μ. This increase in stored energy with increasing x results in an attractive force, pulling the armature into the gap. The specific value of the force depends on the relative size of the residual gap g_μ and the scaled magnetic path length $\mu_0 L_m / \mu$.

This actuator involves three energy domains, as illustrated in Fig. 6.12. Current in the electrical domain establishes a magnetomotive force in the magnetic domain, leading to stored magnetic potential energy in the two-port magnetic "capacitor." The second port of this capacitor is in the mechanical domain, and, in the example shown here, this port is connected to a mechanical capacitor representing the spring.

The equilibrium displacement of the armature depends on the stiffness of the spring. The magnetic force is independent of displacement, so the equilibrium displacement is simply

$$x_{\text{eq}} = \frac{1}{k} \left[\frac{F_{MM}^2}{2x_0} \left(\frac{\mu_0 A}{g_\mu + \frac{\mu_0 L_m}{\mu}} \right) \right] \qquad (6.72)$$

If we now use the gyrator relation

$$F_{MM} = nI \qquad (6.73)$$

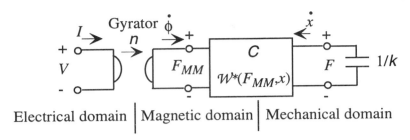

Electrical domain | Magnetic domain | Mechanical domain

Figure 6.12. Equivalent circuit for the magnetic actuator. The inertial mass of the armature and the damping elements are not included.

we can express the entire behavior using only variables from the electrical and mechanical domains:

$$x_{eq} = \frac{n^2 I^2}{2kx_0} \left(\frac{\mu_0 A}{g_\mu + \frac{\mu_0 L_m}{\mu}} \right) \tag{6.74}$$

Note that while we are able to express this behavior entirely with the electrical and mechanical domains, it was necessary to explore the magnetic domain in order to understand how the device operates. That is why it is good practice to make each energy domain explicit in representing multi-energy-domain devices.

As in our previous electrostatic actuator example, we can also add the inertial mass of the armature and a dashpot damping element in series with the mechanical capacitor to create a more complete model of the actuator. The important lesson here is that we have been able to use a model of the stored energy in a two-port magnetic "capacitor" to develop an understanding of a subtle actuation mechanism. This is a useful general approach for gaining insight into transducers and actuators that are energy-conserving.

6.6   Equivalent Circuits for Linear Transducers

The examples of transducers of the previous two sections were both *nonlinear*. The capacitive transducer was nonlinear both in voltage and displacement, and the magnetic transducer was nonlinear in current (although not in displacement). But there are also *linear* transducers, such as the piezoelectric transducer used to build the rate gyroscope of Chapter 21. Furthermore, even nonlinear transducers, when operated over a sufficiently small variation of state variables, behave locally as linear devices.[5] It is useful, therefore, to set up

---

[5] The *linearization* of nonlinear devices is discussed in Section 7.3.2.

the framework for modeling linear transducers in anticipation of using this framework for describing a variety of transducer behaviors.

We already understand that transducers have (at least) two ports. Therefore, in even the simplest case, there are four variables to consider: two efforts and two flows. When formulating the equations that describe a linear transducer, any two of the variables can be considered as "independent," the other two being "dependent." Because of the linearity of the system, it is easy to convert from any one set of independent variables to any other. There are six ways of choosing the two independent variables from a set of four,[6] leading to a potentially bewildering array of possible linear models. But because linearity guarantees easy conversion from one type to another, we choose to introduce here only one of the many possible models. Readers interested in the full array of two-port models are referred to the references [35, 36, 37, 40, 41, 42].

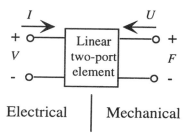

Figure 6.13. A general linear electromechanical two-port transducer.

Figure 6.13 illustrates a linear electromechanical two-port transducer. The left-hand domain is the electrical domain, with voltage as across variable and current as through variable. The right-hand domain is the mechanical domain, with force as the across variable and velocity as the through variable. We choose in this case to use U to represent velocity, both to be consistent with the notation of the cited references and to avoid any potential confusion with voltage.

The specific choice of independent variables will be the two flows, I and U. With this choice, the standard form for the linear equations describing a linear transducer is

$$\begin{pmatrix} V \\ F \end{pmatrix} = \begin{pmatrix} Z_{EB} & T_{EM} \\ T_{ME} & Z_{MO} \end{pmatrix} \begin{pmatrix} I \\ U \end{pmatrix} \qquad (6.75)$$

where the elements of the matrix have the following definitions:

---

[6]An exercise in combinatorial mathematics.

$$Z_{EB} = \frac{V}{I}\bigg|_{U=0} \qquad \text{Blocked electrical impedance}$$

$$Z_{MO} = \frac{F}{U}\bigg|_{I=0} \qquad \text{Open-circuit mechanical impedance}$$

$$T_{EM} = \frac{V}{U}\bigg|_{I=0} \qquad \text{Open-circuit electromechanical transduction impedance}$$

$$T_{ME} = \frac{F}{I}\bigg|_{U=0} \qquad \text{Blocked mechanical-electro transduction impedance}$$

"Blocked" means that in the mechanical domain, there is no motion. "Open-circuit" means that in the electrical domain, there is no current. This formulation is called the *impedance representation*. The units of Z_{EB} are Ohms, just like any electrical impedance. The units of Z_{MO} are (Newton-sec)/meter, the same units as a mechanical damping constant. The units of T_{EM} are (Volt-sec)/meter, a new combination not encountered before. And the units of T_{ME} are Newtons/Ampere, another new combination.

If the transducer is energy-conserving, then both Z_{EB} and Z_{MO} will be combinations of capacitors and inductors, not resistors. However, if there are damping mechanisms in the transducer (as in the squeeze-film damping introduced in the capacitive transducer presented earlier), then there can be resistive components as well as capacitors and inductors in these impedances. The two cross-terms, T_{EM} and T_{ME}, capture the transduction, the transition between the two energy domains. This transduction is said to be *reciprocal* if T_{EM} equals T_{ME}. All linear transducers based on energy-conserving physical behavior such as electromagnetism and linear elastic behavior are reciprocal. Therefore, for the devices we shall encounter, we can assume $T_{EM} = T_{ME}$.[7] We can then define the *impedance transformation factor* φ:

$$\varphi = \frac{T_{EM}}{Z_{EB}} \tag{6.76}$$

and the resulting governing equations become

$$\begin{pmatrix} V \\ F \end{pmatrix} = \begin{pmatrix} Z_{EB} & \varphi Z_{EB} \\ \varphi Z_{EB} & Z_{MO} \end{pmatrix} \begin{pmatrix} I \\ U \end{pmatrix} \tag{6.77}$$

We can make an equivalent circuit that represents this general linear reciprocal transducer with two impedances and a transformer, as shown in Fig. 6.14.

It is readily shown from this circuit and from the definition of a transformer that in order for this circuit to represent the original linear equations, the impedance denoted Z_{MS} must be given by

---

[7]This means that these two new units, Newtons/Ampere and (Volt-sec)/meter, must be the same. They are!

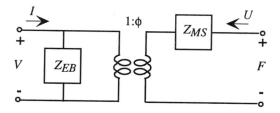

Figure 6.14. Equivalent circuit for a linear reciprocal transducer using impedance parameters.

$$Z_{MS} = Z_{MO}(1 - k_e^2) \tag{6.78}$$

where Z_{MS} is defined as the *short-circuit mechanical impedance*, given by F/U when V is zero, and the *electromechanical coupling constant* k_e is given by

$$k_e^2 = \frac{T_{EM}^2}{Z_{EB}Z_{MO}} \tag{6.79}$$

Thus far, it seems we have manipulated a set of equations and then forced a circuit model to match them. That is exactly what we have done! When we examine the linearization of nonlinear systems in the next chapter, we shall apply this formulation to the electrostatic transducer of this chapter, which will provide some intuition about the meaning of these symbols. In the mean time, this exercise provides good opportunity for practice in manipulating circuits that contain two-port elements.

6.7 The Position Control System – Revisited

We now have in hand two examples of actuators that could be used to control the position of a moveable object. If the object of interest is attached to either the moveable plate of a parallel plate capacitor or to the laterally moveable armature of our magnetic actuator, we could, with proper position information, apply the necessary voltage or current to establish the level of force needed to assure that the equilibrium position agrees with the desired position. And there are many other transducer configurations, such as lateral moveable electrostatic comb-drive actuators, that, like the magnetic example, exhibit a force that is independent of displacement. But before we can complete our analysis of the position control system, we require more tools:

- Dynamics: We need to understand the dynamical behavior of systems, with and without feedback

- Domain-specific knowledge: We need to learn how to determine the lumped-element values for springs, dashpots, and other lumped elements used in our models, based on the actual structure of the device.

- Measurement methods: We need to understand methods for using transducers for measurement as well as for actuation.

Dynamics is the general subject of the next chapter, and feedback is the subject of Chapter 15. Domain-specific knowledge is the subject of all of Part III. Measurement methods appear as examples throughout the rest of the book, but especially in Chapter 16 on noise, and in the Case Studies of Part V.

Related Reading

H. A. C. Tilmans, Equivalent circuit representation of electromechanical transducers: I. Lumped-parameter systems, *J. Microeng. Micromech.*, vol. 6, pp. 157-176, 1996.

L. Beranek, *Acoustics*, New York: McGraw-Hill, 1954.

C. A. Desoer and E. S. Kuh, *Basic Circuit Theory*, New York: McGraw-Hill, 1969.

F. V. Hunt, *Electroacoustics: the Analysis of Transduction, and Its Historical Background*, New York: American Institute of Physics, 1982.

M. Rossi, *Acoustics and Electroacoustics*, Norwood, MA: Artech House, 1988.

Problems

6.1 Add an inertial mass, a mechanical damping element, and a voltage source with source resistor to the magnetic actuator equivalent circuit of Fig. 6.12, and develop a set of state equations that describe the dynamical behavior.

6.2 Determine whether the magnetic actuator of Fig. 6.12 exhibits a pull-in instability.

6.3 Actuator Treasure Hunt: Find a recently published paper that describes an electrostatic or magnetic actuator. Develop the form of a basic lumped-element model for the device using the two-port capacitor elements introduced in this chapter. A list of candidate journals can be found in Problem 1.1.

6.4 Prove that the units of T_{EM} and T_{ME} in the linear transducer model are the same.

6.5 Use Kirchhoff's Laws and the characteristic equation for a transformer in the circuit below to prove Eq. 6.78.

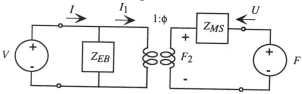

Chapter 7

LUMPED-ELEMENT SYSTEM DYNAMICS

7.1   Introduction

The choice of methods for analyzing the dynamics of a lumped-element system depends critically on whether or not the system is *linear*. A linear system is characterized by state equations that depend only on the first powers of the various state variables. The spring-mass-dashpot system we encountered in the position-control system of Section 2.3 was linear and was described by the following set of state equations:

$$\dot{x}_1 = x_2 \tag{7.1}$$
$$\dot{x}_2 = -\frac{k}{m}x_1 - \frac{b}{m}x_2 + \frac{1}{m}F \tag{7.2}$$

where x_1 is the position of the object, x_2 is the velocity of the object, m is the mass, k the spring constant, and b the damping constant. F in this case is an external force acting on the object, which may be time dependent. From the point of view of the position-control-system block diagram, F is the "input" and the position state variable x_1 is the "output." However, we can also consider the entire state vector, in this case a column vector with two entries, as the output, or we could imagine using some linear function of these two state variables as the output, and the overall system would still be called linear.

We have also encountered a *nonlinear* system, the voltage-controlled parallel-plate electrostatic actuator of Section 6.4. The three state equations for this system were:

$$\dot{x}_1 = \frac{1}{R}\left(V_{\text{in}} - \frac{x_1 x_2}{\varepsilon A}\right) \tag{7.3}$$

$$\dot{x}_2 = x_3 \qquad (7.4)$$

$$\dot{x}_3 = -\frac{1}{m}\left(\frac{x_1^2}{2\varepsilon A} + k(x_2 - g_0) + bx_3\right) \qquad (7.5)$$

where x_1 is the capacitor charge, x_2 is the capacitor gap, x_3 is the velocity of the moveable plate, ε is the permittivity of the air, A is the plate area, g_0 is the at-rest gap, and R is the source resistance of the (possibly time-dependent) voltage source V_{in}. The "input" in this case is the voltage V_{in} and the output can considered to be any desired function of the three state variables. For example, if one is using this actuator for position control, the outputs of interest are x_2 and, possibly, x_3.

If a system is linear, a host of powerful and quite general analytical techniques are available: Laplace transform, s-plane analysis with poles and zeros, Fourier transform, convolution, superposition, and eigenfunction analysis, to name several. If the system is nonlinear, the approach becomes much more problem-specific, and, in general, much more difficult to bring to rigorous closure. Some general techniques include examining the system trajectories in its phase space or using linearization to evaluate dynamic behavior in the vicinity of an operating point. In both types of systems, we shall see that direct numerical integration of the state equations is a very useful simulation method.

An additional issue is whether or not the system involves feedback. Feedback has a profound effect on system dynamics. In linear systems with feedback, we can do extensive analysis by extension of the tools used for linear systems without feedback. Nonlinear systems with feedback can exhibit new and complex behavior, including sustained oscillations called *limit cycles*. In devices involving resonant behavior, such as gyroscopes, some kind of sustained oscillation is needed, so it is essential to examine this particular aspect of nonlinear systems.

This subject of system dynamics is so huge, that we can hope only to touch the highlights. The emphasis in this chapter is on linear system dynamics, plus a few specific cases of nonlinear system analysis. The behavior of nonlinear feedback systems and their associated limit cycles are treated in Chapter 15.

7.2   Linear System Dynamics

We begin with the state equations for a linear system. By assumption, the parameters in these equations (such as resistances, masses, and spring constants) will be *time-independent*. If there is a time-varying element within the device, such as a moving capacitor plate, the energetically correct way to handle it is to assign a mechanical port to the device and to include additional state variables associated with the mechanical motion. That is, we do not represent the effect of the motion with a time-dependent capacitor, but with a *position-dependent capacitor* with explicit dynamic state variables associated with the position.

A typical set of linear state equations can be written in compact matrix form,

$$\dot{x} = Ax + Bu \qquad (7.6)$$
$$y = Cx + Du \qquad (7.7)$$

where x is a column vector of state variables, u is a column vector of system inputs, y is a column vector of outputs, and A, B, C, and D are the time-independent matrices that constitute the system.

For the spring-mass-dashpot system already examined, there is one scalar input F, and the output is equal to x_1. The matrices for this case are, for the state equations,

$$A = \begin{pmatrix} 0 & 1 \\ -\frac{k}{m} & -\frac{b}{m} \end{pmatrix} \text{ and } B = \begin{pmatrix} 0 \\ \frac{1}{m} \end{pmatrix} \qquad (7.8)$$

and, for the output equations,

$$C = \begin{pmatrix} 1 & 0 \end{pmatrix} \text{ and } D = \begin{pmatrix} 0 \\ 0 \end{pmatrix} \qquad (7.9)$$

If we want to have both the position and velocity available as system outputs, then the C-matrix becomes

$$C = \begin{pmatrix} 1 & 0 \\ 0 & 1 \end{pmatrix} \qquad (7.10)$$

Equations in this form can always be developed from the equivalent circuit for a linear system. We now examine a variety of ways of analyzing system behavior.

7.2.1 Direct Integration

One approach, numerical "brute force," involves integrating these first-order differential equations in a matrix-oriented mathematical environment such as MATLAB. The MATLAB commands `ode`, `step`, and `initial`, to name several, are quite useful. Or, one can create a state-space model in SIMULINK, entering the A, B, C and D matrices into the `State-Space` SIMULINK block, or by assigning the entries within such a block to variables whose values are set in the MATLAB environment. Then, with a suitable SIMULINK signal source, one can perform a numerical simulation.

Numerical methods are extremely valuable when seeking such details as the amount of overshoot in a step response. For example, Fig. 7.1 shows the unit step response of a spring-mass-dashpot system with $m = 1$, $k = 1$, and $b = 0.5$, in which the $\mathbf{C}$ matrix of Eq. 7.10 is used, providing both the position and velocity as system outputs. For the general purpose of visualizing the time-evolution of the system, graphs produced from explicit numerical integration of the governing dynamical equations are powerful and easy to obtain.

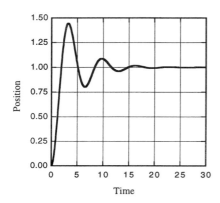 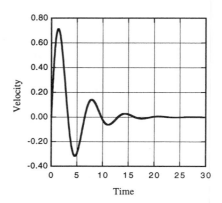

Figure 7.1.   Step response of the spring-mass-dashpot system.

However, as is true with all numerical approaches, deep insight into the underlying dynamics can be difficult to achieve from strictly numerical results. In the case of linear systems, we have many other tools that are quite general and enormously powerful.

7.2.2   System Functions

We can compute the single-sided Laplace transform[1] of the state equations to obtain:

$$sX(s) - x(0) = AX(s) + BU(s) \tag{7.11}$$

where $\mathbf{X}(s)$ is the Laplace transform of the vector $\mathbf{x}(t)$, $\mathbf{x}(0)$ is the initial state for the system, and $\mathbf{U}(s)$ is the Laplace transform of the input. This leads directly to the solution for $\mathbf{X}(s)$

$$\mathbf{X}(s) = (s\mathbf{I} - \mathbf{A})^{-1}(\mathbf{x}(0) + \mathbf{B}\mathbf{U}(s)) \tag{7.12}$$

---

[1]The single-sided Laplace transform of a function $f(t)$ is

$$F(s) = \int_0^\infty f(t)e^{-st}\,dt$$

where s is a complex number. Good presentations can be found in Siebert [43], Bracewell [44], and Desoer and Kuh [40].

where $\mathbf{I}$ is the identity matrix.

We can split this system response into two parts. The first is the *zero-input response* $\mathbf{X}_{zir}(s)$, whose source is the initial state with zero applied input, $\mathbf{U}(s) = 0$ (also called the "natural response" or the "transient response," although these usages have a risk of imprecision). It is the first term in the $\mathbf{X}(s)$ expression above:

$$\mathbf{X}_{zir}(s) = (s\mathbf{I} - \mathbf{A})^{-1}\mathbf{x}(0) \tag{7.13}$$

The second term is the *zero-state response* $\mathbf{X}_{zsr}(s)$ whose source is the applied inputs assuming one started in the zero state, $\mathbf{x}(0) = 0$.

$$\mathbf{X}_{zsr}(s) = (s\mathbf{I} - \mathbf{A})^{-1}\mathbf{B}\mathbf{U}(s) \tag{7.14}$$

Both parts of the response require that the matrix $(s\mathbf{I} - \mathbf{A})$ be an invertable matrix; that is, it cannot be singular. However, as is well-known, there are isolated values of s at which this matrix will have a vanishing determinant, and hence will be singular. The s-values for which this occurs are said to be the *natural frequencies* of the system. The s-values are also called the *poles* of the system.

A partial-fraction expansion of $\mathbf{X}_{zir}(s)$ leads, via the inverse Laplace transform, to a sum of terms dependent on the initial state, each term having a characteristic time dependence of the form[2]

$$e^{s_i t}$$

where s_i is one of the natural frequencies. When s_i is real and negative, the result is a decaying exponential. When s_i is complex with a negative real part, the waveform is a damped sinusoid. If s_i is purely imaginary, the waveform is an undamped sinusoid, but this is never encountered in real systems because real systems are either dissipative, with negative real parts of their natural frequencies, or nonlinear, in which case other methods of analysis must be used. Because real linear systems are dissipative, we know that the zero-input response will die out in time. The zero-state response, however, can persist for as long as the input is applied to the system.

We can write the Laplace transform of the system output as

$$\mathbf{Y}(s) = \mathbf{C}\left[\mathbf{X}_{zir}(s) + \mathbf{X}_{zsr}(s)\right] + \mathbf{D}\mathbf{U}(s) \tag{7.15}$$

---

[2]The exponential time-dependence only appears when each natural frequency is distinct. If there is a double root in the system, additional time dependences of the form $t \exp(s_i t)$ appear; and so on for higher-order roots. A critically damped system has a double root, hence both $\exp(s_i t)$ and $t \exp(s_i t)$ time-dependences must be considered when evaluating the zero-input response. These issues, including methods for inverting the Laplace transform, are thoroughly discussed in the cited references.

We see the output consists of three parts: one part due to the zero-input response, one due to the zero-state response, and one due to direct feedthrough from input to output via the $\mathbf{D}$ matrix. In most systems we shall encounter, the $\mathbf{D}$ matrix is zero.

We can rewrite the zero-state response part of the output in the form

$$\mathbf{Y}_{zsr}(s) = \mathbf{H}(s)\mathbf{U}(s) \tag{7.16}$$

where

$$\mathbf{H}(s) = \mathbf{C}(s\mathbf{I} - \mathbf{A})^{-1}\mathbf{B} \tag{7.17}$$

$\mathbf{H}(s)$ is called the *system function* or *transfer function* for the system.

$\mathbf{H}(s)$ is actually a matrix with the number of rows equal to the number of state variables, and the number of columns equal to the number of inputs. We know from circuit analysis using Laplace-transform-based complex impedances that each entry in the $\mathbf{H}(s)$ matrix will, for lumped-element networks, always be a ratio of polynomials, where the numerator varies from entry to entry, but the denominator is the same for all entries. That is, a general component of the $\mathbf{H}(s)$ matrix has the form

$$H_{ij}(s) = \frac{\mathrm{Num}_{ij}(s)}{\mathrm{Den}(s)} \tag{7.18}$$

where $\mathrm{Num}_{ij}(s)$ is a numerator polynomial and $\mathrm{Den}(s)$ is the denominator polynomial. The roots of $\mathrm{Den}(s)$ are identical to the natural frequencies or poles of the system. The roots of $\mathrm{Num}_{ij}(s)$ are called the *zeros* of the system. They correspond to specific input time-dependences that yield no response at the output.[3]

Once the poles and zeros of a transfer function are specified, the only remaining feature of $\mathbf{H}(s)$ is an overall multiplicative constant for each entry that sets the correct units and magnitude. Thus, *systems with the same pole-zero combinations have the same basic dynamic behavior*. This fact allows powerful reasoning about a wide variety of systems based on a rather small set of thoroughly understood examples, for example, systems with one pole, with one zero, with one pole and one zero, two poles, two poles and one zero, etc.

We now continue with our spring-mass-dashpot example. In this case, the outputs $\mathbf{Y}(s)$ are the state variables themselves. The $\mathbf{H}(s)$ matrix becomes

$$\mathbf{H}(s) = \begin{pmatrix} \dfrac{1}{ms^2+bs+k} \\[2ex] \dfrac{s}{ms^2+bs+k} \end{pmatrix} \tag{7.19}$$

---

[3]Understanding zeros from a physical point of view is often disturbingly difficult. It is suggested that, for each physical example in which a zero in the transfer function is encountered, some time be spent thinking about the specific behavior corresponding to that value of s.

where the first entry is the input-output response with F as the input and position $X_1(s)$ as the output, and the second entry is the input-output response with F as the input and velocity $X_2(s)$ as the response.

We can also write $H(s)$ in a factored form that emphasizes its pole-zero structure:[4]

$$H(s) = \begin{pmatrix} \left(\frac{1}{m}\right) \frac{1}{(s-s_1)(s-s_2)} \\ \left(\frac{1}{m}\right) \frac{s}{(s-s_1)(s-s_2)} \end{pmatrix} \qquad (7.20)$$

where the two poles, s_1 and s_2 are given by

$$s_{1,2} = -\frac{b}{2m} \pm \sqrt{\left(\frac{b}{2m}\right)^2 - \frac{k}{m}} \qquad (7.21)$$

This factored form has a pre-factor, in this case $1/m$, followed by a ratio of factors exhibiting the roots of the numerator and denominator explicitly. We see that this system has two poles, and the velocity response has a zero at $s = 0$. The location of the poles depends on the *undamped resonant frequency* ω_o, given by

$$\omega_o = \sqrt{\frac{k}{m}} \qquad (7.22)$$

and on the *damping constant* α, given by

$$\alpha = \frac{b}{2m} \qquad (7.23)$$

For the numerical values used to create Fig. 7.1, the values are $\omega_o = 1$, and $\alpha = 0.25$. Figure 7.2 shows the *pole-zero diagram* for the velocity response using these values. Note the standard electrical-engineering labeling of the real axis with σ and the imaginary axis with $j\omega$, where j is the square root of -1. Note also that the pole locations are at

$$s_{1,2} = -\alpha \pm \sqrt{\alpha^2 - \omega_o^2} \qquad (7.24)$$

which, for the underdamped example shown here, is equivalent to

$$s_{1,2} = -\alpha \pm j\omega_d \qquad (7.25)$$

---

[4]MATLAB has commands that convert between the state-space, polynomial-ratio system function, and factored pole-zero system function representations of a linear dynamical system. Care must be used, however, because the MATLAB standard ordering of state variables used when constructing the **A** matrix can differ from the ordering of the rows of the corresponding system-function matrix. Row conversion procedures are provided in the MATLAB documentation.

where ω_d is called the *damped resonant frequency*, and for $\omega_o > \alpha$, is given by

$$\omega_d = \sqrt{\omega_o^2 - \alpha^2} \tag{7.26}$$

In Fig. 7.2, the crosses mark the pole locations and the circle marks the zero. There is no indication in the standard pole-zero diagram of the value of the scale factor.

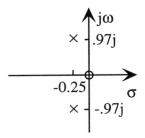

Figure 7.2.   Pole-zero diagram for the velocity response of the spring-mass-dashpot system

We define the *quality factor* or Q of this system as

$$Q = \frac{\omega_o}{2\alpha} = \frac{m\omega_o}{b} \tag{7.27}$$

For this example, $Q = 2$. This gives rise to a moderately *underdamped* step response which exhibits a few cycles of overshoot and ringing after the step (see Fig. 7.1). When $\alpha = \omega_o$, the response is *critically damped*, with both poles occurring at $s_{1,2} = -\alpha$. The Q value of a critically damped system is 1/2. For larger values of α and smaller values of Q, the system is *overdamped*, with both poles occurring on the negative real axis.

When a system is driven with an input of the form e^{st}, then the response consists of a sum of terms, one which is $H(s)e^{st}$, the remaining terms having time dependences determined by the poles of the system. The term that has the same e^{st} time dependence as the input is called the *forced response* , and, provided that the transient pole-related terms die out faster than the e^{st} input, the forced response is the persistent response observed at long times.

We can use this fact to gain insight into the interpretation of a zero. The complex frequency $s = 0$ corresponds to a constant applied force, which will clearly outlast any transient. Thus, the forced response of the velocity to a steady $s = 0$ input is found by substituting $s = 0$ into the velocity component of $H(s)$. It is indeed zero. In contrast, the position forced response is finite at $s = 0$ and has value F/k, just has one would expect from the mechanics of the problem.

7.2.3   Superposition in Linear Time-Invariant Systems: Convolution and the Fourier Transform

The zero-state response of a linear system obeys superposition. That is, the zero-state response to a sum of excitations is the sum of the zero-state responses to each excitation applied separately. Furthermore, because we are insisting that time-dependences in system parameters be explicitly represented with additional state variables, we can assert that our systems (i.e., the **A**, **B**, **C** and **D** matrices) are *time-invariant*, meaning that a shift in the time origin does not affect the response. This property is called *shift invariance*. Superposition and shift invariance in linear-time-invariant systems lead us to many important analytical methods.

7.2.3.1   Convolution and the Fourier Transform

For each component of the input-to-output transfer function, the zero-state response to an impulse input is called the *impulse response*, usually denoted $h(t)$. By definition of the sifting feature of the impulse $\delta(t)$, a single input to a system $u(t)$ can be written in the form of an integral superposition of impulses:

$$u(t) = \int_{-\infty}^{\infty} u(\tau)\delta(t-\tau)\, d\tau \qquad (7.28)$$

Because of the validity of superposition and shift invariance in linear time-invariant systems, we can write the output that results from $u(t)$ as a similar integral superposition of impulse response. The result is the *convolution integral*:

$$y(t) = \int_{-\infty}^{\infty} u(\tau)h(t-\tau)\, d\tau \qquad (7.29)$$

We can also use Fourier decomposition in linear systems. The Fourier transform and its inverse transform (using the so-called electrical-engineering convention) is written[5]

$$F(\omega) = \int_{-\infty}^{\infty} f(t)e^{-j\omega t}\, dt \qquad (7.30)$$

$$f(t) = \int_{-\infty}^{\infty} F(\omega)e^{j\omega t}\, \frac{d\omega}{2\pi} \qquad (7.31)$$

The Fourier transform and the Laplace transform are clearly connected, and the standard notations offer a risk of confusion. Given a time function $x(t)$, we have used $X(s)$ to denote the Laplace transform and $X(\omega)$ to denote the

---

[5]More on the Fourier transform, the impulse response, and convolution can be found in the references [43, 44].

Fourier transform. But at least for functions that are zero for $t < 0$, it is trivially true that

$$X(s)|_{s=j\omega} \quad \text{(Laplace)} \quad = X(\omega) \quad \text{(Fourier)} \tag{7.32}$$

When evaluating the Laplace transform of $x(t)$ at $s = j\omega$, we will use the notation $X(j\omega)$, whereas when evaluating the Fourier transform of $x(t)$ at frequency ω we will use the notation $X(\omega)$. The use of j in the argument is how we will clarify which functional form is being used.

The zero-state response can be found directly from a Fourier superposition of sinusoids if the corresponding system function is known. This is because the impulse response and the zero-state response are related by the Fourier transform. That is, for a system with input $u(t)$ and zero-state-response output $y_{zsr}(t)$, the system function $H(s)$ equal to $Y_{zsr}(s)/U(s)$, when evaluated at $s = j\omega$, is related to the impulse response $h(t)$ by

$$H(j\omega) = \int_{-\infty}^{\infty} h(t)e^{-j\omega t} dt \tag{7.33}$$

Thus, we can find the Fourier transform of $y_{zsr}(t)$ from

$$Y_{zsr}(\omega) = H(j\omega)U(\omega) \tag{7.34}$$

We can then calculate $y_{zsr}(t)$ with the inverse Fourier transform of $Y_{zsr}(\omega)$. Example 7.1 illustrates the details.

The above discussion of superposition deals only with the zero-state response. The zero-input response does not obey superposition. Therefore, if at the time of the application of a signal, the system is not in the zero state, then the zero-input response must be determined, and added to the zero-state response.

7.2.4   Sinusoidal Steady State

An important signal domain is the *sinusoidal steady state*, in which all inputs are sinusoids and all transients are presumed to have died out. The zero-input response and any transient components of the zero-state response can be ignored. Only the forced response to the sinusoidal input need be considered. This forced response must itself be sinusoidal, consisting of a sum of terms (or, as with the Fourier transform, a continuous superposition) at the various frequencies of the inputs. Specifically, for a sinusoidal input $u(t)$ given by

$$u(t) = U_o \cos(\omega t) \tag{7.35}$$

the sinusoidal-steady-state output $y_{sss}(t)$ is given by

$$y_{sss}(t) = Y_o \cos(\omega t + \theta) \tag{7.36}$$

Example 7.1

Let $u(t) = U_o \cos \omega_o t$. The Fourier transform of $u(t)$ is

$$U(\omega) = \frac{U_o}{2} \int_{-\infty}^{\infty} \left(e^{j\omega_o t} + e^{-j\omega t} \right) e^{-j\omega t} \, dt$$

Each of the terms in parentheses gives rise to 2π times an impulse, yielding the result

$$U(\omega) = U_o \pi \left[\delta(\omega + \omega_o) + \delta(\omega - \omega_o) \right]$$

For a system function, we choose a first-order system with

$$H(s) = \frac{1}{1 + s\tau}$$

The zero-state response of this system to $u(t)$, according to Eq. 7.34, thus has Fourier transform

$$Y_{\text{zsr}}(\omega) = \frac{U_o \pi [\delta(\omega + \omega_o) + \delta(\omega - \omega_o]}{1 + j\omega \tau}$$

When these impulse functions appear inside the inverse Fourier transform, the integrals yield

$$y_{\text{zsr}}(t) = \frac{U_o}{2} \left[\frac{e^{j\omega_o t}}{1 + j\omega_o \tau} + \frac{e^{-j\omega_o t}}{1 - j\omega_o \tau} \right]$$

This can be simplified to

$$y_{\text{zsr}}(t) = U_o \frac{\cos \omega_o t - \omega_o \tau \sin \omega_o t}{1 + (\omega_o \tau)^2}$$

---

where

$$Y_o = |H(j\omega)| U_o \qquad (7.37)$$

and where the phase angle θ is given by

$$\tan \theta = \frac{\text{Im}\{H(j\omega)\}}{\text{Re}\{H(j\omega)\}} \qquad (7.38)$$

Graphs of Y_o and θ versus ω are frequently used to show the full *frequency response* of a system in the sinusoidal steady state. When plotted on logarithmic axes for H and ω and a linear axis for θ, these plots are referred to as *Bode plots*.

MATLAB uses the `freqs` or `bode` commands for creating Bode plots from the coefficients of the polynomials $Num(s)$ and $Den(s)$. Figure 7.3 shows the position and velocity Bode plots for the spring-mass-dashpot system with the values $m = 1$, $k = 1$, and $b = 0.5$. Note that the effect of the zero in the velocity transfer function (compared to the position transfer function) is to increase each asymptotic slope on log-log axes by +1 and to add +90° to the phase angle.

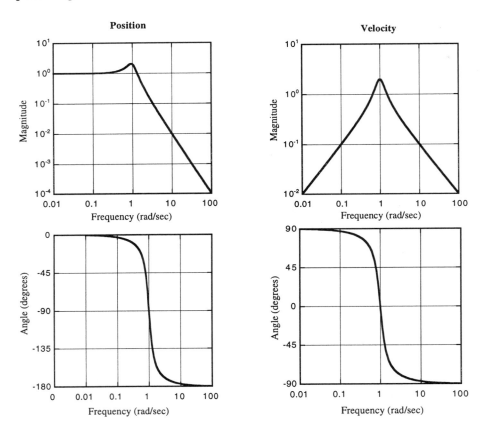

Figure 7.3.   Position and velocity Bode plots for the spring-mass-dashpot system.

7.2.5   Eigenfunction Analysis

We have seen that the zero-input response of a linear system consists of a sum of terms, each term proportional to a time-dependence of the form $e^{s_i t}$, where s_i is the ith natural frequency (and, of course, for double roots, there is a $t e^{s_i t}$ time dependence). These same kinds of terms, in addition to the forced response, occur in the zero-state response. For each term involving a natural frequency, there is a characteristic relation among the amplitudes of the different state variables. This relation is usually expressed as a characteristic vector called an *eigenvector*, and the corresponding natural frequency is called the *eigenvalue*.

The definitions of an eigenvalue and its associated eigenvector are expressed in the following equation: $\mathbf{v}_i$ is an eigenvector of square matrix $\mathbf{A}$ provided it satisfies

$$\mathbf{A}\mathbf{v}_i = \lambda_i\mathbf{v}_i \tag{7.39}$$

where λ_i is a constant and is the eigenvalue associated with the eigenvector. Note that eigenvectors do not have any particular length. Multiplying this equation by a numerical constant does not affect it. Therefore, we should expect that some kind of *normalization* of eigenvectors to a standard scale will occur.

Let us examine the eigenvectors of the $\mathbf{A}$ matrix of a linear system. The procedure for finding eigenvalues is to find the roots of $\mathbf{A} - \lambda\mathbf{I}$, which we recognize as identical in form to the $\mathbf{A} - s\mathbf{I}$ matrix whose inverse was needed to solve the system response using the Laplace transform. Thus, the eigenvalues and natural frequencies are clearly the same.

For each eigenvalue found from the roots of the determinant of $\mathbf{A} - \lambda\mathbf{I}$, substitution into Eq. 7.39 creates a set of constraints among the components of $\mathbf{v}_i$. In our 2×2 spring-mass-dashpot example, we obtain two homogeneous equations involving the two vector components, but only one of these equations can be linearly independent because the set of equations is homogeneous. The selected equation (either row will work) sets the ratio of the components $\mathbf{v}_{1i}$ to $\mathbf{v}_{2i}$. The overall magnitude of $\mathbf{v}_{1i}$ and $\mathbf{v}_{2i}$ is determined by the normalization convention used. Example 7.2 (following page) illustrates these calculations.

MATLAB has a several commands that greatly simplify this extraction. In particular, the command

$$[\mathbf{V}, \Lambda] = \texttt{eig(A)} \tag{7.40}$$

returns a matrix $\mathbf{V}$ whose columns are the eigenvectors and a diagonal matrix Λ whose diagonal entries are the eigenvalues.[6] When MATLAB is used, the

---

[6]When actually using MATLAB, variables must carry text names. Therefore, the Λ used here would have to be given a different name, but we use the Greek symbol here to comply with standard texts on system theory.

Example 7.2

We will analyze the spring-mass-dashpot system, using the same set of numerical values as in our previous examples ($m = k = 1, b = 0.5$).

$$\mathbf{A} = \left(\begin{array}{cc} 0 & 1 \\ -1 & -.5 \end{array} \right)$$

The *characteristic equation* for this matrix is the determinant of $\mathbf{A} - \lambda \mathbf{I}$, which has the value

$$\lambda^2 + 0.5\lambda + 1 = 0$$

which, as we have already seen, has two roots

$$\lambda_{1,2} = -.25 \pm .97j$$

If we substitute the value of s_1 back into Eq. 7.39, and collect terms, we obtain

$$\left(\begin{array}{cc} 0.25 - 0.97j & 1 \\ -1 & -0.25 - 0.97j \end{array} \right) \left(\begin{array}{c} v_{11} \\ v_{21} \end{array} \right) = \left(\begin{array}{c} 0 \\ 0 \end{array} \right)$$

where we use the first index in the components of $\mathbf{v}$ to identify the row and the second index to identify the corresponding natural frequency. Examination of the two equations above shows that they are, indeed, identical to each other and require that for this eigenvector,

$$v_{21} = (-.25 + .97j)v_{11}$$

The same operation can be performed by substituting s_2 for λ, with the result that

$$v_{12} = (-.25 - .97j)v_{22}$$

---

eigenvectors are returned normalized to unit Euclidean length.[7] That is, for a column vector $\mathbf{v}_i$ with transpose $\mathbf{v}_i^{\mathrm{T}}$, the normalization is

$$\mathbf{v}_i^{\mathrm{T}} \mathbf{v}_i = 1 \tag{7.41}$$

---

[7]Other normalizations are in common use. MAPLE, for example, uses a normalization that sets one of the components of each eigenvector to unity. We use the MATLAB normalization in this book.

To show how eigenvectors can be used to represent system behavior, we return to the original state equations:

$$\dot{\mathbf{x}} = \mathbf{A}\mathbf{x} + \mathbf{B}\mathbf{u} \qquad (7.42)$$

and substitute

$$\mathbf{x} = \mathbf{V}\mathbf{z} \qquad (7.43)$$

where $\mathbf{z}$ is a column vector of time-dependent coefficients $z_i(t)$, one for each eigenvector, and $\mathbf{V}$ is a matrix whose columns are the eigenvectors. With this substitution, we obtain

$$\mathbf{V}\dot{\mathbf{z}} = \mathbf{A}\mathbf{V}\mathbf{z} + \mathbf{B}\mathbf{u} \qquad (7.44)$$

If we multiply by the inverse of $\mathbf{V}$, we obtain

$$\dot{\mathbf{z}} = \left(\mathbf{V}^{-1}\mathbf{A}\mathbf{V}\right)\mathbf{z} + \left(\mathbf{V}^{-1}\mathbf{B}\mathbf{u}\right) \qquad (7.45)$$

The first term in parentheses is simply $\boldsymbol{\Lambda}$, since from the definition of the eigenvectors

$$\mathbf{A}\mathbf{V} = \boldsymbol{\Lambda}\mathbf{V} \qquad (7.46)$$

Thus the first term is a diagonal matrix. The second term in parentheses is a time-dependent column vector that depends on the inputs, which we shall denote by $\mathbf{p}(t)$. Thus, *for each row of this matrix equation,*

$$\dot{z}_i(t) = s_i z_i + p_i(t) \qquad (7.47)$$

where we have substituted the natural frequency s_i for λ_i. This differential equation has a well-known solution:

$$z_i(t) = e^{s_i t} z_i(0) + \int_0^t e^{s_i(t-\tau)} p_i(\tau)\, d\tau \qquad (7.48)$$

where $z_i(0)$ is the ith component of $\mathbf{V}^{-1}\mathbf{x}(0)$. The corresponding output of the system can be found from

$$\mathbf{y}(t) = \mathbf{C}\mathbf{V}\mathbf{z}(t) + \mathbf{D}\mathbf{u} \qquad (7.49)$$

Clearly, the two matrices $\mathbf{V}^{-1}\mathbf{B}$ and $\mathbf{C}\mathbf{V}$ are important. If $\mathbf{V}^{-1}\mathbf{B}$ has a row that is all zeros, then the eigenvector and its characteristic time dependence cannot be affected by the input. Such an eigenvector (also called an *eigenmode*, or just a *mode*) is called *uncontrollable*. Similarly, if $\mathbf{C}\mathbf{V}$ has a column that is all zeros, then the corresponding eigenmode does not appear in the system output. Such a mode is called *unobservable*.

We shall see when we examine feedback systems in Chapter 15 why we care about whether modes are controllable and observable. In normally dissipative physical systems, we expect the natural frequencies to be in the left-hand s-plane; hence the time dependence associated with every eigenvector dies out in time. Therefore, as long as we wait long enough, we expect to see only forced responses (the terms due to $\mathbf{p}(t)$). However, once we add feedback to a system, it is possible to create modes that are unstable. Such modes give rise to oscillations, ultimately resulting in some kind of nonlinear behavior. The ability to observe and control such modes can be critical to the overall stability of a complex system.

7.3   Nonlinear Dynamics

When the system is nonlinear, the most general form of the state and output equations is

$$\dot{\mathbf{x}} = \mathbf{f}(\mathbf{x}, \mathbf{u}) \tag{7.50}$$

and

$$\mathbf{y} = \mathbf{g}(\mathbf{x}, \mathbf{u}) \tag{7.51}$$

where $\mathbf{f}$ and $\mathbf{g}$ are nonlinear functions of the state and the inputs. We expect the functions themselves to be time-independent (because we insist, for energy reasons, on capturing time-dependences explicitly into the state equations), but we expect the inputs and the state variables themselves to be time-dependent.

The behavior of nonlinear systems can be very complex, especially when systems have nonlinearities that lead to so-called *jump-phenomena* between different states.[8] We will examine only three specific issues using the parallel-plate electrostatic actuator as a common example:

- Fixed-point analysis

- Linearization about an operating point

- Numerical integration of the state equations

7.3.1   Fixed Points of Nonlinear Systems

The *fixed points* of a nonlinear systems are the solutions of

$$\mathbf{f}(\mathbf{x}, \mathbf{u}) = 0 \tag{7.52}$$

We examine two types of fixed points, global fixed points and operating points. A *global fixed point* is a fixed point of the system when all inputs are zero. It

---

[8] Strogatz [45] provides a good and very accessible introduction to the rich subject of nonlinear system behavior.

corresponds to the system at rest. A nonlinear system can have multiple global fixed points, some of which might be stable, others of which might be unstable. For example, consider a pendulum that is mounted so it can rotate by 360°. A stable global fixed point is the state in which the pendulum hangs down. An unstable global fixed point is the state in which the pendulum is inverted at the precise point of balance.

An *operating point* is a fixed point that is established by non-zero but constant inputs. When we refer to an operating point, we also imply a *stable* fixed point set up by constant inputs. Where a system has more than one stable fixed point for a given input (such as a system with hysteresis), the history of the applied inputs determine which fixed point serves as the operating point.

We already performed fixed-point analysis for a nonlinear system in Section 6.4 when we examined the equilibrium points of the the parallel-plate electrostatic actuator, whether charge-driven or voltage-driven. For the charge-driven case, we found a well-behaved single fixed point for the gap at each value of charge. For the voltage-driven case, we found two fixed points for voltages below pull-in, one of which was stable, the other unstable, and no fixed points for voltages above pull-in. We noted that the stability of each fixed point was important. In the following section, we examine a general method that can, among other things, assist in the evaluation of fixed-point stability.

7.3.2 Linearization About an Operating Point

In many system examples, we are interested in a small domain of state space near an operating point. In this section we examine a technique called *linearization* that creates a linear model of the system for states near the operating point. Suppose that

$$\mathbf{x}(t) = \mathbf{X}_0 + \delta\mathbf{x}(t) \tag{7.53}$$

and

$$\mathbf{u}(t) = \mathbf{U}_0 + \delta\mathbf{u}(t) \tag{7.54}$$

where $\mathbf{X}_0$ and $\mathbf{U}_0$ are constants (the operating point), and where, on an appropriate scale, both $\delta\mathbf{x}$ and $\delta\mathbf{u}$ are small. Substitution into the state equations followed by the use of Taylor's theorem leads to

$$\delta\dot{\mathbf{x}}(t) = \left(\frac{\partial\mathbf{f}}{\partial\mathbf{x}}\bigg|_{\mathbf{X}_0,\mathbf{U}_0}\right)\delta\mathbf{x}(t) + \left(\frac{\partial\mathbf{f}}{\partial\mathbf{u}}\bigg|_{\mathbf{X}_0,\mathbf{U}_0}\right)\delta\mathbf{u}(t) \tag{7.55}$$

The two matrices in this equation are the *Jacobian's* of the original function $\mathbf{f}(\mathbf{x}, \mathbf{u})$. In full matrix form, these equations become

$$
\begin{pmatrix} \delta \dot{x}_1 \\ \vdots \\ \delta \dot{x}_n \end{pmatrix} = \begin{pmatrix} \frac{\partial f_1}{\partial x_1} & \cdots & \frac{\partial f_1}{\partial x_n} \\ \vdots & & \vdots \\ \frac{\partial f_n}{\partial x_1} & \cdots & \frac{\partial f_n}{\partial x_n} \end{pmatrix}\bigg|_{X_0, U_0} \begin{pmatrix} \delta x_1 \\ \vdots \\ \delta x_n \end{pmatrix}
$$

$$
+ \begin{pmatrix} \frac{\partial f_1}{\partial u_1} & \cdots & \frac{\partial f_1}{\partial u_m} \\ \vdots & & \vdots \\ \frac{\partial f_n}{\partial u_1} & & \frac{\partial f_n}{\partial u_m} \end{pmatrix}\bigg|_{X_0, U_0} \begin{pmatrix} \delta u_1 \\ \vdots \\ \delta u_m \end{pmatrix}
\tag{7.56}
$$

where the first Jacobian is an $n \times n$ matrix, where n is the number of state variables, and the second Jacobian is an $n \times m$ matrix, where m is the number of inputs.

This is now a linear problem, which, for small amplitudes of input and output, can be analyzed with all the powerful tools available for linear systems. The advantages of linearization are so great that very often we will intentionally linearize a system even when we know that $\delta \mathbf{x}$ and $\delta \mathbf{u}$ are not all that small, and accept some degree of error in our calculations rather than confront the fully nonlinear system. But, as we shall see in the actuator example later in this section, when a system makes sufficiently large excursions, the full set of state equations must be confronted.

One side-benefit of linearization is that it makes the assessment of fixed-point stability quite easy. If one sets the input to zero in Eq. 7.56 and finds the natural frequencies of the resulting zero-input-response problem, stable fixed points will lead to left-half-plane poles, and unstable fixed points will lead to right-half-plane poles. An example follows.

7.3.3   Linearization of the Electrostatic Actuator

The Jacobian J of the set of three state equations for the parallel-plate electrostatic actuator is given by

$$
J = \begin{pmatrix} -\dfrac{X_2}{R\varepsilon A} & -\dfrac{X_1}{R\varepsilon A} & 0 \\ 0 & 0 & 1 \\ -\dfrac{X_1}{m\varepsilon A} & -\dfrac{k}{m} & -\dfrac{b}{m} \end{pmatrix}
\tag{7.57}
$$

where operating point values for the charge X_1 and position X_2 have been substituted into the Jacobian. In order to evaluate the Jacobian, it is necessary to find the equilibrium operating point values for a specified value of the input. This is done by taking the original three state equations, setting the time derivatives to zero, and solving. The resulting equations are cubic, and algebraically unpleasant. However if we use some numerical values for the parameters, things simplify quite a lot.

Table 7.1. Parameters used in the modeling of the electrostatic actuator.

Parameter	Symbol	Value
Area	A	100
Permittivity	e	1
Initial gap	g_0	1
Minimum gap	g_{min}	.01
Mass	m	1
Damping constant	b	0.5
Spring constant	k	1
Resistance	R	.001

With the component values in Table 7.1, the state equations reduce to

$$\dot{x}_1 = 1000V_{in} - 10x_1x_2 \tag{7.58}$$

$$\dot{x}_2 = x_3 \tag{7.59}$$

$$\dot{x}_3 = -\frac{1}{200}x_1^2 - x_2 + 1 \tag{7.60}$$

The operating-point solutions are found by setting the three equations to zero, and solving. Clearly, the operating-point velocity X_3 should be zero, and we see this directly from setting the equation for $\dot{x}_2$ to zero. Mathematically, there are three possible solutions for the equilibrium charge X_1, the three roots of

$$X_1^3 - 200X_1 + 20000V_0 = 0 \tag{7.61}$$

The corresponding value of the equilibrium gap is then

$$X_2 = 1 - \frac{1}{200}X_1^2 \tag{7.62}$$

We can use MATLAB's roots function to find the roots of the polynomial in Eq. 7.61. One of the three roots turns out to be negative for positive values of V_0 and is discarded. The other two roots have positive values of both charge and gap. The locations of these roots in the voltage-gap plane are plotted in Fig. 7.4.

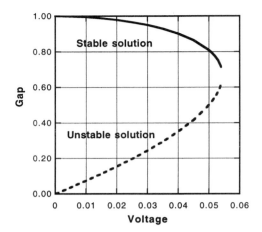

Figure 7.4. The two equilibrium positions for the electrostatic actuator as a function of DC voltage.

We recognize our previously encountered pair of solutions from our discussion of pull-in in Section 6.4.3, the values above $2g_0/3$ being stable and the values below $2g_0/3$ being unstable. The two solutions approach each other, and actually meet exactly at pull-in, which for this system occurs at $V_{in} = 0.05443$. Any point on the dashed curve, when substituted into the Jacobian, yields a right-half-plane natural frequency, indicating an unstable operating point. Every point on the solid curve yields a left-half-plane natural frequency, meaning that the set of operating points along the solid curve are locally stable, and are what would be observed physically as the voltage is increased gradually from zero. Thus we see that our previous force-based analysis of stability agrees with this more general and abstract linearized-Jacobian approach to stability.

We can also use the linearized system to study the small-amplitude behavior about the operating point. In particular, it is very interesting to examine the small-amplitude undamped resonant frequency for this system as a function of an applied operating-point DC voltage. Figure 7.5 shows, on the left, the equilibrium position and charge as a function of DC operating-point voltage and, on the right, the undamped natural frequency for small amplitude vibration about the operating point.

The undamped resonance frequency is found as the imaginary part of the oscillatory natural frequencies obtained from the Jacobian evaluated at the operating point. We note an interesting phenomenon called *spring softening*: as pull-in is approached, the resonant frequency shows a dramatic drop. It actually goes all the way to zero exactly at pull-in. The "softening" of a resonance frequency is one artifact often associated with points of incipient instability. Physically, the downward force from the capacitor opposes the

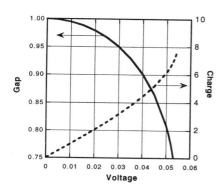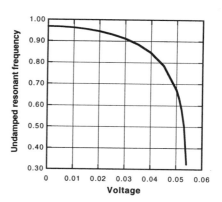

Figure 7.5. Operating-point position and charge as a function of DC voltage together with the undamped resonant frequency, illustrating the spring-softening effect of the nonlinear electrostatic actuator.

spring, and their balance produces equilibrium. Stability is determined by how these forces vary with gap. When the increase in downward force for a small decrease in gap is larger than the increase in upward force from the spring, pull-in occurs. But well before pull-in, the effect of the nonlinear electrostatic force in reducing the net restoring force is evident and shows up directly in the spring softening behavior observed here.

7.3.4 Transducer Model for the Linearized Actuator

With the concept of linearization in hand, we can now apply the linear transducer model of Section 6.6 to a linearized model of the electrostatic actuator. In doing so, we shall gain insight into the meanings of the various terms within the transducer model.

Figure 7.6 shows a redrawn version of the electrostatic actuator circuit of Fig. 6.9. The elements have been gathered into three groups: a source network consisting of the V_{in} voltage source and its source resistance R, a load network consisting of the inertial mass m and dashpot b, and the transducer that connects them. The transducer consists of the two-port capacitor representing the parallel plate capacitor with moveable plate plus the capacitor representing the spring. This grouping is somewhat arbitrary. It would be equally correct to include the mass and dashpot within the model of the transducer, and then connect an additional inertial mass in series with these elements to represent the load. We have also changed the name of the velocity variable from $\dot{g}$ to U to match the standard transducer notation.

For purposes of finding the operating point, this is a voltage-controlled system because, at the operating point, $I = 0$ and V must equal V_{in}. We recall from Section 6.4.2 that for a voltage-controlled representation of the transducer,

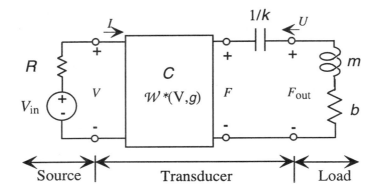

Figure 7.6.   A redrawn version of the electrostatic actuator circuit of Fig. 6.9 grouping the elements into a source network, the transducer, and a load network.

$$F = \frac{\varepsilon A V^2}{2g^2} \tag{7.63}$$

and

$$g = g_0 - \frac{\varepsilon A V^2}{2kg^2} \tag{7.64}$$

where g_0 is the gap with no applied voltage. To linearize these equations, we can assume that V_{in} has a DC value $V_{in,0}$ plus some variation δV_{in}. The DC value sets the operating point of the nonlinear system while the effects of the variation in V_{in} are what will be captured in the linearized model. Using V equal to $V_{in,0}$, we can find the operating-point force, which we denote F_0, and the operating point gap, which we denote $\hat{g}_0$ to distinguish it from the gap with zero applied voltage. The operating-point output force $F_{out,0}$ is zero, since U must be zero. We now assume that we have solved this problem, and have the operating point values in hand. Our goal is to construct a linearized model of the transducer in the form of Fig. 7.7.

To linearize the model so it can be matched to the form of Fig. 7.7, we need expressions for δV and δF. It turns out to be easier to work with the energy formulation rather than the co-energy formulation. (We are allowed to choose which to use in this case because the presence of the source resistor means that this network is not strictly voltage controlled. Only in DC steady state is V required to equal V_{in}.) Therefore, we need one additional operating point value, the charge on the capacitor at the operating point, denoted Q_0. It has the value

$$Q_0 = \frac{\varepsilon A}{\hat{g}_0} V_{in,0} \tag{7.65}$$

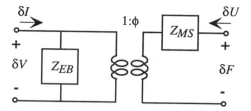

Figure 7.7. The equivalent circuit for the linearized actuator in the form of the transducer model of Section 6.6.

The equations governing the charge-controlled representation of the transducer, now including both the electrostatic force and the force on the spring, are

$$V = \frac{Qg}{\varepsilon A} \qquad (7.66)$$

and

$$F_{\text{out}} = \frac{Q_2}{2\varepsilon A} - k(g_0 - g) \qquad (7.67)$$

To linearize these equations about the operating point, we form the Jacobian, and find

$$\begin{pmatrix} \delta V \\ \delta F \end{pmatrix} = \begin{pmatrix} \frac{g_0}{\varepsilon A} & \frac{Q_0}{\varepsilon A} \\ \frac{Q_0}{\varepsilon A} & k \end{pmatrix} \begin{pmatrix} \delta Q \\ \delta g \end{pmatrix} \qquad (7.68)$$

We note, as expected, that this is a reciprocal transducer, as evidenced by the equality of the off-diagonal elements of the Jacobian. But we have a problem. The linearized variables on the right-hand side are δQ and δg, whereas the variables we will need to form the linear transducer model are δI and δU. The relation between δQ and δI is simply integration. That is,

$$\delta Q = \int \delta I dt \qquad (7.69)$$

Since the linearized equations are now *linear*, we can use the Laplace transform to obtain

$$\delta Q = \frac{\delta I}{s} \qquad (7.70)$$

Similarly,

$$\delta g = \frac{\delta U}{s} \qquad (7.71)$$

Therefore, after substituting, the linearized model becomes

$$
\begin{pmatrix} \delta V \\ \delta F \end{pmatrix} = \begin{pmatrix} \dfrac{\hat{g}_0}{s\varepsilon A} & \dfrac{Q_0}{s\varepsilon A} \\ \dfrac{Q_0}{s\varepsilon A} & \dfrac{k}{s} \end{pmatrix} \begin{pmatrix} \delta I \\ \delta U \end{pmatrix}
\tag{7.72}
$$

From these equations, we can identify the elements of the transducer model:

$$
Z_{EB} = \frac{\hat{g}_0}{s\varepsilon A}
\tag{7.73}
$$

This impedance is a capacitor whose value is that of the parallel-plate capacitance at the operating-point gap.

The other impedance in the model, Z_{MO} is given by

$$
Z_{MO} = \frac{k}{s}
\tag{7.74}
$$

This is simply the capacitance that represents the spring.   The coupling impedances are equal to each other, and have value

$$
T_{EM} = T_{ME} = \frac{Q_0}{s\varepsilon A}
\tag{7.75}
$$

To form the equivalent circuit of Fig. 7.7 we need to compute φ, which is given by

$$
\varphi = \frac{T_{EM}}{Z_{EB}} = \frac{Q_0}{\hat{g}_0}
\tag{7.76}
$$

The apparent units of φ are Coulombs/meter. These units are the same as the Newtons/Volt we would expect on the basis of a transformer that couples voltage to force.

We also need the electromechanical coupling constant k_e given by

$$
k_e^2 = \frac{T_{EM}^2}{Z_{EB}Z_{MO}} = \frac{Q_0^2}{\varepsilon A k \hat{g}_0}
\tag{7.77}
$$

with the result that the final element in the model Z_{MS} can be found:

$$
Z_{MS} = \frac{k}{s}\left(1 - \frac{Q_0^2}{\varepsilon A k \hat{g}_0}\right)
\tag{7.78}
$$

We see that Z_{MS} represents a spring that at zero applied voltage (*i.e.*, $Q_0 = 0$) has value k, but as voltage is applied, shifting the operating point, the value of the spring constant decreases. This shows that the transducer model effectively captures the spring-softening that is intrinsic to these types of electrostatic actuators. The stiffness in the mechanical domain is affected by the forces set up in the electrical domain.

7.3.5   Direct Integration of State Equations

When the state variables of a system undergo large variations, too large to be represented accurately by a linearized model, it is necessary to use the full set of state equations to examine dynamic behavior. Nonlinear state equations can be integrated numerically, using, for example, the ode command in MATLAB. Alternatively, a system block-diagram can be constructed in SIMULINK, and simulations can be performed in that environment. We recall the three state equations for the parallel-plate electrostatic actuator:

$$\dot{x}_1 = \frac{1}{R}\left(V_{\text{in}} - \frac{x_1 x_2}{\varepsilon A}\right) \tag{7.79}$$

$$\dot{x}_2 = x_3 \tag{7.80}$$

$$\dot{x}_3 = -\frac{1}{m}\left(\frac{x_1^2}{2\varepsilon A} + k(x_2 - g_0) + bx_3\right) \tag{7.81}$$

where x_1 is the capacitor charge, x_2 is the capacitor gap, x_3 is the velocity of the moveable plate, ε is the permittivity of the air, A is the plate area, g_0 is the at-rest gap, and R is the source resistance of the time-dependent V_{in} voltage source. A SIMULINK model that implements these equations is shown in Fig. 7.8.

The SIMULINK model has three integrators, one which integrates acceleration to find velocity, one which integrates velocity to find position, and one which integrates current to find charge. There are three nonlinear symbols in the diagram. The electrostatic-force block is a SIMULINK function that can perform nonlinear operations on a scalar input. This input is denoted as $u[1]$ in SIMULINK's required notational format, but the input in this case is the charge Q. The electrostatic force block implements the $Q^2/2\varepsilon A$ part of the total force. The position integrator is a saturating integrator, an integrator with internal limits that prevent the outputs from exceeding specified values. In this model, the integrator has been set so that it's output saturates when the gap g reaches a lower limit of g_{min}. This is necessary to avoid a singularity in the charge which would occur if the gap were allowed to go all the way to zero. The third nonlinear element is the block that computes the Qg product. Product blocks are labelled in SIMULINK with a dot in the middle.

Altogether, this model has eight parameters. The definition of the parameters and the values used for the simulations to follow are given in Table 7.1 (page 167).

With this model, one can apply various voltage waveforms in SIMULINK to the V_{in} input and observe how the system evolves. One immediate result is that the pull-in voltage is in very good agreement with our analytical result. For the parameters selected, the analytical pull-in voltage is 0.05443. As soon as one tries to use a voltage step larger than this, one observes the saturation of the

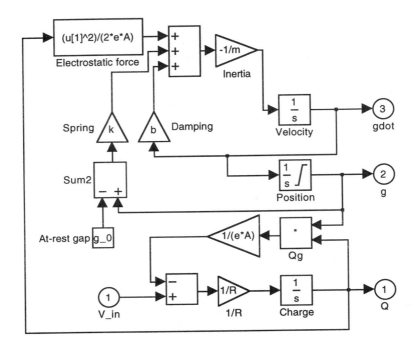

Figure 7.8. SIMULINK implementation of state equations for the parallel-plate electrostatic actuator. The labeling of the various blocks, including the use g_0 to represent g_0, is characteristic of SIMULINK block diagrams.

position integrator at g_{min}, indicating pull-in. Characteristic waveforms will be shown below.

Before examining detailed results, it is useful to confront the state-description of the pulled-in state. Equations 7.79-7.81 describe the dynamics of the system only when it is not pulled in. When it is pulled in, the gap is fixed at g_{min}. This means that the velocity, which might be substantial just before pull-in, must go to zero after pull-in. The plate "crashes" into the lower electrode and, in this model, comes to rest, losing all of its kinetic energy in the crash event.[9] SIMULINK has a number of nonlinear comparison and switch elements that enable the model to be reconfigured to describe different modes of a complex system. Figure 7.9 shows an enhanced SIMULINK model that permits the pulled-in state to be represented, and therefore also allows one to examine the release of the pulled-in plate after the drive voltage is removed.

---

[9] It might also be possible to model some amount of "bouncing" of the plate when it crashes down, allowing the kinetic energy to be dissipated in a few bounces. We choose to avoid this level of complexity, since the stable point is when the plate is pulled in at zero velocity.

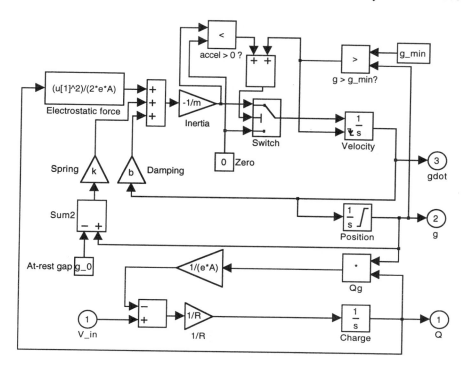

Figure 7.9. SIMULINK implementation of the state equations for the parallel-plate electrostatic actuator, enhanced to include the pulled-in state and capability for release.

There are several new elements in this figure. The velocity integrator has been modified with a "reset" input that is driven from a comparator that compares g with g_{min}. When g reaches g_{min}, the output of the velocity integrator is reset to zero (this is the "crash"). In addition, a switch sets the input to the velocity integrator to zero when g reaches g_{min} so that the velocity remains zero as long as the plate is pulled in. But in order to model the release once the charge has left the plate, we also sense the sign of the net force on the plate (shown as a test on the sign of the acceleration). If it becomes positive, indicating that the capacitor holding the plate down has discharged sufficiently, we re-enable the input to the velocity integrator, and the plate can move upwards.

We now examine a number of simulation results using this more extended model, with the numerical parameters in Table 7.1 (page 167). Figure 7.10 shows, on the left, the drive waveform used in this example, together with the solution for the charge Q. Because of the finite resistance R, there is a time lag between the applied voltage and the appearance of the charge on the plate. The corresponding gap as a function of time is shown on the right. We note the very clear pull-in event and then, once the charge has reduced again to a small value, the release. While it is not evident in this figure, there is some oscillatory motion ("ringing") during the initial pull-down. At the release, there is very strong ringing as the plate pops up and oscillates about its rest position.

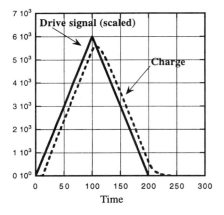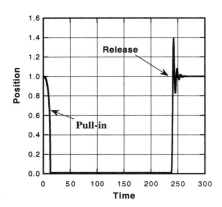

Figure 7.10.   Dynamics of the electrostatic actuator with a single triangular pulse.

The corresponding velocity as a function of time is shown on the left of Fig. 7.11. Here, the "crash" reset of the velocity to zero at the moment of pull-in is evident. The ringing on release is also very clear. On the right is a superposition of the drive signal and the position. Notice that the plate does not release until well after the drive signal has gone to zero. The reason is that when the plate is pulled in, its capacitance is huge and it can take a long time to discharge through the resistor. That is why it is essential to include the source resistance.[10] Models that use an ideal voltage source for the drive will not exhibit this release delay.

Graphs that plot one state variable against another during a dynamic event, with time as a parameter, are very useful. They are called *phase-plane plots* and are routinely used to capture the behavior of systems with several state variables. Our system has three state variables, so we need two phase-plane plots to display the behavior. These are shown in Fig. 7.12.

On the left of Fig. 7.12 is the phase-plane plot of position versus charge. We see that the system has hysteresis. Such hysteresis is observed in most nonlinear actuators that exhibit switching of states. On the right, the more usual phase-plane plot of velocity *vs.* position is shown. The cycle of pull-in, crash, and release are clearly evident. The ringing on release appears as cyclic motion about the final rest position. In system dynamic terms, this rest position

---

[10]It is also important when considering the energy required to achieve pull-in. This subject has been explored in [46].

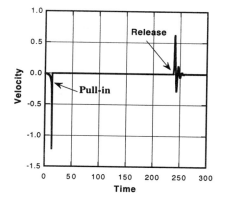 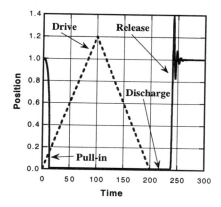

Figure 7.11. Velocity of the plate during the transient of Fig. 7.10, together with the gap and drive signal superimposed. Note the delay between the drive signal going to zero and the release of the plate due to the need to discharge the pulled-in capacitor through the source resistor.

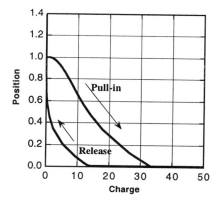 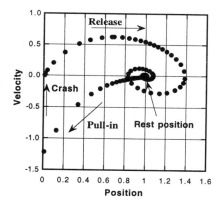

Figure 7.12. Phase-plane plots for the transient of Fig. 7.10.

is a globally stable fixed point of the system. It is the point to which the system returns when all excitations have been removed.

There are many other interesting nonlinear elements and systems encountered in MEMS. A particularly important one is the *Duffing nonlinearity*, characterized by a spring with a nonlinear force-displacement characteristic. The Duffing nonlinearity can be written as

$$F = k_1 x + k_3 x^3 \qquad (7.82)$$

If k_3 is positive, we have what is called an *amplitude-stiffened spring*. The spring gets incrementally stiffer as it stretches. Such behavior is very common in suspended elastic structures, as we shall see in Chapter 10. A characteristic of systems with amplitude-stiffened springs is that if one applies a sinusoidal drive signal below the resonance frequency, then slowly increases the frequency of the sinusoid, the amplitude that builds up as resonance is approached further increases the resonant frequency. This can create an an asymmetry in the apparent frequency response. If the effect is extreme, it can even lead to a jump phenomenon in which, at a critical value of the applied frequency, the amplitude of the response suddenly drops to a low value. In contrast, sweeping through the resonance from high to low frequencies does not produce this asymmetry. Therefore, an important test of whether Duffing-spring effects are present is to sweep through a resonance in both directions. In the absence of nonlinear springs, the two responses should be identical.

If k_3 is negative, we have an *amplitude-softened spring*. The electrostatic spring-softening effect we observed in the linearized electrostatic actuator is an example of this kind of nonlinearity, although we did not examine whether the behavior matches the analytical form of Eq. 7.82. The analysis of systems that include amplitude-stiffened and amplitude-softened Duffing nonlinearities is well documented in texts on nonlinear dynamics [45, 47].

7.3.6   Resonators and Oscillators

A *resonator* is a device with a vibratory natural response. A *linear resonator* is a resonator that can be described in terms of a linear transfer function. That is, it has a complex pole pair near the imaginary axis of the s-plane. There are also *nonlinear resonators*, for example, a mass-spring-dashpot system with a nonlinear spring.

An *oscillator* is a resonator plus an external circuit that provides the energy to sustain steady-state oscillation. The resulting dynamic behavior is called a *limit cycle*, a continuous closed curve when plotted in the phase plane of the dynamical system.

Many people confuse resonators with oscillators. The behavior of an oscillator is always a combination of the characteristics of the resonator and the drive circuit. We will take up this subject in Chapter 15 when we discuss feedback in nonlinear systems.

7.3.7   And Then There's Chaos. . .

There is a class of nonlinear systems that have an even more complex behavior, called *chaotic* behavior. A characteristic of chaotic behavior is that the state of a system after an interval of time t_c is highly sensitive to the exact initial state. That is, a chaotic system, if started from two initial states very close to each other, after time t_c, can be in states that are arbitrarily far away in the accessible phase space of the system. Chaotic systems can also undergo complex jumps between regions of phase space. We will not encounter any intentionally chaotic systems in this book. The interested reader is referred to Strogatz [45].

Related Reading

R. N. Bracewell, *The Fourier Transform and its Applications*, New York: McGraw-Hill, 1978.

C. A. Desoer and E. S. Kuh, *Basic Circuit Theory*, New York: McGraw-Hill, 1969.

A. H. Nayfeh *Nonlinear Oscillations*, New York: Wiley, 1979.

A. V. Oppenheim, A. S. Willsky and I. T. Young, *Signals and Systems*, Englewood Cliffs, NJ: Prentice-Hall, 1983.

A. Papoulis, *Circuits and Systems: A Modern Approach*, New York: Holt, Rinehart & Winston, 1980.

W. McC. Siebert, *Circuits, Signals, and Systems*, Cambridge, MA: MIT Press, 1986.

S. H. Strogatz, *Nonlinear Dynamics and Chaos*, Reading, MA: Addison Wesley, 1994.

M. W. Van Valkenburg, *Network Analysis*, Third Edition, Englewood Cliffs, NJ: Prentice-Hall, 1974.

Problems

7.1 MATLAB practice: Use the MATLAB command ode to find the step response to the linear system of Eqs. 7.1 and 7.2 using the parameter values of Section 7.2.1. This should reproduce Fig. 7.1. Then adjust the value of b to find the smallest value that produces no overshoot in the position response. What is the Q of the system at this value of b?

7.2 The electrostatic actuator system of Eqs. 7.3 - 7.5 exhibits an interesting step response, called dynamic pull-in. Use the MATLAB command ode to set up these equations with the parameter values of Table 7.1 except for b, which you should reduce to 0.1. Study the transient position response to a step voltage. Try using voltages that approach the pull-in voltage. You should find that the system reaches a pull-in instability at a voltage that is less than the quasi-static pull-in voltage for the system. Explain this behavior on the basis of potential and kinetic energy arguments.

7.3   In Problem 5.2, we introduced a two-mass two-spring system for which the form of equivalent circuit is that of Problem 5.3. Using complex impedances, find the natural frequencies (poles) of this system under the assumption that b is small, and provide a physical interpretation of the motion corresponding to each pair of complex poles.

7.4   For the circuit of Problem 5.3, with values $C_1 = 1$, $m_1 = 1$, $C_2 = 10$, $m = .01$, and $b=.01$, make the Bode plot for the transfer function $I_1(s)/V(s)$. Identify the key features of the plot, and show how they change with changes in circuit parameter values.

7.5   A system has an impulse response $h(t)$ given by $u_o(t) \exp -(t/\tau)$, where $u_o(t)$ is the unit step function. Use the convolution method to find the response to a square pulse of amplitude A_P and length t_P. Then, by finding the system function that corresponds to this $h(t)$, find the sinusoidal-steady state response of the system.

7.6   Complete the normalization of the eigenfunctions of Example 7.2, and use these eigenfunctions to find the zero-input response to an initial state $x_1 = 1$, $x_2 = 1$.

7.7   For the magnetic actuator of Section 6.5, using the values $x_o = 50$ μm, $A = 2500$ μm^2, $g = 2$ μm, $g\mu = 0.1\mu$m, $N = 50$, $L_m = 1$ mm, and $\mu = 100$ μ_0, determine the spring-constant k that will permit full travel of the armature with a current I of 100 mA. For this value of k, and using $R = 50$ Ω for the source resistance and a value of damping constant that gives a slightly underdamped response, determine the time required for the armature to move from its rest position to full displacement.

7.8   For the magnetic actuator of Section 6.5 using the component values of Problem 7.7, develop a linearized transducer model for this device at an operating point of 50 mA. What is the small-signal resonant frequency and Q at the operating point?

7.9   SIMULINK practice: Make a SIMULINK model for the magnetic actuator of Section 6.5 using the component values of Problem 7.7.

III
DOMAIN-SPECIFIC DETAILS

Chapter 8

ELASTICITY

Je plie et ne romps pas. (Tr: I bend and I break not.)

—Jean de la Fontaine

8.1 Introduction

We now begin our examination of the physical level of device modeling and analysis, one energy domain at a time. We begin with a three-chapter segment on structural mechanics which includes elasticity, elementary structures, and energy-based methods for analyzing mechanical devices.

We think of things that are *elastic* as having the ability to deform in response to applied forces and recover their original shape when the force is removed. Indeed, all solid materials can, to some extent, deform elastically without breaking or developing a permanent shape change, but all will break if the applied forces are too large. The elastic behavior of materials is a very large field, and we can only hope to sample the key ideas. Our specific goal is to develop enough information to enable the creation of accurate and useful lumped-element models of mechanical microstructures. Much more detailed presentations can be found in the list of Related Reading at the end of this chapter.

We begin with the basic definitions of stress and strain and the linear relations between stress and strain that apply in all solid materials for sufficiently small deformations.

8.2   Constitutive Equations of Linear Elasticity

8.2.1   Stress

Stress is defined microscopically as the *force per unit area* acting on the surface of a differential volume element of a solid body. Figure 8.1 shows a differential volume element of dimensions Δx, Δy, and Δz, and illustrates the reference directions for the definitions of stress. We assume the stress acts uniformly across the entire surface of the element, not just at a point.

Stresses perpendicular to a differential face are called *normal stresses* and are denoted here by the symbols σ_x, σ_y, and σ_z. Forces acting along the faces are called *shear forces*. The corresponding shear stresses are labeled with double subscripts, for example, τ_{xz}, the first subscript identifying the face and the second identifying the direction.

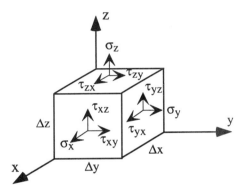

Figure 8.1.   Stresses acting on a differential volume element

An important assumption is that the differential volume element is in *static equilibrium*, which implies that there are no *net* forces or torques (rotational moments) on it. In order to satisfy the zero-net-force requirement, there can be no net body force such as gravity present or, if there is, this body force is balanced by an added normal force on the lower face of the volume element. Further, on the three invisible faces of the differential volume element in Fig. 8.1, there must be stresses acting directed opposite to those shown on the visible faces. That assures that the net force on the differential volume element is strictly zero. In order to assure that there is also no net torque, there must be a relation between the shear forces on neighboring faces. In particular, we require that

$$\tau_{xy} = \tau_{yx}$$
$$\tau_{xz} = \tau_{zx}$$

$$\tau_{yz} = \tau_{zy}$$

The units for stress are, to put it mildly, distressing. The standard MKS unit is the Pascal (symbol Pa), which is what we shall use throughout this book. The Pascal is defined as one Newton of force per square meter of area. However, the literature also uses the dyne per square centimeter, which equals 0.1 Pa, and the pound per square inch, or psi, equal to about 69,000 Pa, or 6.9 kPa. A convenient way to remember the psi unit is that one atmosphere is about 14 psi, which is about 100 kPa. For solid materials, the relevant stresses turn out to be in the mega-Pascal to giga-Pascal range, abbreviated MPa and GPa, respectively. So, for example, 1 MPa equals 10^7 dynes/cm^2 and 1 GPa equals 10^{10} dynes/cm^2.

8.2.2   Strain

When forces are applied to a solid body, it can deform. The differential deformation is called the *strain*, expressed as change in length per unit length. Figure 8.2 illustrates in two dimensions a small area element being displaced and distorted in shape. Each corner point is moved to a new location, and one can assume smooth interpolation to find the new location of any interior point. The net motion can be broken up into three parts: rigid motion of the original body (described as center-of-mass motion), rigid-body rotation about the center of mass, and deformation relative to this displaced and rotated frame of reference. To extract the deformation part of the motion requires consideration of the variation of displacement with position. This leads us to the differential definition of strain.

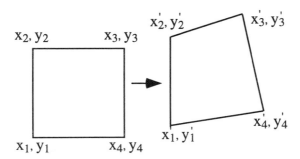

Figure 8.2.   Illustrating displacement and deformation of an area element.

Figure 8.3 illustrates the two basic types of strain. In Fig. 8.3a, the *uniaxial* strain of a differential element with initial length Δx is shown. In general, the displacement **u** is a vector function of the original position x. Considering only the x-component of the displacement vector, the change in length is $u_x(x + \Delta x) - u_x(x)$. (Note that $u_x(x)$ in the figure is actually negative.) Since

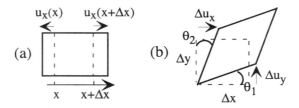

Figure 8.3. (a) Uniaxial normal strain and (b) pure shear strain without rotation ($\theta_1 = \theta_2$). Displacements are greatly exaggerated for clarity.

strain is the change in length per original unit length, the axial strain at point x in this example is

$$\epsilon_x = \frac{u_x(x + \Delta x) - u_x(x)}{\Delta x} = \frac{\partial u_x}{\partial x} \tag{8.1}$$

Note that if all parts of the body moved by the same amount in the x direction (*i.e.*, rigid-body translation), so that $u_x(x)$ were independent of x, there would be no strain.

For shear strains, illustrated in Fig. 8.3b, it is necessary to remove any rigid-body rotation that accompanies the deformation. This is done for small-amplitude deformations by using a symmetric definition of shear strain:

$$\gamma_{xy} = \left(\frac{\Delta u_x}{\Delta y} + \frac{\Delta u_y}{\Delta x}\right) = \left(\frac{\partial u_x}{\partial y} + \frac{\partial u_y}{\partial x}\right) \tag{8.2}$$

The shear strain, for small displacements, is equivalent to the sum of the two angles, θ_1 and θ_2. Thus, if the deformed shape in Fig. 8.3b were to be rigidly rotated such that θ_1 were zero, then the value of θ_2 would double, but the value of the shear strain γ_{xy} would remain unchanged.[1]

Strain, being the ratio of a change of length to a length, is dimensionless. Some authors, however, use a name for this ratio, such as the *microstrain*. One microstrain is a relative change in length of one part in 10^6.

8.2.3   Elastic Constants for Isotropic Materials

Isotropic materials are those with no internal ordering or structure that would make the stress-strain responses depend on direction. For these materials, a uniaxial stress results in a uniaxial strain that is proportional to the stress. The

---

[1]Some authors define the shear strain with a pre-factor of 1/2, so that the shear strain is 1/2 the sum of the angles θ_1 and θ_2, but this definition requires insertion of compensating factors of 2 in later parts of the formulation. The definition we have used is from Timoshenko and Goodier, *Theory of Elasticity* [48]. It has the advantage of simplifying the formulation of strain energy when we get to energy methods.

proportionality constant is called *Young's modulus* and is denoted by E. A normal stress in the x direction is linearly related to the uniaxial strain in the x direction,

$$\sigma_x = E\epsilon_x \tag{8.3}$$

with equivalent expressions for the other normal-stress directions.

Since strain is dimensionless, the units of E are the same as those for stress, i.e., Pa, dynes/cm^2, or psi. Typical values of E range from a few GPa for soft materials to several hundred GPa for stiff materials. Data on selected microelectronic materials can be found in Table 8.1.

When a free-standing object is deformed by a normal stress, resulting in a uniaxial strain, there is a contraction in the directions transverse to the uniaxial strain, as illustrated in Figure 8.4.

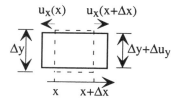

Figure 8.4. When area element of square cross-section (the dashed line) is deformed with a stress in the x-direction, there is an accompanying Poisson contraction in the y-direction. Deformations are shown greatly exaggerated.

The transverse contraction created by a uniaxial extension can be written in terms of the displacements. For example, in the y-direction, if the original length is Δy, the deformed length is written $\Delta y + \Delta u_y$, demonstrating that Δu_y in this example must be negative. Recalling that strain is change in length per unit length, we see that this transverse contraction creates a negative strain. For linear elastic isotropic materials, we write:

$$\epsilon_y = -\nu\epsilon_x \tag{8.4}$$

where ν is the *Poisson ratio*. The Poisson ratio is dimensionless, and typically has a value between 0 and 0.5.[2] Values for inorganic solids are in the range 0.2 - 0.3, while elastomers such as rubber have values very near 0.5. Insight into the Poisson ratio can be obtained from considering the change in volume of

---

[2]It is possible to create structural elements that deform with an apparently *negative Poisson ratio*. That is, when axially strained in one dimension, they expand in the transverse direction [49].

a three-dimensional differential volume element subjected to a uniaxial stress σ_x. In this case, the dimensions of the deformed volume element are

$$\Delta x \quad \rightarrow \quad \Delta x\,(1 + \epsilon_x) \tag{8.5}$$
$$\Delta y \quad \rightarrow \quad \Delta y\,(1 - \nu\epsilon_x) \tag{8.6}$$
$$\Delta z \quad \rightarrow \quad \Delta z\,(1 - \nu\epsilon_x) \tag{8.7}$$

The net result is that the change in volume ΔV is given by:

$$\Delta V = \Delta x \Delta y \Delta z (1 + \epsilon_x)(1 - \nu\epsilon_x)^2 - \Delta x \Delta y \Delta z \tag{8.8}$$

For the case of small strains, this becomes

$$\Delta V \approx \Delta x \Delta y \Delta z (1 - 2\nu)\epsilon_x \tag{8.9}$$

Thus, for a Poisson ratio that approaches 0.5, there is no change in volume accompanying a uniaxial strain. The transverse contractions exactly compensate for the extension. However, for Poisson ratios less than 0.5, there is some net volume expansion that accompanies a uniaxial strain; this is the situation for most solid materials.

There is also a linear relation between the shear stress and resulting shear strain, written in terms of the *shear modulus*, denoted by G. For example, if a shear stress σ_{xy} creates a shear strain γ_{xy}, the two quantities are related by:

$$\tau_{xy} = G\gamma_{xy} \tag{8.10}$$

It can be shown[3] that the shear modulus is related to the Young's modulus and Poisson ratio by the following equation:

$$G = \frac{E}{2(1 + \nu)} \tag{8.11}$$

Thus, if one knows two out of the three quantities, E, ν, and G, one can determine the third.

8.2.4   Other Elastic Constants

Over the long development of solid mechanics, several other ways of defining elastic relations between stress and strain have been created. Many texts use the *Lamé* constants, usually denoted by the Greek symbols μ and λ. In terms of the quantities we have already defined, the Lamé constants are written

---

[3] Timoshenko and Goodier, *Theory of Elasticity*, pp. 9-10.

$$\mu = G \tag{8.12}$$

$$\lambda = \frac{\nu E}{(1 + \nu)(1 - 2\nu)} \tag{8.13}$$

The first constant is just the shear modulus, while the second is a measure of the stresses needed to create volume changes. Note the factor $(1 - 2\nu)$ in the denominator. As $\nu \rightarrow 0.5$, the value of λ blows up, meaning that very large stresses are required to create volume changes. A related quantity, called the *bulk modulus*, also called the *modulus of volume expansion*, is the analog of Young's modulus for an object subjected to hydrostatic pressure (identical normal stresses in all directions). The bulk modulus is defined as

$$K = \frac{E}{3(1 - 2\nu)} \tag{8.14}$$

It relates the volume strain $\Delta V/V$ to the applied pressure. Note, again, the factor $(1 - 2\nu)$ in the denominator. The bulk modulus approaches infinity when the Poisson ratio approaches 0.5. The reciprocal of the bulk modulus is called the *compressibility*. Clearly, the compressibility of a material goes to zero when ν approaches 0.5. Materials with Poisson ratios very close to 0.5 are referred to as *incompressible*.

8.2.5   Isotropic Elasticity in Three Dimensions

Combining the results for normal and shear stresses and strains in three dimensions, we can write the complete stress-strain relations for an isotropic elastic solid (a generalized Hooke's Law) as:

$$\epsilon_x = \frac{1}{E}\left[\sigma_x - \nu\left(\sigma_y + \sigma_z\right)\right] \tag{8.15}$$

$$\epsilon_y = \frac{1}{E}\left[\sigma_y - \nu\left(\sigma_z + \sigma_x\right)\right] \tag{8.16}$$

$$\epsilon_z = \frac{1}{E}\left[\sigma_z - \nu\left(\sigma_x + \sigma_y\right)\right] \tag{8.17}$$

$$\gamma_{xy} = \frac{1}{G}\tau_{xy} \tag{8.18}$$

$$\gamma_{yz} = \frac{1}{G}\tau_{yz} \tag{8.19}$$

$$\gamma_{zx} = \frac{1}{G}\tau_{zx} \tag{8.20}$$

8.2.6   Plane Stress

Plane stress is a special case that occurs very frequently in thin-film materials used in MEMS devices. It refers to a thin film attached to a (relatively) rigid substrate (see Figure 8.5).

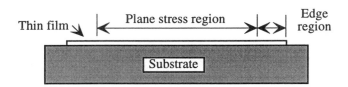

Figure 8.5.   Illustrating the state of plane stress, which occurs away from the edges of a thin film attached to a substrate.

It is typically true that a thin film deposited or formed on a substrate has some in-plane stress, arising either from the details of the deposition process or from mismatches in thermal expansion between the film and the substrate. In regions that are more than about three film thicknesses from the edge of the film, all of the stresses lie in the plane, since the top surface is stress-free. In the edge regions, however, the situation is more complex (see below).

It can be shown from the fundamentals of elasticity[4] that it is always possible, in the case of plane stress, to define a coordinate system in which there are two components of in-plane normal stress, and no in-plane shear stresses. These coordinates are called the *principal axes*. Expressed in principal-axis coordinates, the constitutive equations of plane stress are

$$\epsilon_x = \frac{1}{E}(\sigma_x - \nu\sigma_y) \tag{8.21}$$

$$\epsilon_y = \frac{1}{E}(\sigma_y - \nu\sigma_x) \tag{8.22}$$

with all other stress components equal to zero. A further special case, called *biaxial plane stress* occurs when the two in-plane stress components are equal to each other. In this case, the x- and y- components of strain are also equal to each other. That is

$$\sigma_x = \sigma_y = \sigma \rightarrow \epsilon_x = \epsilon_y = \epsilon \tag{8.23}$$

and the relation between biaxial stress and biaxial strain becomes

---

[4] See, for example, [48, 50, 51].

$$\sigma = \left(\frac{E}{1 - \nu} \right) \epsilon \tag{8.24}$$

The quantity $E/(1 - \nu)$ is called the *biaxial modulus*. It is, unfortunately, occasionally referred to by some authors as the "bulk modulus," which is not precise (the standard definition of bulk modulus is in Eq. 8.14). Therefore, when encountering the phrase "bulk modulus," it is wise to check the exact meaning being used.

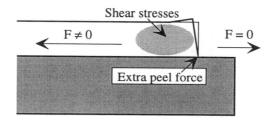

Figure 8.6. The edge region of a tensile film attached to a substrate.

In the edge regions of a film attached to a substrate, there are several important effects. Figure 8.6 illustrates the issues. If we assume that, in the region away from the edge, the film has a tensile stress ($\sigma > 0$), than there is a net non-zero in-plane force to the left of the shaded region, but by the time the right-hand free edge is reached, the in-plane force must be zero. The fact that there is no in-plane normal force at the edge, through the release of the vertical Poisson contraction, means that the film is slightly thicker, and must also be bent back from the originally vertical edge. The shaded edge region has a complex distribution of shear stresses making the transition from the right-hand free edge to the plane stress region on the left. Finally, the discontinuity in the stresses at the attachment point of the film leads, through a phenomenon called *stress concentration*, to extra forces called *peel forces* that tend to detach the film from the substrate. This has important implications in the detailed designs of structures. Debonding of tensile films tends to occur at the edges of patterned features.

8.2.7 Elastic Constants for Anisotropic Materials

When we cannot assume that a material is isotropic, for example, in any crystalline material, then more than two elastic constants are needed to describe the stress-strain relation. Since both stress and strain are second rank tensors, the most general linear relationship between stress and strain is a fourth rank tensor, with 81 components. However, because of various symmetry arguments that must hold (for example, no net force or torque in equilibrium), there are many constraints among the 81 components, and, in every real material, there is a maximum of only 21 parameters to contend with, and these 21 parameters can be written as the elements of a square 6×6 symmetric matrix. As a result, a compact notation has been developed for elasticity. The six independent components of stress and strain, generally written in a coordinate system that has axes along the symmetry axes of the material, are organized into a column-vector array, and the elastic constants are written in a symmetric matrix:

$$
\begin{pmatrix} \sigma_x \\ \sigma_y \\ \sigma_z \\ \tau_{yz} \\ \tau_{zx} \\ \tau_{xy} \end{pmatrix} = \begin{pmatrix} C_{11} & C_{12} & C_{13} & C_{14} & C_{15} & C_{16} \\ C_{12} & C_{22} & C_{23} & C_{24} & C_{25} & C_{26} \\ C_{13} & C_{23} & C_{33} & C_{34} & C_{35} & C_{36} \\ C_{14} & C_{24} & C_{34} & C_{44} & C_{45} & C_{46} \\ C_{15} & C_{25} & C_{35} & C_{45} & C_{55} & C_{56} \\ C_{16} & C_{26} & C_{36} & C_{46} & C_{56} & C_{66} \end{pmatrix} \begin{pmatrix} \epsilon_x \\ \epsilon_y \\ \epsilon_z \\ \gamma_{yz} \\ \gamma_{zx} \\ \gamma_{xy} \end{pmatrix} \quad (8.25)
$$

The matrix elements, C_{IJ} are referred to as the *stiffness coefficients* for the material, where upper case subscripts are used as a reminder that these indices run over the integers (1,2,3,4,5,6). A compact way of writing this is with matrix notation

$$
\sigma_I = \sum_J C_{IJ} \epsilon_J \quad (8.26)
$$

where it is understood that for index values of 4, 5, or 6, the corresponding shear stress τ or shear strain γ is implied. This matrix equation can be inverted, so that strain can be expressed as a function of stress:

$$
\epsilon_I = \sum_J S_{IJ} \sigma_J \quad (8.27)
$$

where the matrix $\mathbf{S}$ is the inverse of the matrix $\mathbf{C}$. The matrix elements S_{IJ} are called the *compliance coefficients* for the material. It is one of the crueler jokes of notational history that the "Stiffness coefficients" usually have symbol C_{IJ} and the "Compliance coefficients" usually have symbol S_{IJ}. The C_{IJ} have the same dimensions as Young's modulus, that is, Pascals. The S_{IJ} have the dimensions Pascals^{-1}.

While $\mathbf{C}$ and $\mathbf{S}$ each have 21 independent components, only a few of them are nonzero and independent in materials of practical interest. For example, in cubic materials such as single-crystal silicon, there are only three independent quantities, and a great deal of symmetry. The stiffness coefficients of silicon [52] are:

$$\mathbf{C} = \begin{pmatrix} C_{11} & C_{12} & C_{12} & 0 & 0 & 0 \\ C_{12} & C_{11} & C_{12} & 0 & 0 & 0 \\ C_{12} & C_{12} & C_{11} & 0 & 0 & 0 \\ 0 & 0 & 0 & C_{44} & 0 & 0 \\ 0 & 0 & 0 & 0 & C_{44} & 0 \\ 0 & 0 & 0 & 0 & 0 & C_{44} \end{pmatrix} \quad (8.28)$$

where

$$\begin{aligned} C_{11} &= 166 \quad \text{GPa} \\ C_{12} &= 64 \quad \text{GPa} \\ C_{44} &= 80 \quad \text{GPa} \end{aligned}$$

Formulae for determining the effective Young's modulus and Poisson ratio for arbitrary orientations and directions for loading of cubic materials can be found in Appendix C.

In materials with less than cubic symmetry, more elastic constants are needed. In Chapter 21, we shall encounter two piezoelectric materials: quartz, which has six elastic constants, and zinc oxide, which has five. A good reference on the form of the stiffness and compliance matrices for different crystal classes is Nye [53].

8.3   Thermal Expansion and Thin-Film Stress

An important thermomechanical effect is *thermal expansion*, the tendency of a free body to increase in size as it is heated. The *linear thermal expansion coefficient* of a material (presumed for the moment to be isotropic), is denoted by the symbol α_T, and is defined in terms of the rate of change of uniaxial strain with temperature:

$$\alpha_T = \frac{d\epsilon_x}{dT} \quad (8.29)$$

where T is temperature. The units for α_T are Kelvins^{-1}. However, since α_T is describing a dimensionless strain per Kelvin, and since the range of numerical values of α_T tends to be 10^{-6} - 10^{-7}, another common unit for α_T is μstrain/Kelvin, capturing the factor of 10^{-6} into the dimensions for α_T.

For moderate temperature excursions (and, in an approximate sense, even for large temperature excursions), we treat α_T as a constant of the material. Thus, for a finite temperature change ΔT, we would write

$$\epsilon_x(T) = \epsilon_x(T_0) + \alpha_T \Delta T \quad (8.30)$$

where $\epsilon_x(T_0)$ is the strain at the original temperature T_0, and ΔT equals $T - T_0$.

In three dimensions, assuming no constraints that prevent expansion, the material expands in each dimension by the same strain. Thus we can determine the *volume thermal expansion coefficient*, as follows, for a differential element of original volume V:

$$\frac{\Delta V}{V} = (1 + \epsilon_x)(1 + \epsilon_y)(1 + \epsilon_z) - 1 \tag{8.31}$$

leading to the relation

$$\frac{\Delta V}{V} = 3\alpha_T \Delta T \tag{8.32}$$

In cubic materials, the linear expansion coefficient must be the same in all three principal cubic directions; hence, for linear expansion purposes, we can treat cubic materials as having isotropic thermal expansion properties.

When a thin-film material is attached to a substrate, the thermal behavior is more complex. Refer back to Fig. 8.5, ignoring the edge regions. Assume that the film is much thinner than the substrate and is deposited onto the substrate in a stress-free state at temperature T_d. The sample is then cooled to room temperature T_r. Because the substrate is so thick compared to the film, a good approximation to assume that the substrate contracts according to its own thermal expansion coefficient, and the thin film, being attached to the substrate, must contract by the same amount as the substrate. The thermal strain of the substrate (in one in-plane dimension) is

$$\epsilon_s = -\alpha_{Ts} \Delta T \tag{8.33}$$

where

$$\Delta T = T_d - T_r \tag{8.34}$$

and where α_{Ts} is the linear expansion coefficient of the substrate. If the film were not attached to the substrate, it would experience a thermal strain given by

$$\epsilon_{f,\text{free}} = -\alpha_{Tf} \Delta T \tag{8.35}$$

where α_{Tf} is the linear thermal expansion coefficient of the film. However, because of its attachment to the substrate, the actual strain in the film must equal that of the substrate. Hence

$$\epsilon_{f,\text{attached}} = -\alpha_{Ts} \Delta T \tag{8.36}$$

The extra strain, the difference between the actual strain and the strain the film would have if free, is called the *thermal mismatch strain*, and is given by

$$\epsilon_{f,\text{mismatch}} = (\alpha_{Tf} - \alpha_{Ts}) \Delta T \tag{8.37}$$

The only way the film can achieve this state of (biaxial) strain is to develop an in-plane biaxial stress, given by

$$\sigma_{f,\text{mismatch}} = \left(\frac{E}{1-\nu}\right) \epsilon_{f,\text{mismatch}} \tag{8.38}$$

An example will illustrate the effect. If a thin polyimide film, with thermal expansion coefficient $\alpha_{Tf} = 70 \times 10^{-6}$ K^{-1}, is deposited with zero initial stress at a temperature of 250°C onto a thick silicon wafer which has thermal expansion coefficient $\alpha_{Ts} = 2.8 \times 10^{-6}$ K^{-1}, and the sample is then cooled to room temperature (25°C), the thermal strain in the polyimide is $\epsilon_{f,\text{mismatch}} = 2.6 \times 10^{-3}$. Assuming the biaxial modulus of the polyimide is 4 GPa, the thermal-mismatch stress that develops is 60 MPa. The stress is positive, hence tensile.

If the thermal expansion coefficient of the film is smaller than that of the substrate, the net thermal mismatch stress has a negative sign, and is compressive. This is the case for silicon dioxide formed on silicon at temperatures above about 950°C, and then cooled to room temperature.

It is interesting to examine the vertical strain in a thin film that experiences a thermal-mismatch in-plane strain. The vertical strain that would occur without the attachment constraint would be $-\alpha_{Tf}\Delta T$, the same as the in-plane strain. However, because of the Poisson effect, there is an additional out-of-plane strain set up by the in-plane thermal-mismatch strain. Hence, the total out-of-plane strain is:

$$\epsilon_{z,\text{total}} = -\alpha_{Tf}\Delta T - 2\nu\epsilon_{x,\text{mismatch}} \tag{8.39}$$

which, after substitution becomes

$$\epsilon_{z,\text{total}} = -\left[\alpha_{Tf} + 2\nu\left(\alpha_{Tf} - \alpha_{Ts}\right)\right]\Delta T \tag{8.40}$$

The attached film has an effective out-of-plane thermal expansion that differs from α_{Tf} by the additional term $2\nu\left(\alpha_{Tf} - \alpha_{Ts}\right)$, which can have either sign, depending on the relative values of α_{Ts} and α_{Tf}.

8.3.1   Other Sources of Residual Thin-Film Stress

Thermal expansion mismatch is only one possible source of in-plane stress. Chemical reactions occurring far from equilibrium, such as the oxidation of silicon and the LPCVD deposition of silicon nitride, can result in films that are highly stressed, far beyond what one would calculate from thermal expansion mismatch. In addition, modifications to a film's equilibrium structural properties by substitutional doping, addition of atoms by ion implantation, mismatch of lattice spacing during epitaxial growth, and rapid deposition processes such as evaporation or sputtering can all result in thin-film stress. We call any stress in a thin film that is present after deposition a *residual stress*. We tend to decompose it into two components: the thermal mismatch stress, already discussed, and anything not explained by thermal mismatch, which is called *intrinsic* stress.

Residual stresses can be modified by thermal processes such as annealing. For example, in polycrystalline silicon, annealing at temperatures above the deposition process can result in grain growth and a corresponding reduction in compressive stresses, even conversion to tensile stress. In addition, residual stresses can result from yield phenomena, as explained in Section 8.5 below.

8.4   Selected Mechanical Property Data

Table 8.1 contains selected mechanical property data for a few microelectronic materials. For the thin-film materials, the data must be considered as approximate, because material properties of thin films do depend in detail on the processing conditions. This is especially true of residual stress. Symbols used in the table are ρ_m for mass density, E for Young's modulus, ν for the Poisson ratio, α_T for the coefficient of thermal expansion, and σ_o for biaxial residual stress.

Table 8.1.   Mechanical property data for selected microelectronic materials. (Sources: [52, 54, 55, 56])

Material	ρ_m kg/m^3	E GPa	ν	α_T μstrain/K	σ_o MPa	Comment
Silicon	2331	page 193		2.8		Cubic
α-Quartz	2648	page 573		7.4, 13.6		Hexagonal
Quartz (fused)	2196	72	.16	0.5		Amorphous
Polysilicon	2331	160	~0.2	2.8	Varies	Random grains
Silicon dioxide	2200	69	.17	0.7	-300	Thermal
Silicon nitride	3170	270	.27	2.3	+1100	Stoichiometric
	3000	270	.27	2.3	-50 – +800	Silicon rich
Aluminum	2697	70	~.3	23.1	varies	Polycrystalline

8.5   Material Behavior at Large Strains

All materials will break at sufficiently large strains. We classify two general types of materials based on their large-strain behavior (see Fig. 8.7). The dashed line illustrates the stress-strain behavior of a *brittle* material. It exhibits linear elastic behavior until some maximum strain, at which point it breaks. The solid line illustrates a *ductile* material, although here, there can be great variations in the exact shape of the plot, and additional details that depend on how fast the deformation is applied. The features shown here include a linear stress-strain curve at low strains (with a slope equal to the Young's modulus), a maximum at what is called the *yield stress* followed by an extension regime characterized by a nearly constant stress called the *flow stress*. At sufficiently large strains, the stress once again increases (the *hardening* region), after which fracture occurs. Some materials do not show any peak at the yield point, but instead exhibit a flattening that could be described by saying the yield stress and the flow stress are about equal (see Fig. 8.8). Furthermore, the terms "yield" and "flow" do not apply to all materials that show nonlinear stress-strain behavior. In particular, rubber and other *elastomers* have linear behavior at small strains, then show a broad elastomeric flat region without a pronounced yield peak, then hardening and fracture.

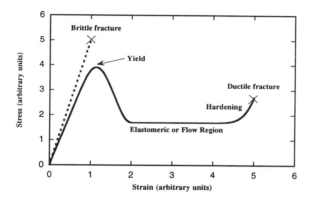

Figure 8.7.   Illustrating brittle and ductile materials.

If one computes the areas under the stress-strain curves up to the fracture point, it is seen that the brittle material has a relatively small area, while the ductile material shows a much larger area. This area is proportional to the total work done in deformation up to the fracture point, and in an engineering sense, tells us how much energy is required to break the material. By this metric, the ductile material would be called *tough* when compared with the brittle material, because it takes more energy to break it.

8.5.1   Plastic Deformation

There is significant variation in the behavior of materials when strained into the nonlinear region, and then unloaded. Figure 8.8 illustrates various the types of unloading behavior, using a stress-strain curve without a pronounced yield peak.

Materials that exhibit so-called *plastic* behavior, like metals, can remain in a bent state. They are typically unloaded by reducing the applied forces to zero. As can be seen from the dashed unloading curve in Fig. 8.8, there is a remanent

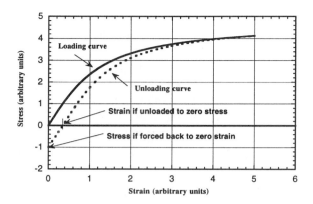

Figure 8.8.   Illustrating plastic deformation.

strain as a result of the loading and unloading operation. The area *between* the loading and unloading curves measures the energy expended to create this change of shape. In metals, as is familiar to anyone who has repeatedly bent a spoon, this energy is converted to heat. The metal gets hot. This ability to bend metals into new shapes is of immense technological importance in a host of fields.

In contrast with metals, elastomeric materials will follow an unloading curve that pretty closely tracks the loading curve. In other words, there is little net work done in loading and unloading; the energy put into the material during loading is returned to the external world during unloading.

A third type of behavior, called *viscoelasticity* is encountered in many plastic materials[5]. Viscoelasticity, in simplest terms, is time-retarded elasticity. When a stress is applied to a material, it responds elastically for short times, but as the load is maintained over time, gradually changes its shape (as if it were flowing – hence the *visco*elastic characterization). The internal viscous friction produces heat, just like in a metal that is bent. When the stress is removed, the material unloads into a deformed shape. However, in many plastics, some of the energy expended in deforming the material is stored internally in the free energy of the plastic,[6] and this can lead to interesting shape-memory effects: when the plastic is heated, the sample tends to return to its original undeformed shape. To the extent that polymeric materials are used in microsystems, especially in packages, the viscoelastic properties of these materials must be kept in view.

---

[5] Another cruel trick: metals exhibit "plastic" behavior while plastics exhibit "viscoelastic" behavior. The definitions simply don't line up nicely.

[6] Strictly speaking, the large-strain deformation of a polymeric material lowers its entropy, which increases its free energy. This stored free energy is available for reshaping the material when the temperature is raised.

The final behavior to consider is the case of a thin film or coating on a substrate that is bent. As we will discover in the next chapter, bending can produce large stresses at the surface of a sample. Thus it is possible for a thin film on a substrate to experience plastic (or viscoelastic) deformation during processing, after which the substrate once again becomes flat. If that is the case, the film is forced to zero strain rather than to zero stress, and thus can end up in a state of residual stress. In the example of Fig. 8.8, the material is deformed in a state of tensile stress, and undergoes plastic deformation that makes it longer. When forced back to a state of zero strain, the material must have a compressive residual stress. Many microelectronic materials, including both silicon and silicon dioxide, can have their stress state changed through this type of plastic deformation cycle.

Related Reading

J. M. Gere and S. P. Timoshenko, *Mechanics of Materials*, Second Edition, Monterey, CA: Brooks/Cole Engineering Division, 1984.

W. M. Lai, D. Rubin, and E. Krempl, *Introduction to Continuum Mechanics*, Third Edition, Oxford: Pergamon Press, 1993.

J. F. Nye, *Physical Properties of Crystals*, Oxford: Oxford University Press, 1960.

S. P. Timoshenko and J. N. Goodier, *Theory of Elasticity*, Third Edition, New York: McGraw-Hill, 1970; reissued in 1987.

Problems

8.1   Derive the bulk modulus (Eq. 8.14) of an isotropic material from the generalized Hooke's Laws.

8.2   Use MATLAB to calculate the compliance coefficients of silicon from the stiffness coefficients. Demonstrate that the compliance coefficients have the same symmetry as the stiffness coefficients.

8.3   Using the results of Problem 8.2 and referring to Appendix C, plot the variation of Young's modulus of silicon for directions within a (100) plane.

8.4   Using the results of Problem 8.2 and referring to Appendix C, plot the variation of the Poisson ratio of silicon as a function of the orientation of two mutually perpendicular directions that lie within a (100) plane.

8.5   Estimate the thermal stress in a silicon dioxide film formed on silicon at 950°C in a stress-free state and then is cooled to room temperature.

8.6   Material Property Treasure Hunt:[a]   For one of the materials listed below, fill in a new row of Table 8.1: LPCVD SiO_2, PECVD SiO_2, PECVD Si_3N_4, tungsten, gold, nickel, permalloy (Fe-Ni alloy), shape-memory alloy (Ti-Ni alloy), polyimide, and SU-8 resist after hardening.

---

[a]Instructors: you may wish to assign one material to each student and share the results with all students in order to expand everyone's access to material properties.

Chapter 9

STRUCTURES

Hence no force however great can stretch a cord however fine into an horizontal line which is accurately straight; there will always be a bending downward.
—William Whewell in *Elementary Treatise on Mechanics* (1819)

9.1 Overview

Microsystems contain many different kinds of structural elements. The devices and the packages that hold them are solid objects governed by the laws of continuum solid mechanics, and the moving parts within the microsystem are usually critical to overall operation. Unlike Mr. Whewell, whose cord always bends downward under the force of gravity, we can almost always ignore the inertial force of gravity in devices scaled to the dimensions of MEMS.[1]

In this chapter, we examine basic structural elements: axially loaded beams, and transversely loaded beams, cantilevers, membranes and plates. We also examine selected composite structures, for example, a thin film on a substrate, and consider the effects of residual stresses, stress gradients, and temperature variations on structural behavior. Our goal is to connect the detailed mechanical behavior of structural elements to equivalent lumped-circuit representations of these elements for inclusion in device-level models.

9.2 Axially Loaded Beams

The simplest structure is a slender beam with rectangular cross section loaded with a uniform stress on its end faces. Figure 9.1 illustrates such a case. The length of the beam is L, the width W, and the height or thickness is H. A force

---

[1]The exception, of course, is if we are trying to build a high-sensitivity accelerometer, in which case the gravitational acceleration figures significantly.

F acts on the end of the beam, and we assume (for the moment) that this total force is distributed as a uniform tensile stress acting on the ends of the beam. The value of this stress is

$$\sigma = \frac{F}{WH} \tag{9.1}$$

Note that equilibrium requires an equal force (and stress) at both ends of the beam.

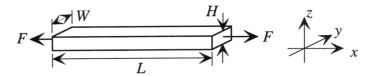

Figure 9.1.   A long slender beam with an axial load.

We know from our basic definitions of elasticity that, under the action of this axial load, the beam will extend. The strain created by the axial stress is

$$\epsilon = \frac{1}{E}\sigma = \frac{F}{EWH} \tag{9.2}$$

Note that because the sides of the beam are free, there are no transverse stresses in the beam and, therefore, no additional contributions to the axial strain. The corresponding change in length of the beam is

$$\delta L = \epsilon L = \frac{FL}{EWH} \tag{9.3}$$

If this beam were functioning as a spring element in a mechanical structure, we would want to determine the *spring constant* of the spring. An ideal linear spring exhibits a proportionality between force and extension of the form

$$F = k\delta x \tag{9.4}$$

where k is the spring constant and δx is the extension. For the axial loaded beam, therefore, we conclude that the equivalent spring constant is

$$k_{\text{axialbeam}} = \frac{EWH}{L} \tag{9.5}$$

Example 9.1 illustrates this result.

Example 9.1

Find the spring constant for an axially loaded silicon beam of length 100 μm, and square cross-section of 2 μm on a side. We use a value of 160 GPa for the Young's modulus of silicon.

$$k = \frac{(160 \times 10^9)(2 \times 10^{-6})^2}{100 \times 10^{-6}} \frac{\text{Pa}-\text{m}^2}{m} = 6400 \frac{\text{N}}{\text{m}}$$

As we shall see when we examine beams that bend, this is a relatively stiff spring. An interesting related point is to ask how much the beam would extend if hung vertically in a gravitational field. This is explored in Problem 9.1.

9.2.1   Beams With Varying Cross-section

If a beam has a nonuniform cross-section, again with a total force acting on the two ends, and the variation of cross-section is gradual enough so that we can assume that the force is still distributed uniformly across the area as a uniform axial stress, then we can consider slicing the beam into differential elements of length dx, finding the extension in length of each segment according to Eq. 9.3, and integrating along the end of the beam. Specifically,

$$\Delta(dx) = \frac{F\,dx}{EA(x)} \tag{9.6}$$

where $A(x)$ is the position-dependent cross-sectional area of the beam, leading to

$$\delta L = \int_0^L \Delta(dx) = \int_0^L \frac{F}{EA(x)}\,dx \tag{9.7}$$

Thus the spring constant for the non-uniform beam is

$$k = \frac{F}{\delta L} = \left[\int_0^L \frac{dx}{EA(x)} \right]^{-1} \tag{9.8}$$

9.2.2   Statically Indeterminate Beams

What happens when an irresistible force meets an immovable object? This is an overconstrained problem. Something must happen. Either the force is not irresistible, or the object is not immovable.

Figure 9.2 illustrates such a situation. A beam is clamped between two immovable ends, shown as hash-marked supports. The beam is also being heated, which we know makes it expand. What happens?

Assuming the supports are really fixed, the length of the beam cannot change. This means that the total strain in the beam must be zero. But there are two possible sources of strain: thermal expansion and an axial stress. Therefore, a possible resolution of the constraints is that an axial stress develops which produces an axial strain that exactly cancels the thermal expansion strain. Let us calculate this effect.

Figure 9.2.   A clamped beam subjected to thermal stress.

Assuming that the effect of the candle is to heat the beam uniformly by a temperature ΔT, the thermal strain is

$$\epsilon_{\text{thermal}} = \alpha_T \Delta T \qquad (9.9)$$

where α_T is the coefficient of thermal expansion. But the total strain must be zero, so we add in the effect of an axial stress σ_{thermal} in the beam such that the total strain is zero:

$$0 = \alpha_T \Delta T + \frac{\sigma_{\text{thermal}}}{E} \qquad (9.10)$$

We conclude that the thermally induced stress is compressive, and has the value

$$\sigma_{\text{thermal}} = -E\alpha_T \Delta T \qquad (9.11)$$

Example 9.2
  Suppose the beam of our previous example is clamped between fixed ends and heated by 100°C. The resulting compressive stress is 52.8 MPa. This is large enough to affect the bending stiffness of the beam, as we will see in Section 9.6.2.

The force that causes this stress comes from the supports. These forces are called *reaction forces*. They occur whenever a problem is overconstrained by the supports. Structural elements subjected to such constraints are called *statically indeterminate* because the state of the structural element cannot be resolved until the reaction forces and elastic deformation of the structure are taken into account. However, once these are taken into account, the state of the structure is, indeed, unique and well-determined.

If, instead of being immovable, the supports were part of a large structure with a much lower thermal expansion coefficient than the beam material, heating the structure and the beam together would also create compressive stresses in the beam. An example is a surface-micromachined beam attached to a substrate of dissimilar material, such as a metal surface-micromachined beam on a silicon substrate. Alternatively, if the beam material had a lower thermal expansion coefficient than the substrate, it would develop a tensile stress when the structure and beam are heated.

9.2.3   Stresses on Inclined Sections

Figure 9.3 shows at the top a beam with an axial force F, with a slanted dashed line that designates a plane through the structure inclined at an angle θ.

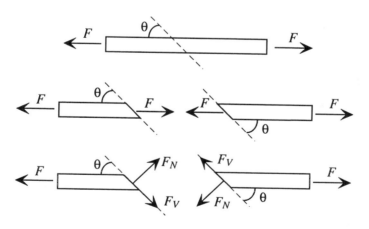

Figure 9.3.   Forces on an inclined section of an axially loaded beam.

The middle and lower parts of Fig. 9.3 show the beam as if it were cut into two sections. Since the original beam was in static equilibrium, with zero net force, both sections must also be in static equilibrium with zero net force. Therefore, we can think of each section as having an axial force F on it. Considering in-plane and normal components of force with respect to that inclined plane, the total force F can be resolved into two components, as in the bottom part of the figure, a normal force F_N and a shear force F_V, where

$$F_N = F \cos \theta \qquad (9.12)$$

$$F_V = F \sin \theta \qquad (9.13)$$

Every incremental inclined section along the beam is subjected to the same normal stress, determined by F_N divided by the cross-sectional area of the section, and shear stress, determined by dividing F_V by the cross-sectional area of the section.

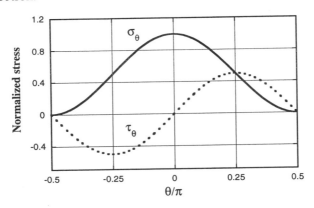

Figure 9.4.   Variation of stress components on an inclined section.

The area of the inclined cross-section is $A/\cos \theta$, where A is the cross-section of the beam end. Therefore, the resolution of forces into stresses yields a normal stress σ_θ and a shear stress τ_θ given by:[2]

$$\sigma_\theta = \frac{F}{A} \cos^2 \theta \qquad (9.14)$$

$$\tau_\theta = \frac{F}{A} \cos \theta \sin \theta \qquad (9.15)$$

Figure 9.4 shows the variation of σ_θ and τ_θ as a function of the angle of the inclined section. It is evident that the shear stress is highest at a $\pm45°$ angle. Materials that are weaker in shear than in tension will tend to fail by yielding along a plane inclined by about 45° to the beam axis. We will build on this relation between normal and shear stresses on inclined sections when we analyze the piezoresistive pressure sensor (Section 18.3.3).

---

[2]There are two possible sign conventions governing the shear stress in such a case, the so-called static sign convention, in which a positive shear stress is in the direction of increasing θ and hence would require a negative sign for τ_θ, and the deformation sign convention which would require a positive sign. Even though this is a static case, we use the positive sign here, so as not to confuse the discussion of the deformation sign convention in the following section. Clear pictures can always resolve questions of signs. . .

9.3   Bending of Beams

We now consider bending-type deformations that result from the transverse loading of beams. This requires that we first examine supports and loads on structures that can bend. There are several definitions and notational standards that must be introduced. We assume that the beams are much longer than either of their transverse dimensions and, for the analyses we shall perform, that the beams have rectangular cross-sections.

9.3.1   Types of Support

When analyzing beam behavior, great care must be taken with the details of the supports. Figure 9.5 shows two beams with four different kinds of supports. The left-hand end of the left-hand beam is *fixed* (also called *clamped* or *built-in*). The behavior of a fixed support is that the beam cannot move vertically or horizontally at the support, nor can it have a non-zero slope at the support. In contrast, the right-hand end of the same beam is free, with no constraint on either its displacement or its slope. A beam with a free end is called a *cantilever* beam. The right-hand beam is *doubly-supported* at its ends. The left-hand *pinned* support (also called a *simple* support) fixes both the vertical and horizontal position of the beam end, but does not restrict its slope. The right-hand *pinned support on rollers* is a construct meant to capture a situation in which the vertical position of a beam end is fixed, but both the horizontal position and slope can vary. The functional equivalent of all four types of supports are encountered in MEMS devices.

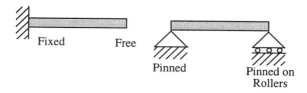

Figure 9.5.   Fixed, free, and pinned supports.

9.3.2   Types of Loads

Transverse loading of beams causes them to bend. The two basic types of external transverse loads that can be applied to beams are illustrated in Fig. 9.6. The left-hand cantilever beam is shown with a *point load F* applied to its end. Such a load can be applied at any position along the length of beam. The right-hand doubly-supported beam is shown with a *distributed* load q applied to a portion of the beam length.

The units and dimensions of these loads must be considered with care. As we shall see below, many properties of beams are expressed *per unit of beam width*, while some other properties include the beam width explicitly. In order to be clear about this, we shall specify F to mean the *total force* acting at a position x along the length of a beam, implicitly assuming that this force is uniformly distributed across the width of the beam. We will denote the *force per unit width* as F', such that the total force would be $F = F' \times W$. Similarly, we must be clear about the distributed load q. Conventionally, the symbol q is used for the *force per unit length of beam*, again assuming that it is distributed uniformly across the width of the beam. In this case, the distributed load per unit width has the dimensions of a *pressure*, so we will use the symbol P for such a load. Thus, the equivalent beam load q when there is a uniform pressure load on the beam would be $q = P \times W$.

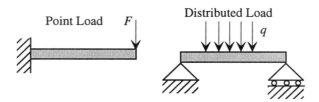

Figure 9.6.   Point and distributed loads.

9.3.3   Reaction Forces and Moments

The beams in Fig. 9.6 cannot be in equilibrium unless we include *reaction forces* from the support, as shown in the top figure of Fig. 9.7. The vertical reaction force F_R is necessary so that the total force on the beam is zero, and its value must equal the external load F using the reference direction shown. The reaction moment M_R is necessary to assure no net rotation at the support, which could lead to a non-zero slope for the beam, a violation of the fixed-support conditions. We recall from elementary mechanics that the total moment acting about a point is the sum of all forces weighted by their distance from that point. If that distance is zero, the force does not contribute to the moment, but a moment at that point does contribute. So if we refer to the top drawing in Fig. 9.7 and compute the total moment acting on the beam M_T using the free end as a reference, we obtain

$$M_T = M_R - F_R L \tag{9.16}$$

In order for M_T to be zero, we require that

$$M_R = F_R L = F L \tag{9.17}$$

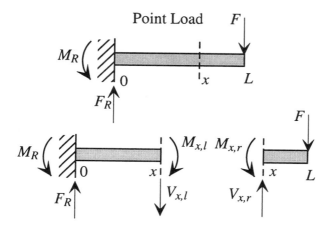

Figure 9.7.   Reaction forces, shear forces, and moments.

To check this result, we can also calculate moments about the fixed end of the beam. In this case, the force F_R does not contribute. The total moment in this case is

$$M_T = M_R - FL \qquad (9.18)$$

But this equals zero because of the value of M_R. Thus we have established the values of the reaction forces and moments needed to assure static equilibrium.

The lower part in Fig. 9.7 illustrates a hypothetical division of the beam into two parts at position x. In order for both parts to be in static equilibrium, the division point can be thought of as having equal and opposite internal forces and moments acting on the left-hand and right-hand parts. In order for the right-hand part to be in equilibrium, its left-hand face must experience a shear force $V_{x,r}$ which has value F, and must also experience a moment $M_{x,r}$ which has the value $F(L - x)$.

Since the total force and moment at position x must be zero, the magnitude of $V_{x,l}$ must be the same as $V_{x,r}$, equal to F, and the two moments $M_{x,l}$ and $M_{x,r}$ must also be of equal magnitude, with value $F(L - x)$. Therefore, we can now find the total force and moment acting on the left-hand part. The total force is $F_R - V_{x,l}$, which equals zero, and the total moment with respect to the fixed support is $M_R - M_{x,l} - xV_{x,l}$. Substitution of values for all three quantities shows that the total moment is zero.

We have learned several things from this example. First, we now have three types of loads to consider: external transverse loads, shear forces, and moments. Second, we have seen that the constraint requiring the total force and total moment to be zero in equilibrium can be applied anywhere in the beam using appropriate definitions for internal moments and shear forces. We now consider the differential beam elements of Fig. 9.8 to show how these three types of loads are related.

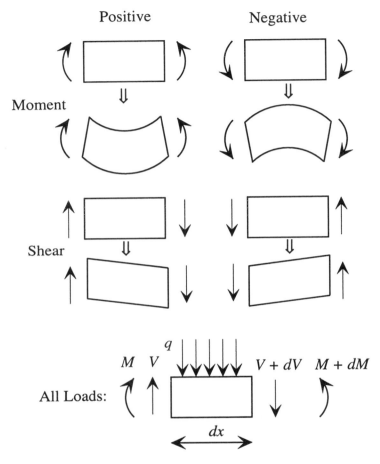

Figure 9.8.   Sign conventions for moments and shear forces and for a fully loaded differential beam element of length dx.

The sign conventions used for definitions of moments and shear forces are shown in the top figures of Fig. 9.8. A positive moment leads to deformation with a positive (upward) curvature, while a negative moment leads to deformation with a negative (downward) curvature. Positive shear forces produce net clockwise rotation; negative shear forces produce net counterclockwise rotation. The differential element at the bottom is shown with all sign conventions carefully observed. Unlike in the upper figures, however, it is not presumed that the moments and shear forces are independent of position.

We can apply the requirement of total static equilibrium (zero force and zero moment) to this differential element in order to obtain differential relations among the different loads. The total force F_T on the differential element is

$$F_T = q\,dx + (V + dV) - V \tag{9.19}$$

If we require this force to be zero, we conclude that

$$\frac{dV}{dx} = -q \tag{9.20}$$

The total moment M_T acting on the differential element with respect to the left-hand edge is

$$M_T = (M + dM) - M - (V + dV)dx - \frac{q\,dx}{2}dx \tag{9.21}$$

If we neglect terms that involve products of differentials and require the total moment to be zero, we obtain

$$\frac{dM}{dx} = V \tag{9.22}$$

We will use these relations to help establish the differential equations governing the bending of beams.

9.3.4 Pure Bending of a Transversely Loaded Beam

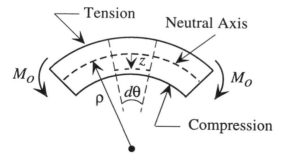

Figure 9.9. A segment of a beam in pure bending. Vertical displacements and angles are greatly exaggerated.

Figure 9.9 illustrates a small section of a beam that has become bent in response to the application of transverse loads. Note that the positive direction of z is downward in this figure.[3] The applied moment M_0 has a positive sign,

---

[3]This use of downward positive axes is common in elementary mechanics texts.

as does the radius of curvature ρ. The dashed arc in the middle of the beam is called the *neutral axis*. It is the axis whose length is unchanged by bending. Portions of the beam below the neutral axis are forced into compression by the bending, while portions above the neutral axis are extended in tension.

A differential angular segment $d\theta$, which is presumed to have axial length dx when not bent, is examined in detail. When the beam is bent, the length of an arc segment that subtends an angle $d\theta$ depends on how far it is from the neutral axis because there is a small variation in radius of curvature from the top to the bottom of the beam. Specifically, the length of the dashed segment at position z is

$$dL = (\rho - z)\, d\theta \qquad (9.23)$$

At the position of the neutral axis, the length of the corresponding segment is equal to dx the differential length of the segment when the beam is not bent. That is,

$$dx = \rho\, d\theta \qquad (9.24)$$

from which we obtain

$$dL = dx - \frac{z}{\rho}\, dx \qquad (9.25)$$

Since the unbent length of the segment is dx, we conclude that the axial strain ϵ_x at position z is

$$\epsilon_x = -\frac{z}{\rho} \qquad (9.26)$$

Because stress and strain are proportional, there is also an axial stress in the beam segment whose value depends on z. The expression is (see Fig. 9.10):

$$\sigma_x = -\frac{zE}{\rho} \qquad (9.27)$$

Note carefully the axes in Fig. 9.10. The z axis is pointing down, in agreement with Fig 9.9, and the horizontal axis is stress, not position. The upper part of the beam (negative z) is in tension, while the lower part of the beam (positive z) is in compression.

This structure has a distribution of internal stresses. We can find the total *internal bending moment* M by calculating the first moment of the distributed internal stress:

$$M = \int_{-H/2}^{H/2} W z \sigma_x\, dz \qquad (9.28)$$

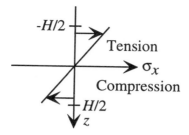

Figure 9.10.   Plot of axial stress in pure bending as a function of position relative to the neutral axis. The beam thickness is H.

which, after substitution, becomes:

$$M = -\int_{-H/2}^{H/2} \frac{EWz^2}{\rho} \, dz = -\left(\frac{1}{12}WH^3\right)\frac{E}{\rho} \qquad (9.29)$$

The quantity in parentheses is the *moment of inertia* of the beam cross-section, taken with respect to the neutral axis, and is usually denoted with the symbol I. We conclude that the internal bending moment is a simple function of the radius of curvature of the beam segment:

$$\frac{1}{\rho} = -\frac{M}{EI} \qquad (9.30)$$

Note the sign of M. A positive radius of curvature (as in Fig. 9.9) creates a negative internal bending moment. Equilibrium requires that the total moment be zero. Hence the external moment M_0 that created the bending in the first place must equal $-M$; M_0 is positive, in agreement with the sign convention of Fig. 9.7. The beam bends exactly to the point where the internal moment balances the externally applied moment. So we could equally write

$$\frac{1}{\rho} = \frac{M_0}{EI} \qquad (9.31)$$

When we formulate the differential equation for beam bending, we have a choice of expressing the radius of curvature in terms of the externally applied bending moment or the internal bending moments. We shall follow the generally used convention, and express this curvature in terms of the *internal* bending moment, M, using Eq. 9.30.

9.3.5   Differential Equation for Beam Bending

Figure 9.11 shows a bent cantilever beam. The beam is drawn as a thin line representing the position of the neutral axis $w(x)$ as a function of position along the original beam x. We temporarily ignore the details of internal stress and moment distributions associated with bending, but will return to them later.

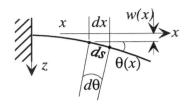

Figure 9.11.   A bent cantilever beam.

An increment of beam length ds along the neutral axis is related to dx by

$$ds = \frac{dx}{\cos \theta} \tag{9.32}$$

and the slope of the beam at any point is

$$\frac{dw}{dx} = \tan \theta \tag{9.33}$$

Furthermore, for a given radius of curvature ρ at position x, the relation between ds and the incremental subtended angle $d\theta$ is

$$ds = \rho d\theta \tag{9.34}$$

For virtually all MEMS devices, the angles encountered are very small – a few degrees. Therefore, we are virtually always justified in making the small-angle approximation, with the result that we can assume $ds \approx dx$. This further implies that

$$\frac{d\theta}{dx} \approx \frac{1}{\rho} \tag{9.35}$$

and

$$\theta \approx \frac{dw}{dx} \tag{9.36}$$

If we differentiate this expression with respect to x and substitute, we find

$$\frac{1}{\rho} = \frac{d^2w}{dx^2} \tag{9.37}$$

which, from Eq. 9.30, gives the basic differential equation for small-angle bending of slender beams:

$$\frac{d^2w}{dx^2} = -\frac{M}{EI} \qquad (9.38)$$

where M is the *internal* bending moment that, in equilibrium, is matched by externally applied loads. Because we typically work with point loads or distributed loads, it is often useful to work with the first or second derivatives of Eq. 9.38:

$$\frac{d^3w}{dx^3} = -\frac{V}{EI} \qquad (9.39)$$

or

$$\frac{d^4w}{dx^4} = \frac{q}{EI} \qquad (9.40)$$

In the examples to follow, we shall examine how beams respond to various combinations of applied loads. Before moving on, we note that if the angle θ becomes large enough to invalidate the small-angle approximations made above, a more exact expression for the radius of curvature can be obtained as follows:

We note that

$$\theta = \arctan w' \qquad (9.41)$$

where

$$w' = \frac{dw}{dx} \qquad (9.42)$$

Thus

$$\frac{1}{\rho} = \frac{d\theta}{ds} \qquad (9.43)$$

and using the relation

$$ds^2 = dx^2 + dw^2 \qquad (9.44)$$

we find

$$\frac{1}{\rho} = \frac{w''}{\left[1 + (w')^2\right]^{3/2}} \qquad (9.45)$$

9.3.6   Elementary Solutions of the Beam Equation

Figure 9.7 (page 209) shows a cantilever beam with a point load at its end. We can readily solve the beam equation for this case. The internal moment at any position x within the beam is $M = -F(L - x)$, which leads to the differential equation

$$\frac{d^2 w}{dx^2} = \frac{F}{EI}(L - x) \tag{9.46}$$

This is a second-order linear equation, so we expect a solution with two constants that are fixed by the boundary conditions:

$$w(0) = 0 \tag{9.47}$$

$$\left. \frac{dw}{dx} \right|_{x=0} = 0 \tag{9.48}$$

Because this is a linear equation, we can solve it many ways: (a) by finding a particular solution and then adding a homogeneous solution of the form $A + Bx$, solving for A and B to satisfy the boundary conditions; (b) using Laplace or Fourier transform methods; (c) finding the eigenfunctions and using an eigenfunction expansion; or (d) guessing a trial solution and seeing if it works. In this case, approaches (a) and (d) are equivalent and straightforward. We note that the right-hand-side is a polynomial of degree one, and it must be the second derivative of w. Therefore, if we assume that w is a polynomial of degree three, we know we can solve the problem:

$$w = A + Bx + Cx^2 + Dx^3 \quad \text{(trial solution)} \tag{9.49}$$

Note that this form already includes the homogeneous solution that might be needed to match boundary conditions. In this case, however, both w and dw/dx are zero at $x = 0$, which requires that both A and B be zero. Hence, the problem reduces to finding C and D such that Eq. 9.46 is solved. After substitution, we find

$$C = \frac{FL}{2EI} \tag{9.50}$$

$$D = -\frac{F}{6EI} \tag{9.51}$$

with the final result that

$$w = \frac{FL}{2EI} x^2 \left(1 - \frac{x}{3L} \right) \tag{9.52}$$

The maximum deflection at the end of the cantilever is

$$w_{max} = \left(\frac{L^3}{3EI}\right) F \tag{9.53}$$

If this maximum deflection were a key displacement in a device, we could assign a spring constant to this cantilever of

$$k_{\text{cantilever}} = \frac{3EI}{L^3} \tag{9.54}$$

or, substituting for I,

$$k_{\text{cantilever}} = \frac{EWH^3}{4L^3} \tag{9.55}$$

Example 9.3

Let us compare the spring constant of the bent cantilever to that of the axially loaded beam in Section 9.2. The 100 μm long beam with the 2 μm square cross-section has a spring constant of

$$k = \frac{(160 \times 10^9)(2 \times 10^{-6})^4}{4(100 \times 10^{-6})^3} \frac{\text{Pa} - \text{m}^4}{\text{m}^3} = 0.64 \frac{\text{N}}{\text{m}} \tag{9.56}$$

This spring constant is 10,000 times smaller than that of the same beam when loaded axially. The obvious conclusion is that relatively thin structures, such as this beam, are much more easily bent than either stretched or compressed.

It is also interesting to determine the stress in the bent cantilever. The curvature is given by

$$\frac{1}{\rho} = \frac{d^2w}{dx^2} = \frac{F}{EI}(L - x) \tag{9.57}$$

The maximum bending stresses occur where the inverse radius of curvature is maximum. This maximum occurs right at the support ($x = 0$), at which point the radius of curvature has the value

$$\frac{1}{\rho} = \frac{L}{EI}F \tag{9.58}$$

The maximum tensile strain is at the top surface, and the maximum compressive strain is at the bottom surface, both having magnitude

$$\epsilon_{max} = \frac{H}{2\rho} = \frac{LH}{2EI}F \tag{9.59}$$

which, after substitution, becomes

$$\epsilon_{max} = \frac{6L}{H^2WE}F \tag{9.60}$$

The maximum strain is the maximum stress multiplied by Young's modulus; hence

$$\sigma_{max} = \frac{6L}{H^2W}F \tag{9.61}$$

Example 9.4

Using the same beam as in Example 9.3, the maximum stress at the top surface of the support of the beam is

$$\sigma_{max} = \frac{6(100 \times 10^{-6})}{(2 \times 10^{-6})^3}F = (7.5 \times 10^{13})(\frac{Pa}{N}) \times F \tag{9.62}$$

We can now ask (and answer) a variety of questions about this cantilever based on Examples 9.3 and 9.4. How large a force would be required to produce a beam deflection at the tip of 1 μm? The answer is 0.64 μN. What is the maximum stress at this deflection? The answer is 72.7 MPa. If this force were to come from the gravitational weight of a silicon proof mass attached to the end of the beam, how large a volume would the mass require? The mass would have to be $(0.64 \times 10^{-6})/9.8 = 6.5 \times 10^{-8}$ kg, or 65 μg. With a silicon density of 2330 kg/m^3, the required volume is 2.8×10^7 μm^3. For a 500 μm thick silicon wafer, this would require a square area of about 237 μm on a side, significantly larger than the beam. Other examples are examined in the Problems at the end of the chapter.

9.4   Anticlastic Curvature

When a beam is bent to a radius of curvature ρ, the axial strain varies with position through the beam thickness. Whenever there is an axial strain, we know from the Poisson effect that there must also be a transverse strain. This gives rise to an effect called *anticlastic curvature*, illustrated in Figure 9.12. The axial strain is

$$\epsilon_x = -z/\rho \tag{9.63}$$

and by the Poisson effect, the transverse strain, ϵ_y is then

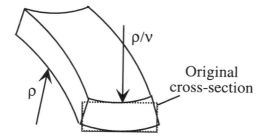

Figure 9.12. Anticlastic curvature. The beam has a radius of curvature ρ along its axial direction, and an anticlastic radius of curvature of magnitude ρ/ν transverse to the beam axis.

$$\epsilon_y = -\nu\epsilon_x \qquad (9.64)$$

Therefore, the transverse strain becomes

$$\epsilon_y = \frac{\nu}{\rho}z \qquad (9.65)$$

which describes pure bending in the opposite direction to the axial bending, with a larger radius of curvature ρ/ν. This effect is usually too small to see in MEMS devices built with slender beams because the radii of curvature tend to be vary large compared with transverse beam dimensions. Nevertheless, when designing elements that must bend, one should check whether or not this anticlastic curvature will affect device behavior.

9.5 Bending of Plates

Our discussion to this point has emphasized beams, with transverse dimensions much smaller than the length. When a structure has transverse dimensions comparable to the length, the elastic behavior must be described with full three-dimensional theories. This is an enormous and mathematically intricate field from which we will extract only a few key results. We begin with the concept of a *plate*.

In contrast to a slender beam, for which we assumed perfectly free Poisson contraction in the transverse direction, a plate is presumed to be wide enough and sufficiently well supported that no transverse Poisson contraction is allowed. Thus, instead of assuming that the transverse stress is zero, as in a bent or axially elongated beam, we will assume that the transverse strain is zero when a plate is stretched. If we assume that a plate has an applied stress σ_x and is constrained such that $\epsilon_y = 0$, the relation between σ_x and ϵ_x must include the effects of the non-zero σ_y required to offset the normal Poisson contraction:

$$\epsilon_x = \frac{\sigma_x - \nu\sigma_y}{E} \qquad (9.66)$$

and

$$\epsilon_y = 0 = \frac{\sigma_y - \nu\sigma_x}{E} \tag{9.67}$$

Combining these equations, we obtain

$$\sigma_x = \left(\frac{E}{1 - \nu^2}\right)\epsilon_x \tag{9.68}$$

The quantity $E/(1 - \nu^2)$ is called the *plate modulus*, and is generally about 10% larger than E.

9.5.1   Plate in Pure Bending

The bending behavior of plates can be understood with a direct extension of what we have already learned about the bending of beams. Instead of a single axial radius of curvature, plates have two principal radii of curvature, expressed in principal axis coordinates for the plate.[4] Assuming the deflection $w(x, y)$ is small, these radii can be written as

$$\frac{1}{\rho_x} = \frac{\partial^2 w}{\partial x^2} \tag{9.69}$$

$$\frac{1}{\rho_y} = \frac{\partial^2 w}{\partial y^2} \tag{9.70}$$

and, in principal axis coordinates, the cross term, $\partial^2 w/\partial x \partial y$ is zero. If we consider a unit width of plate, we can compute the bending moments per unit width in the two principal directions, as follows:

The bending strains are

$$\epsilon_x = -\frac{z}{\rho_x} \quad \text{and} \quad \epsilon_y = -\frac{z}{\rho_y} \tag{9.71}$$

From the constitutive equations, we find

$$\sigma_x = -\frac{Ez}{1 - \nu^2}\left(\frac{1}{\rho_x} + \frac{\nu}{\rho_y}\right) \tag{9.72}$$

$$\sigma_y = -\frac{Ez}{1 - \nu^2}\left(\frac{1}{\rho_y} + \frac{\nu}{\rho_x}\right) \tag{9.73}$$

---

[4] In most MEMS structures, objects with plate-like behavior tend to have rather obvious symmetry directions, so the identification of principal axes is straightforward. The full mathematical description for general structures is found in Timoshenko and Woinowsky-Krieger, *Theory of Plates and Shells* [57].

We can compute moments per unit width by integrating the z-weighted stresses through the thickness of the plate to obtain

$$M'_x = -D\left(\frac{1}{\rho_x} + \frac{\nu}{\rho_y}\right) \tag{9.74}$$

$$M'_y = -D\left(\frac{1}{\rho_y} + \frac{\nu}{\rho_x}\right) \tag{9.75}$$

where D, called the *flexural rigidity* of the plate, is given by

$$D = \frac{1}{12}\left(\frac{EH^3}{1-\nu^2}\right) \tag{9.76}$$

and where H is the thickness of the plate. We note the analogy with the moment of inertia. D is calculated per unit width of plate, hence lacks the factor of W, and substitutes the plate modulus for Young's modulus. Otherwise, the equations are the same.

A special case occurs when the moments in the two principal directions are equal. In this case, the radii of curvature become equal, and we find

$$\frac{1}{\rho_x} = \frac{1}{\rho_y} = -\frac{M'}{D(1+\nu)} \tag{9.77}$$

Equivalently, we can write for the small-deflection case

$$\frac{\partial^2 w}{\partial x^2} = \frac{\partial^2 w}{\partial y^2} = -\frac{M'}{D(1+\nu)} \tag{9.78}$$

which has as a solution

$$w = -\frac{M'(x^2 + y^2)}{2D(1+\nu)} \tag{9.79}$$

which is a paraboloid approximation to a sphere, consistent with our small-displacement assumption that the curvature can be represented by the second derivatives of the displacement without the additional correction terms of Eq. 9.45.

By extension, analogous to what we found for beams, we can write a fourth-order differential equation for the small-amplitude bending of a plate subjected to a two-dimension distributed load $P(x,y)$ as

$$D\left(\frac{\partial^4 w}{\partial x^4} + 2\frac{\partial^4 w}{\partial^2 x \partial^2 y} + \frac{\partial^4 w}{\partial y^4}\right) = P(x,y) \tag{9.80}$$

This equation, together with appropriate boundary conditions along the edges of the plate, can be solved by assuming polynomial solutions in simple cases, or by using eigenfunction expansions for more general cases.

9.6   Effects of Residual Stresses and Stress Gradients

We recall from Section 8.3 that thin films deposited onto substrates can have residual stresses. These stresses can have three important effects on beams. Nonuniform residual stresses in cantilevers, due either to a gradient in the material properties through the cantilever thickness or to the deposition of a different material onto a structure, can cause the cantilevers to curl. In doubly-supported beams, residual stresses modify the bending stiffness, and lead to important nonlinear spring effects when the deflections become comparable to the beam thickness. And for compressive residual stresses, it is possible for the beam to bend out of plane spontaneously, through a mechanism called buckling. We consider these in turn.

9.6.1   Stress Gradients in Cantilevers

Figure 9.13 illustrates a surface-micromachined cantilever beam, on the left still attached to the sacrificial layer, and on the right, fully released. The stress state of the beam in each case is shown beneath the side view. Prior to release, we assume that the beam material has an average compressive stress σ_0, and also a *stress gradient* that results from the deposition, with the bottom part of the film having a more compressive stress than the top part. Once the beam is released, the beam length increases slightly, relieving the compressive stress so that the *average* stress goes to zero; however, the stress gradient is still present, as illustrated in the center of Fig. 9.13. This stress gradient creates a moment that bends the beam, transferring the original stress-gradient-imposed external bending moment into an internal bending moment of the same magnitude.

We assume that, prior to release, the axial stress in the beam is

$$\sigma = \sigma_0 - \frac{\sigma_1}{(H/2)}z \tag{9.81}$$

where the z-axis is directed downward. From the definition of the internal moment

$$M_x = \int_{-H/2}^{H/2} W z \sigma \, dz \tag{9.82}$$

we obtain

$$M_x = -\frac{1}{6}W H^2 \sigma_1 \tag{9.83}$$

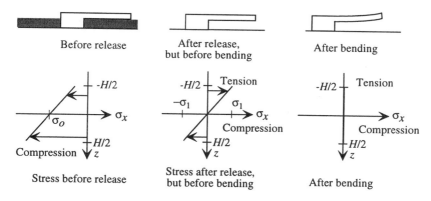

Figure 9.13.   Cantilever with a stress gradient before and after release. A hypothetical intermediate state is shown, after release but before bending, to illustrate the change in bending moment
that results from release. After bending, the stress is zero throughout the beam.

Once the beam is released, it bends, which produces a decrease in the tensile
stress at the top of the beam and simultaneously a decrease in the compressive
stress at the bottom of the beam. The bending creates an internal moment,
yet we know that in equilibrium, the total moment must be zero. The stress
created by bending varies linearly through the thickness of the beam. For the
case of an initial linear residual stress gradient, the stress variation created by
bending exactly cancels the initial stress variation. After bending, the stress is
zero everywhere in the beam, but the beam is bent.

If we neglect transverse Poisson effects (see the following paragraph for how
this is treated), we can calculate that the beam bends to a radius of curvature
ρ_x given by

$$\rho_x = -\frac{1}{12}\frac{EWH^3}{M_x} \qquad (9.84)$$

which, when we substitute for M_x yields

$$\rho_x = \frac{1}{2}\frac{EH}{\sigma_1} \qquad (9.85)$$

If we use this value of ρ_x in Eq. 9.29, we obtain

$$M_x = -\frac{1}{6}WH^2\sigma_1 \qquad (9.86)$$

in exact agreement with the original moment in Eq. 9.83. It all seems rather
magical, but when we examine the strain energy, we will see that this outcome
is exactly what one would expect.

The stress gradient in this example is completely biaxial. In addition to
an x-directed moment, there is also a transverse moment of equal magnitude.

Therefore, for the purposes of finding the radius of curvature, we must actually treat this beam as a plate and include the Poisson effects. One way to understand this reasoning is to note that if we allow free lateral Poisson contraction, the beam would develop an anticlastic curvature. However, because there are identical transverse and axial moments, there are external loads opposing anticlastic curvature in both directions. This has the effect of making the beam stiffer. Hence, we find, substituting into Eq. 9.77 for the case of a bent plate using the moment per unit width as the load,

$$\rho_x = \frac{1}{2} \left(\frac{E}{1 - \nu} \right) \frac{H}{\sigma_1} \tag{9.87}$$

where we recognize the factor $E/(1 - \nu)$ as the biaxial modulus of the beam material.

An obvious application of this result is that the curvature of a cantilever beam can be used to determine the original applied bending moment. The curvature can be measured either by optical profilometry along the beam, or by measuring the tip deflection vs. length for a series of cantilevers.

A more challenging example is a tensile thin film deposited on an originally unstressed beam, as shown in Fig. 9.14. The left hand diagram indicates the stress distribution prior to release. We will assume that the film is very thin compared to the thickness of the cantilever. This greatly simplifies the integrals needed to compute bending moments because we can neglect the small shift in the position of the neutral axis caused by adding the film. However, because the film might be a high-modulus material, we will include the effects of the added stiffness when computing the bending.

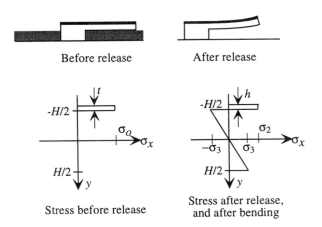

Figure 9.14.   A tensile thin film deposited on a cantilever, then released.

Prior to release, the film has a tensile stress σ_0. After release, two things happen. First, there is a net biaxial contraction of the beam until the average stress is zero. Because the contraction occurs in both directions, we must use the biaxial modulus for both the film and the beam. To keep the algebraic expressions tractable, we use the symbol $\tilde{E}$ for the biaxial modulus.

It can be readily shown that the axial contraction strain is

$$\epsilon_{\text{ax}} = \frac{\sigma_0 h}{\tilde{E}_0 h + \tilde{E}_1 H} \tag{9.88}$$

where $\tilde{E}_0$ is the biaxial modulus of the film, h is the film thickness, and $\tilde{E}_1$ is the biaxial modulus of the beam.

The net tensile stress in the film after this contraction (but before calculating the effects of bending) is

$$\sigma_{0,\text{relaxed}} = \frac{\tilde{E}_1 H \sigma_0}{\tilde{E}_0 h + \tilde{E}_1 H} \tag{9.89}$$

and the corresponding compressive stress in the beam is

$$\sigma_{\text{beam,relaxed}} = -\frac{\tilde{E}_1 h \sigma_0}{\tilde{E}_0 h + \tilde{E}_1 H} \tag{9.90}$$

Since $h \ll H$, the strain is small, and the stress change in the film and the stress level in the beam are also small. For now, we neglect this effect in favor of algebraic simplicity. (In Example 9.5 below, we will calculate the effect of including this relaxation.) Therefore, as an approximation, we will assume that even after release (but before calculating the bending), the film has stress σ_0 and the beam is stress-free.

After release, and after the small axial contraction, there is a net bending moment on the beam created by the tensile film (and the slightly compressed beam). We calculate this moment as

$$M = \int_{-H/2}^{H/2} \sigma_{\text{beam,relaxed}} z \, dz + \int_{-H/2-h}^{-H/2} \sigma_{0,\text{relaxed}} z \, dz \tag{9.91}$$

Because the film is very thin, we can treat the entire film stress as occurring at $z = -H/2$, so that the second integral becomes

$$\sigma_{0,\text{relaxed}} (H/2) h$$

and if we further ignore the small axial relaxation on release, the result simplifies to

$$M = \frac{1}{2}\sigma_0 H h \tag{9.92}$$

Because this is a composite structure, we can no longer separate the Young's modulus effect from the geometric effects, but must perform a integral over the position-dependent modulus, yielding an effective $\tilde{E}I$ product. For this structure

$$\tilde{E}I = \tilde{E}_1 \int_{-H/2}^{H/2} z^2 dz + \tilde{E}_0 h \left(\frac{H}{2}\right)^2 \tag{9.93}$$

which yields

$$\tilde{E}I = \frac{1}{12} H^2 \left(\tilde{E}_1 H + 3\tilde{E}_0 h\right) \tag{9.94}$$

The radius of curvature is then found from

$$\rho = \frac{\tilde{E}I}{M} \tag{9.95}$$

and, using this radius of curvature and the effects of beam bending on stress, we can compute the new stress distribution. In the film, the tensile stress is reduced by the bending to the value shown as σ_2 in Fig. 9.14:

$$\sigma_2 = \sigma_{0,\text{relaxed}} - \frac{\tilde{E}_0 H}{2\rho} \tag{9.96}$$

while the compressive stress in the beam after bending becomes $-[\sigma_3/(H/2)]z$, where (neglecting the effects of the axial contraction)

$$\sigma_3 = -3 \frac{\tilde{E}_1 \sigma_0 h}{H \left(\tilde{E}_1 H + 3\tilde{E}_0 h\right)} \tag{9.97}$$

As a check, we can compute the final moment in the beam using these stress distributions, and we do indeed find that the final moment is zero, in agreement with the requirement of equilibrium.

9.6.2   Residual Stresses in Doubly-Supported Beams

Doubly supported beams with axial constraints at both ends can have net residual stresses, as we already have observed for the case of thermal stresses in a doubly fixed-end beam Whether due to thermal effects , residual stress effects, or external axial loads, these stresses can have a profound effect on beam-bending behavior. Figure 9.15 illustrates a thin beam that has an axial stress σ_0 and has also been bent with a radius of curvature ρ. While the axial stresses produce no net horizontal force, as soon as the beam is bent, there is a

Example 9.5

Consider a 100 μm long by 2 μm square silicon cantilever beam with a fixed end, and deposit a 10 nm film with a tensile stress of 200 MPa and a Young's modulus of 250 GPa onto the top surface of an initially unstressed beam. For simplicity, we assume a Poisson ratio of 0.3 for both materials. By direct substitution, we find that the axial relaxation strain is 6×10^{-6}, which is quite small. After this relaxation, the thin-film stress has reduced to 198 MPa, and the compressive stress in the beam is only 1 MPa. So our neglect of this axial relaxation is justified. If we use these stress values to calculate the initial bending moment and then calculate the resulting radius of curvature, we obtain $\rho = 79$ mm. After bending to this radius of curvature, the thin-film stress has decreased slightly, to 194 MPa, while the maximum stress in the beam (at its two surfaces) has magnitude 1.9 MPa. Thus, it does not take very much bending stress in the beam to compensate for the large load from the tensile film. After release, the cantilever is bent into the arc of a circle. The tip height after bending is $\rho(1 - \cos\theta)$, where θ is the angle subtended by the arclength L, and is given by $\theta = L/\rho$. This angle is 0.0013 radians, and the tip elevation is 64 nm, which could be enough to affect an optical device, but might not affect an accelerometer. However, if the film thickness is increased to 0.1 μm, the radius of curvature drops to 10.1 mm, and the tip elevation is 0.49 μm, or 1/4 of the original beam thickness. The lesson here is that thin-film stress can have a significant effect on the shape of released suspended structures.

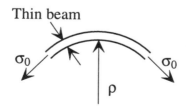

Figure 9.15.   A bent thin beam with axial stress.

net downward force on the beam. The only way to maintain equilibrium is to postulate some kind of upward load on the beam.

The easiest way to analyze this effect is shown in Figure 9.16, in which the thin beam is shown in pure bending through a net angle of π. We postulate

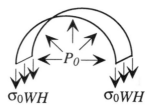

Figure 9.16.   Construction to analyze the effect of residual stress on bending.

a uniformly distributed effective pressure load P_0 acting on the lower surface of the beam so as to cancel the downward force due to the axial tension. The horizontal components of both the axial stress and the uniform pressure load sum to zero. The downward vertical force is $2\sigma_0 W H$, while the net upward force due to the pressure load is equivalent to the pressure times the area of a rectangle of length 2ρ and width W. Hence, the vertical force-balance equation is

$$2\rho W P_0 = 2\sigma_0 W H \qquad (9.98)$$

from which we conclude that

$$P_0 = \frac{\sigma_0 H}{\rho} \qquad (9.99)$$

If we recognize that a uniform pressure load corresponds to a beam load per unit length of $q_0 = P_0 W$, and also replace ρ with the beam curvature, we obtain an equivalent beam load given by

$$q_0 = \sigma_0 W H \frac{d^2 w}{dx^2} \qquad (9.100)$$

where w is now the vertical displacement of the beam. While this example that used a full bend angle of π established the conditions for load balance, once we replace the inverse of the radius of curvature with $d^2 w/dx^2$, we are returning to the case of small displacements and small angles.

With this distributed load needed to compensate the bending when stress is present, the differential equation governing beam bending becomes

$$EI \frac{d^4 w}{dx^4} = q + q_0 \qquad (9.101)$$

where q is some external load, and q_0 is the equivalent load that accounts for the axial stress contribution to the bending stiffness. We can substitute for q_0, and rearrange the equation to yield the famous *Euler beam equation*:

$$EI\frac{d^4w}{dx^4} - (\sigma_0 W H)\frac{d^2w}{dx^2} = q \qquad (9.102)$$

where the quantity in parentheses, $\sigma_0 W H$, has the dimensions of a force, and is called the *tension* in the beam, which we shall denote by N. In effect, the new term works against the d^4w/dx^4 term, making the beam appear stiffer in its response to the external load q.

To illustrate the effect of this stress term, let us find the deflection of a doubly fixed-end beam for small deflections with and without an axial stress component. We will assume that the beam is now loaded with a uniform external load q (equivalent to a pressure load multiplied by the beam width W). The beam equation without the axial stress is

$$EI\frac{d^4w}{dx^4} = q \qquad (9.103)$$

We expect a polynomial solution of fourth order, subject to four boundary conditions:

$$w(0) = 0 \qquad (9.104)$$

$$\left.\frac{dw}{dx}\right|_0 = 0 \qquad (9.105)$$

$$w(L) = 0 \qquad (9.106)$$

$$\left.\frac{dw}{dx}\right|_L = 0 \qquad (9.107)$$

The form of the polynomial is

$$w = Cx^2 + Dx^3 + Fx^4 \qquad (9.108)$$

where the constant and linear terms have already been set to zero in anticipation of having to satisfy the first two boundary conditions. Substitution into the differential equation yields

$$F = \frac{1}{24}\frac{q}{EI} \qquad (9.109)$$

and substitution into the two remaining boundary conditions yields

$$C = \frac{1}{24}\frac{L^2 q}{EI} \qquad (9.110)$$

$$D = -\frac{1}{12}\frac{Lq}{EI} \qquad (9.111)$$

with the final solution being, after substitution

$$w = \frac{x^2 \left(L^2 - 2\,Lx + x^2 \right) q}{2EWH^3} \quad (9.112)$$

The maximum deflection occurs at the middle of the beam ($x = L/2$):

$$w_{\text{max}} = \frac{L^4 q}{32EWH^3} \quad (9.113)$$

The corresponding spring constant of this doubly supported beam is found from the total force on the beam (qL) divided by w_{max}.

$$k_{\text{stress-free}} = \frac{qL}{w_{\text{max}}} = \frac{32EWH^3}{L^3} \quad (9.114)$$

Let us now compare this solution with one for which there is an axial tension in the beam $N = \sigma_0 WH$. We will assume that the deflection of the beam is sufficiently small that the axial stress does not change as a result of the length change of the beam.[5] When the tension is non-zero, the polynomial solution is no longer valid. Instead, we must use hyperbolic functions. Exploiting the symmetry of the beam, we can use only terms that are even about $x = L/2$, and guess a trial solution of the form

$$w = A + C\left(x - \frac{L}{2} \right)^2 + D \cosh\left[k_o \left(x - \frac{L}{2} \right) \right] \quad (9.115)$$

By choosing a symmetric trial solution, two of the four boundary conditions are automatically satisfied. That is, if $w(0)$ is zero, then $w(L)$ must also be zero; similarly for the slope. Substitution into the differential equation yields

$$D\left(\frac{1}{12} EWH^3 k_o{}^4 - N k_o{}^2 \right) \cosh[(k_o\,(x - L/2)] - 2NC = q \quad (9.116)$$

The cosh term can only be satisfied for all x if its coefficient is zero. Solving this for k_o, we obtain

$$k_o = \sqrt{\frac{12N}{EWH^3}} \quad (9.117)$$

while C is given by

---

[5]We shall remove this restriction in the next chapter on Energy Methods.

$$C = -\frac{q}{2N} \tag{9.118}$$

Using these values of k_o and C, and substituting into the two boundary conditions at $x = 0$, then solving, we obtain[6]

$$A = \frac{qL}{4N} \left[\frac{L}{2} - \frac{2}{k_o} \coth\left(k_o L/2\right) \right] \tag{9.119}$$

$$\tag{9.120}$$

$$D = \frac{qL}{2k_o N \sinh\left(k_o L/2\right)} \tag{9.121}$$

The maximum displacement for this case is

$$w_{max} = \frac{qL}{4N} \left(\frac{L}{2} - 2\frac{\cosh(k_o L/2) - 1}{k_o \sinh(k_o L/2)} \right) \tag{9.122}$$

so the corresponding spring constant is

$$k_{with-stress} = \frac{4N}{\dfrac{L}{2} - 2\dfrac{\cosh(k_o L/2) - 1}{k_o \sinh(k_o L/2)}} \tag{9.123}$$

In order to visualize the effect of an axial tensile stress on the stiffening of a beam, the displacement versus position has been plotted in Figure 9.17 for various values of tensile stress, using our standard 100 μm long beam with a 2×2 μm cross section and a Young's modulus of 160 GPa as an example. The load q has been chosen so that the stress-free beam bends by 1/2 of its thickness. It is seen that even a stress as small as 10 MPa can change the mid-point deflection by 5%, and at a stress of 100 MPa, the deflection is reduced to two-thirds its stress-free value. If the beam were longer, this effect would be even more pronounced.

The conclusion is that when designing fixed-end beams as part of a structure, stress control is critical to achieving the desired performance. And, as we have already seen for cantilevers, deposited films can have a large effect. For example, a deposited 10 nm film with a stress of 200 MPa can shift the average stress of a 2 μm thick fixed-end beam by 10 MPa.

9.6.3   Buckling of Beams

In this section, we explore the effects of compressive stresses on doubly-fixed-end beams. We show that for a sufficiently large compressive stress, the

---

[6]This is the kind of problem where a symbolic mathematics package is extremely useful. The solution presented here was done on Maple 5, and then checked by resubstitution into all equations and boundary conditions.

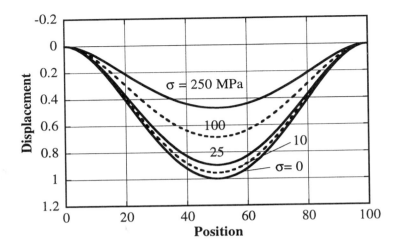

Figure 9.17. Displacement of a $100 \times 2 \times 2\ \mu$m beam with a uniform load for different values of axial tensile stress. Higher stress leads to higher stiffness, hence, to an increased spring constant.

equilibrium position of the beam is no longer a straight position, but a bent one. This phenomenon is called *buckling*.

We shall assume a point load F at the mid-point of a beam with two clamped ends, using Eq. 9.102. We change coordinates so that the midpoint of the beam is in the center and the beam extends from $-L/2$ to $L/2$. We require that the displacement and slope be zero at both ends. The beam equation in this case is

$$EI\frac{d^4 w}{dx^4} - N\frac{d^2 w}{dx^2} = F\delta(x) \qquad (9.124)$$

where N is the tension (equal to $\sigma W H$, where σ is the axial stress), and $\delta(x)$ is the unit impulse at $x = 0$. We shall use this example to illustrate an *eigenfunction solution*. We define the eigenfunctions $\psi_n(x)$ as the solutions to

$$EI\frac{d^4 \psi_n}{dx^4} - N\frac{d^2 \psi_n}{dx^2} = \lambda_n \psi_n \qquad (9.125)$$

where λ_n is the *eigenvalue* of the eigenfunction ψ_n, and n is an index for the various eigenfunctions. Since this is a homogeneous second-order linear differential equation, we know to expect exponentials as solutions. It is readily shown by direct substitution that the eigenfunctions are of the form

$$\psi_n(x) = \sqrt{\frac{2}{L}} \cos(k_n x) \qquad (9.126)$$

where, in order to satisfy the differential equation, we require

$$\lambda_n = EIk_n^4 + Nk_n^2 \tag{9.127}$$

The $\sqrt{(2/L)}$ term is an optional normalization factor, to be explained later.

To determine the values of k_n, we must consider the boundary conditions, and in order to do that, we must actually construct a complete trial solution. We use an eigenfunction expansion plus a solution of the homogeneous solution (*i.e.*, no load):

$$w(x) = A + Bx + \sum_{n=1}^{\infty} C_n \psi_n(x) \tag{9.128}$$

where the $A + Bx$ terms represent the homogeneous solution. In order to satisfy the symmetry condition, B must be zero. And, for each term in the sum, we can identify a constant A_n such that

$$A = \sum_{n=1}^{\infty} A_n \tag{9.129}$$

or, equivalently

$$w(x) = \sum_{n=1}^{\infty} (A_n + C_n \psi_n(x)) \tag{9.130}$$

With this formulation, it is clear that the boundary conditions can be satisfied provided

$$A_n + C_n \psi_n(L/2) = 0 \tag{9.131}$$

$$C_n \left. \frac{d\psi_n}{dx} \right|_{x=L/2} = 0 \tag{9.132}$$

The first boundary condition is trivially satisfied by choosing

$$A_n = -C_n \sqrt{2/L} \cos(nL/2) \tag{9.133}$$

The second boundary condition imposes a restriction on k_n. We require that for any non-zero C_n,

$$k_n \sin\left(\frac{k_n L}{2}\right) = 0 \tag{9.134}$$

If k_n is to be non-zero, then k_n must satisfy

$$\frac{k_n L}{2} = n\pi \tag{9.135}$$

for positive integers n, which yields

$$k_n = \frac{2n\pi}{L} \tag{9.136}$$

Thus the complete solution is of the form

$$w(x) = \sum_{n=1}^{\infty} \sqrt{\frac{2}{L}} C_n \left[\cos\left(\frac{2n\pi x}{L}\right) - (-1)^n \right] \tag{9.137}$$

where we have used the result that $\cos(n\pi) = (-1)^n$.

The question now becomes how to find C_n for the specific applied load. Here we use the *orthonormality* of the eigenfunctions. With our choice of $\sqrt{2/L}$ as a prefactor, then every eigenfunction is *normalized*:

$$\int_{-L/2}^{L/2} |\psi_n(x)|^2 \, dx = 1 \tag{9.138}$$

Further, for any two eigenfunctions ψ_n and ψ_m with $n \neq m$,

$$\int_{-L/2}^{L/2} \psi_n(x)\psi_m(x) \, dx = 0 \tag{9.139}$$

Thus, the eigenfunctions are orthogonal to each other. We exploit this property to find the C_n. After substitution into the differential equation, we obtain

$$\sum_{n=1}^{\infty} \sqrt{\frac{2}{L}} \left(EIk_n^4 + Nk_n^2\right) C_n \left(\cos(k_n x) - (-1)^n\right) = F\delta(x) \tag{9.140}$$

If we multiply this equation by some arbitrary $\psi_m(x)$, integrate from $x = -L/2$ to $L/2$, and use the orthonormality properties of the eigenfunctions, we obtain

$$\left(EIk_n^4 + Nk_n^2\right) C_n = \int_{-L/2}^{L/2} \sqrt{\frac{2}{L}} \cos(k_n x) F\delta(x) \, dx \tag{9.141}$$

which leads to

$$C_n = \sqrt{\frac{2}{L}} \frac{F}{\left(EIk_n^4 + Nk_n^2\right)} \tag{9.142}$$

The complete solution becomes

$$w(x) = \sum_{n=1}^{\infty} \sqrt{\frac{2}{L}} \frac{F}{\left(EIk_n^4 + Nk_n^2\right)} \left[\cos\left(\frac{2n\pi x}{L}\right) - (-1)^n \right] \tag{9.143}$$

The maximum deflection occurs at $x = 0$:

$$w_{max} = \sum_{n=1}^{\infty} \sqrt{\frac{2}{L}} \frac{F}{(EIk_n^4 + Nk_n^2)} (1 - (-1)^n) \qquad (9.144)$$

which reduces to

$$w_{max} = \sum_{n=1,\text{odd}}^{\infty} \sqrt{\frac{2}{L}} \frac{2F}{(EIk_n^4 + Nk_n^2)} \qquad (9.145)$$

Using this solution, the spring constant becomes

$$k_{\text{beam}} = \frac{1}{\sum_{n=1,\text{odd}}^{\infty} \sqrt{\frac{2}{L}} \frac{2}{\left(EIk_n^4 + Nk_n^2\right)}} \qquad (9.146)$$

Let us now examine this result. First, as was already clear from our discussion of axial stress, a positive value of N makes the spring constant larger. But what we can now see for is that for negative values of N, the denominator of the first term in the sum will vanish when

$$N \to -\frac{\pi^2}{3} \frac{EWH^3}{L^2} \qquad (9.147)$$

corresponding to a critical value of stress, called the *Euler buckling limit*, given by

$$\sigma_{\text{Euler}} = -\frac{\pi^2}{3} \frac{EH^2}{L^2} \qquad (9.148)$$

When this value of compressive stress is reached, the first term in the sum diverges, driving the spring constant to zero. Thus the lateral stiffness of the beam goes to zero, and the beam can spontaneously assume the cosine shape of the first eigenfunction. This is called *first-order buckling*. The compressive axial load cannot be supported without the beam bending into a cosine shape.

This analysis was done for perfectly clamped beam ends. However, in real structures, the beam supports tend to be somewhat compliant. The exact details of the support compliance have a strong effect on the details of buckling-type behavior, and the mathematical analysis can get quite intricate. The important lesson, for now, is that compressive stress can lead to buckling, and the physical origin of this buckling is the vanishing of the lateral stiffness at a critical level of stress.

9.7 Plates With In-Plane Stress

We have already demonstrated that axial stresses have a significant effect on the bending behavior of beams. A similar effect occurs when there are stresses in the plane of a plate. The differential equation corresponding to Eq. 9.102 is

$$D \left(\frac{\partial^4 w}{\partial x^4} + 2 \frac{\partial^4 w}{\partial^2 x \partial^2 y} + \frac{\partial^4 w}{\partial y^4} \right) - \left(N_x' \frac{\partial^2 w}{\partial x^2} + N_y' \frac{\partial^2 w}{\partial y^2} \right) = P \qquad (9.149)$$

where N_x' and N_y' are the in-plane tension per unit width in the principal-axis x- and y-directions, respectively, and P is a distributed pressure load. If a plate has a residual biaxial tensile stress σ, then

$$N_x' = N_y' = \sigma H \qquad (9.150)$$

where H is the plate thickness. In direct analogy to what we found for beams, a tensile in-plane stress increases the bending stiffness of the plate, while a compressive in-plane stress reduces the stiffness, and can lead to buckling behavior.

An important subset of plates with in-plane stress are very thin plates, so thin, in fact, that the flexural rigidity (which goes as the cube of H) can be considered negligibly small compared to the rigidity that results from the in-plane stress terms. In this case, the plate is called a *membrane*, and is described by the much simpler differential equation

$$N_x' \frac{\partial^2 w}{\partial x^2} + N_y' \frac{\partial^2 w}{\partial y^2} = -P \qquad (9.151)$$

For the particular case of biaxial tension, then $N_x' = N_y' = N'$, and this equation becomes

$$\frac{\partial^2 w}{\partial x^2} + \frac{\partial^2 w}{\partial y^2} = -\frac{P}{N'} \qquad (9.152)$$

Pressure sensors that involve diaphragm deflection are often thin enough to be considered as membranes, at least to a first approximation. As an example, we find the deflection of a square membrane of length L on a side, subject to the boundary conditions that the deflection is zero at the boundary and that the moment is zero at the boundary (since the membrane has no bending stiffness to withstand a moment). The differential equation is separable, so we use trial solutions that factor

$$w = \sin(k_x x) \sin(k_y y) \qquad (9.153)$$

where, in order that the solution vanish at $x = L$ and also at $y = L$, we require

$$k_x = \frac{n_x \pi}{L} \quad \text{and} \quad k_x = \frac{n_y \pi}{L} \qquad (9.154)$$

for positive integers n_x and n_y. We construct a complete trial solution by superposition:

$$w = \sum_{n_x, n_y} C_{n_x n_y} \sin\left(\frac{n_x \pi x}{L}\right) \sin\left(\frac{n_y \pi y}{L}\right) \quad (9.155)$$

Substituting into the differential equation, we find that

$$- \sum_{n_x, n_y} C_{n_x n_y} (k_x^2 + k_y^2) \sin(k_x x) \sin(k_y y) = -\frac{P}{N'} \quad (9.156)$$

If we multiply by $\sin(k_x x) \sin(k_y y)$ and integrate from 0 to L, we obtain

$$\frac{L^2}{4}(k_x^2 + k_y^2) C_{n_x n_y} = -\frac{P}{N'} \frac{L^2 (\cos(n_x \pi) - 1)(\cos(n_y \pi) - 1)}{n_x n_y \pi^2} \quad (9.157)$$

from which we conclude that

$$C_{n_x n_y} = -\frac{16P}{N'(k_x^2 + k_y^2)} = -\frac{16 L^2 P}{\pi^2 N' \left(n_x^2 + n_y^2\right)} \quad (9.158)$$

for both n_x and n_y odd, leading to the final solution

$$w = -\frac{16 L^2 P}{\pi^2 N'} \sum_{n_x, n_y \text{odd}} \frac{\sin\left(\frac{n_x \pi x}{L}\right) \sin\left(\frac{n_y \pi y}{L}\right)}{(n_x^2 + n_y^2)} \quad (9.159)$$

The maximum deflection occurs in the center of the membrane. We find

$$w_{\max} = -\frac{16 L^2 P}{\pi^2 N'} \sum_{n_x, n_y \text{odd}} \frac{\sin\left(\frac{n_x \pi}{2}\right) \sin\left(\frac{n_y \pi}{2}\right)}{(n_x^2 + n_y^2)} \quad (9.160)$$

Evaluating the first few terms of the sum, we obtain

$$w_{\max} = -.47 \frac{16 L^2 P}{\pi^2 N'} \quad (9.161)$$

We see from this example that the spring constant, which is $PL^2/w_{\max}$, is proportional to the tension.

9.8 What about large deflections?

What happens to all these results if the loads on clamped beams and membranes are sufficient to create a non-negligible change in the in-plane stress or tension? As one might expect, the structures become nonlinear elastic elements, with an incremental stiffness that increases with the load. In order to analyze these examples, we will use some approximate methods based on energy considerations. These are the subject of the following Chapter.

Related Reading

J. M. Gere and S. P. Timoshenko, *Mechanics of Materials*, Second Edition, Monterey, CA: Brooks/Cole Engineering Division, 1984.

W. M. Lai, D. Rubin, and E. Krempl, *Introduction to Continuum Mechanics*, Third Edition, Oxford: Pergamon Press, 1993.

S. P. Timoshenko and J. N. Goodier, *Theory of Elasticity*, Third Edition, New York: McGraw-Hill, 1970; reissued in 1987.

S. P. Timoshenko and S. Woinowsky-Krieger, *Theory of Plates and Shells*, New York: McGraw-Hill, 1959; reissued in 1987.

Problems

9.1 Assume that silicon will fracture when the axial stress reaches ∼1 GPa. Find the maximum length of a vertical silicon beam which, under the action of its own gravitational load, will not exceed this fracture stress.

9.2 A beam has a circular cross section but a conical shape, with a diameter that changes linearly from 100 μm at its support to 500 μm at its free end. The beam length is 1 mm. Calculate the axial stiffness of this beam. What value of constant cross-section would yield the same axial stiffness?

9.3 A silicon cantilever of length 500 μm, width 50 μm, and thickness 2 μm is subjected to a uniform distributed transverse load q_o. Find the tip deflection, and calculate an effective spring constant for this beam. For a load which produces a tip deflection of 2 μm, calculate the maximum stress at the support.

9.4 Derive Eq. 9.88.

9.5 Determine the temperature at which an aluminum beam of length 500 μm and a square 2 × 2 μm cross-section, when held between fixed ends, will buckle. (Since the thermal expansion coefficient of aluminum is much bigger than that of quartz or glass, this problem is a first-order approximation to what happens when surface micromachined aluminum structures are supported on such a substrate.)

9.6 Using the membrane analysis of this chapter, determine the size of an LPCVD silicon-nitride membrane (suspended on a micromachined silicon support) of thickness 1 μm such that it will have a maximum displacement of 1 μm at a differential pressure of one atmosphere.

Chapter 10

ENERGY METHODS

Energy is Eternal Delight.

—William Blake in *The Marriage of Heaven and Hell*

When examining lumped-element device models in Chapters 5 and 6, we had occasion to draw on some basic energy and co-energy concepts to analyze devices in both the electrostatic and magnetic domains. In particular, we noted that for a capacitor with a moveable electrode, the force F on that electrode was obtained from

$$F = \left. \frac{\partial \mathcal{W}(Q, x)}{\partial x} \right|_Q \tag{10.1}$$

where $\mathcal{W}(Q, x)$ is the stored energy on the capacitor, given by

$$\mathcal{W}(Q, x) = \frac{Q^2}{2C} \tag{10.2}$$

where C is the position-dependent capacitance of the device. Alternatively, we could use the co-energy,

$$\mathcal{W}^*(V, x) = \frac{1}{2} C V^2 \tag{10.3}$$

and find the force from

$$F = - \left. \frac{\partial \mathcal{W}^*(V, x)}{\partial x} \right|_V \tag{10.4}$$

Both formulations lead to the same expression for the force. The choice of which energy function to use is entirely a matter of convenience when formulating the problem.

We also used a co-energy formulation when analyzing a magnetic actuator. In that case, we preferred the co-energy, because in the electric domain, the co-energy is formulated in terms of the coil current I which is proportional to the magnetomotive force effort variable, typically a more convenient variable to work with in practical circuits than the total flux linkage ϕ. In that case, the co-energy for a single-coil device was

$$W^*(I, x) = \frac{1}{2} L I^2 \qquad (10.5)$$

and for the force on the moveable permeable material in the gap of the inductor, we obtained

$$F = - \left. \frac{\partial W^*(I, x)}{\partial x} \right|_I \qquad (10.6)$$

We now revisit this subject, for several reasons. First, we have just learned how to calculate the deformation of a variety of elastic structures subject to some rather strict limitations on the amount of allowed deformation. We will discover below that we can extend our analysis to include large-amplitude deformations by using energy methods and some approximations that arise from these methods. This will lead us to understand nonlinear elastic elements and to develop to a much richer set of phenomena with which to build useful devices. Second, we shall, along the way, learn a remarkably powerful approach to using numerical simulation results to create compact but very accurate reduced-order models of complex device behavior. Finally, by extension, we will examine a method for estimating the resonant frequency of dynamical device structures which facilitates the determination of equivalent lumped-element masses for deformable structural elements.

10.1   Elastic Energy

In both the electrostatic and magnetic lumped-element examples we encountered in Chapters 5 and 6, the stored energy was defined in terms of the effort and displacement variables of a lumped element, which are both scalar quantities. In terms of a single scalar displacement q and scalar effort e, the stored energy at displacement value q_1 is

$$W(q_1) = \int_0^{q_1} e(q) \, dq \qquad (10.7)$$

We now extend this concept to a distributed system by defining an *energy density* $\tilde{W}(x, y, z)$, defined such that the total energy W stored in a small volume element of dimensions Δx, Δy, and Δz is

$$W_{\Delta x, \Delta y, \Delta z} = \tilde{W}(x, y, z) \, \Delta x \Delta y \Delta z \qquad (10.8)$$

The concept of an energy density can be applied to many systems that store potential energy. We will begin with elastic deformation, then return to other examples.

Stored energy for a lumped element is the integral of an effort over a displacement. When differential efforts and displacements can be defined locally, then we can find the energy density by the same kind of integral of an effort over a displacement. In the distributed elastic domain, stress is a generalized effort and strain is a generalized displacement. Stress has dimensions N/m^2, while strain is dimensionless. Therefore, the stress-strain product has dimensions of N/m^2, which is equivalent to N-m/m^3, or J/m^3. Thus the stress-strain product has dimensions of energy per unit volume, consistent with an energy density.

The *strain energy density*, in exact analogy to lumped stored energy, is defined as

$$\tilde{W}(x, y, z) = \int_0^{\epsilon(x,y,z)} \sigma(\epsilon) d\epsilon \tag{10.9}$$

where $\epsilon(x, y, z)$ is the value of the strain at (x, y, z), and $\sigma(\epsilon)$ is the function that relates stress to strain at position (x, y, z). Note that this is an integral over the magnitude of the *strain*, not the position, just as when charging a capacitor, the integral is over the increment of charge. For a linear isotropic elastic material with axial strain ϵ,

$$\sigma(\epsilon) = E\epsilon \tag{10.10}$$

Therefore, the strain energy density is

$$\tilde{W}(x, y, z) = \int_0^{\epsilon(x,y,z)} E\epsilon d\epsilon = \frac{1}{2} E \left[\epsilon(x, y, z) \right]^2 = \frac{1}{2} \sigma \epsilon \tag{10.11}$$

The situation for shear strains is the same. The energy density due a shear strain γ_{xy} which develops a shear stress τ_{xy} is

$$\tilde{W} = \frac{1}{2} \tau_{xy} \gamma_{xy} \tag{10.12}$$

We can build on these results to find the *total strain energy* in a deformed body. It is

$$W = \frac{1}{2} \int \int \int \left(\sigma_x \epsilon_x + \sigma_y \epsilon_y + \sigma_z \epsilon_z + \tau_{yz} \gamma_{yz} + \tau_{zx} \gamma_{zx} + \tau_{xy} \gamma_{xy} \right) dx dy dz \tag{10.13}$$

where the integral goes over the entire volume of the elastic body.

A similar set of definitions can be used for distributed electric systems in terms of the fields. With the electric field $\mathcal{E}$ as the effort and the electric displacement $\mathcal{D}$ as the generalized displacement (equal to $\varepsilon_e\mathcal{E}$ in a linear medium[1]), leading to

$$\tilde{\mathcal{W}}(\mathcal{D}) = \frac{\mathcal{D}^2}{2\varepsilon_e} \tag{10.14}$$

To show that this corresponds to our earlier lumped-element approach, we note that for a parallel plate capacitor with voltage V, the fields between the plates are

$$\mathcal{E} = \frac{V}{g} \quad \text{and} \quad \mathcal{D} = \varepsilon_e\mathcal{E} \tag{10.15}$$

where g is the gap. The charge on the capacitor is

$$Q = \frac{\varepsilon_e AV}{g} \tag{10.16}$$

so the lumped-element stored energy is

$$W = \frac{\varepsilon_e AV^2}{2g} \tag{10.17}$$

The field-based energy density in the gap, with fringing fields neglected, is

$$\tilde{\mathcal{W}}(\mathcal{D}) = \frac{1}{2\varepsilon_e}\frac{\varepsilon_e^2 V^2}{g^2} \tag{10.18}$$

This is a constant with respect to position, so we can readily integrate it over the volume of the gap to yield

$$W = \frac{\varepsilon_e V^2}{2g^2} \times gA = \frac{\varepsilon_e AV^2}{2g} \tag{10.19}$$

which is identical to the lumped-element energy.

In distributed magnetic systems, the effort is the $\mathcal{H}$ field, and the generalized displacement is the flux density $\mathcal{B}$. Thus for distributed magnetic systems with linear permeability μ,

$$\tilde{\mathcal{W}}(\mathcal{B}) = \frac{\mathcal{B}^2}{2\mu} \tag{10.20}$$

We now use these energy-density concepts to develop an extremely general and powerful approach to analyzing devices that conserve energy.

---

[1]To avoid notational confusion in this section between strain ϵ and permittivity ε, we add a subscript e to the dielectric permittivity.

10.2   The Principle of Virtual Work

In systems that conserve energy, such as elasticity and electromagnetism, extremely useful insights can be gained from some fundamental properties of the total potential energy of a system in equilibrium. These properties are captured in a concept called the *Principle of Virtual Work.*[2] Simply stated, it says that if a body, initially in equilibrium with zero applied loads, is subjected to a combination of surface forces $\mathbf{F}_s$ and body forces (such as gravity) $\mathbf{F}_b$ and quasi-statically deforms and displaces as a result of those forces, then in equilibrium, the total work done by the external forces in creating the displacement and deformation must equal the internal stored energy in the deformed and displaced state.

This way of stating the principle sounds just like the principle that energy is conserved, since with only quasi-static displacements and deformations to consider, there is no kinetic energy in the system. But the concept is more powerful than that. In effect, it says that one can vary the functions that describe the displacements and deformations arbitrarily, and search for a minimum in the quantity $\mathcal{U}$, given by

$$\mathcal{U} = \text{Stored energy} \quad - \quad \text{Work done} \tag{10.21}$$

Stated more mathematically, if $\delta u(x, y, z)$, $\delta v(x, y, z)$, and $\delta w(x, y, z)$ constitute the x-, y- and z-components of a vector field of imagined or "virtual" displacements of a body from its equilibrium condition, then we can find the virtual strains associated with the virtual displacements, for example, from

$$\delta \epsilon_x = \frac{\partial}{\partial x} \delta u, \quad \ldots, \quad \delta \gamma_{xy} = \frac{1}{2} \left(\frac{\partial}{\partial x} \delta v + \frac{\partial}{\partial y} \delta u \right) \tag{10.22}$$

and the corresponding virtual change in strain energy density is

$$\delta \tilde{\mathcal{W}} = \sigma_x \delta \epsilon_x + \sigma_y \delta \epsilon_y + \sigma_z \delta \epsilon z + \tau_{yz} \delta \gamma_{yz} + \tau_{zx} \delta \gamma_{zx} + \tau_{xy} \delta \gamma_{xy} \tag{10.23}$$

The Principle of Virtual Work states that for any virtual displacement that is compatible with the boundary conditions, because the quantity $\mathcal{U}$ must be at a minimum,

$$\int_{\text{volume}} \delta \tilde{\mathcal{W}} dx dy dz - \int_{\text{surface}} (F_{s,x} \delta u + F_{s,y} \delta v + F_{s,z} \delta w) dS$$
$$+ \int_{\text{volume}} (F_{b,x} \delta u + F_{b,y} \delta v + F_{b,z} \delta w) dx dy dz = 0 \tag{10.24}$$

or, equivalently

---

[2] See Timoshenko and Goodier, *Theory of Elasticity*, pp. 250-252 [48].

$$\delta \left[\int_{\text{volume}} \tilde{\mathcal{W}} dx dy dz - \int_{\text{surface}} (F_{s,x} u + F_{s,y}\, v + F_{s,z} w) dS \right.$$

$$\left. - \int_{\text{volume}} (F_{b,x} u + F_{b,y} v + F_{b,z} w) dx dy dz \right] = 0 \tag{10.25}$$

where it is understood that the differential operator δ acts only on the displacement fields u, v, and w and and their associated strains, not on the external forces. The first integral is the total strain energy. The second integral is the negative of the total work done by the surface forces, while the third integral is the negative of the total work done by the body forces. We recognize the second and third integrals as potential energy functions, in that their negative gradient with respect to the displacements gives back the forces. Thus, the bracketed term is the total potential energy of the system $\mathcal{U}$.

$$\mathcal{U} = \int_{\text{volume}} \tilde{\mathcal{W}} dx dy dz - \int_{\text{surface}} (F_{s,x} u + F_{s,y}\, v + F_{s,z} w) dS$$

$$- \int_{\text{volume}} (F_{b,x} u + F_{b,y} v + F_{b,z} w) dx dy dz \tag{10.26}$$

The Principle of Virtual Work says that, in static equilibrium, the total potential energy is stationary with respect to any virtual displacement.

10.3   Variational Methods

We will now show how the Principle of Virtual Work can be used in an incredibly powerful array of applications by using approximations (even rather poor guesses) for the displacement fields, calculating the total potential energy, and then varying the approximate or guessed solution to find the minimum potential energy.

We will formulate the method rather abstractly, then examine some examples. Let us define trial displacement functions $\hat{u}(x, y, z; c_1, c_2, \ldots c_n)$, $\hat{v}(x, y, z; c_1, c_2, \ldots c_n)$, and $\hat{w}(x, y, z; c_1, c_2, \ldots c_n)$, where $c_1, c_2, \ldots c_n$ are a set of n parameters that appear in the trial functions. For a given set of external surface and body forces, we can use Eq. 10.26 and find the total potential energy of the system. Clearly, it will depend on the values of the parameters $c_1, c_2, \ldots c_n$. But since $\mathcal{U}$ must be stationary with respect to any virtual displacement, we can require that

$$\frac{\partial \mathcal{U}}{\partial c_j} = 0 \tag{10.27}$$

for every c_j. This leads to a set of n simultaneous equations:

$$\frac{\partial \mathcal{U}}{\partial c_1} = 0$$

$$\vdots \qquad \vdots$$

$$\frac{\partial \mathcal{U}}{\partial c_j} = 0 \qquad\qquad (10.28)$$

$$\vdots \qquad \vdots$$

$$\frac{\partial \mathcal{U}}{\partial c_n} = 0$$

Solving these n equations yields the n values of the c_j that minimize the total potential energy and, hence, represent the best approximation to the equilibrium displacements we can make with the set of trial functions we started with.

The above analysis plays a central role in simulation and modeling. For example, this variational approach is the basis of all *finite-element numerical methods*, where for a meshed structure, the variational parameters are the nodal displacements of the various mesh elements.[3] The set of variational equations is formulated with a set of local basis functions that interpolate between the nodal positions so that the strain energy and work done by body forces can be calculated throughout the volume. The result is a set of $3N$ simultaneous equations, where N is the number of nodes in the mesh. While we will not be examining the theory of finite-element methods in this book, the reader should be aware of the relation between finite-element methods using meshes and the general variational methods presented here.

We will now use variational methods in conjunction with the Principle of Virtual Work to find an approximate small-displacement solution to the problem of a doubly-clamped beam of length L with a fixed load F at some position x_o along its length, not necessarily at the center. The boundary conditions are that the displacement and slope be zero both at $x = 0$ and $x = L$. We will try a polynomial trial solution:

$$\hat{w}(x) = c_0 + c_1 x + c_2 x^2 + c_3 x^3 + c_4 x^4 \qquad (10.29)$$

From the boundary conditions, we can immediately conclude that

$$c_0 = c_1 = 0 \qquad\qquad (10.30)$$

and we can further require, from the boundary conditions at $x = L$, that

$$c_2 L^2 + c_3 L^3 + c_4 L^4 = 0 \qquad\qquad (10.31)$$
$$2c_2 L + 3c_3 L^2 + 4c_4 L^3 = 0 \qquad\qquad (10.32)$$

---

[3]For a discussion of finite-element methods, see Bathe [58].

These equations constrain two of the three constants. We can elect to eliminate c_2 and c_3, using

$$c_2 = c_4 L^2 \tag{10.33}$$

$$c_3 = -2c_4 L \tag{10.34}$$

to yield

$$\hat{w} = c_4 \left(L^2 x^2 - 2Lx^3 + x^4 \right) \tag{10.35}$$

We now have a trial solution that is consistent with the boundary conditions. It has one parameter c_4, which we will find by calculating the total potential energy, then minimizing with respect to c_4. Since we assume small displacements, we can neglect axial tension and find the strain energy due to bending alone. The radius of curvature is

$$\frac{1}{\rho} = \frac{d^2\hat{w}}{dx^2} = c_4 \left(2L^2 - 12Lx + 12x^2 \right) \tag{10.36}$$

The bending strain is

$$\epsilon = -\frac{z}{\rho} = -c_4 z \left(2L^2 - 12Lx + 12x^2 \right) \tag{10.37}$$

The total strain energy is found from

$$W = \frac{EW}{2} \int_0^L \int_{-H/2}^{H/2} \epsilon^2 dx dz \tag{10.38}$$

which reduces to

$$W = \frac{4}{5} EW H^3 L^5 c_4^2 \tag{10.39}$$

In order to find the potential energy associated with the external force F, we need to find $\hat{w}(x_o)$ the displacement of the beam at position $x = x_o$. The work done by force F in displacing through a distance $\hat{w}(x_o)$ is then $\hat{w}(x_o)F$. Substituting $x = x_o$ into Eq. 10.35 yields

$$\hat{w}(x_o) = c_4 \left(L^2 x_o^2 - 2Lx_o^3 + x_o^4 \right) \tag{10.40}$$

Therefore, the total potential energy, expressed with c_4 as a parameter, is

$$\mathcal{U} = \frac{4}{5} EW H^3 L^5 c_4^2 - \left(L^2 x_o^2 - 2Lx_o^3 + x_o^4 \right) F c_4 \tag{10.41}$$

To find the value of c_4 that minimizes $\mathcal{U}$, we solve

$$\frac{\partial \mathcal{U}}{\partial c_4} = 0 \qquad (10.42)$$

to obtain

$$c_4 = \frac{5}{8} \frac{x_o^2 \left(L^2 - 2Lx_o + x_o^2 \right)}{EWH^3L^5} F \qquad (10.43)$$

Thus, the final solution starting from this particular trial solution is

$$w = \frac{5}{8} \frac{\left(L^2 x^2 - 2Lx^3 + x^4 \right) x_o^2 \left(L^2 - 2Lx_o + x_o^2 \right)}{EWH^3L^5} F \qquad (10.44)$$

10.3.1   Properties of the Variational Solution

We can now ask several interesting questions about this variational solution:
  The first question: Does the trial solution satisfy the beam-bending differential equation? The answer is no. The point load F is represented by an impulse at $x = x_o$. The trial solution is a continuous fourth-order polynomial which cannot have the third-derivative discontinuity required to produce an impulse in the fourth derivative. Thus, we are clearly dealing with an approximate solution.

The second question: Is the point of maximum displacement located at x_o? Again, the answer is no. It is trivial to show that the maximum displacement for this trial solution occurs at the middle of the beam, $x = L/2$. The reason for this is subtle. As we have seen, the fourth-order polynomial solution has only one degree of freedom, the value of c_4, which determines the overall amplitude of the deformation in response to the applied load. There is no parameter available to affect the *shape*. The shape was fixed by the selection of the trial function, and the selected shape has its maximum deflection in the center of the beam, regardless of where the load is actually applied.

The third question: How can we determine how accurate the variational solution is? The answer here is discouraging. The only way to determine the accuracy is to try a different trial solution with more degrees of freedom, and see whether the answer changes. If it does, then the variational method has not yet converged to a close approximation to the true equilibrium state. In finite-element methods, adding degrees of freedom means using a finer mesh, so this process is called *mesh refinement*. One makes an increasingly fine mesh, and tests whether the resulting displacements have changed significantly. In more general variational methods, we select a more complex trial function and derive a new solution.

As it turns out, our fourth-degree polynomial trial solution is a *very bad trial solution*, precisely because it constrains the shape of the beam, even for an off-center load. If, instead, we use a trial solution of the form

$$\hat{w} = c_2 x^2 + c_3 x^3 + c_4 x^4 + c_5 x^5 \qquad (10.45)$$

and go through the exact same procedure, we get a much better solution. At the minimization step, there are two free parameters, c_4 and c_5. Thus, there are two simultaneous equations from which to determine their values. Because of the x^5 term, it is now possible for the point of maximum deflection to deviate from the center of the beam. A comparison of the two solutions is shown in Fig. 10.1.[4]

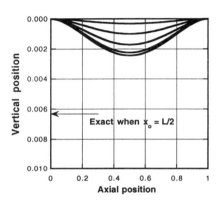 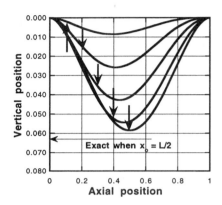

Figure 10.1. Comparison of the fourth-degree (on the left) and fifth-degree (on the right) polynomial trial solution for the point-loaded beam. The horizontal arrow shows the exact solution for the case where the point load is applied at the center of the beam. The vertical arrows in the right-hand figure show the points at which the loads are applied for the various curves. They are also the loading points in the corresponding curves in the left-hand graph.

Several things are immediately obvious from Fig. 10.1. First, the point of maximum deflection is always in the middle for the fourth-degree polynomial trial function, but shifts with loading point for the fifth-degree polynomial trial function. Also, the fifth-degree solution gets much closer to the exact solution. The horizontal arrow shows the exact solution for the maximum deflection when the point load is at the center of the beam. The fourth-degree polynomial, by comparison, does a very poor job.

We can extract several important conclusions from this example. First, variational solutions are relatively easy to produce, provided one is supported with good symbolic math tools to prevent algebraic disasters. Second, it is important to ensure that the trial solution have enough degrees of freedom to capture the essential behavior of the problem. It is clear in hindsight that the fourth-degree polynomial had only one undetermined coefficient, and that was

---

[4]The solution of the variational problem with the fifth-degree polynomial is algebraically daunting for hand solution, but is perfectly tractable with the aid of a symbolic mathematics package such as Maple 5.

fixing the amplitude. Therefore, by implication, the shape of the trial function was fixed. The fifth-degree polynomial, on the other hand, could capture the essential shape change when the load moved off center. Finally, we have seen that a poor choice of trial function gives a very poor answer, while a better choice can yield a much better answer.

An obvious question, at this point, is where do we get the "exact" solution for a general problem, so we can bound the errors in our approximate variational solutions? Highly meshed simulations with state-of-the-art finite-element and boundary-element modeling programs are, in fact, sufficiently accurate that they can serve as a "first-order" exact solution.[5] Strategically, this means that one is well advised to use fairly simple variational solutions at first to get a feeling for how a problem behaves, but when one needs highly accurate solutions (on the order of 1% accuracy in deflections), one then goes to an appropriate meshed simulation. The advantage of the analytical variational solution is that it makes explicit how the solution varies with device geometry and material properties. Typical finite-element simulations cannot provide this insight. Therefore, this author strongly recommends using analytical variational methods in combination with numerical simulation to build accurate and insightful device models. This is explored further in the following sections.

10.4   Large Deflections of Elastic Structures

We now remove the restriction imposed in Chapter 9 on the size of deformations of elastic structures, and use variational methods to get approximate expressions for their large-amplitude load-deflection behavior. After examining an example, we will comment on how variational solutions can be combined with numerical simulations to yield extremely useful and accurate models of device behavior. Then further examples are explored.

10.4.1   A Center-Loaded Doubly-Clamped Beam

We will once again deal with a doubly clamped beam, but with a point load exactly at the middle of the beam (see Fig. 10.2). This time, we shall use a trial function in the form of a cosine:

$$\hat{w} = \frac{c}{2}\left(1 + \cos\frac{2\pi x}{L}\right) \tag{10.46}$$

---

[5]The correctness of meshed models depends on the accuracy with which such effects as real geometric variations, boundary conditions, and constitutive properties, especially residual stress variations, are captured. Thus, in a sense, the only true "exact" solution is the result of fabrication followed by careful experiment, but it is often slow and expensive to obtain. That is why we need good models.

Like our fourth-degree polynomial example, this solution has only one free parameter, and thus the shape is totally constrained. However, the purpose of this exercise is to examine what happens when the deflections get large compared to the thickness of the beam. We already know that the solution may not be numerically accurate, even for small deflections, but there is merit in carrying this example through. After going through the calculation, we will examine the benefits of this kind of analysis.

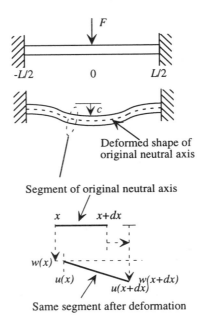

Figure 10.2. Doubly clamped beam with a point load in the center. The center deflection c can be comparable to or larger than the beam thickness. The expanded view of a segment of the original neutral axis is used to show the deformation components in detail.

Because we are allowing the deflections to be large, we must include two sources of strain energy, bending and stretching. That is, if the beam bends, its length must increase. This length increase produces some axial stress, and we have already seen that axial stress increases the stiffness of the structure. To solve the beam differential equation while taking into account the increase in length produced by bending is very difficult. However, to develop an approximate solution with variational methods is quite straightforward.

We already know how to evaluate bending energy. The axial strain energy requires some care. When a structure undergoes large deflections, we must include both transverse displacements w and axial displacements u. This is indicated at the bottom of Fig. 10.2, which shows a differential segment of the original neutral axis before and after deformation. There is a vertical deflection

$w(x)$ and also an axial deflection $u(x)$ of the left-hand end of the segment, and corresponding deflections of the right-hand segment. The original length of the segment is dx. The deformed length is

$$ds = \sqrt{[dx + u(x + dx) - u(x)]^2 + [w(x + dx) - w(x)]^2} \qquad (10.47)$$

Using the result that for small δ

$$\sqrt{1 + \delta} \approx 1 + \frac{\delta}{2} \qquad (10.48)$$

we obtain, to lowest order,

$$ds = dx \left[1 + \frac{du}{dx} + \frac{1}{2}\left(\frac{dw}{dx}\right)^2\right] \qquad (10.49)$$

The axial strain of the differential element is given by

$$\epsilon_x = \frac{ds - dx}{dx} \qquad (10.50)$$

Hence,

$$\epsilon_x = \frac{du}{dx} + \frac{1}{2}\left(\frac{dw}{dx}\right)^2 \qquad (10.51)$$

The total change in length is given by

$$\delta L = \int_{-L/2}^{L/2} \epsilon_x \, dx \qquad (10.52)$$

Thus, the aggregate axial strain of the original neutral axis, $\epsilon_a = \delta L / L$ is

$$\epsilon_a = \frac{1}{L}\int_{-L/2}^{L/2}\left[\frac{du}{dx} + \frac{1}{2}\left(\frac{dw}{dx}\right)^2\right] dx \qquad (10.53)$$

which becomes

$$\epsilon_a = \frac{1}{L}\left[u\left(\frac{L}{2}\right) - u\left(-\frac{L}{2}\right)\right] + \frac{1}{L}\int_{-L/2}^{L/2}\frac{1}{2}\left(\frac{dw}{dx}\right)^2 dx \qquad (10.54)$$

The first term in brackets is zero, since the beam is clamped at the ends. Thus only the second term survives. We conclude that the aggregate axial strain that results when a clamped beam is bent is

$$\epsilon_a = \frac{1}{L}\int_{-L/2}^{L/2}\frac{1}{2}\left(\frac{dw}{dx}\right)^2 dx \qquad (10.55)$$

We can now proceed with our variational analysis. We will use the fact that linear elasticity obeys superposition, so that we can think of the total deformation as consisting of two virtual steps. The first step is bending the beam to the desired shape with the right-hand end free to move in the x direction.[6] The second step is to move the right-hand support back to the original fixed position, which creates the axial stress. The bookkeeping method we choose is to follow the *original* neutral axis of the beam. During the first step, there is no axial stress, and the neutral axis remains at its original length. However, after the second step, this original neutral axis has been stretched. There is a new neutral axis location after deformation, but we are not interested in its location. Instead we use the change in length of the original neutral axis to measure aggregate axial strain.

The total axial strain ϵ_T can be written as

$$\epsilon_T = \epsilon_{\text{bending}} + \epsilon_{\text{stretching}} \tag{10.56}$$

or,

$$\epsilon_T = -z\frac{d^2\hat{w}}{dx^2} + \frac{1}{L}\int_{-L/2}^{L/2}\frac{1}{2}\left(\frac{d\hat{w}}{dx}\right)^2 dx \tag{10.57}$$

The second term is the axial strain ϵ_a. If we substitute the trial solution of Eq. 10.46, we obtain

$$\epsilon_a = \left(\frac{\pi^2}{4L^2}\right)c^2 \tag{10.58}$$

The total strain energy can now be found from

$$W = \frac{EW}{2}\int_{-H/2}^{H/2}\int_{-L/2}^{L/2}\epsilon_T^2\,dzdx \tag{10.59}$$

The work done by the external load is simply Fc. Performing the indicated integrals, we find the total potential energy to be

$$U = \frac{EWH\pi^4(8H^2 + 3c^2)c^2}{96L^3} - Fc \tag{10.60}$$

Taking the derivative with respect to c and setting the result to zero yields the *load-deflection characteristic* for this beam

$$F = \left(\frac{\pi^4}{6}\right)\left[\frac{EWH^3}{L^3}\right]c + \left(\frac{\pi^4}{8}\right)\left[\frac{EWH}{L^3}\right]c^3 \tag{10.61}$$

---

[6]This is equivalent to a built-in support on rollers (see Section 9.3.1).

This is a very interesting result, worthy of close examination. The first thing we notice is that the relation between the applied load F and the resulting deflection c is *nonlinear*. This is not unexpected, but it is our first robust example of a nonlinear mechanical spring. In particular, this is an *amplitude-stiffened Duffing spring*.

The second important aspect of this solution is that it can be broken down into pieces. Each term in the response can be uniquely associated with one of the deformation modes. The first term, linear in c, is the small-deflection bending result, proportional to the moment of inertia (*i.e.*, to WH^3). The second term, proportional to c^3, is the stretching term. It is proportional to WH.

Next, we note that each term consists of three parts: a numerical constant in front, a square-bracketed factor that depends on structural dimensions and material constants, and a factor containing the displacement c to some power. We expect that our very simple trial solution might not give use the correct numerical constants, but it has the very important virtue of giving us the square-bracketed terms that provide the dependences on dimensions and material properties. This is a major advantage of variational methods over purely numerical simulations. A designer looking at this equation can immediately see that thicker beams will be more likely to be dominated by the bending term, thinner beams by the stretching term. Further, the transition from bending-dominated behavior to stretching-dominated behavior occurs when $c \approx H$. Thus, we conclude that the thickness of the beam is the correct metric to determine whether the deformation is "small" or "large."

10.4.2 Combining Variational Results with Simulations

It must be emphasized that all of these insights were obtained with what might be a very poor trial solution. In order to improve the accuracy of this solution, we need to work harder. One approach is to repeat this variational solution with increasingly complex trial solutions. The disadvantage of that approach is that the algebraic complexity grows exponentially with the number of terms in the trial solution. A better alternative is to use the *functional form* of the load-deflection solution obtained from a simple variational trial solution, and use meshed numerical simulations both to verify that the form is sufficiently accurate to cover the design space of interest, and to determine better values for the numerical constants in each term.

Specifically, we have seen that the load-deflection behavior for a doubly-clamped beam should follow the form

$$F = C_\mathrm{b} \left[\frac{EWH^3}{L^3} \right] c + C_\mathrm{s} \left[\frac{EWH}{L^3} \right] c^3 \qquad (10.62)$$

where the numerical constants C_b and C_s may need to be determined from careful numerical simulation. If we wish to apply this result to a range of beam lengths and widths that cover a geometric design space we are interested in and a range of beam materials with different values of E, we can do a set of finite-element simulations that sample this design space, then fit the results to the form of Eq. 10.62. This will establish whether or not the functional form is a good representation for the design space, and will, from the fitting, provide good values for the constants C_b and C_s.

This strategy is widely used in model building. The only significant risk in this approach is that the selected trial function may omit some important physical effect that figures significantly in the numerical modeling. In that case, one would expect a poor fit of the numerical results to the analytical form obtained from the trial solution. Assuming that a good fit is obtained, this approach draws out the best aspects of both methods – an analytical functional form with the correct dependence on device geometry and material properties, which comes directly from the variational solution, and the numerical accuracy of highly-meshed finite-element simulation (which is, after all, the ultimate form of variational solution). We shall encounter several specific examples later in the book.

10.4.3   The Uniformly Loaded Doubly-Clamped Beam

We can use the results of the previous section to find very quickly the load-deflection behavior of a beam uniformly loaded with pressure P. The strain energy W depends only on the assumed trial function, and we have already done that calculation. The only change required for a different form of loading is to calculate the work done by the loading force in establishing the deformation. This work is

$$W P \int_{-L/2}^{L/2} \frac{c}{2} \left(1 + \cos \frac{2\pi x}{L} \right) \, dx = \frac{WLPc}{2} \tag{10.63}$$

The resulting total potential energy is

$$\mathcal{U} = \frac{EWH\pi^4(8H^2 + 3c^2)c^2}{96L^3} - \frac{WLPc}{2} \tag{10.64}$$

Differentiating with respect to c, setting the result to zero, and solving yields

$$P = \left(\frac{\pi^4}{3} \right) \left[\frac{EH^3}{L^4} \right] c + \left(\frac{\pi^4}{4} \right) \left[\frac{EH}{L^4} \right] c^3 \tag{10.65}$$

Once again, we have an amplitude stiffened Duffing spring, but now with a pressure-deflection characteristic whose square-bracketed terms lack the factor of W (a result of the distributed load), and have an extra factor of L in the denominator. Otherwise, the behavior is the same as for the point-loaded beam.

10.4.4   Residual Stress in Clamped Structures

Of immediate concern for MEMS devices is how to handle clamped structures that already have some residual axial stress σ_0, and then are subjected to bending that modifies the axial stress. There are several ways of dealing with this. The easiest is to go back to the fundamental definition of the strain energy density

$$\tilde{W} = \int_0^{\epsilon_a} \sigma \, d\epsilon \tag{10.66}$$

We recall that when we apply variational methods, we are concerned with the variation in strain energy that results from the deformation. Therefore, the most direct way of handling residual stress is to use it as the starting point for the integral, and calculate the change in strain energy resulting from the deformation:

$$\tilde{W} = \int_0^{\epsilon_a} (\sigma_0 + E\epsilon) \, d\epsilon \tag{10.67}$$

The second term inside the integral has already been evaluated, but there is now a new residual-stress contribution to the total strain energy of the form:

$$W_r = \sigma_0 W \int_{-H/2}^{H/2} dz \int_{-L/2}^{L/2} \epsilon_a dx \tag{10.68}$$

Using the trial solution of Eq. 10.46, this term equals

$$\sigma_0 (WLH) \left(\frac{\pi^2}{4L^2} \right) c^2 \tag{10.69}$$

When added to $\mathcal{U}$ prior to differentiation with respect to c, the modified load-deflection characteristic becomes

$$F = \left\{ \left(\frac{\pi^2}{2} \right) \left[\frac{\sigma_0 W H}{L} \right] + \left(\frac{\pi^4}{6} \right) \left[\frac{EW H^3}{L^3} \right] \right\} c + \left(\frac{\pi^4}{8} \right) \left[\frac{EW H}{L^3} \right] c^3 \tag{10.70}$$

The corresponding calculation for the uniformly loaded beam with residual stress yields

$$P = \left\{ (\pi^2) \left[\frac{\sigma_0 H}{L^2} \right] + \left(\frac{\pi^4}{3} \right) \left[\frac{EH^3}{L^4} \right] \right\} c + \left(\frac{\pi^4}{4} \right) \left[\frac{EH}{L^4} \right] c^3 \tag{10.71}$$

We see that the presence of residual stress directly affects the linear stiffness term, a result we already encountered in Section 9.6.2. The general form for the load-deflection characteristic, using the uniformly loaded beam as an example, becomes

$$P = \left\{ C_r \left[\frac{\sigma_0 H}{L^2} \right] + C_b \left[\frac{EH^3}{L^4} \right] \right\} c + C_s \left[\frac{EH}{L^4} \right] c^3 \qquad (10.72)$$

now with three constants to be determined from simulation. Furthermore, assuming, based on the trial-solution results, that the constants C_r and C_b are of comparable magnitude, we can estimate that the linear term becomes stress-dominated when

$$\sigma_0 \approx \frac{EH^2}{L^2} \qquad (10.73)$$

The longer the beam, the more important is the residual stress in determining its bending stiffness. In the limit of large residual stress, the beam behaves like a tensioned wire, with a deflection characteristic much more like the hyperbolic cosine solution to the beam equation in Section 9.6.2. In such a case, our assumed cosine trial function is very bad indeed. What is interesting, though, is that the general form of the load-deflection characteristic is correct even for this very approximate trial function.

10.4.5 Elastic Energy in Plates and Membranes

When considering elastic energy in beams, it was sufficient to consider only axial strains, a single degree of freedom. In two-dimensional plates and membranes, the situation is more complex. It is necessary to include the Poisson effects that couple strains in different directions. The subject is covered fully in standard texts.[7] For a fully three dimensional problem, the energy density can be written in terms of the various stress components:

$$\tilde{W} = \frac{1}{2E} \left(\sigma_x^2 + \sigma_y^2 + \sigma_z^2 \right) - \frac{\nu}{E} \left(\sigma_x \sigma_y + \sigma_y \sigma_z + \sigma_x \sigma_z \right)$$

$$+ \frac{1}{2G} \left(\tau_{xy}^2 + \tau_{yz}^2 + \tau_{xz}^2 \right) \qquad (10.74)$$

where the shear modulus G is

$$G = \frac{E}{2(1+\nu)} \qquad (10.75)$$

---

[7]See, for example, Timoshenko and Goodier, *Theory of Elasticity*, pp. 244-249 [48].

Plates and membranes are in a state of *plane stress*, with σ_z, τ_{yz}, and τ_{xz} all equal to zero. Thus, in plane stress, the energy density becomes

$$\tilde{W} = \frac{1}{2E}\left(\sigma_x^2 + \sigma_y^2\right) - \frac{\nu}{E}\sigma_x\sigma_y + \frac{1}{2G}\tau_{xy}^2 \qquad (10.76)$$

Using the equations of elasticity for an isotropic material, this equation can be recast in terms of strains:

$$\tilde{W} = \frac{E}{2(1-\nu^2)}\left(\epsilon_x^2 + \epsilon_y^2 + 2\nu\epsilon_x\epsilon_y\right) + \frac{G}{2}\gamma_{xy}^2 \qquad (10.77)$$

We now use this result to analyze the large-deflection behavior of plates and membranes.

10.4.6 Uniformly Loaded Plates and Membranes

We have already seen that even a relatively poor choice of trial function gives useful analytical insights into the behavior of structures. In this section, we shall examine the large-deflection behavior of uniformly loaded plates and membranes, and for the sake of algebraic simplicity, we will work with square structures. The extension to rectangular structures is conceptually easy, but algebraically more intricate.

We assume a trial solution which is the two-dimensional equivalent of the cosine solution we used for the bent doubly-clamped beam:

$$\hat{w} = \frac{c_1}{4}\left[1 + \cos\left(\frac{2\pi x}{L}\right)\right]\left[1 + \cos\left(\frac{2\pi y}{L}\right)\right] \qquad (10.78)$$

The constant c_1 is the deflection of the plate at the center $x = 0$, $y = 0$. Because this is a large-amplitude solution, it improves the solution if we include in-plane displacements as well. We know that the in-plane displacements must be zero at the clamped edges, and also zero at the center. Therefore, a suitable trial solution describing the in-plane displacements of the original neutral axis is

$$\hat{u} = c_2 \sin\left(\frac{2\pi x}{L}\right)\left[1 + \cos\left(\frac{2\pi y}{L}\right)\right] \qquad (10.79)$$

$$\hat{v} = c_2 \sin\left(\frac{2\pi y}{L}\right)\left[1 + \cos\left(\frac{2\pi x}{L}\right)\right] \qquad (10.80)$$

However, we must add to this neutral-axis displacement the displacement that results from bending at positions away from the neutral axis. Hence, a complete trial solution including the bending terms is

$$\hat{u} = c_2 \sin\left(\frac{2\pi x}{L}\right)\left[1 + \cos\left(\frac{2\pi y}{L}\right)\right] - z\frac{d\hat{w}}{dx} \qquad (10.81)$$

$$\hat{v} \;=\; c_2 \sin\left(\frac{2\pi y}{L}\right)\left[1 + \cos\left(\frac{2\pi x}{L}\right)\right] - z\frac{d\hat{w}}{dy} \qquad (10.82)$$

From these displacement functions, we can readily calculate the strains ϵ_x, ϵ_y and γ_{xy}, substitute them into Eq. 10.77, and integrate over the volume of the plate to yield the elastic energy W due to bending and stretching of the plate. In addition, we must include the strain energy due to any biaxial residual stress σ_0, given by

$$W_r = \int_{\text{area}} \sigma_0 \left(\epsilon_x + \epsilon_y\right) dx dy dz = \left(\frac{3\pi^2}{16}\right)\sigma_0 H c_1^2 \qquad (10.83)$$

The work done by the external uniformly distributed load P is given by

$$W_e = P \int_{\text{area}} \hat{w}\, dx dy \qquad (10.84)$$

The total potential energy is then

$$U = W + W_r - W_e \qquad (10.85)$$

The variational solution is found by differentiating U with respect to c_1 and c_2, setting the results to zero, and solving the two equations. The results are

$$c_2 = -\frac{\pi \nu}{32L}c_1^2 \qquad (10.86)$$

which, when substituted into the equation for c_1 gives the pressure-deflection relation:

$$P = \left\{ C_r\left[\frac{\sigma_0 H}{L^2}\right] + C_b\left[\frac{EH^3}{(1-\nu^2)L^4}\right] \right\} c_1 + C_s f_s(\nu)\left[\frac{EH}{(1-\nu)L^4}\right] c_1^3 \qquad (10.87)$$

where, for this particular choice of trial solution

$$C_r \;=\; \frac{3\pi^2}{2L^2} \qquad (10.88)$$

$$C_b \;=\; \frac{2\pi^4}{3} \qquad (10.89)$$

$$C_s \;=\; \frac{\pi^4}{4} \qquad (10.90)$$

$$f_s(\nu) \;=\; \frac{(7-2\nu)(5+4\nu)}{32(1+\nu)} \qquad (10.91)$$

This result is very similar to what we found for the beam in the large-deflection regime. First, there are three terms, a linear stiffness term due to the residual stress which is proportional to the stress-thickness product $\sigma_0 H$, a second linear stiffness term due to bending, proportional to H^3 and to the plate modulus $E/(1 - \nu^2)$, and a cubic stiffness term due to stretching, proportional to H and to the Young's modulus E. But the stretching term also involves some additional factors that depend on the Poisson ratio. How they are grouped is arbitrary. However, to capture the primary dependence on ν, we choose to gather a factor $(1 - \nu)$ into the square-bracketed term, so that it depends on the biaxial modulus $E/(1 - \nu)$, and collect the remaining factors together with a numerical scale factor of 32 into a function $f_s(\nu)$. This factor is quite a weak function of Poisson ratio, changing from .934 to .975 over the range of Poisson ratio from 0.25 to 0.35. Thus a "guess" of Poisson ratio of 0.3, will give a value of $f_s(\nu)$ good to a few percent over the typical range of Poisson ratios in non-elastomeric materials.

We can now be quite precise about the difference between a "plate" and a "membrane." If the stress term is dominant at small deflections, the structure behaves like a membrane. If the bending term is dominant at small deflections, it behaves like a plate.

10.4.7 Membrane Load-Deflection Behavior

When a membrane is sufficiently thin that the bending term can be neglected (as is often the case in MEMS structures), the load-deflection characteristic simplifies to

$$P = C_r \left[\frac{\sigma_0 H}{L^2} \right] c_1 + C_s f_s(\nu) \left[\frac{EH}{(1 - \nu)L^4} \right] c_1^3 \qquad (10.92)$$

The specific values of the constants C_r and C_s and the specific form of the function $f_s(\nu)$ will depend on the trial solution. Therefore, a good way to model membranes is to use the functional form of Eq. 10.92, and with a few finite-element simulations in the design space of interest, capture good values for the constants. This has been done in the membrane limit (assuming C_b is zero) [59], with the result that

$$C_r = 13.64 \qquad (10.93)$$
$$C_s = 21.92 \qquad (10.94)$$
$$f_s(\nu) = 1.446 - .427\nu \qquad (10.95)$$

The numerical scale factors for C_s and $f_s(\nu)$ are somewhat arbitrary; only their product appears in the full load-deflection characteristic.

This membrane load-deflection characteristic together with the constants of Eqs. 10.93, 10.94, and 10.95 can be used to measure the residual stress and biaxial modulus of suspended thin films [59, 60]. A membrane attached to a suitable supporting frame is placed beneath a microscope with a shallow depth of field and a calibrated z-axis motion, and pressure P is applied to one side of the membrane. The deflection c_1 is measured with the shift of focus point in the microscope.

The load-deflection data can be fit with the expression of Eq. 10.92; alternatively, a straight-line method can be used by recasting the equation into the following form:

$$\frac{PL^2}{Hc_1} = C_r\sigma_0 + C_s f_s(\nu)\frac{E}{(1-\nu)L^2}c_1^2 \qquad (10.96)$$

If the left-hand-side is plotted against c_1^2, the result is a straight line whose intercept is $C_r\sigma_0$ and whose slope is $C_s f_s(\nu)E/(1-\nu)L^2$. Thus the value of the residual stress is obtained from the intercept, and the value of the biaxial modulus is obtained from the slope. This method has been widely used to measure material properties. The accuracy of the method depends on the dimensional accuracy of the sample and on the measurement errors in determining the deflection.

10.5   Rayleigh-Ritz Methods

We have already seen that we can estimate both linear and nonlinear stiffnesses of structures using energy methods. This enables us to develop models for the "spring" parts of elastic mechanical systems. We will now see that it is also possible to estimate the resonant frequency of such flexible systems by a direct extension of these energy methods. From the resonant frequency, we can obtain the equivalent lumped "mass" for the structure.

10.5.1   Estimating Resonant Frequencies

When a structure deforms, for example by bending, the displacements vary with position. If we were to excite the structure with a sinusoidal force at its resonant frequency, the resulting resonant waveshape would also have maximum displacements that vary with position. As a result, the kinetic energy of the system at resonance (hence the "mass" part of an equivalent lumped circuit) must somehow incorporate this waveshape, with more kinetic energy in the parts that move a lot and less in the parts that move less.

A very nice way of estimating the kinetic energy is to assume that under sinusoidal excitation, the system motion is equal to the quasistatic trial function multiplied by a sinusoidal time dependence. That is, a time-dependent trial function becomes

$$\hat{w}(x, t) = \hat{w}(x)\cos(\omega t) \qquad (10.97)$$

From this trial function, we can use the methods developed in this chapter to calculate the elastic strain energy $\mathcal{W}_e$ and the corresponding total potential energy $\mathcal{U}$ using the maximum deflection of the trial solution (for example, $t = 0$). On the other hand, when $t = \pi/2\omega$, the deflection is everywhere zero, but the velocity is maximum. We can use this velocity profile to calculate the estimated maximum kinetic energy, as follows.

The velocity of the beam when $t = \pi/2\omega$ is

$$\left.\frac{\partial\hat{w}(x,t)}{\partial t}\right|_{t=\pi/2\omega} = -\omega\hat{w}(x) \tag{10.98}$$

Since the kinetic energy of a lumped mass m is

$$\mathcal{W}_k = \frac{1}{2}mv^2 \tag{10.99}$$

where v is the velocity, we can define a local *maximum kinetic energy density* $\tilde{\mathcal{W}}_k$ as

$$\tilde{\mathcal{W}}_k = \frac{\omega^2}{2}\rho_m(x)\hat{w}(x)^2 \tag{10.100}$$

where $\rho_m(x)$ is the mass density of the beam at position x. For a homogeneous beam, ρ_m is constant, but for a compound beam, it can be position-dependent. To find the total estimated maximum kinetic energy, we integrate $\tilde{\mathcal{W}}_k$ over the volume of the beam:

$$\mathcal{W}_k = \frac{\omega^2}{2}\int_{\text{volume}} \rho_m(x)\hat{w}(x)^2\,dxdydz \tag{10.101}$$

If we assume that the beam is being driven at its resonant frequency ω_o, then it is well known from analysis of linear second-order systems (like the resonant R-L-C circuit analyzed in Chapter 7) that the maximum kinetic energy and the maximum potential energy are equal at resonance. Therefore, by equating the maximum potential energy, which occurs at $t = 0$, to the maximum kinetic energy, which occurs at $t = \pi/2\omega_o$, we can obtain an estimate of ω_o:

$$\omega_o^2 = \frac{\mathcal{W}_e}{\displaystyle\int_{\text{volume}} \frac{1}{2}\rho_m(x)\hat{w}(x)^2\,dxdydz} \tag{10.102}$$

This procedure is called the *Rayleigh-Ritz Method* for estimating resonant frequencies.[8] What is interesting is that even rather poor trial functions give reasonably good estimates of the resonant frequency.

---

[8] See, for example, Bathe, pp. 586-593 [58].

As an example, we estimate the resonant frequency of a very thin doubly-clamped beam of length L with an axial residual stress σ_0, subjected to a uniform load P. We will use two different trial solutions, and compare the results. By assuming a very thin beam and assuming small-amplitude motion, we can neglect the bending and stretching terms and consider only the residual stress term in the elastic energy.

The first trial solution we will use is the the correct solution to the vibration of a tensioned wire:

$$\hat{w}_1(x) = c_1 \cos\left(\frac{\pi x}{L}\right) \tag{10.103}$$

The second trial solution is the same bent beam we have already used in several examples. We know this will be a relatively poor trial solution because it is the appropriate shape when the beam is bending dominated, not stress dominated. The bending trial solution is:

$$\hat{w}_2(x) = \frac{c_2}{2}\left[1 + \cos\left(\frac{2\pi x}{L}\right)\right] \tag{10.104}$$

If we go through the variational calculation for these two trial solutions, we find that

$$c_1 = \left(\frac{4}{\pi^3}\right)\left[\frac{PL^2}{\sigma_0}\right] \tag{10.105}$$

and

$$c_2 = \left(\frac{1}{\pi^2}\right)\left[\frac{PL^2}{\sigma_0}\right] \tag{10.106}$$

Figure 10.3 shows a comparison of these two solutions for identical values of $P = 1$, $L = 1$, and $\sigma_0 = 1$. The tensioned-wire solution has a larger center deflection, and lacks the inflection point that is characteristic of bending.

If we compute the estimated resonant frequencies using these two trial solutions, we obtain

$$\omega_{o,1} = \frac{\pi}{L}\sqrt{\frac{\sigma_0}{\rho_m}} \tag{10.107}$$

and

$$\omega_{o,2} = \frac{\pi}{L}\sqrt{\frac{4\sigma_0}{3\rho_m}} \tag{10.108}$$

The ratio of the two frequency estimates is

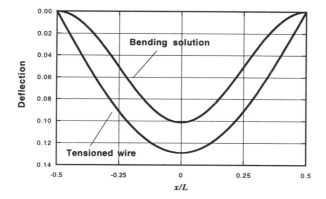

Figure 10.3. Comparison of two trial functions for a uniformly loaded thin beam with axial tensile stress.

$$\frac{\omega_{o,2}}{\omega_{o,1}} = \frac{2}{\sqrt{3}} = 1.15 \tag{10.109}$$

The bending solution gives a resonant frequency that is within 15% of the correct value for the tensioned wire, even with a rather poor shape for the trial function. Further, because the bending solution is stiffer (i.e., it has a smaller deflection for the same load), its estimated resonant frequency is higher.

This result illustrates a general property of variational solutions: The better the trial solution, the larger the deflections, the smaller the stiffness, and the smaller the resonant frequency. In general, an arbitrarily selected trial function will overestimate the stiffness and the resonant frequency.

10.5.2 Extracting Lumped-Element Masses

We can now use the combination of stiffness and resonant frequency to obtain an estimate of the appropriate mass to use when creating a lumped-element model for a structure. We use the fact that the undamped resonant frequency of a spring-mass-dashpot system is

$$\omega_o = \sqrt{\frac{k}{m}} \tag{10.110}$$

where k is the lumped linear spring constant for the structure and m is the lumped mass. By using variational methods to estimate k and the Rayleigh-Ritz procedure to estimate ω_o, we can obtain an equivalent lumped mass

$$m = \frac{k}{\omega_o^2} \tag{10.111}$$

With the exception of the damping effects, which will be taken up in the following chapters, we now have an approach that will permit us to analyze a wide range of elastic structures to obtain their linear (and even nonlinear) stiffness behavior, their small-amplitude resonant frequency, and the corresponding equivalent lumped mass.

Related Reading

K.-J. Bathe, *Finite Element Procedures in Engineering Analysis*, Englewood Cliffs, NJ: Prentice Hall, 1982.

S. P. Timoshenko and J. N. Goodier, *Theory of Elasticity*, Third Edition, New York: McGraw-Hill, 1970; reissued in 1987.

S. P. Timoshenko and S. Woinowsky-Krieger, *Theory of Plates and Shells*, New York: McGraw-Hill, 1959; reissued in 1987.

Problems

10.1 Use a parabolic trial function and find the variational solution for a cantilever beam with a built-in end and a point load at its free end. Compare the result to the exact solution, both in form, and numerically. Repeat for the case of a uniformly loaded cantilever.

10.2 Use the fourth-order trial solution of Eq. 10.29 to calculate the large-deflection behavior of an initially unstressed doubly-clamped beam with a point load at its center (the use of Maple or another symbolic math package is recommended). Compare the result with the cosine solution of Eq. 10.46. Which gives the larger center deflection?

10.3 A silicon pressure sensor is fabricated as a 10 μm thick 500 × 500 μm diaphragm supported by a frame of silicon that, in effect, creates fixed edges. Using a variational solution, estimate the pressure at which the deflection equals the diaphragm thickness and the maximum stress at the top surface of the diaphragm at this load.

10.4 A micromachined quartz load cell consists of a beam 5 mm long with a 600 × 500 μm cross-section supported at both ends with fixtures that permit an axial load to be applied either in tension or compression. Using Rayleigh-Ritz methods, show that for small values of axial load, the lowest resonant frequency is a linear function of the load. (The piezoelectric methods for getting such a beam to vibrate are discussed in Chapter 21.) Determine the lumped equivalent spring constant and mass for the lowest resonant mode of this device.

Chapter 11

DISSIPATION AND THE
THERMAL ENERGY DOMAIN

*Heat not a furnace for your foe so hot
That it do singe yourself.*

—William Shakespeare, in *Henry VIII*

11.1   Dissipation is Everywhere

In our study of electrostatic, magnetic, and elastic effects, we made effective use of the conservation of energy and its associated concept, the Principle of Virtual Work. In particular, we were able to use the essential *reversibility* of the energy storage process to create energy-based models of transducers. Energy could be put into generalized capacitors or generalized inductors, and all that energy could be extracted again. However, we know that lurking in the background, even for perfectly conservative effects such as elastic and capacitive energy storage, are effects that are *dissipative*, effects that convert energy into heat. In our spring-mass-dashpot model of the basic mechanical element, we introduced a damping element as an *ad hoc* addition to guarantee stability and to mirror reality. In our electric circuit used to drive the electrostatic actuator, we introduced the source resistance of the voltage source, and found that it had a profound effect on the operation of the system, introducing additional time-delays. But it also has an important effect on the entire energy-transfer process, as we shall see using the electrical resistor as an example.

11.2   Electrical Resistance

Ohm's law, in continuum form, is written

$$\mathbf{J}_e = \sigma_e \mathcal{E} \tag{11.1}$$

where $\mathbf{J}_e$ is the current density, with units Amperes/m^2, $\mathcal{E}$ is the electric field, and σ_e is the *electric conductivity*. In a semiconductor material, the conductivity at low electric field is

$$\sigma_e = q_e \mu_n n + q_e \mu_p p \qquad (11.2)$$

where q_e is the electronic charge (1.6×10^{-19} Coulombs), n and p are the concentrations of free electrons and free holes, respectively, and μ_n and μ_p are the electron and hole mobilities, with units of m^2/Volt-second. In doped semiconductors, only the majority carrier will contribute significantly to the conductivity.[1]

The electric field is the gradient of the electrostatic potential ϕ

$$\mathcal{E} = -\nabla \phi \qquad (11.3)$$

We usually design resistors to be in the form of a long bar, perhaps folded for compactness. With this design, the electric field inside the bar is uniform, and has magnitude

$$\mathcal{E} = \frac{V}{L} \qquad (11.4)$$

where V is the potential difference applied between the ends of the resistor and L is the resistor length. The current density inside the resistor is then uniform (neglecting detailed turning effects where the path may be folded), and has magnitude

$$J = \frac{\sigma_e V}{L} \qquad (11.5)$$

The total current I is the integral of J over the cross section of the conductor, yielding

$$I = \frac{\sigma_e A}{L} V \qquad (11.6)$$

where A is the area. This can be expressed in more usual Ohm's law form as

$$V = \left(\frac{\rho_e L}{A} \right) I \qquad (11.7)$$

where ρ_e is the *resistivity*, the reciprocal of the conductivity σ_e. We recognize the quantity in parentheses as the resistance of the resistor, given by

$$R = \frac{\rho_e L}{A} = \frac{L}{\sigma_e A} \qquad (11.8)$$

---

[1] See Chapter 14 for additional discussion of the physics of semiconductor devices.

When the shape of the resistor is more complex, it is first necessary to solve the governing differential equation for current flow using the methods to be presented in Chapter 12. In DC steady state, the current distribution must satisfy

$$\nabla \cdot \mathbf{J_e} = 0 \qquad (11.9)$$

which can be expanded to

$$\sigma_e \nabla^2 \phi + \nabla \sigma_e \nabla \phi = 0 \qquad (11.10)$$

For a region with uniform conductivity, this reduces to the Laplace equation

$$\nabla^2 \phi = 0 \qquad (11.11)$$

This equation is solved throughout the volume of the conductor, subject to boundary conditions of imposed potentials at the contacting surfaces and zero gradient over the non-contacted surfaces. The current density is found from Eq. 11.1. The current through a contact is then found from the integral of the current density over the contact area. Resistors fabricated in semiconductor technologies usually have non-uniform doping, hence nonuniform conductivities, requiring inclusion of the $\nabla \sigma_e$ term.

The power entering a resistor when it has voltage V and is carrying current I is VI, which, using Ohm's law for a linear resistor, is equal to $I^2 R$. All positive resistors absorb power, regardless of the direction of current.[2] This is an example of an intrinsically dissipative element. Whenever a resistor carries a current, it is absorbing net power, which must be converted to heat. This process is called *Joule heating*. Joule heating is a friend when one is trying to use resistors as heaters in a temperature-control system. It is an enemy when one is trying to make accurate measurements of resistance values, since resistivities tend to vary with temperature. If one uses too large a current to sense a resistor value, the *self-heating* can disturb the measurement. We will model this effect in Section 11.6.

11.3   Charging a Capacitor

We now use a resistor to show the effects of dissipation within the resistor on the energy transfer that occurs when charging a capacitor from a voltage source that has a non-zero source resistance. The circuit is shown in Fig. 11.1.

A capacitor, initially uncharged, is connected to a DC voltage source V_S at time $t = 0$. The current through the resistor is

---

[2] All passive resistors are positive, hence dissipate energy. It is possible, however, to create electronic circuits that behave over a portion of their characteristics as negative resistors. These circuits do supply power to the external circuit. They are used to build oscillators.

$$I = \frac{V_S - V_C}{R} \tag{11.12}$$

The voltage on the capacitor and the current follow the usual exponential charging transients

$$V_C = V_S \left(1 - e^{-\frac{t}{RC}} \right) \quad \text{and} \quad I = \frac{V_S}{R} e^{-\frac{t}{RC}} \tag{11.13}$$

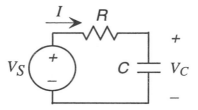

Figure 11.1.   Charging of a capacitor through a resistor

At the end of the transient, the stored energy in the capacitor (expressed as a co-energy) is

$$W_C^* = \frac{1}{2} C V_S^2 \tag{11.14}$$

Let us now calculate W_S, the total energy supplied by the source during the transient. This is found by integrating the power from the source, the negative of its I-V product, from $t = 0$ to the end of the transient.

$$W_S = \int_0^\infty V_S \left(\frac{V_S}{R} e^{-\frac{t}{RC}} \right) dt \tag{11.15}$$

This integral is readily evaluated to yield CV_S^2, which is exactly twice the amount of energy that ends up in the capacitor after the transient. The other half of the energy was dissipated in the resistor during the transient through *Joule heating*, which we can confirm by direct calculation:

The total energy dissipated in the resistor during the transient is the integral of $I^2 R$.

$$W_R = R \left[\int_0^\infty \left(\frac{V_S}{R} e^{-\frac{t}{RC}} \right)^2 dt \right] = \frac{1}{2} C V_S^2 \tag{11.16}$$

This is exactly the difference between what came from the voltage source and what was stored in the capacitor.

We conclude that energy is indeed conserved, but there is a 50% *energy tax* for charging the capacitor through a resistor, *regardless of the value of the resistor.* The source must supply twice as much energy as eventually ends up in the capacitor. And when the source is set to zero volts so that the capacitor can discharge, the rest of the energy is dissipated in the Joule heating of the resistor during the discharge transient. This means, in general, that one cycle of charging and discharging a capacitor costs a total energy of CV_S^2, even though the capacitor itself dissipates no energy. It is the *process of charging and discharging* that creates the energy loss.

Is the energy really "lost?" The answer, of course, is no. The *First Law of Thermodynamics* states that energy is conserved. Indeed, the energy lost from the electric domain when a current flows through a resistor appears in the thermal energy domain as heat. But this conversion to heat is *irreversible.* The heat energy is not readily available for re-conversion into electrical energy. This is a manifestation of the *Second Law of Thermodynamics* which states, in simplified terms, that all energy-transfer processes are inherently dissipative, hence irreversible. Only in carefully constructed ideal systems is it possible to represent perfect energy transfer without loss, and these systems require unphysical elements, such as perfect voltage sources with zero source resistance.

11.4   Dissipative Processes

As stated above, all dissipative processes produce heat and are inherently irreversible. Therefore, the thermal energy domain must play a special role among the array of energy domains encountered in MEMS devices.

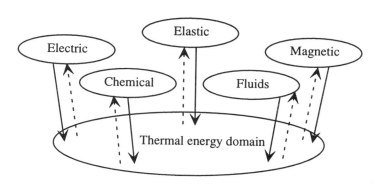

Figure 11.2. Illustrating the relation of the thermal energy domain to other energy domains. Solid arrows represent dissipative processes. Dashed arrows indicate that temperature affects constitutive properties in all other domains.

Figure 11.2 illustrates this schematically. The thermal energy domain is linked to the other energy domains in three ways. First, while each of the other energy domains has some conservation laws that apply, there are always dissipative processes that create heat. The solid arrows indicate these dissipative processes; examples are listed in Table 11.1. Second, all constitutive properties depend on temperature. Hence, the production of heat through dissipative processes always has the possibility of modifying the constitutive properties in the nominally "conservative" energy domain. We mentioned self-heating of a resistor in Section 11.2 as an example of this kind of dependence. The dashed arrows indicate these dependencies. Finally, it is possible to use heat engines or thermoelectric effects to convert some of the heat energy back to other forms of energy, although not with perfect efficiency (again, a consequence of the Second Law of Thermodynamics).

Table 11.1.   Dissipative effects in different energy domains.

Energy Domain	Dissipative Effects
Rigid-body motion	Contact friction
Elastic	Internal friction
Electric	Joule heat, dielectric loss
Magnetic	Eddy currents, hysteresis
Fluidic	Viscosity
Chemical	Diffusion and reaction
Thermal	Heat transfer

At this point, we could plunge off a cliff into chasms of thermodynamic complexity, and while it is a fascinating subject, our target is the modeling of MEMS devices. Therefore, we will take a closely edited walk through the various issues associated with dissipative effects, raising fundamental thermodynamic considerations where appropriate, but mostly concentrating on effects that arise in the MEMS world.

11.5   The Thermal Energy Domain

Heat energy, which we shall represent by the symbol Q, is energy associated with internal degrees of freedom in a material: the motion of individual atoms and molecules. In the macroscopic world, we do not observe these individual motions, only their aggregate effect. The amount of heat energy a material contains is directly related to its temperature T. We define the *heat capacity* at constant volume as

$$C_V = \frac{\partial Q}{\partial T}\bigg|_V \tag{11.17}$$

We can also define a heat capacity at constant pressure. This is important for gases, less so for liquids and solids:

$$C_P = \frac{\partial Q}{\partial T}\bigg|_P \tag{11.18}$$

Both heat capacities have units of Joules/Kelvin. Because they do not differ significantly in solids, we will simply use C to denote heat capacity.

Heat energy is an *extensive* quantity. The amount of heat present depends on how much matter is being considered. It is common practice, therefore, to normalize the heat energy either to a unit mass, a unit volume or a fixed number of molecules, such as a mole. We will use the symbol $\tilde{C}_m$ as the normalized heat capacity per unit mass and $\tilde{C}$ for the heat capacity per unit volume. We will not use the molar normalization in this book. The relation between the total and normalized heat capacities is then

$$C = \mathcal{V}\tilde{C} \tag{11.19}$$

and

$$\tilde{C} = \rho_m \tilde{C}_m \tag{11.20}$$

where ρ_m is the mass density and $\mathcal{V}$ is the volume.[3]

Heat energy obeys a continuity equation that results from the conservation of energy. To express this equation, it is most convenient to normalize the heat energy to a unit volume, denoted by $\tilde{Q}$. The continuity equation is

$$\frac{d\tilde{Q}}{dt} + \nabla \cdot \mathbf{J}_Q = \tilde{P}\bigg|_{sources} \tag{11.21}$$

where $\mathbf{J}_Q$ is the heat flux out of the surface of the material, with units of Watts/meter^2, and $\tilde{P}|_{sources}$, with units of Watts/m^3, represents the generated heat power inside the material due, for example, to any dissipative process from another energy domain as in Fig. 11.2. This source term couples the thermal energy domain to all others.

The heat flux $\mathbf{J}_Q$ through a bounding surface of a sample of material is driven by temperature differences. There are three important mechanisms of heat transfer: conduction, convection, and radiation. Heat conduction is the most

---

[3] We have now entered the notationally treacherous region where the symbol V must be reserved for voltage, so we adopt $\mathcal{V}$ for volume. This notation continues through Chapter 13 on fluids.

important in MEMS devices. The constitutive equation for heat conduction, called Fourier's law, is written

$$\mathbf{J}_Q = -\kappa \nabla T \qquad (11.22)$$

where κ is the *thermal conductivity*, and has units of Watts/Kelvin-meter. A fundamental thermodynamic constraint (from the Second Law) is that $\kappa > 0$. That is, heat never spontaneously flows from a colder body to a hotter body. Heat flow is an *irreversible* energy-transfer process, a one-way street.

Convection is a complex process in which heat conduction across a bounding surface into a fluid occurs, and the motion of the fluid contributes to the net rate of heat transfer. In most MEMS devices, convection is far less important than conduction. The continuous-flow polymerase-chain-reaction system of Chapter 22 is affected by convection and Section 22.5 includes a very brief and very approximate convection analysis. For more thorough treatments, the reader is referred to standard texts on heat transfer [61, 62].

Radiation is a fundamental heat-transfer process involving the quantum statistical mechanics of photons (electromagnetic radiation). The heat transfer occurs over a space between two bodies at different temperatures, T_1 and T_2. Assuming that the electromagnetic radiation does not interact with the medium filling the space (for example, air), the magnitude of the heat flux is

$$J_Q = \sigma_{\text{SB}} F_{12} \left(T_1^4 - T_2^4 \right) \qquad (11.23)$$

where σ_{SB} is the Stefan-Boltzmann constant, with value 5.67×10^{-8} W/m^{2}K^4, and F_{12} is a factor between 0 and 1 that takes account of the relative efficiency of the radiant energy transfer. It includes both geometry effects and an intrinsic property of the material called the *emissivity*. The emissivity of a surface varies between zero and one, depending strongly on the kind of material and its surface finish. For two closely spaced parallel plates of materials with emissivities e_1 and e_2, the rate of radiative heat transfer per unit plate area is[4]

$$J_Q = \frac{\sigma_{\text{SB}} \left(T_1^4 - T_2^4 \right)}{\dfrac{1}{e_1} + \dfrac{1}{e_2} - 1} \qquad (11.24)$$

If both emissivities are unity, the denominator has value unity, and the heat flux is at its maximum. The heat flux is decreased if the either emissivity is less than unity.

---

[4]See Incropera and DeWitt, *Fundamentals of Heat and Mass Transfer*, Chapters 12 and 13, and specifically p. 739 [62].

11.5.1   The Heat-Flow Equation

If we take the divergence of the equation for heat flux (Eq. 11.22), we obtain

$$\nabla \cdot \mathbf{J}_Q = -\nabla \cdot \kappa \nabla T \qquad (11.25)$$

If we substitute this result into the continuity equation (Eq. 11.21), we obtain

$$\frac{\partial \tilde{Q}}{\partial t} = \nabla \cdot \kappa \nabla T + \tilde{P}\Big|_{sources} \qquad (11.26)$$

This is the heat-flow equation, a partial differential equation describing the time and space variation of temperature in the thermal energy domain. If we assume that the specific heat is a constant over the temperature range of interest, and that κ is constant over the spatial region of interest, then this reduces to

$$\frac{\partial T}{\partial t} = \frac{\kappa}{\tilde{C}} \nabla^2 T + \frac{1}{\tilde{C}} \tilde{P}\Big|_{sources} \qquad (11.27)$$

where $\tilde{C}$ is the heat capacity per unit volume and the ratio $\kappa/\tilde{C}$ is called the *thermal diffusivity*. It has the units of m^2/second. This equation can be solved to find the time- and space-dependence of the temperature distribution. This equation is also quite fundamental for distributed dissipative processes in general. We will see later in this chapter that this same equation arises in several other dissipative processes.

11.5.2   Basic Thermodynamic Ideas

Thermodynamics is the science of the thermal energy domain and its interaction with other energy domains.[5] We present here the laws of thermodynamics assuming we are dealing with stationary (non-fluidic) systems. When we discuss fluids in Chapter 13, the necessary revisions to these formulations will be presented.

The First Law of Thermodynamics expresses energy conservation between heat energy and all other forms of energy. In differential form

$$dU = dQ - dW \qquad (11.28)$$

where dU is the change in the *internal energy* of the system, dQ is the differential heat energy entering the system, and dW is the differential work done by the system on the external world. dQ and dW depend on the path the system follows to go from one state to another, while dU depends only on the starting and ending state. For example, a gas in a fixed volume can be initially

---

[5] See, for example, Robert B. Lindsay, *Introduction to Physical Statistics*, Chapter III [63], or Landau and Lifshitz, *Statistical Physics*, Chapter II [64].

compressed to a smaller volume (work is done on the gas). The internal energy goes up and the gas gets hotter. If the gas is then allowed to cool through the walls back to its starting temperature, the final state is the same amount of gas at its initial temperature, but in the smaller volume. Alternatively, the gas could initially be cooled by heat removal, then compressed to the smaller volume. If the amount of heat withdrawn by cooling is matched to the work done by compression to the smaller volume, the final state will be the same amount of gas at its initial temperature, but in the smaller volume. The paths were different, but the internal energy of the gas depends only on the initial and final states expressed in terms of total mass, temperature, and volume.

We can connect the internal energy to our heat-transfer ideas by imagining a process in which there is *reversible* heat transfer, heat transfer that requires an infinitessimal temperature gradient. We can imagine a sequence of state changes produced by reversibly adding heat to a system. This reversibly added heat can be related to the thermodynamic state of a system through the *entropy*.

From a statistical point of view, entropy measures the degree of internal disorder in a system. Ordered systems have low entropy, while disordered systems have high entropy. Random atomic and molecular processes are intrinsically disordering. Hence, from a statistical point of view, we expect systems to evolve in a way that tends to increase entropy. Once they have reached equilibrium, the entropy ceases to change.

In thermodynamic terms, entropy is defined differentially as

$$dS = \frac{\Delta Q_r}{T} \tag{11.29}$$

where T is the temperature at which the reversible heat transfer ΔQ_r takes place. If we restrict our attention to reversible heat transfer, which takes a system from one equilibrium state to another, we can write the First Law as

$$dU = TdS - dW \tag{11.30}$$

How can we understand the entropy if we have an *irreversible* heat transfer process, such as heat conduction through a region with a strong temperature gradient? We can still retain the entropy as a thermodynamic concept, but we cannot calculate it from Eq. 11.29. The details of calculating entropy in non-equilibrium states goes beyond the scope of this discussion. However, we can identify the thermodynamic equivalent of our statistical notion that entropy should increase until a system reaches equilibrium. We consider a *closed system*, one that has no interaction with its environment. It can exchange neither heat nor work with the outside world. Therefore, from the First Law, its internal energy is fixed. However, its entropy is not. The Second Law of Thermodynamics says that *for a closed system, the entropy can only increase or stay the same.* It can never decrease. It is precisely this non-decreasing property of entropy that helps us understand dissipative processes. A very simple example will suffice:

Let us prepare an isolated system by taking a bar of material and heating one end very hot while keeping the other end cold. Then, at time $t = 0$, we insulate the bar from the outside world, so that no heat can enter or leave. The internal energy of the bar is fixed by the total heat content, but the entropy is lower than its equilibrium value because of the ordering (more heat energy at one end than the other). The initial spatial distribution of the heat energy is far from equilibrium. Heat will flow from the hot end to the cold end (*but not the other way!*) until the bar achieves a uniform temperature. This unidirectional requirement on heat flow from hot to cold is what makes the trend toward equilibrium unidirectional. And the amount of heat flow is quantitatively related to the rate of entropy increase. When the bar has reached a uniform temperature, the heat flow goes to zero and the rate of change of entropy does also. The maximum entropy in a closed system occurs when the system reaches equilibrium.

11.5.3   Lumped Modeling in the Thermal Domain

As we have seen in the previous section, we can write the First law as

$$dU = TdS - dW \tag{11.31}$$

Further, we can express the work dW in terms of effort and displacement variables in the various energy domains to which the thermal system is coupled:

$$dU = TdS - \sum_j e_j dq_j \tag{11.32}$$

where e_j and q_j are the effort and displacement coordinates in the jth energy domain. This formulation suggests that if we want to represent the thermal energy domain with the $e \rightarrow V$ convention introduced in Chapter 5, we should use temperature as the generalized effort and entropy as the generalized displacement. However, as we have seen, entropy is an awkward choice for a displacement variable because it is not associated with any conserved quantity (such as charge for the electrical energy domain). It has this nasty habit of spontaneously increasing during irreversible processes.

A more pragmatic (and successful) choice for lumped modeling is the *thermal convention* introduced in Chapter 5. We use temperature difference as the across variable analogous to voltage, and heat current as the through variable analogous to electric current. This choice implies that the heat energy itself is the generalized displacement, analogous to charge. Note that the product of the across variable and the through variable does not have the dimensions of power. Hence, for the sake of building lumped networks in the thermal domain, we must abandon the power-conjugate-variable pairs that work so well in all the other energy domains. Instead, we must explicitly calculate the transfer of energy into and out of the thermal domain. Within the thermal domain, however, the network elements do conserve thermal energy, just as charge is conserved in electric networks.

With this thermal convention, heat capacities are represented by lumped capacitors, and the thermal conductivity κ becomes an exact analogy for the electrical conductivity σ_e. Thus one can create lumped resistors to represent heat conduction paths by exact analogy to electrical resistors for charge conduction paths, even including the solution of the Laplace equation for the temperature distribution in nontrivial geometries. One serious complication here is that unlike electrical currents, which tend to stay rigorously confined to conductors, heat currents can leak out the sides of thermal resistors by conduction, convection, and radiation. A second complication is that because thermal time constants tend to be long, it is often necessary to analyze transient effects as well as steady-state behavior. Therefore, accurate modeling of the temperature distribution in the presence of heat sources is a more complex task than modeling the distribution of quasi-static electric fields in the presence of voltage differences.

Chapter 12 presents some general and systematic methods for building lumped-element models of thermal systems. However, it is always possible to be quite approximate, and create hand-built lumped models using estimates of element values. We shall use this approach here to analyze an example in which the electrical and thermal domains are coupled.

11.6   Self-Heating of a Resistor

In this section, we hearken to Shakespeare's warning (page 267) and study the self-heating of a resistor, that is, the change in temperature and electrical resistance that occurs when passing current through a resistor. This phenomenon is central to many systems, both macro and micro. In the macro world, the simple fuse that protects many electrical circuits from current overloads depends on the change in electrical resistance with temperature to produce a runaway temperature rise when the current exceeds a critical value. The heating element in an electric toaster glows red hot when making toast, but does not produce a runaway thermal rise. We shall see below why this difference in behavior arises. In the micro world, we often use resistors to measure things, such as strain (using the piezoresistance effect, explained in Chapter 18) or temperature (using the intrinsic temperature dependence of the resistor). When using resistors for measurement, we must be concerned with the temperature change that can occur during a measurement. Three examples follow: a self-heated current-driven resistor (the fuse), a self-heated voltage-driven resistor (the toaster), and the self-heating of a semiconductor resistor used for sensing applications.

11.6.1   Temperature Coefficient of Resistance

The starting point for all the examples is the definition of the *temperature coefficient of resistance*, or TCR. All resistors vary with temperature (except those carefully designed for zero TCR over a restricted temperature range). For moderate temperature excursions, we can use a linear variation of resistance with temperature:

$$R = R_0 \left[1 + \alpha_R (T_R - T_0)\right] \tag{11.33}$$

where R_0 is the resistance at reference temperature T_0, and R is the resistance at temperature T_R. The temperature coefficient of resistance α_R is typically positive for metallic resistors, but can be either positive or negative for semiconductors, depending on the region of operation and the specific design of the resistor. Typical values for metals are in the range of 10^{-4} per Kelvin. Values in semiconductors can be larger. In specially designed *thermistor materials*, the resistance-temperature function can be an exponential, with incremental negative TCR values on the order of 10^{-2} per Kelvin or greater.[6]

11.6.2   Current-source drive

The Joule heating that occurs when a current flows in a resistor can be represented with the aid of the lumped-element thermal circuit of Fig. 11.3.

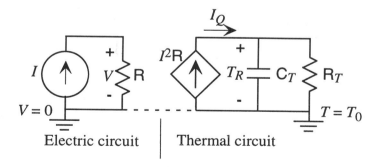

Figure 11.3.   Circuit model for the self-heating of a resistor driven from a current source.

The electric circuit contains the resistor and a current source. Not shown is a voltmeter that would be used to measure the voltage across the resistor. The thermal circuit has three elements: a diamond-shaped *dependent current*

---

[6] See, for example, Khazan, *Transducers and Their Elements*, Chapter 10 [65].

source[7] that supplies the Joule heat power $I^2 R$, a thermal capacitor C_T representing the heat capacity of the resistor, and a thermal resistor R_T representing the heat conduction from the resistor to a thermal reservoir held at temperature T_0. The current variable in the thermal circuit has dimensions of power, and is denoted by I_Q. The ground symbol in the electric circuit denotes the zero of electric potential. The ground symbol in the thermal circuit denotes the reference temperature T_0. The voltage across C_T is the temperature difference $T_R - T_0$, but for convenience in the following analysis, we set T_0 to zero, making the voltage across C_T simply T_R.

The two grounds are shown connected to each other with a dashed line. This connection is not necessary; however, when using circuit simulators to analyze mixed electrical-thermal circuits, it is essential that there be a common ground for the entire mixed-domain network. The connection between the two grounds is included here in anticipation of the transfer of such circuits to circuit-simulation environments.

This is a linear first-order system, with a rather simple governing equation:

$$C_T \frac{dT_R}{dt} = -\frac{T_R}{R_T} + I^2 R_0 \left(1 + \alpha_R T_R\right) \tag{11.34}$$

We can collect terms to yield

$$\frac{dT_R}{dt} = -\frac{1}{R_T C_T} \left(1 - \alpha_R R_0 R_T I^2\right) T_R + \frac{I^2 R_0}{C_T} \tag{11.35}$$

We recognize this as a first-order system with input $I^2 R_0 / C_T$. The factor that occupies the usual place for the time constant of a first-order system depends on I^2, and thus might be time dependent. However, if I is a constant, then this system has a time constant τ_I given by

$$\tau_I = \frac{R_T C_T}{1 - \alpha_R R_0 R_T I^2} \tag{11.36}$$

The steady-state temperature rise $T_{\text{ss,I}}$ for a steady current I is[8]

$$T_{\text{ss,I}} = \frac{R_0 R_T I^2}{1 - \alpha_R R_0 R_T I^2} \tag{11.37}$$

---

[7]A dependent current source is a current source whose value depends on some other circuit variable, in this case, the electric current I.

[8]Readers may recognize this form as that of a linear system with positive feedback. The fuse blows when the magnitude of the loop gain reaches unity. We will revisit this example from the point of view of feedback systems in Chapter 15.

This equation explains how a metal fuse works. The TCR is positive. For a sufficiently large current, the denominator vanishes, the temperature rise diverges, and the fuse melts.

11.6.3 Voltage-source drive

Figure 11.4 shows the same thermal circuit, but driven with a voltage source V. Not shown is the ammeter or other circuit that would be used to measure the resistor current I.

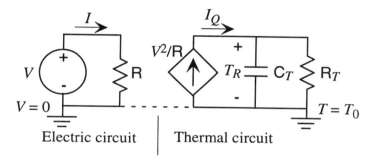

Figure 11.4. Circuit model for the self-heating of a resistor driven from a voltage source.

This is still a first-order system, but it is now nonlinear because the input term proportional to V^2 has the resistor value in the denominator:

$$C_T \frac{dT_R}{dt} = -\frac{T_R}{R_T} + \frac{V^2}{R_0 \left(1 + \alpha_R T_R\right)} \tag{11.38}$$

If we assume that the total change in resistance is small, we can linearize the model by expanding the denominator to yield

$$\frac{V^2}{R_0 \left(1 + \alpha_R T_R\right)} \approx \frac{V^2}{R_0} \left(1 - \alpha_R T_R\right) \tag{11.39}$$

Substituting and collecting terms yields

$$\frac{dT_R}{dt} = -\frac{1}{R_T C_T} \left(1 + \frac{\alpha_R R_T V^2}{R_0}\right) T_R + \frac{V^2}{R_0} \tag{11.40}$$

Once again, the factor that corresponds to the time constant contains V^2, which might be time-varying. However, if V is constant, the time constant τ_V for this system is

$$\tau_V = \frac{R_T C_T}{1 + \frac{\alpha_R R_T V^2}{R_0}} \tag{11.41}$$

and the steady-state temperature rise is[9]

$$T_{ss,V} = \frac{R_T V^2 / R_0}{1 + \alpha_R R_T V^2 / R_0} \tag{11.42}$$

This system is perfectly stable for a positive TCR. It never reaches a runaway situation. Our toaster, which has a positive TCR heating element connected to the power-main voltage source, does not experience thermal runaway, and our toast does not burn up. However, if we built our toaster with a negative TCR resistor, it could experience runaway for sufficiently large drive voltages, in exact analogy to the behavior of a positive-TCR fuse with a current drive.

11.6.4   A Self-Heated Silicon Resistor

When using resistors as transducers, it is necessary to dissipate Joule heat during the resistor measurement. The simplest method is to use a steady applied voltage or current, and measure the other variable with a suitable meter or circuit. The self-heating that results can produce errors.

In a typical piezoresistive strain measurement,[10] diffused semiconductor resistors may have a total change of only 1% at the maximum strain. If it is desired to measure this change to better than 1%, then the total accuracy required of the resistance measurement is 1 part in 10^4. We now use an example to see under what circumstances self-heating can produce a resistance change on this order.

Figure 11.5 shows on the top left a schematic of an n-type resistor formed in a p-type silicon substrate by ion implantation and diffusion. An expanded view of just the resistor shown at the right. We will use rather coarse approximations in order to estimate the thermal behavior of this resistor. As will be explained in detail in Section 14.4, the implanted resistor is electrically isolated from the substrate by the pn junction that is formed between the n-type implant and the p-type substrate. Assuming effective junction isolation between the resistor and the substrate, so that the electrical leakage current to the substrate can be ignored, the electrical resistance is approximated by[11]

$$R = \frac{L}{aW\sigma_e} \tag{11.43}$$

where σ_e is the electrical conductivity of the resistor, presumed uniform for this example. We assume a value of 1500 Siemens/meter for this example,

---

[9]This is a linear system with negative feedback. See Chapter 15.

[10]Piezoresistance is discussed in Chapter 18.

[11]Contact and end effects are ignored.

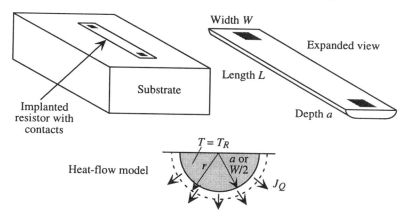

Figure 11.5.   An implanted resistor embedded in a silicon substrate, presumed to be very thick compared to the resistor. On the right is an expanded view, showing dimensions. At the bottom is a simplified heat-transfer model used to estimate the thermal resistance between the resistor and the substrate, which functions as a thermal reservoir.

corresponding roughly to n-type silicon doped to a concentration of 10^{17} cm^{-3}. The TCR of such a resistor is about 2500 ppm per Kelvin.

The heat capacity of the resistor portion of the sample is found from

$$C_T = \rho_m a W L \tilde{C}_m \tag{11.44}$$

where ρ_m is the mass density of silicon (2330 kg/m^3) and $\tilde{C}_m$ is the specific heat per unit mass [712 J/(kg-Kelvin)].

To estimate the thermal resistance, we use the highly simplified model shown at the bottom of Fig. 11.5. We assume the resistor is very long compared to its width or depth, and we assume that the heat flow from the resistor to the substrate is uniform around a semicircle of radius of order a or $W/2$, which will be chosen as comparable numbers in our example. The shaded region inside the heated resistor is assumed to have a uniform temperature T_R. Outside this region there is a radial temperature gradient giving rise to a heat flux J_Q

$$J_Q = -\kappa \nabla T \tag{11.45}$$

If we imagine a semicircular cylindrical boundary corresponding to the dashed arc in Fig. 11.5, the total heat current I_Q through that surface is

$$I_Q = \pi r L \kappa \frac{\partial T}{\partial r} \tag{11.46}$$

Using $\partial T/\partial r$ from this equation, the resistor temperature T_R takes the form

$$T_R = \int_a^{r_o} \frac{I_Q}{\pi r L \kappa} \, dr = \frac{I_Q}{\pi L \kappa} \ln\left(\frac{r_o}{a}\right) \qquad (11.47)$$

where r_o is a characteristic depth beyond which the temperature is presumed not to vary significantly. We will use a value of $r_o = 10a$ in our example, recognizing that a more precise estimate requires solution of the heat-flow partial differential equation using the methods of Chapter 12. With this assumption, the thermal resistance between the resistor and the thermal reservoir is about

$$R_T = \frac{T_R}{I_Q} = \frac{1}{\pi L \kappa} \ln(10) \qquad (11.48)$$

We now have all the elements of the equivalent circuit, and can evaluate some interesting quantities. Parameter values for this example are shown in Table 11.2.

Table 11.2.   Numerical values used for self-heated resistor example.

Parameter	Symbol	Value	Units
Length	L	300×10^{-6}	m
Width	W	4×10^{-6}	m
Depth	a	2×10^{-6}	m
Electrical conductivity	σ_e	1500	Siemens/m
Electrical resistance	R	2.5×10^4	Ohms
TCR	α_R	2500×10^{-6}	K^{-1}
Density	ρ_m	2330	kg/m^3
Specific heat	$\tilde{C}_m$	712	J/(kg-K)
Thermal capacitance	C_T	4×10^{-9}	J/K
Thermal conductivity	κ	148	W/(K-m)
Reference length	r_o	10*a	m
Thermal resistance	R_T	16.5	K/W
Thermal time constant	$R_T C_T$	6.6×10^{-8}	sec

The first thing we notice is that the thermal time constant at low current for this device $R_T C_T$ is extremely short, on the order of 60 ns. This means that this resistor is in very good thermal contact with the reservoir. A transient thermal rise will spontaneously die out in less than 1 μs.

We are interested primarily in the steady-state temperature rise, $T_{ss,I}$ as a function of drive current. If our criterion that the self-heating-induced error be less than a part in 10^4, then the maximum permitted value of T_R is

$$T_{R,\text{max}} = \frac{10^{-4}}{\alpha_R} = 0.04 \text{ K} \qquad (11.49)$$

Because we are interested in such small temperature rises, the current- or voltage-dependent terms in the denominator of Eqs. 11.37 and 11.42 can be ignored. This means that the current-drive and voltage-drive cases become identical.

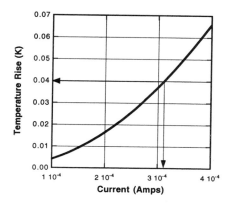

Figure 11.6.   Temperature rise vs. drive current for the self-heated silicon resistor. The arrows designate the targeted maximum allowable temperature rise, and the corresponding drive current.

Using the values in Table 11.2, $T_{ss,I}$ was calculated for a range of currents. The results for the region of greatest interest are shown in Fig. 11.6. We see that the temperature rise for this example reaches the maximum allowable value of 0.04 K for a drive current of about 0.3 mA, corresponding to a voltage across the resistor of 7.5 Volts. This is a very large voltage compared to the tens-to-hundreds of millivolts typically used when reading resistor values. Hence, we conclude that this extremely well heat-sunk resistor will not suffer appreciably from self-heating effects at normal voltages used for readout.

Let us consider a more challenging example. If this same resistor is mounted on the thin neck of a silicon accelerometer, displaced from the substrate by a beam of length 50 μm, width 20 μm and thickness 5 μm. Suddenly the thermal resistance to the substrate has gone from 16.5 K/W up to about 300 K/W, a factor of 20, and the allowable current level drops to about 25 μA, corresponding to a voltage drop across the resistor of only 0.6 V. If this resistor is used as part of a Wheatstone bridge, then there can be only 1.2 V across the bridge.

Finally, if such a resistor is used on a thermal isolation platform of the kind used in infrared cameras or chemical sensors, the thermal resistance can be much greater, meaning that even lower currents and voltages are required to avoid self-heating effects. At some point, the available signal level becomes comparable to the random noise generated by the resistor, a subject we shall explore in Chapter 16.

11.7   Other Dissipation Mechanisms

Table 11.1 (page 272) lists additional dissipation mechanisms. They all have one thing in common: the dissipation depends on the time-evolution of the system, either through a velocity, or through an irreversible flow process involving electric currents, heat energy, or matter (diffusion). Short comments on each[12] are provided below.

11.7.1   Contact Friction

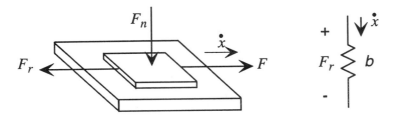

Figure 11.7.   Illustrating the frictional retarding force F_r that opposes motion in response to an applied tangential force F in the presence of a normal force F_n. The lumped circuit element used to represent frictional forces in the $e \rightarrow V$ convention is a damping resistor b shown on the right.

Contact friction occurs whenever two bodies must move in contact with one another (see Fig. 11.7). In linear models of contact friction, the *coefficient of friction* is the ratio of the retarding force to the applied normal force holding the two surfaces together.

$$F_r = \mu_f F_n \tag{11.50}$$

where μ_f is the coefficient of friction. Coefficients of friction range from the ideal of zero to values of order unity. It is important to recognize that the retarding force F_r is present only when there is relative motion between the two surfaces driven by some other external force F. That is, just as heat flow occurs only from hot to cold, frictional forces only retard, they do not accelerate. Hence, the equivalent lumped circuit element is a resistor. The power dissipated by frictional forces is the product $\dot{x}F_r$. This energy is converted to heat directly at the contacting surfaces, giving rise to irreversible heat flow and a corresponding entropy production.

Coefficients of friction depend on the surface morphology and roughness and on the speed of the motion, typically decreasing for sliding motions compared to the frictional forces that occur when initiating the motion from a stopped

---

[12]Except for fluid viscosity, which is discussed in Chapter 13.

position (so-called *static* friction). There are no good ways of calculating frictional forces from first principles. The value of the equivalent circuit element must be inferred from measurement.

There is also a nonlinear type of friction, called *Coulomb friction*, in which there is no motion until the force reaches a threshold level, beyond which there is a retarding frictional force proportional to velocity (see Fig. 11.8).

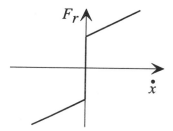

Figure 11.8.   Illustrating Coulomb friction.

The concept of Coulomb friction can also be used to model a normal-force type of friction, called "stiction." The origin of stiction forces can be chemical bonding, dispersion forces (van der Waals forces), hydrogen bonding, or Coulomb attraction between charged surfaces. The idea is that when surfaces are brought into intimate contact, there is an attractive force that must be overcome before the surfaces can be separated. While some finite work must be done to separate the surfaces (this work is called the *work of adhesion*), there is no dissipation of power once stiction has been overcome.[13]

11.7.2   Dielectric losses

Dielectric materials can have internal losses, not only from Joule heating due to any small conductivity the material might have, but also from internal friction that retards the orientation of molecular dipoles within the material. In the context of Maxwell's Equations, the relevant terms are found in equation for the curl of the magnetic field[14]

$$\nabla \times \mathcal{H} = \mathbf{J} + \frac{\partial \mathcal{D}}{\partial t} \qquad (11.51)$$

---

[13]This is a highly oversimplified way to look at the problem of adhesion between surfaces. A more fundamental approach involves fracture mechanics, which is the examination of the energy balance between deformation of elastic elements and the energy required to create surfaces by separation of the elements. The reader is referred to basic books on adhesion, such as Chapter 7 of Kinloch [66].
[14]Electromagnetic fields are reviewed in Appendix B.

In the sinusoidal steady state, all fields are represented as a complex amplitude multiplied by a time dependence of the form $e^{-j\omega t}$, where $j = \sqrt{-1}$. If the dipoles cannot orient instantly, there is a phase delay in the polarization that results from the sinusoidal electric field. Therefore, using j to represent a 90° phase shift, we can write

$$\mathcal{D} = (\varepsilon' + j\varepsilon'')\mathcal{E} \tag{11.52}$$

where ε' is the normal dielectric constant at frequency ω, the response that is in phase with the applied field, and ε'' is the dielectric *loss factor*, the response that is 90° out of phase. If, in addition, the material has conductivity σ_e, so that

$$\mathbf{J} = \sigma_e\mathcal{E} \tag{11.53}$$

we obtain a compact expression for the equivalent total dielectric and conductivity response:

$$\nabla \times \mathcal{H} = -j\omega \left[\varepsilon' + j\left(\varepsilon'' + \frac{\sigma_e}{\omega} \right) \right] \mathcal{E} \tag{11.54}$$

The term in square brackets is called the *complex permittivity* . If this quantity is used as the permittivity in a parallel plate capacitor operated in the sinusoidal steady state, the equivalent circuit model is a parallel *RC* circuit, with a capacitor of value

$$C = \frac{\varepsilon' A}{g} \tag{11.55}$$

and a resistor given by

$$\frac{1}{R} = \frac{(\sigma_e + \omega\varepsilon'') A}{g} \tag{11.56}$$

The out-of-phase dielectric response ε'', scaled by the frequency ω, enters the equivalent loss model in a fashion identical to conductivity, and is thus dissipative. In fact, insulating materials that have a high dielectric loss factor at certain frequencies can be efficiently heated by exciting them with a sinusoidal electric field.

11.7.3   Viscoelastic losses

Viscoelastic materials have an internal friction at work, creating internal stresses that oppose the motion. As a material deforms, power is dissipated by the velocity (i.e., the strain rate) working against these retarding stresses. Internal friction produces heat, irreversible heat flow, and entropy production. Linear viscoelastic losses can be modeled in the sinusoidal state using a complex elastic modulus, in exact analogy to the complex dielectric permittivity introduced above. The equivalent mechanical model, called a Maxwell-Voigt model, consists of a spring-dashpot combination, with the lossless spring representing the in-phase elastic part of the viscoelastic response and the dashpot representing the out-of-phase, hence lossy, part of the viscoelastic response. When translating these models to lumped electrical equivalents in the $e \rightarrow V$ convention, the spring becomes a capacitor and the dashpot becomes a resistor.

11.7.4   Magnetic Losses

There are several sources of dissipation in magnetic systems. First, currents are required to establish magnetic fields, and in all systems at non-cryogenic temperatures, the current-carrying elements have resistance, hence dissipation and Joule heating. There are two other dissipation sources: eddy currents, and magnetic hysteresis.

We will treat magnetic eddy currents in the magnetoquasistatic approximation (MQS), which neglects the displacement current in Eq. 11.51 compared to the conduction current $\mathbf{J}$.[15] Faraday's law of magnetic induction states that whenever there is a changing magnetic flux, an electromotive force (EMF) is established which drives current in conductors in a direction to oppose the flux change. In the context of Maxwell's equations, this is written as

$$\nabla \times \mathcal{E} = -\frac{\partial \mathcal{B}}{\partial t} \tag{11.57}$$

If we use Ohm's law to substitute for $\mathcal{E} = \mathbf{J}/\sigma_e$ and use Eq. 11.51 to substitute for $\mathbf{J}$, we obtain

$$\nabla \times \left(\frac{\nabla \times \mathcal{H}}{\sigma_e}\right) = -\frac{\partial \mu \mathcal{H}}{\partial t} \tag{11.58}$$

Using the result that

$$\nabla \cdot \mu \mathcal{H} = 0 \tag{11.59}$$

we obtain

$$\frac{1}{\mu \sigma_e} \nabla^2 \mathcal{H} = \frac{\partial \mathcal{H}}{\partial t} \tag{11.60}$$

This is the equation for *magnetic diffusion*. It is a vector equivalent of the same diffusion equation (Eq. 11.27) we discovered for heat flow. It states that changes in magnetic fields inside conductors do not take place instantaneously, but must build up in the conductors by a diffusion-like process. While this change is taking place, there are *eddy currents* inside the conductors set up by

---

[15]Electromagnetic fields are reviewed in Appendix B.

Faraday's law of induction in combination with Ohm's law, and these currents dissipate power and create heat.

Magnetic hysteresis is the magnetic analogy to internal friction in viscoelasticity. When a magnetic field is applied to a permeable material, the magnetic flux density B attempts to align with the $\mathcal{H}$-field. The microscopic process involves the orientation of atomic magnetic dipoles with the external field. This occurs primarily by the motion of the bounding regions of magnetic domains that have their dipoles oriented in various directions. Domains aligned with the applied field tend to grow at the expense of domains aligned in other directions.

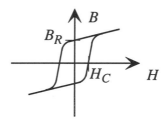

Figure 11.9.   A typical B-H curve for a permeable material, showing magnetic hysteresis.

Domain wall motion is a frictional process. In addition, there is a certain amount of magnetic field called the *coercive force H_C* that must be applied before domain walls will move (somewhat like internal Coulomb friction). This leads to hysteresis in the permeability response, as shown in Fig. 11.9. When the applied magnetic field is removed, the domains do not spontaneously return to their original position. Instead, there is net alignment, leading to a *remanent flux density B_R*.

Permanent magnets, such as samarium-cobalt alloys, are characterized by high coercive forces, and large remanent flux densities. Such materials are called *magnetically hard*. Other materials have very low coercive force, such as soft iron and iron-nickel alloys (permalloy). These materials are called *magnetically soft*.

11.7.5   Diffusion

Imagine a box with ten black and ten white marbles arranged so that all the black marbles are on the left side and all the white marbles on the right. Now imagine a process by which random pairs of adjacent marbles can exchange places. Over time, there will be a net flux of black marbles from the left toward the right and a net flux of white marbles from right to left because, until the distribution becomes random, there is a *concentration gradient* of both white and black marbles. Random motion in the presence of a concentration gradient leads to net particle transport. This process is called diffusion.

The equations governing diffusion are identical in their form to the equations of heat flow. If $[X]$ is the concentration of species X in a material (whether solid, liquid, or gas), the diffusion behavior of the species can be written

$$\frac{\partial[X]}{\partial t} = D_X \nabla^2[X] \tag{11.61}$$

We recognize this equation as identical to the equation for heat flow, and the equation for magnetic diffusion. The quantity D_X is called the *diffusion constant* for species X, and has units of m^2/second. We already encountered diffusion in Chapter 3 as a method for intentionally moving dopants introduced in to semiconductors. What is now clear is that this is an irreversible dissipative process, just like heat flow.

11.8 Irreversible Thermodynamics: Coupled Flows

In systems that contain temperature gradients and well as concentration gradients, both heat flow and diffusion can take place simultaneously. In addition, the two flows become coupled. That is, a flux of particles simultaneously produces a flux of heat energy because each particle carries some kinetic energy with it (this is in addition to the volume generation of heat energy resulting from the dissipative flow). Similarly, a temperature gradient can produce a flux of particles. The general topic of these coupled flows is called *irreversible thermodynamics.*[16] Basically, one must construct appropriate sets of driving forces and flows, and when one does, the coupling coefficients between the flows are symmetric. What defines "appropriate" is determined by the requirement that the complete set of forces and flows must account for all of the entropy production in the system.

Important examples of these coupled flows occur in electrical conductors. If a conductor contains n charge carriers per unit volume, and these carriers obey Boltzmann statistics[17], then under isothermal conditions, the relation between the electric current density $\mathbf{J}_e$ and the gradients of electric potentials and concentration can be written

$$\mathbf{J}_e = -z_n q_e D_n \nabla n - q_e n \mu_n \nabla \phi \tag{11.62}$$

where z_n is the charge of the charge carrier expressed in units of the electronic charge q_e, D_n is the diffusivity of the charge carrier, with dimensions m^2/sec, μ_n is the electric mobility of the species, with dimensions m^2/Volt-second, and

---

[16]See, for example, van der Ziel, *Physical Electronics*, pp. 491-500 [67].

[17]Boltzmann statistics is an equilibrium relation governing the probability of finding a carrier in a state with energy E. This probability is proportional to exp(-E/k_BT), where T is the absolute temperature in Kelvins and k_B is Boltzmann's constant, equal to 1.38×10^{-23} J/K.

ϕ is the electrostatic potential in Volts. Thus, an electric current can arise either from a concentration gradient or an electric field. In more advanced treatments, the electric current is described as due to the gradient of a new potential called the *electrochemical potential*, which combines the thermodynamic chemical potential, which is related to the concentration, with the electrostatic potential into a single quantity. In our treatment, we shall avoid cases which require the use of the electrochemical potential.

Under equilibrium conditions, with $\mathbf{J}_e$ equal to zero, there is a relation between the concentration n and the electrostatic potential:

$$\frac{1}{n}\nabla n = -\frac{z_n \mu_n}{Dn}\nabla\phi \tag{11.63}$$

There is also a fundamental relationship between the mobility and the diffusivity, called the *Einstein relation*. It can be derived from the basic statistical mechanics of carriers that obey Boltzmann statistics, and it states that

$$\frac{\mu_n}{Dn} = \frac{q_e}{k_B T} \tag{11.64}$$

The quantity $k_B T/q_e$ has the dimensions of voltage, and at room temperature equals about 26 mV. It is often called the *thermal voltage*. Substituting the Einstein relation, we conclude that the equilibrium relation between concentration and potential is

$$n = n_r e^{\frac{-z_n q_e(\phi - \phi_r)}{k_B T}} \tag{11.65}$$

where n_r is the concentration at some reference point which has potential ϕ_r. Thus, if we intentionally fabricate a semiconductor to have a variation in the concentration of electrons by changing its doping, there is an internal electric field in equilibrium that results from this concentration gradient. A way to think about it is that the internal electric field is exactly what is required to cancel the diffusion current that would result from the concentration gradient.

The same kind of thing happens when there are simultaneous gradients of temperature and concentration. Even for neutral species, in which there is no electric current but there can be a particle current. If there is a temperature gradient in a system, there must be a compensating concentration gradient if the particle current is to be zero. In the absence of such a concentration gradient, a thermal gradient will produce a net particle current. The underlying mechanism is that the particles in the hotter region have, on average, a higher kinetic energy than the particles in the colder region. Hence, there is slightly more tendency to find particles moving from hot to cold than from cold to hot. This creates a net particle current from hot to cold.

We now consider a homogeneous conductor, with n a constant, but with gradients of both temperature and electrostatic potential. We can write a pair of linear coupled flow equations:

$$\mathbf{J}_e = -L_{11}\nabla\phi - L_{12}\nabla T$$

$$\frac{\mathbf{J}_Q}{T} = -L_{21}\nabla\phi - L_{22}\nabla T \qquad (11.66)$$

The $1/T$ factor in the second equation converts the heat current into an "entropy current," which is the flux required by irreversible thermodynamics. We recognize L_{11} as the electrical conductivity σ_e and L_{22} as the thermal conductivity κ divided by T, but we need new information to interpret the cross terms L_{12} and L_{21}.

With this choice of fluxes and driving forces, according to one of the major the results of irreversible thermodynamics called the *Onsager relations*, the system is reciprocal. That is

$$L_{12} = L_{21} \qquad (11.67)$$

Thus a total of three coefficients are required to described the coupled electric-current/heat-current problem. If we manipulate these equations, we obtain the pair of relations

$$-\nabla\phi = \rho_e\mathbf{J}_e + \alpha_S\nabla T$$

$$\mathbf{J}_Q = \Pi\mathbf{J}_e - \kappa\nabla T \qquad (11.68)$$

where α_S is called the *Seebeck coefficient*, equal to L_{12}/L_{11}, and Π is called the *Peltier coefficient*, equal to TL_{21}/L_{11}. By the reciprocity relation, we conclude that the Seebeck coefficient and the Peltier coefficient are related:

$$\alpha_S = \frac{\Pi}{T} \qquad (11.69)$$

Let us now examine the implications of these cross terms.

11.8.1   Thermoelectric Power and Thermocouples

If we imagine a sample of conducting material with its two ends at different temperatures, but no closed circuit so that there can be no electric current flowing, then we immediately conclude that there must be an electric field in the sample:

$$\mathcal{E} = -\nabla\phi = \alpha_S\nabla T \qquad (11.70)$$

If we integrate this potential gradient from one end of the bar to the other, we obtain for the voltage V_{ab} that appears between the ends of the bar (assuming a one-dimensional gradient)

$$V_{ab} = \int_{T_a}^{T_b} \alpha_S(T)\, dT \qquad (11.71)$$

where the possibility of a temperature dependence for α_S has been explicitly included.

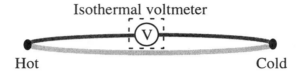

Isothermal voltmeter

Hot   Cold

Figure 11.10.   Illustrating the operation of a thermocouple.

The potential difference due to the Seebeck coefficient occurs inside the sample, and cannot be measured with an external voltmeter, because as soon as we contact the voltmeter lead to the sample, we create a *contact potential* that disturbs the measurement. Furthermore, the contact potentials at the two different temperatures might not be the same. Therefore, to measure this potential difference, we use a *pair* of materials with *different* Seebeck coefficients joined at junctions at both the hot and cold ends. Such an arrangement is called a *thermocouple*, illustrated in Fig. 11.10. Examples include platinum with rhodium, or iron with constantan. We create a closed loop with the voltmeter in the middle, as shown in Fig. 11.10. The region around the voltmeter is presumed to be locally isothermal at temperature T_V, while the hot junction is at temperature T_H and the cold junction at temperature T_C. Since the ideal voltmeter draws no current, we can integrate around the loop from one voltmeter contact to the other to find the thermocouple voltage V_{TC} measured by the voltmeter.

$$V_{TC} = \int_{T_V}^{T_C} \alpha_{S,1}\, dT + \int_{T_C}^{T_H} \alpha_{S,2}\, dT + \int_{T_H}^{T_V} \alpha_{S,1}\, dT \qquad (11.72)$$

This can be simplified to

$$V_{TC} = \int_{T_C}^{T_H} (\alpha_{S,2} - \alpha_{S,1})\, dT \qquad (11.73)$$

We see that the voltage measured by the voltmeter depends on the temperature difference, and on the *difference* in the Seebeck coefficients of the two materials. Because the Seebeck coefficient is an intrinsic property of a material, thermocouples are quite reliable methods of making temperature-difference measurements. Typical thermocouple-voltage magnitudes are 40 μV/K. To generate larger voltages, it is possible to construct a series connection of alternating materials, attaching every other junction to the hot region, the others to the cold region. Such an arrangement is called a *thermopile*.

11.8.2   Thermoelectric Heating and Cooling

Let us now imagine a sample whose temperature we wish to control connected to a heat sink through two conductors with different Seebeck coefficients (see Fig. 11.11). If we hold the heat sink at temperature T_0, and if the temperature of the sample is T_S, then the temperature gradient in the two conductors is the same.

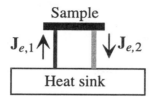

Figure 11.11.   Illustrating Peltier heating and cooling.

We now pass an electric current through this loop circuit. The rate of net heat delivery to the sample $\mathcal{P}$ (or removal from the sample, depending on the direction of the current) is found from the heat-flow equation in the two materials, evaluated at the position of the sample:

$$\begin{aligned} \mathcal{P} &= A_1 \mathbf{J}_{Q,1} - A_2 \mathbf{J}_{Q,2} \\ &= A_1 \mathbf{J}_{e,1} \Pi_1 - A_2 \mathbf{J}_{e,2} \Pi_2 - (A_1 \kappa_1 + A_2 \kappa_2) \nabla T|_{\text{sample}} \end{aligned}$$

(11.74)

where A_1 and A_2 are the areas of the two conductors. But for a closed electric circuit, it must be true that

$$A_1 \mathbf{J}_{e,1} = A_2 \mathbf{J}_{e,2} = I_e \qquad (11.75)$$

Therefore, the final result is

$$\mathcal{P} = I_e(\Pi_1 - \Pi_2) - (A_1 \kappa_1 + A_2 \kappa_2) \nabla T|_{\text{sample}} \qquad (11.76)$$

We conclude that this system can operate as either a heater or a refrigerator by supplying heat to the sample for one direction of current, removing it with the other. The effectiveness of the heating and cooling depends on the difference in Peltier coefficients between the two materials at the temperature of the sample.

Once the Peltier effect establishes a temperature difference between the sample and the heat sink, then there is a temperature gradient in each conductor and net heat conduction back toward the substrate. This means that there is a maximum temperature difference that can be created by the Peltier effect. Commercial Peltier heater/refrigerators with temperature swings of $\pm 30^\circ$ are readily available. With cascading, it is possible to achieve greater temperature differences.

11.8.3   Other Coupled-Flow Problems

We shall encounter other problems of coupled dissipative flows in Chapter 13 when we discuss electrokinetic phenomena in liquid electrolytes. The subject can become very complex because, as we have seen, one type of driving force can establish several different flows. The more components and phenomena to consider, the more complex the formulation. Fortunately, most of the cases that are important in MEMS can be handled with relatively few coupled flows.

11.9   Modeling Time-Dependent Dissipative Processes

Finally, it would be nice to have a general procedure for modeling dissipative processes that is as direct and understandable as the energy methods and variational calculations we used for energy-conserving systems. But because dissipation involves time-dependences, the problem is intrinsically harder. Nevertheless, there are good procedures available. These are the subject of the following chapter.

Related Reading

H. S. Carslaw and J. C. Jaeger, *Conduction of Heat in Solids*, Second Edition, New York: Oxford University Press, 1959, reprinted in 1986.

William M. Deen, *Analysis of Transport Phenomena*, New York: Oxford University Press, 1998.

Frank P. Incropera and David P. DeWitt, *Fundamentals of Heat and Mass Transfer*, Fourth Edition, New York: Wiley, 1996.

L. D. Landau and E. M. Lifshitz, *Statistical Physics*, Reading, MA: Addison-Wesley, 1958.

Robert B. Lindsay, *Introduction to Physical Statistics*, New York: Wiley, 1941.

Problems

11.1 A silicon plate, 100 μm square and 2 μm thick is suspended 2 μm above a silicon wafer with four tethers, each 100 μm long with a 2 μm square cross section. If this plate is heated 50°C above the wafer (held at room temperature), compare the heat lost by conduction along the tethers to radiative heat loss from the plate to the substrate. Assume unity emmissivities.

11.2 Use lumped-element arguments to estimate the thermal time constant of the heated silicon plate of Problem 11.1. How would this time constant change if the plate and supports were made of silicon nitride?

11.3 A silicon nitride plate and suspension with the dimensions given in Problem 11.1 has an aluminum resistor deposited on it. The resistor has a meander pattern. The aluminum thickness is 0.1 μm, and the width is 1 μm. The total meander length is 1 mm. Contacts to this resistor are made by running the aluminum lines down two of the supports. How does the presence of the aluminum affected the thermal capacity, the thermal resistance and the thermal times constant of the structure as calculated in Problem 11.2? How much current will be needed in the aluminum resistor to raise the temperature of the plate by 50°C above the substrate temperature? Is self-heating of the aluminum important in this application?

11.4 A 300 μmlong silicon beam with a 5 μmsquare cross section is supported on rigid raised supports attached to a silicon wafer whose temperature is maintained at room temperature. Using the resistivity and TCR from Table 11.2, determine at what electrical current this beam will buckle due to thermal expansion effects driven by Joule heating.

11.5 A Thermocouple Treasure Hunt: Locate in the literature a MEMS device that incorporates a microfabricated thermocouple or thermopile. What combination of materials was used, with what net thermocouple sensitivity?

11.6 A Thermal Platform Treasure Hunt: Locate in the literature a MEMS device that uses a suspended structure to create thermal isolation. Determine the lumped element thermal model for the device (or, if the authors provide such a model, verify the reported model).

Chapter 12

LUMPED MODELING
OF DISSIPATIVE PROCESSES

If you can't stand the heat, get out of the kitchen.

—Harry S. Truman

12.1 Overview

Unlike energy-conserving systems, for which quasi-static stored energy functions can be used for analysis and for which good variational methods are at hand to estimate the quasi-static energy functions, dissipative processes require us to confront the time-dependence of the system from the outset. We have seen in the previous chapter that a variety of dissipative processes share some features, such as a continuity equation and a flow equation driven by gradients of an intensive quantity, such as temperature. These, in turn, lead to mixed space-and-time partial differential equations governing physical behavior. Our goal here is to find good methods for building lumped-element approximate models that capture the essential features described by these partial differential equations.

12.2 The Generalized Heat-Flow Equation

The heat-flow equation is the most fundamental equation describing dissipative processes. It has already turned up in heat flow, particle diffusion, and eddy currents (magnetic diffusion), and it will turn up again in Chapter 13 in the linearized Reynolds equation describing squeeze-film damping. Understanding the basic properties of this equation permits the modeling of a wide variety of dissipative processes.

The origin of the heat-flow equation is the combination of a continuity equation together with a linear flux equation. The general continuity equation for a variable X can be written

$$\tilde{C}(\mathbf{r})\frac{\partial X(\mathbf{r},t)}{\partial t} + \nabla \cdot \mathbf{J}_X = \tilde{S}(\mathbf{r},t) \tag{12.1}$$

where $\tilde{C}(\mathbf{r})$ is a (possibly) position-dependent "capacity" coefficient per unit volume, the flux $\mathbf{J}_X$ depends linearly on the gradient of X

$$\mathbf{J}_X = -\kappa(\mathbf{r})\nabla X \tag{12.2}$$

$\tilde{S}(\mathbf{r},t)$ is a distributed source per unit volume term, and $\kappa(\mathbf{r})$ is a (possibly) position-dependent "conductivity" coefficient.

The interpretation of the variables depends on the energy domain. In describing heat flow, for example, X is the temperature T, $\tilde{C}$ becomes the heat capacity per unit volume, κ is the thermal conductivity, and $\tilde{S}$ is the time- and position-dependent heat generation rate $\tilde{\mathcal{P}}$ (with units of power per unit volume) due to local dissipative processes, such as Joule heating or viscous friction.

If we combine the two equations, we obtain

$$\tilde{C}(\mathbf{r})\frac{\partial X(\mathbf{r},t)}{\partial t} - \kappa(\mathbf{r})\nabla^2 X(\mathbf{r},t) - \nabla\kappa(\mathbf{r}) \cdot \nabla X(\mathbf{r},t) = \tilde{S}(\mathbf{r},t) \tag{12.3}$$

In a homogeneous region, where $\tilde{C}$ and κ are not dependent on position, the equation simplifies to the general heat-flow equation, also called the *diffusion equation*:

$$\frac{\partial X(\mathbf{r},t)}{\partial t} - D\nabla^2 X(\mathbf{r},t) = \frac{1}{\tilde{C}}\tilde{S}(\mathbf{r},t) \tag{12.4}$$

where D is the *diffusivity*, and equals κ/C.

This is a linear partial differential equation with constant coefficients. There are many excellent approaches to solving this equation for a variety of source conditions and boundary conditions. The goal of this chapter is to illustrate several methods that are particularly well-suited for creating lumped-element equivalent circuit models of the dissipative processes.

12.3   The DC Steady State: The Poisson Equation

When the source terms are constant in time, and one waits until all transients have died out, one reaches the *DC steady state*. In the DC steady state, the heat-flow equation simplifies into another fundamental equation, the *Poisson equation*.

$$\nabla^2 X(\mathbf{r}) = -\frac{S(\mathbf{r})}{\kappa} \tag{12.5}$$

The solution to this equation depends, of course, on the spatial variation of the source term, but also on the boundary conditions. A *Dirichlet boundary condition* is one which fixes the value of X on some portion of the surface. A *Neumann boundary condition* is one which fixes dX/dn, the normal derivative of X, on some portion of the surface. It is also possible to have *mixed boundary conditions* which specify a functional relationship between X and dX/dn.

The Poisson equation is a linear equation, and thus can be solved by superposition. For example, one can find the response to a spatial impulse source term (called the *impulse response* or *Green's function*), and then use the convolution integral to find the response to the actual source. This approach, while guaranteed to be successful, does not lend itself easily toward making lumped-element models. Instead, we shall examine two methods: the lumped-element equivalent of finite-difference methods, and eigenfunction methods.

Because this equation applies to many energy domains, there is a risk of terminology overload. In order to keep things crisp, we shall work exclusively with heat-flow examples, and will now replace X with T in Kelvins, J_X with J_Q in W/m^2, and $\tilde{S}$ with $\tilde{\mathcal{P}}$ in W/m^3. The units for κ are W/(Kelvin-meter). For the correct analogy to another energy domain, one simply uses the appropriate constants and units.

12.4 Finite-Difference Solution of the Poisson Equation

If we have a sample within which we must solve the Poisson equation, we can imagine dividing it up by meshing into N smaller volume elements and assign a point at the center of that volume element as a node in the network we shall build. The DC steady state requires that no net flux can enter or leave any volume element. Therefore, one can create a resistor network from node to node to capture the flow between volume elements, and the fact that Kirchhoff's Current Law (KCL) must be satisfied at every node captures the "no net flux" condition on the volume element. We complete the lumped-element model by attaching a flow source to each node to capture the part of the distributed source $\tilde{\mathcal{P}}$ that falls within that volume element.

To create the correctly sized resistors, we will create a discretized approximation to ∇^2. We assume that the nodes are equally spaced by distance h. The volume associated with each node is then h^3. For the sake of algebraic simplicity, we will assume that the boundary conditions are such that the only variation in T is along the x-direction. Then the operator ∇^2 reduces to d^2/dx^2, which can be approximated by the finite-difference expression

$$\left. \frac{d^2T}{dx^2} \right|_{x_n} \approx \frac{T(x_n + h) + T(x_n - h) - 2T(x_n)}{h^2} \tag{12.6}$$

The discretized form of the Poisson equation becomes

$$\frac{T(x_n + h) + T(x_n - h) - 2T(x_n)}{h^2} = -\frac{\tilde{P}(x_n)}{\kappa} \tag{12.7}$$

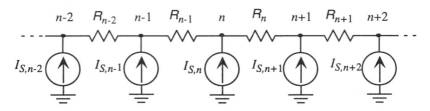

Figure 12.1.   A one-dimensional resistor-equivalent circuit for the Poisson equation.

To transform the Poisson equation into an equivalent circuit, we create an array of resistors between nearest-neighbor nodes. Figure 12.1 shows such a one-dimensional resistor network, with a current source from ground at each node to represent the input. KCL applied at node n yields

$$I_{S,n} + \frac{T_{n+1} - T_n}{R_n} + \frac{T_{n-1} - T_n}{R_{n-1}} = 0 \tag{12.8}$$

This equation has the same general form as the discretized Poisson equation. To link the two, we must first assign the "across" and "through" variables. For a heat-flow problem, we use the generalized temperature T as the across variable, and the generalized heat current I_Q as the through variable, having units of Watts. To connect the two formulations, the current source $I_{S,n}$ must capture all of the heat-current input within one node spacing of node n. Therefore, for a volume element of size h^3, we identify

$$I_{S,n} = h^3 \tilde{P}(x_n) \tag{12.9}$$

With this identification, the equation for the resistor network becomes

$$\frac{T_{n+1} - T_n}{R_n} + \frac{T_{n-1} - T_n}{R_{n-1}} = -h^3 \tilde{P}(x_n) \tag{12.10}$$

We can make the resistor network KCL equation identical to the finite-difference Poisson equation by choosing each resistor to have the value

$$R_n = \frac{1}{\kappa h} \tag{12.11}$$

If there is a three-dimensional mesh, with spacings h_x, h_y, and h_z, then using three indices to index each node, the current source at each node has value

$$I_{S,\mathbf{n}} = h_x h_z h_z S\tilde{\mathcal{P}}(\mathbf{r}_{n_x,n_y,n_z}) \tag{12.12}$$

where $\mathbf{r}_{n_x,n_y,n_z}$ is the position of node $\mathbf{n}$, and the resistor between adjacent nodes in the x-direction has the value

$$R_x = \frac{h_x}{\kappa h_y h_z} \tag{12.13}$$

Resistors in the other directions use appropriate permutations of the node spacings.

12.4.1 Temperature Distribution in a Self-Heated Resistor

We can use this formulation to set up the problem of finding the temperature distribution in a one-dimensional electrical resistor, with length L and cross-sectional area A. If we assume such a resistor is carrying a uniform current, then there is uniform Joule heating at the rate of $\tilde{\mathcal{P}}_o$ W/m^3. For simplicity, we will assume a zero TCR for the electrical resistor.[1] We will assume heat-sink boundary conditions at the two ends of the electrical resistor such that the temperature falls to its reference value of zero at both ends. This is a Dirichlet boundary condition. When using resistor networks to solve the Poisson equation, it is necessary to have at least one Dirichlet boundary condition somewhere in the problem. Otherwise, the resulting set of equations is singular. Physically, it means that one must establish a reference for T somewhere in the problem.

We will assume that the length of the resistor is divided into N equal intervals, corresponding to $N + 1$ nodes. Thus, h_x has the value L/N. There is no need to specify the values of h_y and h_z, since all of the elements will have cross-sectional area A.

To apply the boundary conditions, we fix the zeroth node and the $N + 1$st nodes at $T = 0$. At every other node, there is a current source with value $hA\tilde{\mathcal{P}}_o$.

We can assemble a set of $N+1$ simultaneous equations, as shown in Fig. 12.2, using the conductance $G = 1/R = \kappa A/h$.

For solving this equation, we have two good choices. One is to set up this conductance matrix and source vector in MATLAB, and solve the matrix equation $\mathbf{G} \cdot \mathbf{T} = \mathbf{P}$ numerically.[2] An alternative is to put the equivalent circuit

---

[1]It is possible to use the methods of this section to deal with resistors with non-zero TCR, but this requires simultaneous solution of two coupled networks, one electrical and one thermal. This issue is examined at the end of Chapter 23.

[2]It is advisable to use L-U factorization method rather than the `invert` command to solve this matrix equation when using a large number of nodes. Additionally, MATLAB has special commands for efficient solutions with *sparse matrices*, matrices with only a few non-zero entries per row. Consult MATLAB Help for further information.

$$
\begin{bmatrix}
1 & 0 & 0 & \cdots & & & & & & & & 0 \\
-G & 2G & G & \ddots & & & & & & & & 0 \\
0 & -G & 2G & \ddots & & & & & & & & 0 \\
\vdots & \ddots & \ddots & \ddots & \ddots & & & & & & & \vdots \\
0 & & & \ddots & 2G & -G & 0 & & & & & 0 \\
0 & & & \ddots & -G & 2G & -G & \ddots & & & & 0 \\
0 & & & & 0 & -G & 2G & \ddots & & & & 0 \\
\vdots & & & & & \ddots & \ddots & \ddots & \ddots & \ddots & & \vdots \\
0 & & & & & & & \ddots & 2G & -G & 0 \\
0 & & & & & & & & \ddots & -G & 2G & -G \\
0 & & & & & & & & & \cdots & 0 & 0 & 1
\end{bmatrix}
\begin{pmatrix}
T_0 \\ T_1 \\ T_2 \\ \vdots \\ T_{n-1} \\ T_n \\ T_{n+1} \\ \vdots \\ T_{N-1} \\ T_N \\ T_{N+1}
\end{pmatrix}
$$

$$
= -hA\tilde{\mathcal{P}}_o
\begin{pmatrix}
0 \\ 1 \\ 1 \\ \vdots \\ 1 \\ 1 \\ 1 \\ \vdots \\ 1 \\ 1 \\ 0
\end{pmatrix}
$$

Figure 12.2.   Matrix equation for finite-difference solution of the Poisson equation with a resistor network.

into a simulator such as SPICE, and use the SPICE solver. In either case, the accuracy of the solution will depend on the mesh, and if one needs accurate solutions, one must conduct a mesh refinement study to ensure that the results of the discretized problem are not highly sensitive to mesh size.

To create small analytical lumped-element models that can be solved by hand, one can use the methodology described here, but with a relatively coarse subdivision of the sample, creating a model with only a few resistor elements. An extreme example of this approach was presented in Section 11.6.4 when we analyzed the self-heating of a silicon resistor with a single equivalent thermal resistance. Such hand-built models have the virtue of providing analytical expressions for the dependences of behavior on device dimensions and material properties, but do not necessarily provide accurate quantitative results. In that sense, relatively coarse lumped models are analogous to relatively simple trial functions when using variational methods in energy conserving systems. They have the same virtue, namely, good algebraic insight into the functional form of the solution, but the same drawback: uncertain accuracy.

12.5   Eigenfunction Solution of the Poisson Equation

In the previous section, we divided the volume of the sample up into a set of local regions, built a local model for each region, and then connected the local models to create a lumped-element model for the entire structure. One disadvantage of this method is that it may require a large number of elements to create a good model.

An alternative is to seek a set of *global basis functions* that satisfy the boundary conditions of the problem, and then to determine how much of each of these basis functions is excited by the given source term. A particularly good set of functions is the set of eigenfunctions of the ∇^2 operator. In Cartesian coordinates, these eigenfunctions in one dimension have the form

$$e^{\pm jkx} \quad \text{or} \quad \cos(kx) \quad \text{and} \quad \sin(kx) \tag{12.14}$$

with obvious extension to three dimensions. The allowed values of the parameter k are determined by the boundary conditions.

The advantage of using eigenfunctions is that they are orthogonal to each other. That is, if ψ_i and ψ_j are different eigenfunctions, then

$$\int_{\text{sample}} \psi_i^* \psi_j dx = 0 \tag{12.15}$$

Furthermore, the eigenfunctions can be normalized such that

$$\int_{\text{sample}} \psi_i^* \psi_i dx = 1 \tag{12.16}$$

where the complex conjugate is required only when using the complex-exponential form of the eigenfunctions.

Let us use this approach to set up the same problem as in the previous section, namely, the temperature rise of a Joule-heated one-dimensional resistor with heat sinks at both ends. If we select the origin at one end of the resistor, the appropriate set of eigenfunctions are the $\sin(kx)$ set. The values of k are determined by the requirement that

$$\sin(kL) = 0 \tag{12.17}$$

which yields

$$k_n = \frac{n\pi}{L} \quad \text{for } n = 1, 2, 3 \ldots \tag{12.18}$$

The normalized set of eigenfunctions is

$$\psi_n(x) = \sqrt{\frac{2}{L}} \sin\left(\frac{n\pi x}{L}\right) \tag{12.19}$$

The general form of the solution is

$$T(x) = \sum_{n=1}^{\infty} A_n \psi_n(x) \tag{12.20}$$

If we substitute this form into the Poisson equation, we obtain

$$\sum_{n=1}^{\infty} k_n^2 A_n \psi_n(x) = \frac{\tilde{P}(x)}{\kappa} \tag{12.21}$$

To find the value of A_m, we multiply by $\psi_m(x)$ and integrate over the length of the resistor. Because of orthogonality and normalization, the left-hand-side collapses to $k_m^2 A_m$, and we obtain

$$A_m = \left(\frac{1}{\kappa k_m^2}\right) \sqrt{\frac{2}{L}} \int_0^L \sin\left(\frac{m\pi x}{L}\right) \tilde{P}(x)\, dx \tag{12.22}$$

Thus A_m is the projection of the functional form of $\tilde{P}(x)$ onto the eigenfunction, and we have a complete formal solution after only a few lines of algebra.

The primary virtue of this approach is that depending on the functional form of $\tilde{P}(x)$, only a few of the eigenfunctions may be needed to express the result. Therefore, we may be able to achieve extremely good accuracy with a model that contains only a few terms. But converting that solution into a lumped model is somewhat arbitrary, and it depends on what you wish to call input and output.

To illustrate, let us determine the maximum temperature in the center of the resistor as a function of the amount of Joule heating. We assume $\tilde{P}(x)$ is a constant $\tilde{P}_o$. Under those circumstances

$$A_n = \frac{\tilde{P}_o}{\kappa k_n^3} \sqrt{\frac{2}{L}} [1 - (-1)^n] \tag{12.23}$$

Only the odd values of n contribute, and the final solution is

$$T(x) = \frac{4\tilde{P}_o L^2}{\kappa \pi^3} \sum_{n \text{ odd}} \frac{\sin\left(\frac{n\pi x}{L}\right)}{n^2} \tag{12.24}$$

The maximum temperature at $x = L/2$ is

$$T_{\text{max}} = \frac{4\tilde{P}_o L^2}{\kappa \pi^3} \left[1 - \frac{1}{9} + \frac{1}{25} - \frac{1}{49} \cdots \right] \tag{12.25}$$

With only four terms, we can obtain the maximum temperature to a few percent. To use the result in a system, we can create an input-output lumped-element function block, suitable for insertion into a system level simulator such as SIMULINK, with Joule heat in and maximum temperature out.

12.6   Transient Response: Finite-Difference Approach

We can apply an extension of the finite-difference methods developed for the steady state problem to handle transients. What we must do is attach to each node a capacitive element that captures the heat capacity of the volume element associated with a node. Thus, by adding a capacitor

$$C = h_x h_y h_z \tilde{C} \tag{12.26}$$

between every node in the finite-difference network and ground, the transient behavior will be correctly captured. This approach is most useful for coarse lumped modeling by hand or when using a circuit simulator such as SPICE to find responses. When attempting to build accurate models that are small enough to be modeled by hand, it is often more useful to take the eigenfunction approach, as discussed in the following section.

12.7   Transient Response: Eigenfunction Method

We can extend the eigenfunction method in order to model transient behavior with global basis functions. As a first step, we will find the response to a spatially distributed source that is excited as an impulse in time. The method of separation of variables is extremely powerful for equations of this type. We write a trial solution as

$$T(\mathbf{r}, t) = \hat{T}(\mathbf{r}) e^{\alpha t} \tag{12.27}$$

where $\hat{T}(\mathbf{r})$ depends only on position, and α is a parameter to be determined. By substitution, we obtain

$$\left(\alpha \hat{T} - D \nabla^2 \hat{T} \right) e^{-\alpha t} = \frac{1}{\tilde{C}} \tilde{P}(\mathbf{r}, t) \tag{12.28}$$

Depending on the nature of the source term, there are various paths forward from here. Since we anticipate using Laplace transform methods to build an equivalent circuit, we choose to find the response when $\tilde{P}$ is an impulse in time:

$$\tilde{P}(\mathbf{r}, t) = \tilde{Q}_o(\mathbf{r}) \delta(t) \tag{12.29}$$

It is important to be clear about units. The function $\delta(t)$ has units of sec^{-1}. Therefore, $\tilde{Q}_o(\mathbf{r})$ has different units from $\tilde{P}(\mathbf{r}, t)$. While $\tilde{P}$ has units Watts/m^3, $\tilde{Q}_o$ has units J/m^3. The units of D are always m^2/sec.

The impulse is not a physically realizable input, but it is a mathematically useful construction for extracting an equivalent circuit model from which we will be able to obtain physically meaningful results for fully realizable inputs, including DC sources, sinusoidal sources, or various pulsed sources.

With the impulse input, the right-hand side is zero for all times $t > 0^+$, and we can divide out the exponential time factor to obtain:

$$D\nabla^2 \hat{T} = -\alpha \hat{T} \tag{12.30}$$

Thus $-\alpha/D$ is identified as an eigenvalue of ∇^2, and we obtain the immediate result that

$$\hat{T} = e^{j\mathbf{k}\cdot\mathbf{r}} \tag{12.31}$$

where

$$\mathbf{k} = \left(\pm\sqrt{\frac{\alpha_x}{D}}x, \pm\sqrt{\frac{\alpha_y}{D}}y, \pm\sqrt{\frac{\alpha_z}{D}}z \right) \tag{12.32}$$

The values of α_x, α_y and α_z depend on the spatial boundary conditions. We will typically use linear combinations of the plus and minus exponentials to create sinusoidal eigenfunctions. The full solution will be a superposition of sinusoidal terms, assembled so as to match both the boundary conditions and the initial conditions.

12.8   One-Dimensional Example

To illustrate the method, we select a one-dimensional example in which there is no y- or z-dependence in the result, but there is an x-dependence (see Fig. 12.3). We further assume that the impulsive source at $t = 0$ is spatially uniform; that is, $\tilde{Q}_o(\mathbf{r})$ is a constant.

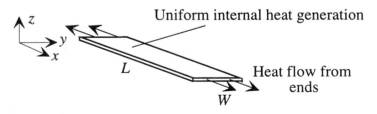

Figure 12.3.   A one-dimensional heat-flow example.

The boundary condition we shall select is that $\hat{T}$ is zero at $x = 0$ and at $x = L$, where L is the length of the sample. This example corresponds to the heat distribution in a long and thin resistor attached to heat sinks at its ends, but insulated at the top and bottom and at its edges. The width of the sample is W and the thickness H.

To match the boundary conditions, it is necessary to combine the plus and minus eigenfunctions to create a spatial sinusoid:

$$\hat{T}_n = \sin\left(\frac{n\pi x}{L}\right) \tag{12.33}$$

where we have omitted the normalization factor, but we have retained the fact that eigenfunctions of different n are orthogonal. This choice of eigenfunction means that

$$\sqrt{\frac{\alpha_n}{D}} = \frac{n\pi}{L} \; ; \quad \text{for } n = 1, 2, 3 \ldots \tag{12.34}$$

Thus, the full solution is of the form

$$T(x,t) = \sum_{n=1}^{\infty} A \left\{ \sin\left[\frac{n\pi x}{L}\right] \right\} e^{-\alpha_n t} \tag{12.35}$$

To find the coefficients A_n, we take the limit at $t = 0$ and require the spatial function to match the spatial distribution of the original impulse $\tilde{Q}_o(x)$. In the example selected here, $\tilde{Q}_o(x)$ is presumed uniform, with amplitude $\tilde{Q}_o$. Thus, the coefficients A_n are the Fourier coefficients of a square wave, found from

$$A_n = \frac{\tilde{Q}_o}{\tilde{C}} \frac{2}{L} \int_0^L \sin\left[\frac{n\pi x}{L}\right] dx = \frac{4}{n\pi} \frac{\tilde{Q}_o}{\tilde{C}} \tag{12.36}$$

for n odd. Thus, the final solution is

$$T(x,t) = \frac{\tilde{Q}_o}{\tilde{C}} \sum_{n \text{ odd}} \frac{4}{n\pi} \sin\left[\frac{n\pi x}{L}\right] e^{-\alpha_n t} \tag{12.37}$$

This is the impulse response. The interpretation of this response is that the system has a variety of spatial modes, and each mode has its own decay constant. When we excite the system with a temporal impulse that is spatially uniform, each mode is excited to the amplitude A_n, and, thereafter, it decays with time with decay constant α_n.

12.9 Equivalent Circuit for a Single Mode

The impulse response is a sum of terms of identical form. Therefore, we can begin to build an equivalent-circuit representation by examining a single term. The pathway to finding an equivalent circuit representation is to use the Laplace transform.

The system function relating the Laplace transform of the time-dependence of $\tilde{P}$ to the Laplace transform of T is the Laplace transform of the impulse response. It is a bit awkward to use system functions when working with combined space and time dependences. However, if we construct a scalar function of interest, such as the total flux out of the edges of the plate, the problem reduces to a lumped problem. To carry the heat-flow example forward, we will use as our scalar function the total heat current leaving the sample along its two edges:

$$I_Q(t) = -\kappa \int_0^W \int_0^H \left(\frac{\partial T}{\partial x}\bigg|_{x=0} - \frac{\partial T}{\partial x}\bigg|_{x=L} \right) dy\, dz \qquad (12.38)$$

This reduces to

$$I_Q(t) = \left[\frac{8\kappa W H}{L} \sum_{n=1}^{\infty} e^{-\alpha_n t} \right] \frac{\tilde{Q}_o}{\tilde{C}} \qquad (12.39)$$

Because this response is a sum of transient terms, we shall examine one typical term and create its equivalent circuit. Then we will return to build the complete equivalent circuit including the rest of the terms. The Laplace transform of a single term in the sum is

$$I_{Q,n}(s) = \frac{8\kappa W H}{\alpha_n L} \left(\frac{1}{1 + \frac{s}{\alpha_n}} \right) \frac{\tilde{Q}_o}{\tilde{C}} \qquad (12.40)$$

This is a single-pole transfer function. We can build an equivalent circuit by assigning T as the across variable and I_Q as the through variable. The equivalent circuit for the source term is a current source, since it supplies an impulse of heat current at $t = 0$ to establish the initial temperature distribution.

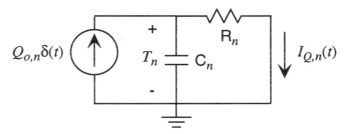

Figure 12.4. An equivalent circuit for a single mode relating input heat current due to Joule heating to output heat current at the ends of the sample.

An equivalent circuit that yields the correct form of the transfer function is shown in Fig. 12.4. It shows an impulse current source and an *RC* current divider, with the voltage across the capacitor labeled as T_n. If we identify T_n

as the amplitude of the excitation of mode n, then we can identify the elements of the model as follows:

The capacitor C_n represents the heat capacity of the mode. Thus, it is the total energy in the mode when the mode has unit amplitude. Therefore, we can write:

$$C_n = C \int_0^W \int_0^L \int_0^H \sin\left[\frac{n\pi x}{L}\right] dx\,dy\,dz \qquad (12.41)$$

This evaluates to

$$C_n = \left(\frac{2}{n\pi}\right) \mathcal{V}\tilde{C} \qquad (12.42)$$

where $\mathcal{V}$ is the volume of the sample, and $\tilde{C}$ is the specific heat per unit volume. This expression provides a reasonably clear explanation of the capacitance. It is proportional to the heat capacity of the entire sample, but scaled down by a factor in parenthesis that derives from the mode shape.

To find the resistor, we note that we must have

$$\frac{1}{R_n C_n} = \alpha_n = \frac{Dn^2\pi^2}{L^2} \qquad (12.43)$$

which leads to

$$R_n = \frac{1}{n\pi} \frac{L/2}{\kappa W H} \qquad (12.44)$$

Thus, the resistance is exactly what one would expect for a sample of length $L/2$ and cross-sectional area WH, but scaled by a factor $1/n\pi$ that comes from the mode shape.

Finally, we must consider the value of the current source. The requirement is that just after the impulse at $t = 0^+$ the value of X_n must equal A_n. This means that

$$Q_{o,n} = C_n A_n = \frac{8}{n^2\pi^2} \mathcal{V}\tilde{Q}_o \qquad (12.45)$$

12.10   Equivalent Circuit Including All Modes

Since each mode is independent, we can combine the results of the last section for each mode to create the partial equivalent circuit is shown in Fig. 12.5, which shows the effect of three modes. Obviously, more could be added.

The principles and methods used to create this equivalent circuit can now be applied to more general problems. For example, if the generation of heat is spatially uniform but time varying in some general manner, one can replace

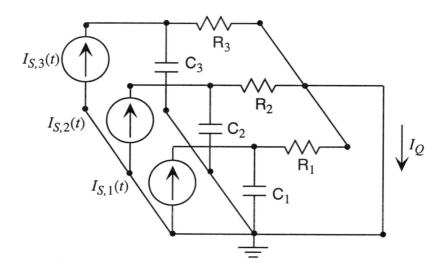

Figure 12.5. Equivalent circuit for three of the modes of the one-dimensional heat-flow example.

the temporal impulse by the appropriate time-dependence and use the circuit to find the net response.

As an example, we will consider the Joule heating of a ribbon-like resistor with heat sinking on its ends. We assume that the resistor carries a uniform current density J_e in the x-direction, so that the Joule heating is spatially uniform. However, we will assume that the current waveform is a sinusoid at frequency ω. That is, we assume an electrical current I_e given by

$$I_e = I_o \cos \omega t \tag{12.46}$$

The Joule heating, which provides the spatially uniform source term for the heat flow equation, is then

$$\tilde{P}(t) = \frac{I_o^2 R_e}{2} (1 + \cos 2\omega t) \tag{12.47}$$

Since the heat source is spatially uniform, we can use for each current source

$$I_{S,n}(t) = \left(\frac{8\mathcal{V}}{n^2 \pi^2} \right) \frac{I_o^2 R_e}{2} (1 + \cos 2\omega t) \tag{12.48}$$

There will be a transient associated with establishing the Joule heating, after which there will be a combined DC and sinusoidal steady state. The DC steady state reduces to the problem we considered in Section 12.5. Now that we have an equivalent circuit, we can find both the DC heat current and the sinusoidal-steady-state heat current at frequency 2ω directly from the circuit. The sinusoidal component has complex amplitude

$$I_Q(2j\omega) = \sum_{n \text{ odd}} \left(\frac{1}{1 + 2j\omega R_n C_n}\right) \left(\frac{8\mathcal{V}}{n^2\pi^2}\right) \frac{I_o^2 R_e}{2} \qquad (12.49)$$

We see that the amplitude of the various modes decreases as $1/n^2$. Thus the second mode ($n = 3$) contributes only 1/9 as much as the first mode, and the third mode ($n = 5$) only 1/25th. Therefore, we are often justified in using only the mode with the largest coefficient for our model, corresponding in this case, to the first term in the sum.

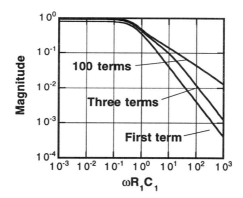

Figure 12.6. Bode plot for the self-heated resistor transfer function.

The transfer function between total heat generated inside the sample and the heat current through the ends can be written

$$\frac{I_Q(2j\omega)}{(I_o^2 R_e \mathcal{V}/2)} = \frac{8}{\pi^2} \sum_{n \text{ odd}} \left(\frac{1}{1 + 2j\omega R_n C_n}\right) \qquad (12.50)$$

Figure 12.6 shows the magnitude Bode plots for this transfer function using only the first term, the first three terms, and the sum to 100 terms, which for the frequency range in the graph, is equivalent to the exact solution. The exact solution has a magnitude of unity at zero frequency. The model with a single term has a zero-frequency magnitude of 0.81, and the three-term solution, .93. All three solutions show a roll-off in response at about the same point, $\omega R_1 C_1 \approx 1/2$. However, the shape of the high-frequency response is very different for the three models. The single-term model, as expected, shows a $1/\omega$ roll-off characteristic of a single-pole response. The exact solution shows a $1/\sqrt{\omega}$ behavior. This is characteristic of a variety of systems controlled by diffusion.[3] The three-term model is in between. It shows a hint of the $1/\sqrt{\omega}$ behavior near the roll-off frequency, but eventually goes over to the $1/\omega$ behavior at high frequencies.

Our conclusion is that the simple lumped model with only one resistor and one capacitor gives a faithful but approximate estimate of the low-frequency response and the roll-off frequency, but fails in the details of the high-frequency response. If the high-frequency behavior is important, a more complete model is needed. Overall, we have achieved our goals of finding a compact circuit model that has reasonable accuracy, and gaining insight into where and how such a model will fail.

Related Reading

H. S. Carslaw and J. C. Jaeger, *Conduction of Heat in Solids*, Second Edition, New York: Oxford University Press, 1959, reprinted in 1986.

J. Crank, *The Mathematics of Diffusion*, Second Edition, Oxford: Clarendon Press, 1975.

Problems

12.1 A silicon plate 100 μm square and 2 μm thick is suspended 2 μm above a silicon substrate by tethers attached to each corner, each of which is 100 μm long, 10 μm wide, and 2 μm thick. Divide this structure into square cells 10 μm on a side, and set up the finite-difference Poisson equation. Assume that this structure is heated with identical Joule heat sources in the four cells at the center of the plate. Determine the steady-state temperature distribution. How valid is it to treat the plate as an isothermal element connected to the substrate with lumped-element thermal resistors determined by the tethers? How does this result change if the Joule heating is more uniformly spread throughout the area of the plate?

---

[3] A distributed R-C transmission line has $1/\sqrt{\omega}$ behavior at high frequencies, and we have already shown in discussing the finite-difference approach in Section 12.6 that all heat-flow problems can be mapped into distributed R-C systems.

12.2 A silicon beam 300 μm long with a 2 μm square cross-section is suspended above a silicon wafer by rigid supports. The beam is heated by Joule heat from a steady current along the beam. Assuming that the temperature of the supports and of the wafer are fixed at room temperature, determine how fine a finite-difference mesh is required to match the accuracy of the temperature distribution calculated from a solution based on the first eigenfunction? What if the first three eigenfunctions are used?

12.3 Use the eigenfunction method to estimate the thermal time constant of a circular silicon nitride plate, 200 μm in diameter and 1 μm thick, assuming the edge of the plate is clamped at the reference temperature. This will require determination of the radial eigenfunctions in polar coordinates.

Chapter 13

FLUIDS

Flow gently, sweet Afton . . .

—Robert Burns

13.1   What Makes Fluids Difficult?

Unlike solids, fluids don't stay where they are put. When subjected to shear forces, they deform without limit, and because of their viscosity, this deformation is inherently dissipative and irreversible.   However, fluids do have elastic properties when subjected to normal forces (pressure). And they have inertia. Thus, just like an elastic medium, fluids can propagate sound waves. There are great differences in fluid behavior depending on the relative importance of viscous forces to inertial forces, on the one hand, and the relative size of the velocity of flow to the speed of sound, on the other. The final difficulty: the governing partial differential equations for fluids, unlike the diffusion and heat flow equations, are intrinsically nonlinear.

The reader will not be surprised to learn that we shall not be covering all these topics within the confined space of one chapter. Our goal is to introduce basic fluidic concepts, and focus on the phenomena that are important in a broad range of MEMS devices. References to more complete treatments can be found in the Related Reading section at the end of this chapter.[1]

---

[1]The recent development of microturbomachinery [29, 68] and microrocket engines [69] has now made *all* of fluid mechanics relevant to MEMS. It is possible that future editions of books like this will have to include a much wider range of topics.

13.2   Basic Fluid Concepts

13.2.1   Viscosity

The primary feature of fluidic behavior is the continuous deformation in the presence of shear forces. Figure 13.1 illustrates the concept.

Figure 13.1. Fluid between two plates. The upper plate moves to the right with velocity U, setting up shear forces τ.

When the upper plate in Fig. 13.1 moves to the right at velocity U, the friction between the fluid and the wall applies a shear force to the fluid. This shear force sets up a transverse velocity gradient in the fluid, with the fluid velocity increasing from zero at the lower plate to U at the upper plate. The fluid flows! Such flow is described in terms of continuous shear deformation of the fluid, the shear rate increasing with increasing shear force. A *Newtonian fluid* is one for which the shear rate is proportional to the shear force. The proportionality constant is called the *viscosity*, denoted here[2] by the symbol η. Under steady-state conditions in the geometry of Fig. 13.1, the relation between shear force and viscosity is

$$\tau = \eta \frac{U}{h} \tag{13.1}$$

In the limit of a differentially thin layer of fluid, this becomes a differential relation:

$$\tau = \eta \frac{\partial U_x}{\partial y} \tag{13.2}$$

This transverse gradient in velocity describes the continuous shearing of the fluid in the presence of shear force. The units for viscosity are Pascal-seconds.

A related quantity is called the *kinematic viscosity*, which we shall denote by η^*. It is the ratio of the viscosity to the density of the fluid.

$$\eta^* = \frac{\eta}{\rho_m} \tag{13.3}$$

Shear flow of a fluid is inherently dissipative. The dissipated power density $\tilde{\mathcal{P}}$ (in Watts/m³) is

---

[2]Many authors use the symbol μ for viscosity.

$$\tilde{\mathcal{P}} = \tau \frac{\partial U_x}{\partial y} \tag{13.4}$$

The total power dissipation in a specified volume of fluid is found from integration of $\tilde{\mathcal{P}}$ over that volume.

13.2.2   Thermophysical Properties

A fluid can never be in static equilibrium in the presences of shear forces. However, it can be in static equilibrium with pressure forces. The density of a fluid, in general, depends on the temperature and the pressure. We write

$$\frac{\delta \rho_m}{\rho_m} = \left(\frac{\partial \ln \rho_m}{\partial P} \right)_T \delta P + \left(\frac{\partial \ln \rho_m}{\partial T} \right)_P \delta T \tag{13.5}$$

The first term in parentheses is the bulk compliance, which is the reciprocal of the bulk modulus, while the second term in parentheses is the negative of the volume coefficient of thermal expansion. For water, the bulk modulus is about 2 GPa, comparable to a "soft" solid. For gases such as air, the bulk modulus is on the order of 0.1 MPa, many orders of magnitude more compressible.

In most MEMS applications involving liquids, the finite compressibility of the liquid can be neglected compared to flow phenomena set up by pressure gradients and shear forces. Therefore, it is common practice to treat liquid flow as *incompressible* to first order, except that if one wishes to study sound waves in liquids, one must include the compressibility. Gases, on the other hand, are very compressible. We shall find that in some device regimes, the compressibility of air has important implications for operation of MEMS devices.

The thermal expansion coefficient of water is about 1.5×10^{-4} K^{-1}, about 10 times (or more) greater than that of typical solids. The thermal expansion coefficient of air is more than 10 times larger than that of water.

The relation between density, pressure, and temperature for a fluid is expressed through an *equation of state*. If we treat liquids as incompressible, we shall not need an equation of state, since we will treat the density as constant with respect to pressure, and it will not change very much with temperature. For gases, however, we will use the *Ideal Gas Law*, which is written

$$P = \rho_m \left(\frac{R}{M_W} \right) T \tag{13.6}$$

where R is the universal gas constant, 8.3×10^3 J/kg-K, and M_W is the molecular weight of the gas. We shall use this equation of state when analyzing the effects of compressibility on the squeezed-film damping of mechanical motions of structures.

13.2.3   Surface Tension

Molecules at the surface of a liquid sample experience different forces than molecules in the interior. In general, surface locations are energetically unfavorable compared to interior positions. This creates a distributed force called *surface tension* that tends to minimize the free surface of a liquid. Hence, liquids tend to form spherical drops.

We can define the surface tension quantitatively by considering exactly one half of a spherical drop, as shown in Fig. 13.2.

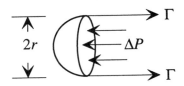

Figure 13.2.   Illustrating the force balance between the surface tension of a hemispherical liquid surface and the pressure difference across the surface.

A hemispherical segment of fluid with radius r can be in equilibrium provided that the total force on it is zero. If we postulate a surface-tension force per unit length Γ, with units N/m, we must then also postulate a pressure difference ΔP between the two sides of the surface to balance this surface tension. The force-balance equation is

$$(2\pi r)\Gamma = \Delta P \left(\pi r^2\right) \tag{13.7}$$

Solving for ΔP, we obtain

$$\Delta P = \frac{2\Gamma}{r} \tag{13.8}$$

When the surface is flat (infinite r), there is no pressure difference, but just as we found for fixed-end beams with residual stress, when a curved surface is created, there must be a balancing pressure load in equilibrium. The surface tension units of N/m is equivalent to J/m^2. Thus the surface tension can be interpreted as the energy required to create a unit of surface. An interesting calculation is to compare the surface energy of a drop of radius r, equal to $4\pi r^2\Gamma$, to the volume of the drop $(4\pi r^3/3)$. This ratio is $3\Gamma/r$. The surface energy per drop volume is inversely proportional to drop radius. Thus small drops are energetically unfavorable compared to larger drops.

The opposite of a drop is a bubble of vapor inside a liquid, but the concept is the same. There must be pressure inside the bubble to balance the surface tension trying to collapse the bubble. Small bubbles can have very large internal pressures, and when bubbles attach to surfaces, the large localized forces associated with bubble formation and collapse (generally called *cavitation*) can cause damage.

Surface tension can cause fluid transport and can affect states of equilibrium. The simplest example is the static (equilibrium) rise of liquid in a capillary, as illustrated in Fig. 13.3. The fluid in the meniscus forms an acute angle θ with the wall of the capillary, called the *contact angle*. The contact angle is a characteristic of the various interfaces involved, and can vary from convex for liquids that wet the surface to concave for nonwetting liquids.

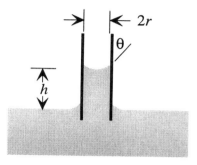

Figure 13.3.   Illustrating the rise of a wetting liquid in a capillary due to surface tension forces.

The net upward force on the liquid due to surface tension is $2\pi r\Gamma\cos\theta$. There must be a pressure difference between the atmosphere above the meniscus and the liquid just beneath the meniscus. If we take the atmosphere at a reference pressure of zero, it means that just inside the meniscus, the fluid pressure is actually negative. The mass of fluid in the capillary column, in combination with the acceleration of gravity g, creates a gradually increasing pressure as one moves from the meniscus down to the reference level of the liquid. The height of the column in equilibrium is just that needed for the pressure at the bottom of the column to equal the atmospheric reference pressure. The total force on the body of fluid in the capillary is zero, since it has atmospheric pressure both at the top and the bottom. Therefore, the upward force due to the meniscus and the downward force due to the inertial mass must balance:

$$\rho_m g(\pi r^2 h) = 2\pi r\Gamma\cos\theta \tag{13.9}$$

or

$$h = \frac{2\Gamma\cos\theta}{\rho_m g r} \tag{13.10}$$

Contact angles depend both on the wetting liquid and the chemistry of the surface and can be sensitive to contamination and the effects of roughness. Generally, when water wets a compatible surface such as glass, contact angles are small, typically tens of degrees. In the case of non-wetting liquids, the contact angle is obtuse, with $\cos\theta$ negative, and the meniscus is depressed below the reference level.

In MEMS devices, we very often make fluid channels that are small enough for surface tension forces to have a large effect on device operation. For example, the surface tension of water is about 72×10^{-3} N/m at room temperature. If it wets the wall of a 20μm diameter channel with a contact angle of $30°$, the pressure drop across the meniscus is 12.5 kPa, or about 1/8 atmosphere.

13.2.4   Conservation of Mass

Because fluids are typically in motion, we must be careful when constructing conservation laws. We must decide whether to follow a specific block of fluid as it deforms and moves (the Lagrangian approach), or follow a particular volume of space as the fluid moves through it (the Eulerian approach). Both have their place in treatments of fluid mechanics, but for the sake of compactness, we present only the Eulerian approach.

The starting point for an Eulerian description is a *control volume*, a region fixed in space through which the fluid can move. If we imagine a volume $\mathcal{V}$, and compute the total mass m inside that volume, we obtain

$$m = \int_{\mathcal{V}} \rho_m \, d\mathcal{V} \qquad (13.11)$$

The time rate of change of mass inside the control volume must equal the negative of the outflow of mass. We find this outflow by a surface integral over the bounding surface $\mathcal{S}$. Thus

$$\frac{dm}{dt} = -\int_{\mathcal{S}} \rho_m \mathbf{U} \cdot \mathbf{n} \, d\mathcal{S} \qquad (13.12)$$

Since the control volume is fixed in space, we can take the time derivative inside the integral to obtain

$$\int_{\mathcal{V}} \frac{\partial \rho_m}{\partial t} d\mathcal{V} + \int_{\mathcal{S}} \rho_m \mathbf{U} \cdot \mathbf{n} \, d\mathcal{S} = 0 \qquad (13.13)$$

and we can convert the surface integral to a volume integral using the divergence theorem, to obtain a mass conservation law:

$$\int_{\mathcal{V}} \left(\frac{\partial \rho_m}{\partial t} + \nabla \cdot (\rho_m \mathbf{U}) \right) d\mathcal{V} = 0 \qquad (13.14)$$

Since the control volume is arbitrary, the integrand must vanish uniformly. The resulting equation is called the *continuity equation*.

$$\frac{\partial \rho_m}{\partial t} + \nabla \cdot (\rho_m \mathbf{U}) = 0 \qquad (13.15)$$

We can rewrite this expression by expanding the divergence expression to yield

$$\frac{\partial \rho_m}{\partial t} + (\mathbf{U} \cdot \nabla)\rho_m + \rho_m \nabla \cdot \mathbf{U} = 0 \tag{13.16}$$

By convention, the first two terms are combined into a form called the *material derivative*, using the upper-case D to distinguish it from normal or partial differentiation

$$\frac{D\rho_m}{Dt} = \frac{\partial \rho_m}{\partial t} + (\mathbf{U} \cdot \nabla)\,\rho_m \tag{13.17}$$

More generally, we write the operator identity for the material derivative as

$$\frac{D}{Dt} = \frac{\partial}{\partial t} + \mathbf{U} \cdot \nabla \tag{13.18}$$

The material derivative measures the time rate of change of a property for an observer who moves with the fluid. Using the material derivative, we can rewrite the continuity equation as

$$\frac{D\rho_m}{Dt} + \rho_m \nabla \cdot \mathbf{U} = 0 \tag{13.19}$$

13.2.5   Time Rate of Change of Momentum

We can apply the same mathematical approach to calculate the time rate of change of other fluid properties. If we define B as the integral of a property $\rho_m b$ over the volume of the fluid, we can write

$$B = \int_{\mathcal{V}_M} \rho_m b \, d\mathcal{V}_M \tag{13.20}$$

where $\mathcal{V}_M$ is a material volume that moves with the fluid (the Lagrangian description). With the aid of the Reynolds' Transport Theorem[3], we can calculate the time rate of change of B and convert it to an Eulerian description with a fixed control volume:

$$\frac{dB}{dt} = \frac{d}{dt} \int_{\mathcal{V}} \rho_m b \, d\mathcal{V} + \int_{\mathcal{S}} \rho_m b \mathbf{U} \cdot \mathbf{n} \, d\mathcal{S} \tag{13.21}$$

If we use the velocity $\mathbf{U}$ as the quantity b, we obtain the total momentum $\mathbf{p}$ inside the control volume:

$$\frac{d\mathbf{p}}{dt} = \frac{d}{dt} \int_{\mathcal{V}} \rho_m \mathbf{U} \, d\mathcal{V} + \int_{\mathcal{S}} \rho_m \mathbf{U}(\mathbf{U} \cdot \mathbf{n}) \, d\mathcal{S} \tag{13.22}$$

---

[3] See, for example, Fay, *Introduction to Fluid Mechanics*, Section 5.2 [70].

We now use this result to obtain the governing equation of fluid flow.

13.2.6   The Navier-Stokes Equation

We can compute the total forces acting on the control volume and apply Newton's law of motion to the time rate of change of momentum from Eq. 13.22 to obtain an equation of motion governing fluid behavior. There are three components of the total force: (1) the net pressure acting normal to the bounding surface, (2) the net shear force acting tangential to the surface, and (3) any body force, such as gravity, acting throughout the volume. If we sum up all the forces acting on the control volume, and equate that sum to $d\mathbf{p}/dt$, we obtain

$$
\frac{d}{dt} \int_{\mathcal{V}} \rho_m \mathbf{U} d\mathcal{V} + \int_{\mathcal{S}} \rho_m \mathbf{U} (\mathbf{U} \cdot \mathbf{n}) d\mathcal{S}
$$

$$
= \int_{\mathcal{S}} (-P\mathbf{n} + \boldsymbol{\tau}) \, d\mathcal{S} + \int_{\mathcal{V}} \rho_m \boldsymbol{g} d\mathcal{V}
$$

(13.23)

where $\mathcal{S}$ is the surface bounding the volume $\mathcal{V}$ and $\boldsymbol{g}$ is the acceleration of gravity. The negative sign on the pressure term is due to the fact that the pressure force acting on the control volume is in the opposite direction to the surface normal.

We can apply the same kind of manipulations that converted the integral form of the continuity equation into the differential form. The surface integral of pressure times the normal vector $\mathbf{n}$ becomes a volume integral over ∇P. When the surface integral over the shear forces is converted to a volume integral, the result[4] for uniform viscosity is an equivalent body force $\mathbf{f}$ of the form

$$
\mathbf{f} = \eta \nabla^2 \mathbf{U} + \frac{\eta}{3} \nabla (\nabla \cdot \mathbf{U})
$$

(13.24)

In differential form, Newton's second law becomes the famous *Navier-Stokes equation*, describing the motion of a fluid in the presence of external and internal viscous forces:

$$
\rho_m \frac{D\mathbf{U}}{Dt} = -\nabla P + \rho_m \boldsymbol{g} + \eta \nabla^2 \mathbf{U} + \frac{\eta}{3} \nabla (\nabla \cdot \mathbf{U})
$$

(13.25)

We shall use simplified versions of this equation to analyze several examples later in this chapter.

13.2.7   Energy Conservation

Fluids have kinetic energy from their motion and potential energy that arises from the action of conservative forces such as gravity. They are subject to frictional shear forces at bounding surfaces and to internal dissipation due

---

[4]See, for example, Kuethe and Chow, *Foundations of Aerodynamics*, Appendix B [71].

to viscous forces. The dissipative effects generate heat and heat flow. The formulation of the conservation of energy in a moving fluid with heat generation and heat flow is daunting. While we will not be using this equation for the examples in this book, the result is included here for completeness [71]:

$$\rho_m \frac{D\tilde{\mathcal{U}}_m}{Dt} - \frac{P}{\rho_m} \frac{D\rho_m}{Dt} = -\nabla \cdot \mathbf{J}_Q + \Phi \qquad (13.26)$$

where $\tilde{\mathcal{U}}_m$ is the internal energy per unit mass, $\mathbf{J}_Q$ is the heat flux, and Φ is the dissipation function given by

$$\Phi = \frac{2}{3}\eta \left[\left(\frac{\partial U_x}{\partial x} - \frac{\partial U_y}{\partial y}\right)^2 + \left(\frac{\partial U_y}{\partial y} - \frac{\partial U_z}{\partial z}\right)^2 + \left(\frac{\partial U_x}{\partial x} - \frac{\partial U_z}{\partial z}\right)^2 \right]$$

$$+ \eta \left[\left(\frac{\partial U_y}{\partial x} + \frac{\partial U_x}{\partial y}\right)^2 + \left(\frac{\partial U_z}{\partial y} + \frac{\partial U_y}{\partial z}\right)^2 + \left(\frac{\partial U_z}{\partial x} + \frac{\partial U_x}{\partial z}\right)^2 \right] \quad (13.27)$$

$$+ \eta_B \left(\nabla \cdot \mathbf{U}\right)^2$$

The quantity η_B is the *bulk viscosity*, which arises from the internal-friction component of pure compression of a fluid. This term is normally negligible compared to other terms.

13.2.8   Reynolds Number and Mach Number

There is a huge array of dimensionless numbers used to characterize the many different regimes of fluid behavior. Here we present only two: the Reynolds Number and the Mach Number.

The *Reynolds number Re* is a ratio that measures the relative importance of inertial forces to viscous forces. It emerges naturally from the process of converting the Navier-Stokes equation to normalized variables,[5] and is defined as

$$Re = \frac{\rho_m LU}{\eta} = \frac{LU}{\eta^*} \qquad (13.28)$$

where L is a characteristic dimension for the problem and U is a characteristic velocity of the flow. We identify several regimes. For $Re \ll 1$, viscous forces dominate and we can neglect inertial forces by comparison. Most of the fluid flow problems in MEMS devices fall in this regime. For Re roughly less than 1000, the flow is *laminar*, following smooth streamlines. At higher values of

---

[5]See, for example, Deen, *Analysis of Transport Phenomena*, Section 5.10 [61].

Reynolds number, the flow is turbulent, with complex fluctuations in the flow field. We rarely encounter turbulent flow in MEMS devices.

The *Mach Number* M is the ratio of the flow velocity to the local velocity of sound c.

$$M = \frac{U}{c} \tag{13.29}$$

The velocity of sound is a function of the compressibility of the fluid and its density, and can be expressed as the derivative of the pressure with respect to density at constant entropy

$$c = \left(\frac{\partial P}{\partial \rho_m} \right)_S \tag{13.30}$$

In liquids, the speed of sound is quite large, on the order of thousands of meters per second. We virtually never encounter flow velocities in liquid MEMS systems that approach the speed of sound. However, gases, especially passing through narrow orifices under the influence of large pressure drops, can accelerate to speeds on the order of the sound velocity. An array of complex phenomena, falling under the general headings of compressible flow and rarefaction, result. We shall not explore this high-velocity regime in this book. All of our examples will involve Mach numbers considerably smaller than unity.

13.3   Incompressible Laminar Flow

An *incompressible flow* is one for which the time rate of change of density $D\rho_m/Dt$ is negligible. Under these circumstances, the continuity equation for mass conservation becomes:

$$\nabla \cdot \mathbf{U} = 0 \tag{13.31}$$

and the Navier-Stokes equation reduces to

$$\rho_m \frac{D\mathbf{U}}{Dt} = -\nabla P + \rho_m \mathbf{g} + \eta \nabla^2 \mathbf{U} \tag{13.32}$$

We can combine the pressure and gravity body-force terms by defining

$$P^* = P - \rho_m \mathbf{g} \cdot \mathbf{r} \tag{13.33}$$

where $\mathbf{r}$ is the position. With this definition, the Navier-Stokes equation reduces to

$$\rho_m \frac{D\mathbf{U}}{Dt} = \eta \nabla^2 \mathbf{U} - \nabla P^* \tag{13.34}$$

We recognize our old friend, the heat-flow equation of Chapters 11 and 12, recast in terms of the material derivative of a vector velocity **U**. We now examine several basic steady-flow examples.

13.3.1   Couette Flow

Steady viscous flow between parallel plates, one of which is moving parallel to the other, is called *Couette flow*. Figure 13.4 illustrates this situation on the left. On the right is the steady-state velocity distribution.

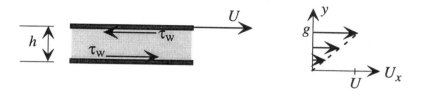

Figure 13.4.   Illustrating Couette flow

Far away from any edges or entrance ports, the flow is perfectly one-dimensional. Further, we know that $\nabla \cdot \mathbf{U} = 0$, so there can be no x-dependence to U_x. Thus, we can write the general form of the velocity distribution as

$$\mathbf{U} = U_x(y)\mathbf{i}_x \tag{13.35}$$

where $\mathbf{i}_x$ is the unit vector in the x direction. *Streamlines* are curves in space that are everywhere tangent to the local velocity field **U**. Thus the arrows on the right of Fig. 13.4 are the streamlines.

Under these steady flow conditions both $\partial \mathbf{U}/\partial t$ and $\mathbf{U} \cdot \nabla \mathbf{U}$ are zero. Thus, $D\mathbf{U}/Dt$ is zero. Further, we shall assume that there is no gradient of P^* created by the moving plate (we shall consider pressure gradients separately). Under these assumptions, the Navier-Stokes equation collapses to

$$\frac{\partial^2 U_x}{\partial y^2} = 0 \tag{13.36}$$

The solution is a linear velocity profile. The coefficients are determined by the boundary conditions. The boundary conditions we shall use are the classical *no-slip* boundary conditions, which state that the relative tangential velocity of a fluid in contact with a wall is zero. This assumption is valid for liquids and for gases in relatively large structures. When the dimensions of the structure become comparable with the molecular mean free path in the gas, this assumption breaks down. We shall discuss this further when we consider squeezed-film damping in Section 13.4. With these no-slip boundary conditions, the solution, shown on the right of Fig. 13.4, is

$$U_x = \frac{y}{h}U \tag{13.37}$$

The shear stress τ_w acting on the plate as a result of the motion is

$$\tau_w = -\eta \left. \frac{\partial U_x}{\partial y} \right|_{y=h} = -\eta \frac{U}{h} \mathbf{i}_x \tag{13.38}$$

The wall shear stress is proportional to the velocity gradient at the wall, and is directed so as to oppose the applied force creating the motion. This is a dissipative stress, proportional to velocity.

To create a lumped-element model of the mechanical effect of Couette flow on the plate, we can compute the total retarding force, which is $\tau_w A$, where A is the wall area. If we are using the $e \rightarrow V$ convention, this wall force is the across variable, and the plate velocity U is the through variable. Therefore, in the mechanical domain using the $e \rightarrow V$ convention, Couette flow appears as a resistor with value

$$R_{\text{Couette}} = \frac{\text{across}}{\text{through}} = \frac{\eta A}{h} \tag{13.39}$$

The power dissipation corresponding to Joule heating in the resistor is

$$\mathcal{P}_{\text{Couette}} = R U^2 \tag{13.40}$$

13.3.2   Poiseuille Flow

We now consider a pressure-driven flow between stationary parallel plates, and we will assume that the flow is horizontal, so there is no contribution from gravity.[6]

Once again, we can assume that the velocity in the x-direction is only a function of y, but now we will assume a uniform pressure gradient in the x-direction.

$$\frac{dP}{dx} = -K \quad \text{(constant)} \tag{13.41}$$

With these assumptions, the Navier-Stokes equation becomes

$$\frac{\partial^2 U_x}{\partial y^2} = -\frac{K}{\eta} \tag{13.42}$$

---

[6]This makes $P^* = P$. For flows where gravity is important, it is necessary to replace P by P^* in the results that follow.

This equation has a quadratic-polynomial solution. With the no-slip boundary condition, the solution becomes

$$U_x = \frac{1}{2\eta}[y(h-y)]K \qquad (13.43)$$

This parabolic flow profile is called *Poiseuille flow*. It is illustrated in Fig. 13.5.

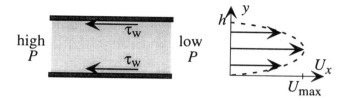

Figure 13.5.   Illustrating Poiseuille flow.

The maximum velocity $U_{\max}$ occurs in the middle of the channel, and has value

$$U_{\max} = \frac{h^2}{8\eta}K \qquad (13.44)$$

The volumetric flow rate Q is found by integrating the velocity profile over the flow cross-section.

$$Q = W \int_0^h U_x \, dy = \frac{Wh^3}{12\eta}K \qquad (13.45)$$

where W is the width of the channel. The mean flow velocity, averaged across the cross-sectional area, is

$$\bar{U} = \frac{Q}{Wh} = \frac{h^2}{12\eta}K = \frac{2}{3}U_{\max} \qquad (13.46)$$

The shear stress on the wall is found from the velocity gradient in the fluid evaluated at the position of the wall.

$$\tau_w = \eta \left.\frac{\partial U_x}{\partial y}\right|_{y=0} = \frac{h}{2}K \qquad (13.47)$$

We note that the shear stress in this case does not depend explicitly on the viscosity. However, we can recast this expression by noting that the pressure gradient K and the flow rate are linked

$$\tau_w = \frac{6\eta}{Wh^2}Q \qquad (13.48)$$

Thus, the wall stress turns out to be proportional to the viscosity and the volumetric flow rate.

The lumped-element model for Poiseuille flow can be expressed in both the mechanical and fluidic domains. In the fluid domain using the $e \to V$ convention, we can use pressure as the across variable and volumetric fluid flow as the through variable. In that case, the pressure gradient must be expressed in terms of the pressure drop between the ends ΔP, spaced by a length L.

$$K = \frac{\Delta P}{L} \tag{13.49}$$

The relation between flow and pressure drop is then

$$\Delta P = \frac{12\eta L}{W h^3} Q \tag{13.50}$$

Thus, in the fluid domain using the $e \to V$ convention, the lumped element model for a Poiseuille flow path is

$$R_{\text{pois}} = \frac{\text{across}}{\text{through}} = \frac{12\eta L}{W h^3} \tag{13.51}$$

In the mechanical domain, there is an interesting force-balance problem. We began by assuming that the plates bounding the fluid channel did not move, which requires the total force on the plates to be zero. We also assumed that the fluid was not accelerating, meaning that the total force on the fluid must also be zero. The fluid experiences a net pressure force of ΔPWh in the $+x$ direction and a shear force at the walls of magnitude $2 \times \tau_w WL$ in the $-x$ direction. In order that the fluid not be accelerated, these two forces should cancel, and they do. However, if the wall exerts a force on the fluid in the $-x$ direction, then, by Newton's laws, the fluid must exert an equal and opposite force in the $+x$ direction on the walls. So in order that the walls not move, they must be restrained by additional external forces that exactly balance the wall shear forces set up by the pressure gradient.

Poiseuille flow can take place in structures of arbitrary cross-section. For a circular cross-section with radius a, the velocity distribution becomes

$$U_x = \frac{a^2 - r^2}{4\eta} K \tag{13.52}$$

The volumetric flow rate becomes

$$Q = \frac{\pi a^4}{32\eta} K \tag{13.53}$$

and the average flow velocity is

$$\bar{U} = \frac{a^2}{8\eta} K \tag{13.54}$$

To find the shear force on the walls, we can use the requirement of force balance to obtain

$$2\pi a L \tau_w = \pi a^2 L K \tag{13.55}$$

or

$$\tau_w = \frac{a}{2} K \tag{13.56}$$

For more general noncircular geometries, we can define the *hydraulic diameter* D_h as

$$D_h = \frac{4 \times \text{Area}}{\text{Perimeter}} \tag{13.57}$$

and, with this definition, write the steady Poiseuille flow expression as

$$\Delta P = f_D \left(\frac{1}{2} \rho_m \bar{U}^2 \right) \frac{L}{D_h} \tag{13.58}$$

where f_D is the *Darcy friction factor*. The friction factor depends on the average flow velocity. This expression is usually written in terms of the Reynolds number. That is, for a fixed cross-sectional geometry,

$$f_D Re_D = \text{Dimensionless constant} \tag{13.59}$$

where

$$Re_D = \frac{\rho_m \bar{U} D_h}{\eta} \tag{13.60}$$

The dimensionless constant is tabulated for various cross-sections.[7] For a circle it is 64; for plane Poiseuille flow, it is 48.

Poiseuille flow and Couette flow can take place simultaneously, with both a pressure gradient and a laterally moving wall. The governing equations are linear, so the results can be simply added to match the conditions of the problem.

13.3.3   Development Lengths and Boundary Layers

When a flow enters a pipe or experiences some other abrupt change in flow profile, a characteristic distance, called the *development length* or *entrance length*, is required before the flow assumes its steady-state profile. A very

---

[7] See, for example, Fay, *Introduction to Fluid Mechanics*, p. 293 [70].

rough estimate of the development length L_d for a fluid entering a pipe with hydraulic diameter D_h is

$$L_d \approx \frac{D_h}{16} Re_D \qquad (13.61)$$

In MEMS devices, hydraulic diameters tend to be sub-millimeter, and Reynolds numbers tend to be small, with the net result that development-length effects are short, and can often be ignored.

In flows over free surfaces, or in large open regions, the development-length effects must be taken into account. The flow itself is often conceptually separated into a boundary region and a free-stream region. Far away from the free surface, viscous effects are minimal, and the fluid is characterized by a free-stream velocity. But close to the surface, viscous shear forces can dominate. The *boundary layer* is a region near a free surface where flow profiles are characterized by large shear effects, such as in Couette flow. The reader is referred to the Related Reading at the end of the chapter for more on this subject.

13.3.4   Stokes Flow

When the Reynolds number is much less than unity, which is a typical situation in MEMS devices, the inertial term $\rho_m D\mathbf{U}/Dt$ can be neglected compared to the viscous term, leading to a vector Poisson equation:

$$\eta \nabla^2 \mathbf{U} = \nabla P^* \qquad (13.62)$$

where, again, $P^* = P - \rho_m g \cdot \mathbf{r}$. This regime is called *Stokes flow*, or *creeping flow*. A good example is the force on a sphere moving slowly through a fluid at constant speed. The total drag force is $6\pi\eta U a$, where a is the radius of the sphere. This drag force is not strongly sensitive to the shape of the object.

13.4   Squeezed-Film Damping

Figure 13.6 illustrates a situation that arises in many MEMS devices. A gas, such as air, fills the space between two parallel plates. The lower plate is fixed, while the upper plate is moveable. The motion does not have to be uniform. The plate can move as a rigid body, or it can rotate, or bend. The net result is a gap h that depends both on space (x and y) and time.

If the upper plate moves downward, either because the plate has velocity imparted to it by some external agent, or because a force is applied directly to the plate, two things happen: (1) the pressure inside the gas increases due to the plate motion, and (2) the gas is squeezed out from the edges of the plate. If the plate is moved up, the situation reverses: the pressure drops between the plates, and gas is sucked back into the space. The viscous drag of the air

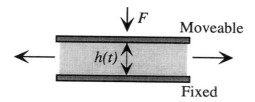

Figure 13.6.   Illustrating the basic idea of squeezed-film damping.

during the flow creates a dissipative mechanical force on the plate, opposing the motion. This dissipative force is called *squeezed-film damping*.

Because we are dealing with a gas, we choose to include the compression of the gas explicitly. However, in analyzing this problem, we will make several simplifying assumptions:

1. The gap h is always much smaller than the lateral extent of the plates.

2. The motion is sufficiently slow that we can treat the gas as moving under Stokes flow.

3. There is no pressure gradient in the normal direction (from plate to plate).

4. The lateral flow has a Poiseuille-like velocity profile, with zero transverse velocity at the upper and lower plates.

5. The gas obeys the ideal gas law.

6. The system is isothermal. That is, any temperature rise due to gas compression or to viscous dissipation or any temperature drop due to gas dilation is quickly compensated by heat flow to or from the walls.

With these assumptions, the continuity equation, the Navier-Stokes equation, and the ideal gas law can be combined to yield the *Reynolds equation* [72, 73]:

$$12\eta\frac{\partial(Ph)}{\partial t} = \nabla \cdot \left[(1 + 6K_n)h^3 P\nabla P\right] \tag{13.63}$$

where the pressure P depends on x and y but not on z, h is time- and position-dependent, and K_n is a new factor, called the *Knudsen number*, which is the ratio of the mean free path of the gas molecules to the gap [74]. When the Knudsen number is small, much less than 0.1, it is reasonable to use the no-slip boundary condition to describe the flow. However, when K_n gets as large as 0.1, then slip-flow begins to become important. A one-micron gap, for example, filled with a gas with a typical room-air mean free path of 0.1 microns, already has a Knudsen number of 0.1. In effect, the structure is too small to permit the "many collisions close to the wall" requirement for non-slip conditions. Molecules can travel a significant fraction of the height of the channel before experiencing a collision. The Knudsen-number correction in Eq. 13.63 is one possible approximate way to take account of the slip. This is a subject of active research, with a rapidly developing literature [75, 76] .

13.4.1   Rigid Parallel-Plate Small-Amplitude Motion

The simplest example of squeezed-film damping is the small-amplitude rigid motion of the upper plate, without rotation. That is, h is a function only of time. Further, we shall assume in this example that the Knudsen number is negligibly small.[8] Under these assumptions, the Reynolds equation becomes:

$$\frac{\partial(Ph)}{\partial t} = \frac{h^3}{12\eta}\left(\frac{1}{2}\nabla^2 P^2\right) \tag{13.64}$$

We recognize this as a distinctly non-linear partial differential equation, with a complicated input, since $h(t)$ appears on the right-hand side to the third power, and on the left-hand-side inside the time derivative. Ultimately, it is the *velocity* of the plate $\partial h/\partial t$ that is the driving force for the damping, but the value of h figures prominently as well.

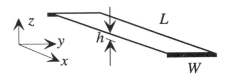

Figure 13.7.   Long narrow beam used for squeeze-film damping calculation.

We can linearize this equation about an assumed operating point, with average gap h_o and average pressure P_o:

$$h = h_o + \delta h \tag{13.65}$$

where $\delta h \ll h_o$, and

$$P = P_o + \delta P \tag{13.66}$$

where $\delta P \ll P_o$. If we now apply this equation to a long narrow beam, as illustrated in Fig. 13.7, we can assume that all the fluid motion is in the y-direction, and normalize the y-coordinate to the beam width:

---

[8]It is not difficult to repeat the various examples in this section with the Knudsen-number factor in place.

$$\xi = \frac{y}{W} \tag{13.67}$$

Applying this normalization and the assumption of small perturbations about an operating point yields the linearized isothermal compressible Reynolds equation:

$$\frac{\partial \hat{p}}{\partial t} = \frac{h_o^2 P_o}{12\eta W^2} \frac{\partial^2 \hat{p}}{\partial \xi^2} - \frac{\dot{h}}{h_o} \tag{13.68}$$

where $\hat{p} = \delta P/P_o$. Once again, we recognize this form as the same heat-flow equation that has appeared in so many dissipative processes. The equation is now linear, and the source term is, as expected, the velocity of the plate $\dot{h}$.

13.4.1.1 Solution for a Suddenly Applied Motion

If we assume that at time $t = 0$, the plate is suddenly displaced vertically by an amount z_o, the effective excitation is simply a velocity impulse, and we can find the transient solution to Eq. 13.68 using the methods of the Chapter 12. We assume a separated solution of the form

$$\hat{p}(\xi, t) = \tilde{p}(\xi)e^{-\alpha t} \tag{13.69}$$

For $t > 0^+$, the velocity of the plate is zero, so the Reynolds equation becomes

$$\frac{h_o^2 P_o}{12\eta W^2} \frac{d^2\tilde{p}}{d\xi^2} + \alpha\tilde{p} = 0 \tag{13.70}$$

The general solution is of the form

$$\tilde{p} = A_n \sin \sqrt{\sigma_n}\xi + B_n \cos \sqrt{\sigma_n}\xi \tag{13.71}$$

where

$$\sigma_n = \frac{12\eta W^2 \alpha}{h_o^2 P_o} \tag{13.72}$$

Assuming that the origin is in the edge of the plate, the boundary conditions that must be satisfied are that $\tilde{p}$ is zero at $\xi = 0$ and $\xi = 1$ and, by symmetry, $d\tilde{p}/d\xi$ is zero at $\xi = 1/2$. This means that the B_m terms are all zero, and that

$$\sqrt{\sigma_n} = n\pi; \quad \text{for } n = 1, 3, 5\dots \tag{13.73}$$

The corresponding values for α are

$$\alpha_n = \frac{h_o^2 P_o n^2 \pi^2}{12\eta W^2} \tag{13.74}$$

The coefficients A_n are found from the requirement that the displacement be uniform over the area of the plate. Therefore, the A_n are the same as for the uniformly heated resistor example of Chapter 12, with the result that

$$\hat{p} = -\frac{z_o}{h_o} \sum_{n \text{ odd}} \frac{4}{n\pi} [\sin n\pi\xi] e^{-\alpha_n t} \tag{13.75}$$

The total time-dependent force $F(t)$ acting on the plate is

$$F(t) = P_o W L \int_0^1 \hat{p}(t, \xi) \, d\xi \tag{13.76}$$

with the result

$$F(t) = -P_o W L \frac{z_o}{h_o} \sum_{n \text{ odd}} \frac{8}{n^2\pi^2} e^{-\alpha_n t} \tag{13.77}$$

The above expression is the *step response* of the squeeze-film damper. To find a more general response, we can take the single-sided Laplace transform, to obtain

$$F(s) = \frac{96\eta L W^3}{\pi^4 h_o^3} z_o \sum_{n \text{ odd}} \frac{1}{n^4} \frac{1}{1 + \dfrac{s}{\alpha_n}} \tag{13.78}$$

This is now the Laplace transform of the response to a step input of magnitude z_o. If we replace by the Laplace transform of the impulse source z_o/s by the Laplace transform $z(s)$ of a more general time-dependent source, we obtain the transfer function (in square brackets below) for the small-amplitude damping for an arbitrary source waveform:

$$F(s) = \left[\frac{96\eta L W^3}{\pi^4 h_o^3} \sum_{n \text{ odd}} \frac{1}{n^4} \frac{1}{1 + \dfrac{s}{\alpha_n}} \right] sz(s) \tag{13.79}$$

We now examine this result.

The first important observation is that the driving term is the velocity of the plate, which has Laplace transform $sz(s)$. This makes sense. Second, the magnitudes of the coefficients decrease very rapidly with increasing n. The second term is 1/81 as large as the first term, and successive terms are, by comparison, completely negligible. Therefore, for small-amplitude excitation, we are justified in using only the first term of the sum. That is, to better than a few percent,

$$F(s) = \frac{b}{1 + \dfrac{s}{\omega_c}} sz(s) \tag{13.80}$$

where the *damping constant*

$$b = \frac{96\eta L W^3}{\pi^4 h_o^3} \tag{13.81}$$

and the *cutoff frequency* ω_c is

$$\omega_c = \frac{\pi^2 h_o^2 P_o}{12\eta W^2} \tag{13.82}$$

The cutoff frequency, when compared to a sinusoidal frequency of motion, is a measure of the relative importance of viscous forces to spring forces. The *squeeze number* σ_d, a quantity widely used in the squeezed-film-damping literature, is defined as

$$\sigma_d = \frac{12\eta W^2}{h_o^2 P_o}\omega \tag{13.83}$$

The squeeze number is equal to $\pi^2 \omega / \omega_c$.

An equivalent circuit for squeezed-film damping in the $e \rightarrow V$ convention, with force as the across variable and plate velocity (with Laplace transform $sz(s)$) as the through variable is shown in Fig. 13.8. We note that both a damping resistor and a capacitor (equivalent to a spring) are required for this model, capturing the fact that there is both viscous drag and gas compression built into our model.

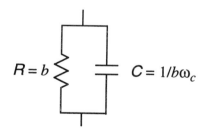

Figure 13.8. Equivalent circuit for small-amplitude squeezed-film damping using the $e \rightarrow V$ convention.

In the limit of low frequencies, where the capacitor is an open circuit, this model reduces to a linear resistive damping element, with a value determined by a combination of geometry and gas composition. Until we reach the limit of Knudsen numbers greater than about 0.1, the viscosity of a gas is only weakly dependent on pressure. Therefore, for slow motions compared to frequency ω_c, that is, for squeeze numbers much smaller than unity, the effect of the squeezed film is a pure damping force, independent of pressure.

The cutoff frequency ω_c is itself proportional to pressure. Therefore, as the pressure is decreased, the value of the capacitor goes up and the spring-force becomes relatively more important, compared to the damping force. We can understand this with an impedance model. Figure 13.9 shows the spring-mass-dashpot linear system we have studied before, but now with the equivalent circuit for squeezed-film damping replacing the simple linear resistor.

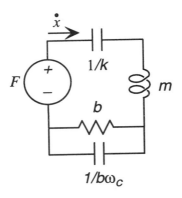

Figure 13.9.   Spring-mass-dashpot model using a squeezed-film damping element.

In the sinusoidal steady state, the complex amplitude of the force-displacement relation can be written

$$\frac{F(s)}{x(s)} = s\left(sm + \frac{b}{1 + s\tau} + \frac{k}{s}\right) \qquad (13.84)$$

where $\tau = 1/\omega_c$ and where the s factor is what converts displacement $x(s)$ into velocity $\dot{x}(s)$. In the sinusoidal steady state at frequency ω, this becomes

$$\frac{F(j\omega)}{x(j\omega)} = j\omega\left\{j\omega m + \frac{b}{1 + (\omega\tau)^2} + \frac{k}{j\omega}\left[1 + \frac{(\omega\tau)^2}{1 + (\omega\tau)^2}\frac{b\omega_c}{k}\right]\right\} \qquad (13.85)$$

We observe that at low frequencies, the added capacitor term has no effect, but as ω approaches ω_c, so that $\omega\tau$ approaches unity, there is a decrease in the net damping force due to the $1 + (\omega\tau)^2$ term in the denominator of the b term, and an increase in the total stiffness given by the term in square brackets. This stiffness increase is a result of the compressibility of the gas in combination with a relatively high frequency, so that the gas does not have time to squeeze out before the motion of the mass is reversed. The effect of this added stiffness is to increase the undamped resonant frequency of the system. This effect is routinely observed in MEMS structures in the presence of squeeze-film damping when $\omega\tau$ approaches unity [77, 78, 79, 80].

This model must be used with caution. As we observed in our self-heated resistor example in Section 12.10, this lumped circuit will only be valid for frequencies up to about ω_c. Beyond that frequency, additional terms in the damping sum will start to become important compared to the first term. Therefore, if the pressure of the system is decreased until ω_c becomes comparable to ω_o, or beyond, then the effect of squeeze-film damping on the resonant behavior must be calculated with more terms in the sum and a correspondingly more elaborate circuit model.[9]

13.5   Electrolytes and Electrokinetic Effects

Electrolytes are solutions of ionic species, and they have special fluidic properties that arise because of the possibility of coupling electric fields with the flow. If C_i is the concentration of ionic species i, and z_i is the charge of that ionic species normalized to the electronic charge q_e, then the total charge density in the solution can be written

$$\rho_e = \sum_i z_i q_e C_i \tag{13.86}$$

In normal electrolytes, far away from bounding surfaces, this charge density is zero, and the electrostatic potential in such a system then obeys the Laplace equation

$$\nabla^2 \phi = 0 \tag{13.87}$$

Near a wall, as will be examined in more detail below, there can be departures from charge neutrality, in which case the electrostatic potential obeys the Poisson equation

$$\nabla^2 \phi = -\frac{\rho_e}{\varepsilon} \tag{13.88}$$

We recall from Section 11.8 that in equilibrium, there is a relation between concentration and potential. Thus if we assume that this relation continues to apply to small departures from equilibrium (a typical assumption when dealing with irreversible thermodynamic concepts), then we can write

$$\nabla^2 \phi = -\frac{q_e}{\varepsilon} \sum_i z_i q_e C_{i,0} e^{-\frac{z_i q_e (\phi - \phi_0)}{k_B T}} \tag{13.89}$$

where $C_{i,0}$ is the concentration at a reference position where the potential is ϕ_0. Since ϕ_0 is a constant, we can rewrite this equation in terms of $\hat{\phi} = \phi - \phi_0$:

---

[9] Veijola has demonstrated this circuit approach for lumped-element modeling of squeeze-film damping. He uses the $f \rightarrow V$ convention for assigning across and through variables [79].

$$\nabla^2 \hat{\phi} = -\frac{q_e}{\varepsilon} \sum_i z_i q_e C_{i,0} e^{-\frac{z_i q_e \hat{\phi}}{k_B T}} \tag{13.90}$$

For small potential variations, we can expand the exponential to obtain

$$\nabla^2 \hat{\phi} \approx -\frac{q_e}{\varepsilon} \sum_i z_i q_e C_{i,0} + \frac{q_e^2}{\varepsilon k_B T} \left[\sum_i z_i^2 C_{i,0} \right] \hat{\phi} \tag{13.91}$$

Assuming that the reference location is a charge-neutral region, then the first term on the right vanishes, and we are left with a very simple expression

$$\nabla^2 \hat{\phi} = \frac{1}{L_D^2} \hat{\phi} \tag{13.92}$$

where L_D is called the *Debye length*, and is given by

$$\frac{1}{L_D} = \sqrt{\frac{q_e^2}{\varepsilon k_B T} \left[\sum_i z_i^2 C_{i,0} \right]} \tag{13.93}$$

Basic solutions in one dimension are exponentials of the form

$$\hat{\phi} = e^{\pm \frac{x}{L_D}} \tag{13.94}$$

This equation tells us that a variation in potential in a solution with a charge-weighted ionic strength $\sum_i z_i^2 C_{i,0}$ has a characteristic exponential dependence with a decay length given by the Debye length. Departures from neutral conditions are screened out over a distance of about 3 Debye lengths. As the ionic strength of the electrolyte increases, the Debye length decreases. The range of Debye lengths in water varies from about 1 μm in pure water down to about 0.3 nm in a one-molar solution of a monovalent salt such as KCl. Thus it is typically true that fluid-flow channels in MEMS devices, which are in the 10 μm- 1 mm range in diameter, are much greater than the Debye lengths encountered in aqueous solutions. This has important implications for the design of "lab-on-a-chip" devices, as will be described later.

13.5.1   Ionic Double Layers

When an electrolyte is placed in contact with a solid surface, as in Fig. 13.10, a combined chemical-electrostatic interaction occurs that leaves the surface of the solid with a net charge. There are two primary mechanisms. On metal surfaces, the electron gas within the metal produces strong image charges that can bind ions of either sign. This tends to produce several very compact layers of adsorbed ions, shown in Fig. 13.10, with the solid black circles representing ions of one charge, the shaded circles the countercharge. These compact layers are called the *inner and outer Helmholtz planes*. Which ion dominates in the inner layer is a function of the specific material and the composition of the electrolyte, but the outer layer tends to have the opposite charge from that of the inner layer. Functionally, the Helmholtz layers behave as a very large electrical capacitance between the metal and the electrolyte. Applying potentials between the metal and the electrolyte can modify the amount of charge in the two layers, and unless very large amounts of current are required by the rest of the electrical circuit, the Helmholtz planes can supply net charge of either sign by adjustment of the relative amount of the two types of ion in the compact double layer.

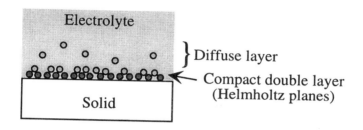

Figure 13.10.   Illustrating the compact and diffuse double layers.

An important detail is that the Helmholtz planes are not exactly charge-balanced. There is some net charge between the inner and outer Helmholtz planes, and this net charge must be balanced by a slight increase in the concentration of the oppositely charged ion in the region near the solid surface. This region of concentration unbalance is called the *diffuse double layer*. Schematically, in Fig. 13.10, there are fifteen dark circles in the inner Helmholtz plane, ten light circles in the outer Helmholtz plane, and five light circles in the diffuse layer. The circles in the diffuse layer represent the extra concentration of that type of ion in the diffuse layer. This extra concentration is typically a very small fraction of the total ion concentration in the solution. Where there is extra concentration, there must be net charge density. Thus, the electrolyte in the immediate vicinity of the wall (within several Debye lengths) has a net charge and can be influenced by externally applied electric fields.

We can analyze the diffuse double layer by assuming that the potentials are small, and that the Debye length is small compared to the dimensions of the sample. Therefore, the potential $\hat{\phi}$ near a wall can be written

$$\hat{\phi} = \phi_w e^{-\frac{z}{L_d}}$$

(13.95)

where the *wall potential* ϕ_w is the extrapolated potential at the outer edge of the compact layer.[10] We require that the total charge in the diffuse layer equal the negative of the net charge in the compact layer, and we can use this fact to determine ϕ_w. The charge density in the diffuse layer in the Debye-length approximation is

$$\rho_e = -\frac{\varepsilon}{L_D^2}\hat{\phi} \tag{13.96}$$

The total charge per unit area in the diffuse layer is then

$$\sigma_d = \int_0^\infty \rho_e \, dz = -\frac{\varepsilon \phi_w}{L_D} \tag{13.97}$$

Thus

$$\phi_w = \frac{\sigma_w L_D}{\varepsilon} \tag{13.98}$$

where σ_w is the net charge per unit area in the compact layer. We can thus write the potential as

$$\hat{\phi} = \frac{\sigma_w L_D}{\varepsilon} e^{-\frac{z}{L_D}} \tag{13.99}$$

On insulating surfaces, the situation is a little different. There is no electron gas in the solid to provide image charges for compact layers of adsorbed ions. However, there are typically surface binding sites for ions, and these binding sites can shift the net charge on the surface depending on the chemistry of the material and the composition of the electrolyte. For example, the pH of the electrolyte can have a strong influence on the sign and magnitude of the surface charge on a dielectric interface, such as silicon dioxide or silicon nitride. However, whatever the net surface charge on the wall, the diffuse layer must have a compensating net charge distributed over several Debye lengths. Thus, again, the electrolyte in the immediate vicinity of the wall has a net charge, and can be influenced by externally applied electric fields.

With an insulating wall, it is possible for the applied electric field to be *tangential* to the surface (this is not possible for a metal). This means that an electrostatic body force can be applied to the diffuse double layer, and this body force can drag the fluid along the wall. This effect gives rise to an very important phenomenon, called *electroosmotic flow*, which is described in the next section.

---

[10]This extrapolated wall potential is also commonly referred to as the *zeta potential* [81].

13.5.2   Electroosmotic Flow

We now consider the situation in Fig. 13.11, in which an electrolyte is bounded by insulating walls, and there is a tangential applied electric field throughout the electrolyte created by voltage applied between electrodes immersed in the electrolyte at either end of the sample. The field strength $\mathcal{E}_x$ is simply the applied voltage divided by the fluidic path length. We assume that this electric field is small enough not to disrupt the double layer, an assumption that is generally well satisfied. The black dots along the walls, in this case, represent only the charge in the diffuse layer; the compact layers are not shown. The Debye length is very small compared to the distance between the walls, so on this scale, the diffuse layer looks like a thin charge sheet at the edge of the fluid.

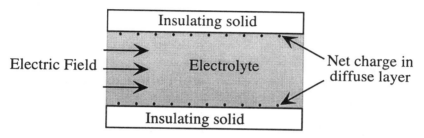

Figure 13.11.   Illustrating electroosmotic flow

The liquid is presumed to be incompressible. The Navier-Stokes equation for this case becomes

$$\rho_m \frac{DU_x}{Dt} = -\frac{dP}{dx} + \eta \nabla^2 U_x + \rho_e \mathcal{E}_x \qquad (13.100)$$

where, for a compact-layer wall charge σ_w per unit area,

$$\rho_e = -\frac{\sigma_w}{L_D} e^{-\frac{z}{L_D}} \qquad (13.101)$$

We shall assume there is no pressure gradient in the system, and that we are in steady state with a uniform cross-section and fully developed flow. With these assumptions, the velocity U_x can depend only on z, the distance from the wall. The Navier-Stokes equation becomes

$$\frac{d^2 U_x}{dz^2} = \frac{\sigma_w \mathcal{E}_x}{\eta L_D} e^{-\frac{z}{L_D}} \qquad (13.102)$$

Requiring that U_x vanish at the wall (the no-slip condition) yields

$$U_x = \frac{\sigma_w \mathcal{E}_x L_D}{\eta} \left(1 - e^{-\frac{z}{L_D}}\right) \qquad (13.103)$$

The flow velocity is maximum in the center of the sample, with flow velocity

$$U_0 = \frac{\sigma_w \mathcal{E}_x L_D}{\eta}$$ (13.104)

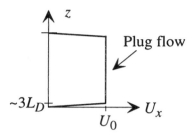

Figure 13.12.   Electroosmotic flow profile.

The flow profile is shown schematically in Fig. 13.12. Within about three Debye lengths of the walls, there is a highly sheared boundary layer, but away from the boundary layers, the velocity is completely uniform across the sample. This type of flow profile is called *plug flow*. Because there is no shear throughout most of the volume, this kind of flow is extremely valuable in chemical analysis systems.

The electroosmotic flow velocity is proportional to the wall charge σ_w. Therefore, control of this charge is important if repeatable and well-behaved flow velocities are to be achieved. Methods for robust control of σ_w for various types of materials is a subject of active research.

13.5.3   Electrophoresis

Let us now imagine that in addition to the background electrolyte, there is an ionic species such as an amino acid or protein segment present in concentrations much less than the background ions. Thus its presence does not perturb the basic electroosmotic flow profile. This species will be carried along in the flow at velocity U_0. However, because this species is charged, it will also drift in the electric field relative to the moving fluid with velocity

$$v_{ep} = \mu_{ep} \mathcal{E}_x$$ (13.105)

where μ_{ep} is called the *electrophoretic mobility* of the species.

Because different chemical species typically have different electrophoretic mobilities, it is possible to separate them within electroosmotic flow channels using the fact that electroosmotic flow velocities can be designed to be greater than the electrophoretic velocities. Microfabrication offers enormous improvements in this type of electrophoresis because flow channels of precise dimensions and remarkable intricacy can be fabricated in a single substrate [82].

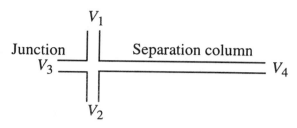

Figure 13.13. A junction and separation column used for electrophoretic separation in combination with electroosmotic flow.

Figure 13.13 shows a typical junction structure of the kind used to perform electrophoretic chemical separations. Typical dimensions are 100 μm wide and 20 μm deep. The channels are etched in glass, and then capped with a flat glass slide containing entry ports for the various fluids (not shown in the Figure). Four flow channels meet at an intersection. The system is filled with an electrolyte buffer solution,[11] and voltages are applied via electrodes immersed in the reservoirs at the ends of each channel. To the right is a long separation column.

The use of this device for chemical separation is illustrated in Fig. 13.14, which shows an injection and separation system [82]. The clear regions are filled with buffer; the shaded regions contain similar buffer in which the sample to be separated is dissolved. In this example, we shall assume that the channel walls have a negative charge, so the diffuse layer in the buffer has a positive charge and the flow is in the direction of the applied electric field.

The first step is to prepare a sample plug. This is shown at the top of Fig. 13.14. A positive voltage is applied to V_1 and V_2 is grounded so that electroosmotic flow carries the sample electrolyte through the junction region as shown. Voltages must also be applied at V_3 and V_4 to prevent the sample from entering the horizontal channels. These voltages are determined in advance from the device design. The applied voltage at V_1 determines the voltage at the junction V_J through an electrical voltage divider:

$$V_J = \frac{R_2}{R_1 + R_2} V_1 \qquad (13.106)$$

where R_1 and R_2 are the electrical resistances between the point where the respective voltages are applied and the junction. R_1, for example, is given by

---

[11] A buffer solution is a mix of electrolytes that results in a stable pH value.

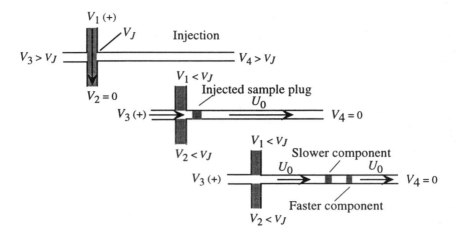

Figure 13.14.   Illustrating electrophoretic separation with electroosmotic flow. The voltages used during the injection and separation sequence are described in the text.

$$R_1 = \frac{L_{1,J}}{\sigma_e A_1} \qquad (13.107)$$

where $L_{1,J}$ is the distance between the junction and the application point for voltage V_1, A_1 is the cross-sectional area of channel 1, and σ_e is the conductivity of the buffer carrying the sample. Note that while this conductivity varies with buffer composition, the conductivity cancels out of the expression for V_J. Voltages slightly greater than V_j are applied to V_3 and V_4, forcing a small amount of buffer toward the sample stream, slightly pinching the sample buffer in the intersection.

Once the sample-filling voltages have been applied long enough for the slowest moving component to reach the intersection, the voltages are switched. A large voltage is applied at V_3 and V_4 is grounded. This creates a rapid electroosmotic flow down the separation channel, fed with buffer from the V_3 reservoir. To keep any further sample out of the separation channel, the voltages at V_1 and V_2 are now adjusted to be slightly smaller than the new junction voltage established by V_3, driving small amounts of buffer toward the V_1 and V_2 reservoirs.

The sample plug is carried down the separation channel by the electroosmotic flow, but within the sample, the different components are separated according to their electrophoretic mobilities. If the ionic species are positively charged, then they move faster than the fluid; if negatively charged, they move slower. And, depending on their respective mobility, the difference chemical species get separated by the electric field while the flow is taking place because their velocity relative to the fluid increases with the magnitude of their respective mobilities. By the time the sample reaches the end of the separation channel, distinct bands of different species are observed, either by flourescence or some other optical technique.

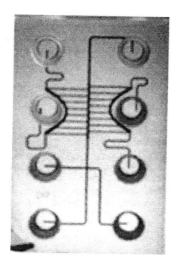

Figure 13.15.   A glass microfluidic chip for biochemical assays. (Source: Caliper Technologies, Inc.; reprinted with permission.)

This basic combination of electroosmotic flow and electrophoretic separation has spawned a rich array of chemical assay and analysis systems. Commercial separation systems are now being sold by several vendors. Figure 13.15 shows a commercial device made of two bonded glass wafers containing a set of flow channels together and several large wells for introduction of liquid reagents. This particular chip is designed for combining multiple samples into a single flow stream. The chip fits into a complete instrument that provides electrical contacts to each well, and a pre-aligned optical detection system.

13.5.4   Diffusion Effects

While electroosmotic flow and electrophoretic separation is taking place, the sample species can also diffuse. An infinitessimal slab of a sample will spread out in width due to diffusion, even while it is being carried by the electro-osmotic flow and being drifted relative to the flow by electrophoresis. If the length of the separation channel is L_S, and the electro-osmotic flow speed is U_0, then each species is in the separation channel for a transit time τ_t of approximately L_S/U_0. During this time, an infinitessimally thin slab of sample will grow to a width of approximately $\sqrt{D\tau_t}$, where D is the diffusivity of the species. Thus the narrowest separated band that can possible be achieved in the system is

$$W_{\min} = \sqrt{\frac{DL_S}{U_0}} = \sqrt{\frac{DL_S\eta}{\sigma_w \mathcal{E}_x L_D}} \qquad (13.108)$$

This says that to get the sharpest separated bands, one needs to use short columns, large electric fields, and relatively low-ionic strength buffers to achieve large Debye length. The sample injection process creates a sample slab of finite width. A good rule of thumb is that one would like the initial sample slab to be smaller than $W_{\min}$, so that the best possible separation, limited only by diffusion, is achieved. This is why microfabrication has been so successful in this arena. Small sample plugs can be formed, and large electric fields can be produced in the separation channels (1000 V/cm are typical field strengths). Thus separations that require several minutes in larger apparatus can be achieved over a length of a few cm in 15 seconds. Because the sample spends less time in the separation channel, diffusive effects are minimized.

13.5.5   Pressure Effects

If buffers of two different ionic strength are used, pressure gradients can develop. This is because the velocity at which a buffer would travel varies with the Debye length, and this depends on ionic strength. A region of high ionic strength will travel more slowly than one of low ionic strength, and this creates pressure differences across the interface between regions of different ionic strength, leading to Poiseuille-like behavior in the flow profile. The subject is quite complex, because as the characteristically curved flow front develops, diffusion of the sample tends to move the sample radially in the channel. Advanced high-throughput systems presently being designed must confront this issue.

13.5.6   Mixing

Figure 13.16 illustrates an effect which is extremely important for microchemical analysis systems. The flow in these small structures is strictly laminar. When two streams are joined, as in the T-junction illustrated in the Figure, there is virtually no mixing of the flow fields. If one wishes to mix a reagent with a sample stream, the only mechanism available is diffusion across the boundary. We have already seen that diffusion effects can be made to be quite small by using large electric fields, short columns, and relatively large Debye length buffers. This is the exact opposite problem – one seeks good mixing, which means that the residence time must be long enough to permit diffusion across the width of the channel. Under these circumstances, one must either use very long columns, or slow the flow down by using smaller electric fields when mixing is desired.

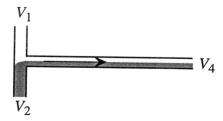

Figure 13.16. Illustrating laminar flow when two streams are combined. Mixing occurs only by diffusion.

13.5.7   Modeling of Electrokinetic Systems

To first order, the models we have already presented are sufficient to get a good estimate of how one of these systems will perform. The geometry of the channels in combination with the applied voltages determines the electric fields everywhere in the system, and flow velocities can then be calculated, presuming one has characterized the wall charge for the material being used for the channels and the specific buffer composition. Thus, unless there are pressure differences that develop because of mismatches in flow speed, the electrical circuit and the fluidic circuit behave identically. However, when pressure and diffusion effects must be taken into account, it is valuable to use numerical simulation to study these systems. Some commercial CAD systems now include modules for modeling electrokinetic effects in fluids.

Related Reading

J. O'M. Bochris and A. K. N. Reddy, *Modern Electrochemistry*, Sixth edition (2 volumes), New York: Plenum Press, 1979.

W. M. Deen, *Analysis of Transport Phenomena*, New York: Oxford University Press, 1998.

J. A. Fay, *Introduction to Fluid Mechanics*, Cambridge, MA: MIT Press, 1994.

A. J. Grodzinksy and M. L. Yarmush, *Electrokinetic Separations*, in H.-J. Rehm and G. Reed (editors), *Biotechnology*, Second Edition, vol. 3: *Bioprocessing*, pp. 680-693, Weinheim: VCH, 1993.

A. M. Kuethe and C.-Y. Chow, *Foundations of Aerodynamics*, Fifth Edition, New York: Wiley, 1998.

Problems

13.1 Use the Ideal Gas Law to calculate the bulk modulus and thermal expansion coefficient of room-temperature helium at one atmosphere pressure. Compare to the results for air.

13.2 Referring to Fig. 13.3, determine the vertical force that must be applied to a 20 μm diameter capillary tube to keep it stationary in the presence of the surface tension forces set up by water, assuming a surface tension of 72×10^{-3} N/m and a contact angle of $30°$. What is this force if the diameter becomes 1 μm?

13.3 A very lightly oxidized polysilicon cantilever beam 100 μm long, 1 μm thick, and 5 μm wide is suspended 1 μm above a lightly oxidized silicon surface. If a water drop with in-plane diameter 3 microns becomes trapped beneath the cantilever near its free end, estimate the amount of cantilever bending due to surface tension forces. Use the data from Problem 13.2. What can happen as the liquid evaporates, shrinking the drop?

13.4 A fluid microchannel has depth 20 μm and width 100 μm. Determine the fluid resistance per unit length of this channel for water at room temperature ($\eta = 10^{-3}$ Pa-sec) and air at room temperature and pressure ($\eta = 1.82 \times 10^{-5}$ Pa-sec). For these two fluids, assuming incompressible flow, how much pressure drop along a 1 cm channel is required to reach a Reynolds number of 1, and what is the volumetric flow rate under these conditions? Is the assumption of incompressible flow justifiable for the case of air?

13.5 A doubly-clamped polysilicon beam 100 μm long, 1 μm thick, and 5 μm wide is suspended 1 μm above a substrate. Assuming an axial tensile stress of 15 MPa in the beam, create a lumped-element model for the vertical motion of this beam under the action of a uniformly distributed load in the presence of squeezed-film damping due to room temperature air at one atmosphere. Is the fundamental resonance underdamped or overdamped?

13.6 An Electrokinetic Treasure Hunt: Go to the literature and find an example of electrophoretic separation in a microfabricated electroosmotic flow cell. What flow velocity was achieved, and what wall charge is needed to explain that flow velocity using Eq. 13.104? Which dominates the separation resolution of the system, the initial width of the sample plug or diffusion during the separation flow?

IV

CIRCUIT AND SYSTEM ISSUES

Chapter 14

ELECTRONICS

Throw physic to the dogs; I'll none of it.

—William Shakespeare, in *Macbeth*

14.1 Introduction

We have now completed our discussion of the various physical energy domains.[1] In this chapter, we begin our study of signal-processing devices and systems. We present a summary of key integrated-circuit electronic components: the semiconductor diode, the semiconductor resistor, the bipolar junction transistor (BJT), the metal-oxide-semiconductor field-effect transistor (MOSFET), and the operational amplifier with emphasis on MOSFET versions of these amplifiers. The treatment is of necessity quite abbreviated. References to more detailed treatments are listed in the Related Reading section at the end of this chapter.

14.2 Elements of Semiconductor Physics

We present here the semi-classical picture for conduction in semiconductors. At the atomic level, we think of semiconductors as solids in which the number of valence electrons exactly fills a set of states (called the *valence band*). For a range of energies above the top of this valence band, there are no allowed quantum states. This range is called the *energy gap*. At higher energies, there is a new set of quantum states called the *conduction band*. A *conduction electron* is an electron occupying a state in the conduction band. Conduction electrons

---

[1] The reader is asked to pardon the author's malapropism in selecting the quotation that begins this chapter. Macbeth is referring to medicine, not physics. But after the intensity of the last six chapters, this quote seemed refreshingly appropriate.

behave like classical particles with charge $-q_e$, where q_e is the electronic charge. A *hole* is the result of a vacant electron state near the top of the valence band. Holes behave like classical particles with charge q_e.

14.2.1   Equilibrium Carrier Concentrations

In equilibrium at temperature T, there is a relation between the concentration of electrons n_0 and the concentration of holes p_0.

$$p_0 n_0 = n_i^2 \tag{14.1}$$

where n_i is called the *intrinsic carrier concentration* and has the value of about 1×10^{10} cm^{-3} in silicon at room temperature.[2] The temperature dependence of n_i is given by

$$n_i \propto e^{-\dfrac{E_G}{2k_B T}} \tag{14.2}$$

where E_G is the band gap of silicon, equal to 1.1 electron-Volts at room temperature.[3]

These equations are consistent with the picture presented during our discussion of transport in Section 11.8 in which it was asserted that there is a relation between equilibrium concentration and potential of the form

$$n = n_r e^{\dfrac{-z_n q_e (\phi - \phi_r)}{k_B T}} \tag{14.3}$$

where z_n is the charge number of the carrier (+1 for holes and -1 for electrons), ϕ is the electrostatic potential, and n_r is the concentration at a reference position where the potential is ϕ_r. In this case, we choose to define the origin of potential such that ϕ_r is zero when n equals n_i. Thus

$$n_0 = n_i e^{\dfrac{q_e \phi}{k_B T}} \quad \text{and} \quad p_0 = n_i e^{-\dfrac{q_e \phi}{k_B T}} \tag{14.4}$$

As discussed in Section 3.2.5, the relative size of p_0 and n_0 can be adjusted by addition of minute quantities of substitutional impurities, called *dopants*. Using silicon with valence 4 as a starting point, a *donor* is an atom of valence 5 such as phosphorus, arsenic or antimony. When substituted for a silicon atom, a donor atom contributes a conduction electron to n_0, and leaves behind a positively charged *ionized donor* with charge q_e. An *acceptor* is an atom with

---

[2]Semiconductor conventions use the centimeter as the standard unit of length instead of the meter. In order to connect to other basic texts, we adopt that convention here.

[3]The electron-Volt, abbreviated eV, is equal to 1.6×10^{-19} Joules. It is the energy of an electron in a potential of one Volt.

valence 3, such as boron or aluminum. When substituted for a silicon atom, an acceptor atom contributes a hole and leaves behind a negatively charged *ionized acceptor* with charge $-q_e$. Overall electrical neutrality requires that

$$p_0 - n_0 = N_A - N_D \tag{14.5}$$

where N_D and N_A are the concentrations of donors and acceptors.[4] A material is called *p-type* if $N_A > N_D$, because then p_0 must be greater than n_0. An *n-type* material is one for which $N_D > N_A$.

Using p-type material as an example, and substituting, we find that

$$p_0 = \frac{N_A - N_D}{2} + \sqrt{\left(\frac{N_A - N_D}{2}\right)^2 + n_i^2} \tag{14.6}$$

It is typically true that the net doping $N_A - N_D$ is much greater than n_i, leading to the simplified result

$$p_0 = N_A - N_D \quad \text{and} \quad n_0 = \frac{n_i^2}{N_A - N_D} \tag{14.7}$$

In n-type material, the corresponding result is

$$n_0 = N_D - N_A \quad \text{and} \quad p_0 = \frac{n_i^2}{N_D - N_A} \tag{14.8}$$

Given these equilibrium concentrations, we can also determine the equilibrium potentials in n-type and p-type regions characterized by net dopings N_D and N_A, respectively:

$$\phi_{n0} = \frac{k_B T}{q_e} \ln \frac{N_D}{n_i} \quad \text{and} \quad \phi_{p0} = -\frac{k_B T}{q_e} \ln \frac{N_A}{n_i} \tag{14.9}$$

There is a relationship between potential and concentration at room temperature such that a factor of ten in concentration corresponds to a change in potential of about 60 mV. This is a very convenient rule of thumb when estimating behavior in semiconductor devices.

14.2.2 Excess Carriers

Non-equilibrium conditions in a semiconductor can be created by temperature gradients, by a process called *carrier injection* from a material that contacts the region of interest, or by direct generation of electron-hole pairs by the absorption of light or other internal excitations. The case of the temperature

---

[4]We assume that all donors and acceptors are ionized, a correct assumption for silicon at ordinary device temperatures and for doping levels below about 10^{20} cm^{-3}.

gradient has already been discussed in Section 11.8. Here we consider what are called *excess carriers*, a departure from equilibrium carrier concentrations under isothermal conditions.

Excess carriers can be written as

$$n' = n - n_0 \quad \text{and} \quad p' = p - p_0 \qquad (14.10)$$

In general, excess carriers are created and destroyed in pairs, so that $n' = p'$. Thus as a percentage of the equilibrium concentration, it is the smaller or *minority carrier concentration* that is most dramatically affected by the creation of excess carriers.

Because excess carriers represent a departure from equilibrium, there must be a dissipative process driving toward equilibrium. This process is called *recombination* and is typically described by a *minority carrier lifetime* τ_m. If we include processes for creating and removing excess carriers, we can write a continuity equation for each type of carrier. For electrons, this continuity equation takes the form:

$$\frac{\partial n}{\partial t} = \frac{1}{q_e} \nabla \cdot \mathbf{J}_n + G - \frac{n}{\tau_m} \qquad (14.11)$$

where G is the generation rate of excess carriers due either to optical absorption or ionization effects produced by impact of highly-excited carriers with background atoms. Charge transport in a semiconductor is an example of coupled transport due to electric-field drift and to diffusion set up by a concentration gradient. The flow equation for electrons is

$$\mathbf{J}_n = q_e \left(\mu_n n \boldsymbol{\mathcal{E}} + D_n \nabla n \right) \qquad (14.12)$$

where μ_n is the electron mobility, $\boldsymbol{\mathcal{E}}$ is the electric field, and D_n is the electron diffusion coefficient. Electron and hole mobilities are about 1000 cm^2/(Volt-sec) and 500 cm^2/(Volt-sec), respectively, in lightly doped material, decreasing markedly when dopings increase above about 5×10^{-16} cm^{-3}. The diffusion constants for both electrons and holes are proportional to their respective mobilities via the Einstein relation.[5]

In a homogeneously doped region of a semiconductor, both p_0 and n_0 are independent of both space and time. Therefore, all of the gradient terms can be re-expressed in terms of only the excess carriers. The continuity equation becomes

$$\frac{\partial n'}{\partial t} = \frac{1}{q_e} \nabla \cdot \mathbf{J}_n + G - \frac{n'}{\tau_m} \qquad (14.13)$$

---

[5]The Einstein relation is discussed on page 292. Additional details on semiconductor mobilities can be found in Howe and Sodini, pp. 34-37 [83], or in Sze, pp. 27-33 [84].

When dealing with the minority carriers, we can usually ignore the electric-field drift term compared with the diffusion term. Thus, after substitution for J_n, we obtain the equation governing the excess minority carriers:

$$\frac{\partial n'}{\partial t} = D_n \nabla^2 n' + G - \frac{n'}{\tau_m} \tag{14.14}$$

Once again, we recognize the heat-flow equation, but this time accompanied by generation and recombination terms.[6] Under DC steady state conditions in a region with uniform doping, this reduces to

$$\nabla^2 n' = \frac{1}{L_n^2} n' \tag{14.15}$$

where L_n is the *diffusion length* for minority-carrier electrons, given by

$$L_n = \sqrt{D_n \tau_m} \tag{14.16}$$

The solutions are exponentials with decay lengths given by L_n.

14.3   The Semiconductor Diode

If a region of p-type material is created in an n-type semiconductor substrate, a *semiconductor diode* is created. A typical fabrication sequence would be, first, to oxidize an n-type silicon wafer, open a window in the oxide with photolithography, introduce boron either by a diffusion process or by ion implantation followed by appropriate annealing steps, and, finally, create metal contacts to the p-type and n-type regions. Figure 14.1 shows the structure without the metal contacts.

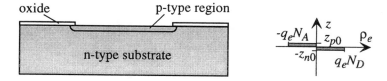

Figure 14.1.   A p-n junction created in an n-type substrate by the introduction of p-type dopants through a window in the oxide. The space-charge in the vicinity of the junction is shown at the right.

In equilibrium, a p-n semiconductor junction has a double layer of charge, illustrated schematically on the right of Fig. 14.1. There is an electric field within this double layer set up by the positive and negative charges, with the

---

[6]Similar transport equations arise in chemical systems which undergo diffusion and simultaneous chemical reactions that convert one type of species into another.

integral of the electric field across the double layer giving the equilibrium difference in equilibrium electrostatic potential that is required by the different carrier concentrations on the two side of the junction. This so-called *built-in potential* is given by

$$\phi_B = \frac{k_B T}{q_e} \ln \frac{N_A N_D}{n_i^2} \tag{14.17}$$

where N_A and N_D are now the net dopant acceptor and donor concentrations on the p and n sides of the junction.

To create the charged double layer, mobile electrons and holes are depleted from the immediate vicinity of the junction, leaving a positive charged region on the n-type side with charge density $q N_D$ and a negatively charged region on the p-type side, with charge density $-q N_A$. A good way to think about it is to recognize that the potential must make a continuous transition between a positive ϕ_{n0} on the n-type side of the junction to a negative ϕ_{p0} on the p-side, and it is the gradient of this potential that corresponds to the electric field within the space-charge region. Linked to this continuous potential change through the junction region is a continuous change of the equilibrium carrier concentrations n_0 and p_0 such that their product is still n_i^2, but the majority carrier concentration in the junction region is far below its value in the bulk region. In the *depletion approximation*, the charge layer is modeled as fully depleted[7] to a depth x_{n0} into the n-type material and x_{p0} into the p-type material. This depleted region is also called the *space-charge layer*. The width of the space-charge region in equilibrium is

$$X_{J0} = x_{p0} + x_{n0} = \sqrt{\left(\frac{2 \varepsilon_s \phi_B}{q_e}\right) \left(\frac{1}{N_A} + \frac{1}{N_D}\right)} \tag{14.18}$$

where ε_s is the permittivity of silicon, equal to $11.7\, \varepsilon_0$. Typical values for X_{J0} are in the range of 1 - 10 μm, depending on the doping profile.

To complete the diode structure, a metal contact (typically aluminum or an aluminum-silicon alloy) can be deposited and patterned on the top surface, and similar metal can be blanket deposited on the bottom surface. Care must be taken to ensure good resistive (ohmic) contacts between the semiconductors and the metals. For aluminum, which is itself a p-type dopant of silicon, the contact to the p-region is straightforward. The n-region might require an n-type implant on the back surface to raise the surface concentration into the 10^{19} cm^{-3} range so that the aluminum forms a good contact. The completed structure is illustrated in Fig. 14.2.

---

[7] The region is "depleted" only in the sense that the concentration of mobile carriers is much less than $q N_D$ or $q N_A$. Both carriers are present, but not in sufficient numbers to affect the macroscopic charge density.

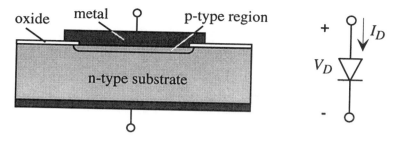

Figure 14.2. A completed semiconductor junction diode, together with its standard circuit symbol.

The effect of an applied voltage between the contacts, to first order, is to change the total potential difference across the space-charge layer, and this has the effect of changing the carrier concentrations in the immediate vicinity of the space charge layer. The *Law of the Junction* expresses the concentration changes in the form

$$\frac{n_p}{n_{p0}} = \frac{p_n}{p_{n0}} = e^{\frac{q_e V_D}{k_B T}} \tag{14.19}$$

where the subscripts p and n refer to positions at the edges of the space-charge layer on the p-side and n-sides of the junction, respectively, and the 0 refers to equilibrium. A positive voltage V_D increases the minority carrier concentrations at the boundary of the space charge layer by a factor $\exp(q_e V_D / k_B T)$, leading to *excess minority carriers* given by

$$n_p' = n_{p0} \left(e^{q_e V_D / k_B T} - 1 \right) \tag{14.20}$$

This increase in minority concentration at the edge of the space-charge layer is called *minority carrier injection* across the *pn* junction. Diffusion of these excess minority carriers away from the junction toward the contacts then leads to the following current-voltage characteristic for an ideal exponential diode:

$$I_D = I_0 \left(e^{q_e V_D / k_B T} - 1 \right) \tag{14.21}$$

where the quantity $k_B T / q_e$ is the thermal voltage, equal to about 25 mV at room temperature, and I_0 is a prefactor, called the *reverse saturation current*, having the form

$$I_0 = q_e A \left(\frac{n_{p0} D_n}{L_n} + \frac{p_{n0} D_p}{L_p} \right) \tag{14.22}$$

where A is the junction area. In forward bias, with V_D positive, there is a large current through the diode in the p-to-n direction. However, in reverse bias, with V_D negative, the minority carrier concentrations at the edges of the space-charge layer are driven to zero, and the reverse current is small.

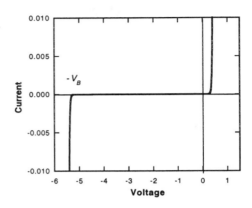

Figure 14.3.   A typical diode current-voltage characteristic.

In practice, real diodes have many detailed characteristics that produce variations from this ideal. The first of these is illustrated in Fig. 14.3. For positive voltages, the characteristic exponential behavior is observed. At moderate sized reverse voltages, the current is small (this is the *reverse blocking region*.) but at sufficiently large negative voltages, characterized by a reverse breakdown voltage V_B, the reverse blocking behavior breaks down and the diode conducts in the reverse direction. There are several possible mechanisms for this reverse breakdown. For diodes with breakdown voltages greater than a few Volts, the dominant mechanism is avalanche breakdown, a runaway process of generation of excess holes and electrons in the space-charge layer by impact from highly accelerated electrons and holes with the atoms.

Another detail that has a significant effect on diode operation is *space-charge-layer generation and recombination* The ideal-diode equation assumes that the space charge-region is so narrow that any kinetic effects within that region can be ignored. However, when a reverse voltage is applied to the diode, the width of the space-charge layer grows. For uniform doping on the two sides of the junction, this variation takes the form

$$X_J = X_{J0}\sqrt{\frac{\phi_B - V_D}{\phi_B}} \tag{14.23}$$

This equation is valid for V_D negative, and also for V_D slightly positive (less than about 0.4 V).[8] Within this space-charge layer, there can be net generation under reverse (negative-voltage) bias. The reverse current is of the form

$$I_D = -I_0 - \frac{q_e A X_J n_i}{\tau_m} \qquad (14.24)$$

where I_0 is the prefactor from the ideal-diode behavior and τ_m is the minority carrier lifetime, which can be in the range of microseconds to nanoseconds, depending on the details of the device. The net result is a reverse-current characteristics of the type shown in Fig. 14.4.

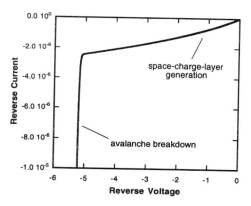

Figure 14.4. Illustrating the reverse-bias characteristic of a typical diode. The growth of reverse leakage current prior to breakdown is due to carrier generation within the space-charge layer. The width of this layer grows with the square-root of the reverse voltage.

The space-charge layer also affects the forward characteristic. As soon as the diode is forward biased by about 60 mV, the exponential term makes the -1 in Eq. 14.21 insignificant. If we plot the log of the forward current versus voltage in this range, a typical diode has the characteristic shown in Fig. 14.5. We see that the characteristic has the ideal-diode exponential slope at the larger voltages, but at lower voltages, has an exponential behavior with a different slope. This is due to recombination of excess carriers in the space-charge layer while they are trying to cross the junction. An equation that models this behavior is

$$I_D = I_0 e^{q_e V_D / k_B T} + I_1 e^{q_e V_D / 2 k_B T} \qquad (14.25)$$

---

[8]For voltages more positive than 0.4 V, the excess electrons and holes in the space-charge layer make the depletion approximation no longer valid. More complete models are required.

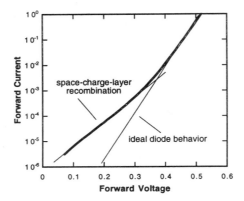

Figure 14.5.   Forward-bias characteristic of a typical junction diode.

The ideal-diode term has has slope q_e/k_BT, or one decade of current for 60 mV of voltage. The space-charge-layer-recombination term, with prefactor I_1, has slope $q_e/2k_BT$, or one decade of current for 120 mV of voltage.

The variation of the space-charge-layer width with applied voltage also creates a nonlinear capacitor across the junction. The magnitude of the charge on this capacitor is

$$Q_J = q_e A N_D x_n \tag{14.26}$$

where x_n is the width of the depletion region on the n-type side of the junction. This width is voltage dependent. Therefore, the charge in the space-charge-layer is also voltage dependent. We can define a *junction capacitance*, which is the incremental capacitance of the junction at bias voltage V_D:

$$C_J = \left.\frac{\partial Q_J}{\partial V_D}\right|_{V_D} \tag{14.27}$$

The reverse-bias variation of junction capacitance, normalized to its value at zero bias, is shown in Fig. 14.6. The decrease with increasing reverse bias is due to the increased width of the space-charge layer. It is the same kind of variation as that of X_J, so that

$$\frac{C_J}{C_{J,0}} = \frac{1}{\sqrt{1 - \frac{V_D}{\phi_B}}} \tag{14.28}$$

Under forward bias conditions, the space-charge layer narrows, causing an increase in C_J.

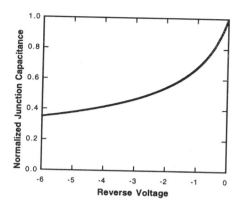

Figure 14.6.　The voltage variation of the junction capacitance under reverse-bias conditions, normalized to its value at zero bias. Under forward bias, the capacitance increases markedly, especially for forward bias values in excess of about 0.4 V.

14.4　The Diffused Resistor

If we take the same structure we used to build the semiconductor diode, and simply make two contacts to the top surface, we get a *diffused resistor* illustrated in Fig. 14.7. Diffused resistors of this type, in addition to being used as components of ordinary circuits, are often used as piezoresistors to sense strain, a subject that will be discussed in Chapter 18.

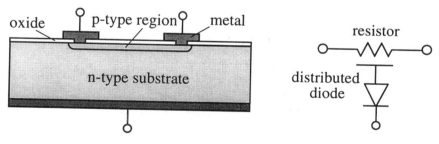

Figure 14.7.　Illustrating a diffused semiconductor resistor.

There is a diode between the p-region and the n-region, illustrated schematically as a distributed diode, but the p-region itself functions as a resistor. As long as the applied potential everywhere in the p-region is less positive than the potential applied to the n-type substrate, this distributed diode is reverse biased, and very little leakage current flows. Under these circumstances, we say that the diffused resistor is *junction isolated* from the substrate. However, if we allow the voltage on the p region to become more positive than the n substrate, significant current can flow in the diode. It is important to keep this distributed diode in mind when designing circuits that use diffused resistors.

When light of sufficient photon energy (2 eV or greater, corresponding to the visible spectrum) impinges on silicon, optical generation of excess carriers occurs. This gives rise to an effect called *photoconductivity*. The steady-state concentration of excess carriers is determined by the product $G\tau_m$, where G is the optical generation rate, giving rise to an electrical conductivity change

$$\sigma_e = q_e[\mu_n(n_0 + n') + \mu_p(p_0 + p')] \tag{14.29}$$

Photoconductivity leads to measurable resistance changes. This is a benefit when one is seeking to detect the presence of light, but it can be a disadvantage when attempting to use a resistor to measure some other quantity, such as strain or temperature, in the presence of external light.

14.5   The Photodiode

When visible light impinges on the space-charge region of a semiconductor diode, the excess carriers generated in that region are separated by the electric field present in the space-charge layer. If the diode is connected to an external circuit, this leads to current. If, instead, the diode is open-circuited, a forward voltage V_{OC} appears across the diode. The simplest model for this *photocurrent effect* is a light-dependent shift in the diode current-voltage characteristic

$$I_D = I_0\left(e^{q_e V_D/k_B T} - 1\right) - I_P \tag{14.30}$$

where I_P is the photocurrent. Figure 14.8 illustrates the effect of the photocurrent on the current-voltage characteristic.

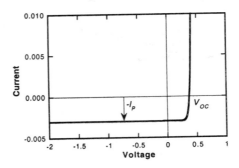

Figure 14.8.   A photocurrent shifts the diode characteristic so that now a portion of the characteristic lies in the fourth quadrant. In this part of the characteristic, the diode can deliver power to an external circuit.

Photodiodes are direct optical-to-electrical energy conversion devices, and are used both as photodetectors and as power-generation devices (for example, solar cells).

14.6   The Bipolar Junction Transistor

If two diodes are constructed back-to-back so as to share a common middle layer, and this layer is thin compared to the minority-carrier diffusion length, the result is a *bipolar junction transistor*, abbreviated BJT. Figure 14.9 illustrates the basic device structure for an npn transistor.

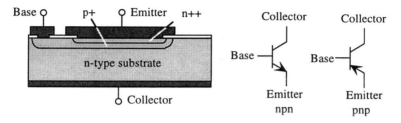

Figure 14.9.   Basic structure of an npn bipolar junction transistor together with standard circuit symbols for both npn and pnp transistors.

The n-type substrate in an npn transistor serves as the collector and has a p+ base region diffused into it, followed by an even more heavily doped n-type emitter, labeled as n++ to distinguish its very heavy doping. By reversing all the dopings, a pnp transistor can be built. Circuit symbols for both devices are shown in the Figure.

The BJT is typically operated with the base-emitter junction forward biased and the base-collector junction reverse biased. The emitter forward bias creates excess minority carriers within the base at the the space-charge-layer edge of the emitter junction, while the collector reverse bias reduces the minority carrier concentration to zero at the space-charge-layer edge of the collector junction. If the distance between base and collector is much less than the minority carrier diffusion length, a linear gradient of minority carriers develops between the emitter side of the base and the collector side of the base. The resulting diffusion current is *collected* by the reverse-biased collector junction, resulting in a current in the external circuit. We shall encounter bipolar transistors briefly in Chapter 18.

14.7   The MOSFET

The Metal-Oxide-Semiconductor Field-Effect Transistor, or MOSFET, is illustrated in Fig. 14.10. The enhancement-mode n-channel MOSFET is constructed on a p-substrate with two heavily doped n-type regions called the *source* and *drain*, denoted n^+ to signify heavy doping, separated by a region called the *channel*. Above the channel is a relatively thin oxide and a gate conductor, typically polysilicon. Because in the absence of gate voltage at least one of the two pn junctions must be reverse biased, there is no conducting path between the source and drain. However, when positive voltage is applied

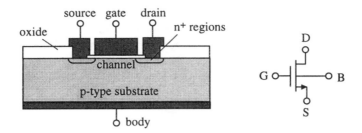

Figure 14.10. An n-channel enhancement-mode MOSFET. The gate conductor is typically polysilicon, while the contacts to the n^+ source and drain regions as well as to the body are typically metal.

between the gate and the substrate (called the *body*), the potential at the surface of the channel is pulled positive, initially depleting the surface of its holes, and eventually, creating a sufficiently positive surface potential that the surface becomes n-type. The voltage needed to create this n-type *inversion layer* is called the *threshold voltage*. In the presence of a gate-to-body voltage greater than the threshold, there now is a conducting path between source and drain, and the amount of conduction is controlled by the gate voltage.

The source and drain are actually identical in their structure. In an n-channel device, we assign the drain to be the terminal with the more positive voltage compared to the source. Before presenting quantitative models of the n-channel enhancement mode MOSFET, it is useful to introduce three other varieties:

The p-channel enhancement mode MOSFET is identical in structure to the n-channel device in Fig. 14.10, but with an n-type substrate and p^+ source and drain regions. To turn on the conducting channel between source and drain, a negative voltage is applied between the gate and the body. In a p-channel device, we assign the drain to be the terminal with the more negative voltage with respect to the source.

There is a second generic type of MOSFET called a *depletion-mode* MOS-FET in which the channel region is lightly doped with the same type as in the source and drain. In these devices, there is a conducting path between source and drain with zero gate-to-body voltage. For the n-channel depletion device, a negative gate-to-body voltage repels the electrons from the channel region and turns the device off. The voltage at which the conducting channel disappears is called the threshold voltage. Similarly, for a p-channel depletion device, a positive gate-to-body voltage is required to turn the device off, and the threshold voltage is positive. Circuit symbols for the four types are shown in Fig. 14.11. We show the p-channel devices with the sources on top since, when we use these devices in circuits, the source is usually connected to the most positive supply voltage, and this, by convention, is usually drawn at the top of the circuit diagram.

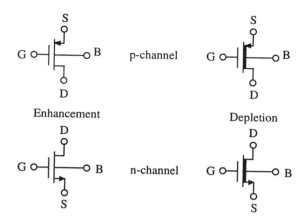

Figure 14.11.   Circuit symbols for the four types of MOSFET.

14.7.1   Large-Signal Characteristics of the MOSFET

There are three operation regimes for a MOSFET.[9] We will use an n-channel enhancement-mode MOSFET as our example. The usual scheme for describing how the drain-to-source current varies with terminal voltages is shown in Fig. 14.12.

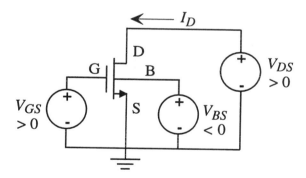

Figure 14.12.   Normal biasing scheme for an n-channel MOSFET.

We use the source terminal as the circuit ground. Let us consider first the case where V_{BS} is zero, that is, the body is connected to the same ground as the source. When V_{GS} is below the threshold voltage V_{Tn}, there is no conducting path, and we say the transistor is in the *cutoff* region of operation. The opposite

---

[9]Our discussion of the MOSFET and MOSFET circuits draws heavily on Howe and Sodini [83]. Readers are referred there for additional details.

case is where V_{GS} is made very positive, greater than $V_{DS} + V_{Tn}$. In this case, there is an n-type inversion layer all along the channel region. This is called the *nonsaturation* or *triode* region. At intermediate values of V_{GS}, there is an inversion layer at the source end of the channel since V_{GS} is greater than V_{Tn}, but this layer disappears at some point along the channel because there is not sufficient gate potential to exceed threshold at the drain end. This intermediate range is called the *saturated* region.

We shall not derive the governing equations for these regions, since the derivations are a bit lengthy. In the triode region, the relation between V_{GS}, V_{DS} and I_D for zero V_{BS} is

$$I_D = \frac{W}{L} \mu_n \hat{C}_{ox} \left[V_{GS} - V_{Tn} - (V_{DS}/2) \right] V_{DS} \tag{14.31}$$

where W is the width of the channel region (into the page in Fig. 14.10), L is the length of the channel between source and drain, μ_n is the electron mobility in the channel (which can be about a factor of two less than the electron mobility in bulk, because of surface-scattering effects), and $\hat{C}_{ox}$ is the oxide capacitance per unit area, given by

$$\hat{C}_{ox} = \frac{\varepsilon_{ox}}{t_{ox}} \tag{14.32}$$

The permittivity of the oxide is $\varepsilon_{ox} = 3.9\,\varepsilon_0$, and t_{ox} is the gate oxide thickness, typically less than 20 nm in modern devices. Of the various parameters involved, the channel width-to-length ratio is controlled by the mask layout used to define the channel region, and $\hat{C}_{ox}$ is determined by the gate-oxide thickness. The threshold voltage is discussed in detail a little later in this section.

For a given V_{GS}, the drain current reaches its maximum value when V_{DS} reaches a value called $V_{DS,\text{sat}}$, given by

$$V_{DS,\text{sat}} = V_{GS} - V_{Tn} \tag{14.33}$$

The corresponding drain current is

$$I_{D,\text{sat}} = \frac{1}{2} \frac{W}{L} \mu_n \hat{C}_{ox} V_{DS,\text{sat}}^2 \tag{14.34}$$

To first order, increasing V_{DS} beyond $V_{DS,\text{sat}}$ produces no increase in drain current. The net result is a set of characteristics shown in Fig. 14.13.

The threshold voltage is determined by the gate-to-body voltage needed to create an inversion layer under the gate. The substrate doping is N_A, so the potential in the substrate has the value

$$\phi_p = -\frac{k_B T}{q_e} \ln \left(\frac{N_A}{n_i} \right) \tag{14.35}$$

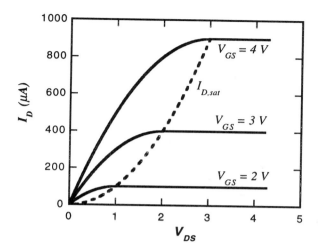

Figure 14.13. Drain current versus drain-to-source voltage with gate-to-source voltage as a parameter for an n-channel MOSFET with $V_{Tn} = 1$ V, $W/L = 4$, and $\mu_n \hat{C}_{ox} = 50$ μA/V^2. The dashed curve plots $I_{D,\text{sat}}$ *vs.* $V_{DS,\text{sat}}$.

Inversion is achieved when the surface potential is made positive by about the magnitude of ϕ_p. If the body-to-source voltage V_{BS} is nonzero and negative, then it takes a more positive gate voltage to pull the surface potential up to $|\phi_p|$. This is called the *body effect*. The details will carry us too far afield, but the result is that the threshold voltage, including the body effect, is defined as

$$V_{Tn} = V_{FB} - 2\phi_p + \frac{1}{\hat{C}_{ox}}\sqrt{(2q_e N_A \varepsilon_s)(-2\phi_p - V_{BS})} \qquad (14.36)$$

where V_{FB}, called the *flat-band voltage*, is the voltage that must be applied between gate and source to bring the total charge on the gate electrode to zero. V_{FB} depends on the potential difference between the n^+-doped polysilicon gate and the p-type substrate, and also on the small residual positive fixed charge per unit area that is present at the oxide-silicon interface, denoted $\hat{Q}_{ox}$. The expression for V_{FB} is

$$V_{FB} = -(\phi_{n+} - \phi_p) - \frac{\hat{Q}_{ox}}{\hat{C}_{ox}} \qquad (14.37)$$

In order to separate the body effect from other effects, it is conventional to express the body effect as a shift of threshold from $V_{Tn,0}$, the threshold voltage when V_{BS} is zero:

$$V_{Tn} = V_{Tn,0} + \gamma_n \left(\sqrt{-2\phi_p - V_{BS}} - \sqrt{-2\phi_p}\right) \qquad (14.38)$$

where $\gamma_n = (\sqrt{(2q_e\varepsilon_s N_A)}/\hat{C}_{ox}$.

There are many more details we could consider, but there is one that is extremely important in the design of amplifiers. In the saturation region, the inversion layer disappears somewhere between the source and the drain, limiting the total current to $I_{D,\text{sat}}$. As the drain voltage is increased, this disappearance point moves slightly toward the source and the value of $I_{D,\text{sat}}$ increases slightly. This effect is called *channel-length modulation*. The simplest model that captures this effect is to modify the saturation-region drain-current expression to be of the form

$$I_D = \frac{1}{2}\frac{W}{L}\mu_n\hat{C}_{ox}\left(V_{GS} - V_{Tn}\right)^2\left(1 + \lambda_n V_{DS}\right) \qquad (14.39)$$

where the parameter λ_n depends inversely on the length of the channel. A typical value is

$$\lambda_n = \frac{0.1}{L}\ \text{V}^{-1} \qquad (14.40)$$

where the channel length L is expressed in microns.

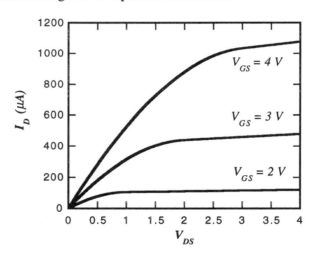

Figure 14.14.   Drain characteristic of the same MOSFET as in Figure 14.13, but now including channel-length modulation. The channel length for this example is 2 μm.

With this channel-length modulation, we must also modify our expression for triode-region behavior so that the characteristics are continuous at the boundary between the triode and saturation regions:

$$I_D = \frac{W}{L}\mu_n\hat{C}_{ox}\left[V_{GS} - V_{Tn} - (V_{DS}/2)\right]\left(1 + \lambda_n V_{DS}\right)V_{DS} \qquad (14.41)$$

The net result is a modified drain current characteristic, as shown in Fig. 14.14. The primary difference between this graph and that of Fig. 14.13 is the non-zero slope of the drain current in the saturation region. When these device characteristics are linearized about an operating point in the saturation region, this slope gives rise to a finite conductance which limits the achievable gain of a single-state of amplification, as we shall see in Section 14.8.

14.7.2   MOSFET Capacitances

The MOSFET has many capacitances that are inherent in its structure. The most obvious one is the gate-to-channel capacitance. In the triode region, with an inversion layer present throughout the channel, this capacitance appears entirely connected to the source, and has a value given by the total area of the channel times $\hat{C}_{ox}$. In the saturation region, however, the channel is not inverted throughout its length. A good rule of thumb is that the effective capacitance between gate and source decreases to 2/3 its triode-region value. In addition, there is a parasitic capacitance between gate and source where the thin oxide and polysilicon gate overlap the source diffusion. Thus, the total capacitance between gate and source in saturation is expressed as

$$C_{GS} = \frac{2}{3} W L \hat{C}_{ox} + W \hat{C}_{ov} \qquad (14.42)$$

where $\hat{C}_{ov}$ is the overlap capacitance per unit channel width.

The gate-to-drain capacitance in saturation is simply the overlap capacitance $W \hat{C}_{ov}$. And there are junction capacitances between the source and body and the drain and body. These depend on the actual areas of the source and drain diffusions, and to a lesser extent, on the perimeter of these diffusions.

14.7.3   Small-Signal Model of the MOSFET

If we linearize the drain-current characteristics about an operating point in the saturation region, and include the various capacitances discussed in the previous section, we obtain the small-signal MOSFET transistor model shown in Fig. 14.15.

In keeping with circuit-analysis conventions, upper-case letters with upper-case subscripts are used for DC quantities, and lower case letters with lower-case letters with lower-case subscripts are used for linearized incremental variables. The left-hand dependent current source, $g_m v_{gs}$ captures the increase in drain current with V_{GS}. The quantity g_m is called the *transconductance*, and (with neglect of the channel-length modulation correction) in the saturation region has the value

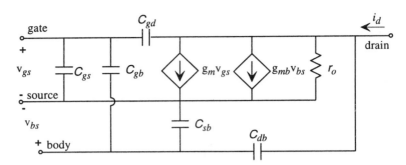

Figure 14.15.   Complete small-signal model for an n-channel MOSFET in the saturation region.

$$g_m = \left.\frac{\partial I_D}{\partial V_{GS}}\right|_{V_{DS}} = \sqrt{2\left(\frac{W}{L}\right)\mu_n \hat{C}_{ox} I_D} \qquad (14.43)$$

It increases as the square-root of the operating-point drain current. The right-hand current source captures a similar but much smaller effect if the body-to-source voltage has a small-signal component. Normally, V_{BS} is a constant, and has no incremental component. Hence, the right-hand current source becomes an open circuit and disappears from the model. The resistor r_o is the reciprocal of the slope of the drain-current characteristic at constant V_{GS}:

$$\frac{1}{r_o} = \left.\frac{\partial I_D}{\partial V_{DS}}\right|_{V_{GS}} \approx \lambda_n I_D \qquad (14.44)$$

Here we see that inclusion of the channel-length modulation term is essential in order to get a first-order estimate of r_o.

This model includes all the capacitances discussed in the previous section plus one more, the capacitance between gate and body. When a transistor is operated with the source and body at fixed potentials, both the body and source are connected to ground in the linearized model, and this additional capacitance gets added to the value of C_{gs}. And if V_{SB} is zero or a constant, the C_{sb} capacitor disappears from the model, since there is no incremental voltage across it.

We are now prepared to consider amplifiers.

14.8   MOSFET Amplifiers

An *amplifier* is a circuit with several ports that must be connected to *energy sources*, and several other ports for, first, one or more *inputs*, and, second, and one or more *outputs*. Except for recent attention to low-power designs, conventional electronic circuit analysis tends to be interested only in the signal-path between input(s) and output(s). This is quite different from the view we have taken of other devices in which we track total energy. It is useful, therefore, before adopting the conventional electronics view, to place amplifiers in the same context as other lumped elements.

We usually think of an amplifier as something that increases signal amplitude, or signal power, or both. But, in the simplest picture, with internal dynamics due to charging and discharging of internal capacitances ignored for the moment, *a transistor is a passive resistive element.* It always dissipates net power.[10] So how can it "amplify?"

Amplification occurs through the *nonlinear* resistive behavior of the transistor. If we consider an amplifier to be a three-port device, one port for a DC power supply, one for an input voltage, and one for an output voltage, we tend to express the relationship between input and output as a graph (the transfer function) *assuming that the power supply is connected.* When we do that (see the following section), we get a transfer characteristic that is nonlinear and which, in a region between the two output limits, shows incremental gain. We shall find that the output can supply more power than comes in at the input. This extra power *must* come from the power supplies.

It is common practice to ignore the power from the power supply, and, as was done in Fig. 14.15, model the transistor with *dependent sources,* non-energy-conserving elements that set either a voltage or a current depending on some other variable in the network. While these are extremely useful for electronic design, they are a potential disaster if one is concerned with total energy management because dependent sources are not required to obey any particular rules about where their energy comes from. Nevertheless, we will continue to follow standard electronics practice, and use dependent sources for transistors, and also for modeling complete transistor amplifiers.

14.8.1   The CMOS Inverter

Figure 14.16 shows a basic MOS amplifier which uses an n-channel enhancement-mode MOSFET as the *driver,* with its gate as the "input," and one p-channel enhancement-mode MOSFET as the *load,*[11] having a fixed voltage V_B at its gate and its source tied to a power supply voltage V_{DD}. The two drains are connected together at the output port, creating what is called a *complementary MOS* circuit, or CMOS. A load capacitance has been added representing, perhaps, the input to another MOS circuit, or the equivalent capacitance of a bonding pad. If this circuit is to deliver DC current to an external circuit, then

---

[10]The entire drain-current characteristic of the MOSFET lies in the first quadrant, and if we were to extend to negative V_{DS}, the characteristic would like in the third quadrant. Hence MOSFETs are strictly passive, always dissipating power.

[11]The term *load* is used in two ways: as an internal part of the circuit, such as the PMOS transistor in this example, and as an external component connected to the output of the amplifier, such as C_L.

a load resistor would be added in parallel with C_L, providing a DC path to ground.

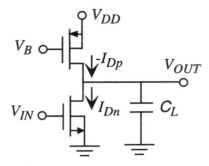

Figure 14.16.   A CMOS inverter circuit.

The CMOS inverter is used both in digital integrated circuits and as one type of analog amplifier circuit. When used in digital circuits, the two gates are connected together, so that $V_B = V_{IN}$. We shall first examine the case where V_B is fixed. We will find the quasistatic large-signal transfer function, estimate the switching speed for digital-circuit applications, and calculate the small-signal input-output characteristics as a linear amplifier when biased to a suitable operating point.

14.8.1.1   Quasi-Static Transfer Function

Under quasistatic conditions, the current into the capacitor is negligible, which means that the PMOS drain current, shown as $-I_{Dp}$, must equal the NMOS drain current I_{Dn}. Since the various characteristics are nonlinear, analytical solutions over the entire range are difficult to work with. Instead we use a graphical method, by plotting I_{Dn} and $-I_{Dp}$ vs. V_{DS} on the same axes. In order to do this, we must use the KVL constraint that

$$V_{SDp} = V_{DD} - V_{DSn} \tag{14.45}$$

With this constraint, the expression for $-I_{Dp}$ becomes, for the triode region,

$$-I_{DP} = \left(\tfrac{W}{L}\right)_p \mu_p \hat{C}_{ox} [V_{DD} - V_B + V_{Tp} - (V_{DD} - V_{DSn})/2]$$

$$\times \ [1 + \lambda_p(V_{DD} - V_{DSn})](V_{DD} - V_{DSn}) \tag{14.46}$$

and, for the saturation region,

$$-I_{DP} = \frac{1}{2} \left(\frac{W}{L}\right)_p \mu_p \hat{C}_{ox} \left(V_{DD} - V_B + V_{Tp}\right)^2 \left[1 + \lambda_p(V_{DD} - V_{DSn})\right]$$

$$(14.47)$$

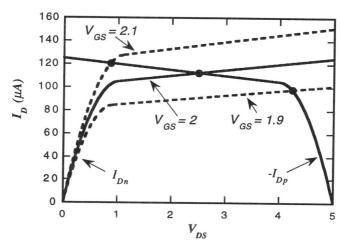

Figure 14.17.   Simultaneous plot of I_{Dn} and $-I_{Dp}$ for the CMOS inverter with $V_{Tn} = -V_{Tp}$ = 1 V, V_{DD} = 5 V, V_B = 3 V, $k_n = k_p = 200 \ \mu A/V^2$, and $\lambda_n = \lambda_p = .05$. The solid circles mark the various operating points for the indicated values of V_{GS}.

Figure 14.17 shows the simultaneous plot of I_{Dn} and $-I_{Dp}$ for transistors with the data shown in the figure caption. The quantities k_n and k_p are defined as

$$k_n = \left(\frac{W}{L}\right)_n \mu_n \hat{C}_{ox} \quad \text{and} \quad k_p = \left(\frac{W}{L}\right)_p \mu_p \hat{C}_{ox} \qquad (14.48)$$

Because hole mobilities are typically a factor of two smaller than electron mobilities, it is common practice to make the PMOS transistor have twice the W/L ratio as the NMOS transistor, in which case $k_n = k_p$. We have also chosen a symmetric design in which the two threshold voltages have the same magnitude. With these choices, the two graphs are symmetric, and for V_{GS} = 2 V, they intersect exactly at the middle, with an operating point value of V_{DS} equal to 2.5 V, half the supply voltage. For slightly larger values of V_{GS}, the operating point swings to the left, entering the triode region of the NMOS device, while for slightly smaller values, the operating point swings to the right, entering the triode region of the PMOS device. Between the two triode-region limits, the circuit can work as an amplifier. Note that a change of only 0.2 V at the input produces a swing of almost 3.5 V at the output. On the other hand, when switched hard between the alternate triode regions using input voltages

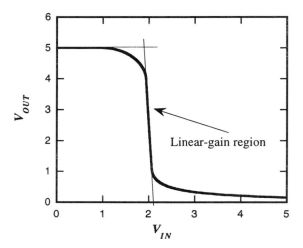

Figure 14.18.   The DC voltage transfer function for the CMOS inverter of Fig. 14.16.

either near zero or V_{DD}, the circuit functions as a digital inverter in the sense that a high input voltage produces a low output voltage, and vice versa.

The complete transfer function for this circuit is shown in Fig. 14.18. This was obtained by solving the equation $I_{Dn} = -I_{Dp}$ across the various regions. For example, for $V_{IN} < VT_n$, the NMOS device is cutoff. For slightly larger V_{IN}, the NMOS device is in saturation but the PMOS device is in the triode region. Then both devices enter the triode region, which corresponds to the linear-gain region in Fig. 14.18. Finally, at large V_{IN}, the NMOS device enters the triode region while the PMOS device is in saturation.[12] We note that the linear-gain region is very narrow and has a steep slope. A small change in input voltage produces a large change in output voltage. The slope is the *voltage gain*, and it is noticeably larger than unity. Thus, we have an amplifier for this restricted range of inputs, but only in the sense that a *change of input* produces a *change of output*. Therefore, to see the amplification most clearly, we should consider the incremental model for this circuit. Before doing so, we shall make an estimate of the large-signal switching speed.

14.8.2   Large-Signal Switching Speed

In the circuit of Fig. 14.16, the gate-to-source capacitances are connected to voltage sources, the drain-to-ground capacitance for both transistors appear in parallel with C_L, and the body-to-source capacitances are held at zero volts.

---

[12] As a practical matter, it is easier to solve for V_{IN} as a function of V_{OUT} because the resulting equations are of lower order.

Therefore, if we temporarily ignore the gate-to-drain capacitances of the two transistors, we can approximate the circuit as a first-order nonlinear *RC* circuit. Because the resulting state equation is nonlinear, it is helpful to use numerical methods to investigate large-signal transients. There are several good methods: (1) use a circuit simulator, such as SPICE, incorporating a standard transistor model for each device; (2) set up the state equation in MATLAB and solve it with a suitable input function; or, (3) set up an equivalent system in SIMULINK, and simulate it there. While the circuit-simulator approach is preferred when available, in this particular example, we will demonstrate how SIMULINK can be set up for this nonlinear transient problem, partly to get the answer, and partly to provide an example of how to use SIMULINK for nonlinear dynamic simulation.

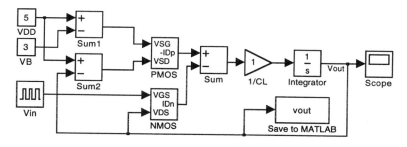

Figure 14.19. A SIMULINK implementation of the nonlinear first-order differential equation describing a switching transient of the CMOS inverter.

Figure 14.19 shows a SIMULINK block diagram that solves the following differential equation:

$$\frac{dV_{OUT}}{dt} = -\frac{1}{C_L}(I_{Dn} + I_{Dp}) \tag{14.49}$$

Note that in DC steady state, $I_{Dn} + I_{Dp} = 0$, which is the condition we used to analyze the static transfer function. The integrator integrates the net current into the capacitor, whose size is set by the "$1/C_L$" gain block, which is set to unity for this example. All other numerical parameters are buried in the transistor models. The blocks labeled 'NMOS' and 'PMOS' are SIMULINK subsystems that implement the cutoff, saturation, and triode regions of the transistor characteristics. Because there are three regions, two switches are needed in each subsystem. Figure 14.20 shows the NMOS subsystem. The two function blocks implement the saturation- and triode-region equations using there parameters, k_n, V_{Tn}, and λ_n. A similar subsystem is used for the PMOS transistor.

Switching transients were calculated for an initial state of zero at the output, and a 4 V square pulse applied to the input. In one simulation, V_B was fixed at three volts. This is the kind of arrangement that would be used in an analog

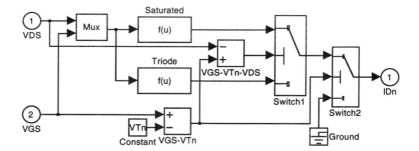

Figure 14.20. A SIMULINK model that implements the quasi-static characteristic of an NMOS transistor. The drain-current equations corresponding to saturation- and triode-region operation are placed in the function blocks. The switches select the value of drain current depending on the relative sizes of V_{GS}, V_{Tn}, and V_{DS}.

amplifier. In the second simulation, the value of V_B was set equal to V_{IN}. This is the arrangement that is used for CMOS inverters in digital circuits. The two transients are plotted in Fig. 14.21.

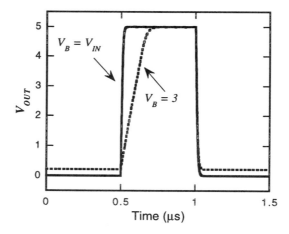

Figure 14.21. Switching transients for a 4 V square pulse for two different choices of V_B. Transistor parameters were identical to those in Fig. 14.17. The value of C_L is 1 pF.

The two output waveforms are similar, in that they have a low value during the first half microsecond, which is when there is a 4V signal applied at the input. At $t = 0.5\mu s$, the input drops to zero, and the output climbs toward the supply voltage of 5 V. When both inputs are tied together, as in the digital inverter, this climb is very rapid, but with V_B fixed at 3 V, the rate of rise is notably slower. This is because the capacitor charging is limited by how much current the PMOS transistor can supply. This current is much greater if its input is pulled to zero than if it is fixed at 3 V. When the 4 V pulse returns at $t = 1\mu s$, the NMOS transistor turns on and quickly discharges the capacitor, with virtually identical behavior in the two circuits. To summarize, the *slew rate* of an output, that is, how fast it can swing from low to high or high to low, depends on the details of the circuit. Faster slew rate is achieved with both inputs tied together.

14.8.3   The Linear-Gain Region

We can find the analog gain of this circuit by examining the small-signal model,[13] linearized about the operating point $V_{IN} = 2$ V, which corresponds to a DC output voltage of 2.5 V. We replace each transistor by its incremental model, DC voltage sources by short circuits, and DC current sources by open circuits. The resulting linear circuit is shown in Fig. 14.22. Note that because V_B is constant, the only elements of the small-signal model of the PMOS transistor are its output resistance and the drain-to-gate and drain-to-body capacitances. The capacitances appear in parallel with the load and, hence, can be lumped into C_L. Similarly, the drain-to-body capacitance of the NMOS transistor appears in parallel with C_L.

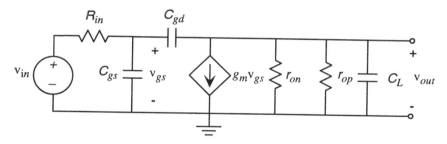

Figure 14.22.   Small-signal model of the CMOS amplifier of Fig. 14.16. The values of C_{gs} and C_L include additional contributions from transistor incremental capacitances that are connected to incremental grounds.

A source resistance R_{in} has been included in order to represent a general drive circuit. For a DC incremental voltage v_{in}, with all capacitors equivalent to open circuits, the incremental output is $v_{out} = -g_m r_o$, where

$$r_o = \frac{r_{on}r_{op}}{r_{on} + r_{op}} = \frac{1}{(\lambda_n + \lambda_p)I_{Dn}} \tag{14.50}$$

---

[13]We have not discussed methods for "biasing" a circuit in order to create a stable operating point. Since understanding operational amplifiers is our primary goal here, we shall skip over this amplifier-design issue, and refer the reader to the references.

Now it is clear why the channel-length modulation had to be included in our model. The gain of this amplifier, indeed, all amplifiers, is limited by the output resistance of the amplifying device in parallel with the load resistance. In this case, the only resistive load is the output resistance of the PMOS transistor. If there were also a resistive external load in parallel with C_L, it would further reduce the gain. Substituting for g_m, we obtain

$$|A_{max}| = \frac{1}{(\lambda_n + \lambda_p)I_{Dn}}\sqrt{2\left(\frac{W_n}{L_n}\right)\mu_n \hat{C}_{ox}I_{Dn}} \qquad (14.51)$$

If we use the values for the transistors, $L_n = 2\ \mu\text{m}$, $k_n = 200\ \mu\text{A/V}^2$, $\lambda_n = \lambda_p = .05\ \text{V}^{-1}$, and $I_{Dn} = 110\ \mu\text{A}$, we find

$$|A_{max}| = 19 \qquad (14.52)$$

We can calculate the frequency response for this amplifier under sinusoidal-steady-state conditions. We note two important effects: the charging of C_{gs} and C_{gd} through R_{in} will produce a roll-off. Also, the capacitor C_L in parallel with the output resistors will produce a roll-off. Thus we expect a two-pole response. The full transfer function is

$$H(s) = \frac{v_{out}(s)}{v_{in}(s)} = -g_m r_0 \frac{1 - sC_{gd}}{D(s)} \qquad (14.53)$$

where the denominator $D(s)$ is

$$D(s) = r_o R_{in}[C_{gd}C_{gs} + C_L(C_{gs} + C_{gd})]s^2$$

$$+ \{R_{in}[C_{gs} + C_{dg}(1 + g_m r_o] + r_o(C_{gd} + C_L]\}\, s + 1 \qquad (14.54)$$

This transfer function has two poles, plus a very-high-frequency right-half-plane zero at $s = g_m/C_{gd}$. If we assume $\hat{C}_{ox} = 2.5\ \text{fF}/\mu\text{m}^2$, and $\hat{C}_{ov} = 0.5\ \text{fF}/\mu\text{m}$ and evaluate the pole locations, we obtain

$$s_1 = -1.1 \times 10^7\ \text{rad/sec} \quad \text{and} \quad s_2 = -6.5 \times 10^9\ \text{rad/sec} \qquad (14.55)$$

The lower-frequency pole at -1.1×10^7 rad/sec, equivalent to about 175 kHz, is due to the combination of r_o and C_L. In fact, one can make a good estimate of this pole location by calculating

$$\frac{1}{r_o(C_L + C_{gd})} = 1.12 \times 10^7 \qquad (14.56)$$

Our conclusion is that capacitance attached to the high-resistance node in an amplifier circuit can dominate the roll-off frequency. Indeed, in operational amplifiers, it is necessary to place an additional amplifier stage after the high-gain amplifier so as to minimize capacitive loading of the high-gain state, and still provide the capability of driving current into an output load.

14.8.4   Other Amplifier Configurations

There are many other transistor combinations that operate as amplifiers. Of special interest in analog circuits are circuits called *source followers*, *differential pairs*, and *cascode* circuits. It is also possible to use combinations of these to create circuits that, at least over a range of outputs, function as good voltage sources and good current sources. CMOS versions can be built in which PMOS transistors create the load elements for NMOS drivers just as in the inverter example analyzed in the previous sections. The details carry us well beyond the scope of this book. Suffice it to say that all of these combinations play a role in the design of operational amplifiers, which are the basic building block of most analog signal-processing circuits. A good treatment can be found in Howe and Sodini [83].

14.9   Operational Amplifiers

Operational amplifiers (also known by their more friendly name: *op-amps*) are two-input amplifier circuits with two power-supply ports (one positive with respect to system ground, the other negative) and, usually, one output (although there are also differential-output operational amplifiers). The basic structure and the standard op-amp symbol are shown in Fig. 14.23. Op-amps contain a *differential amplifier* as the first stage which subtracts the two inputs from each other and amplifies the difference, then a high-gain stage that provides significant voltage gain, and, finally, an output stage that can provide reasonable current to an external load without itself loading down the high-gain stage. Operational amplifiers tend to have a large static (DC) gain and a single-pole frequency response, at least up to relatively high frequencies.

The various terminals of an op-amp have names: One input terminal is called the *non-inverting input*, and is denoted with the $+$ symbol. The other input is called the *inverting input*, and is denoted with the $-$ symbol. These $+$ and $-$ symbols have nothing to do with the actual polarity of the applied voltages. The voltages at these inputs, referred to a system ground, are called v_+ and v_-, respectively. The power supply terminals are usually labeled with upper-case symbols, such as V_{S+} and V_{S-}, which implies that DC power supplies or batteries have been connected between the terminals and the system ground. Except when we are concerned with the allowed output voltage swing of the op-amp, we tend to ignore these terminals, and often do not even show them in the diagram. But remember, *all of the extra signal power* comes from these

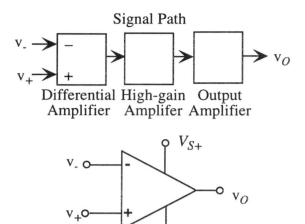

Figure 14.23.   Basic architecture of an operational amplifier together with its standard symbol. The power supply connections are often omitted in circuit diagrams.

terminals. The output terminal, also referenced to system ground, is called the *output*, and is typically labeled v_o.

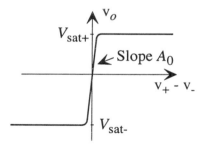

Figure 14.24.   Quasi-static difference-mode transfer characteristic of an op-amp.

The typical quasi-static difference-mode transfer characteristic, plotted in Fig. 14.24 as a function of $v_+ - v_-$, has a slope of A_0 at the origin, called the *DC open-loop gain*, and saturates at voltages 1 or 2 volts below the two power supply voltages. In this model, the amplifier produces no output if $v_+ = v_-$. That is, this model of the amplifier has a non-zero *difference-mode* incremental gain A_0 but zero *common-mode* gain. Of course, real amplifiers will show a variation of output when both inputs are equal to each other and swing over a range of voltages. Thus, the common-mode gain of real amplifiers is not zero, but it is typically much less than the difference-mode gain. If we denote the common-mode gain as A_C, we define the *common-mode rejection ratio*, or CMRR, as the ratio between A_0 and A_C. In our ideal model, this ratio is infinite. In real amplifiers, it is finite, but usually at least 1000, often as large as 100,000. In decibel units,[14] these correspond to CMRR's between 60 and 100 dB.

We can define the *linear region* of op-amp operation as that portion of the transfer characteristic well away from the saturation plateaus. In this region, we can linearize the transfer characteristic, and represent the linearized transfer function as

$$v_o = A_0(v_+ - v_-) + A_C \frac{v_+ + v_-}{2} \qquad (14.57)$$

where the factor of 2 in the common-mode term uses the conventional definition of the common-mode signal as the average of the two inputs. In addition to the difference-mode and common-mode gains included here, there is also some dependence of output on the power supply voltages.

Real op-amps also have *offset voltages* due to small mismatch between the transistors inside. The effects of offset are to make the transfer characteristic miss the origin. An easy way to take this into account is to refer the offset to an equivalent shift in $v_+ - v_-$, thereby creating the linearized transfer characteristic:

$$v_o = A_0(v_+ - v_- - v_{off}) + A_C \frac{v_+ + v_-}{2} \qquad (14.58)$$

where v_{off} is the *input-referred offset voltage*.

When the op-amps are built from bipolar transistors or junction field-effect transistors, there are non-zero currents required for amplifier operation. When the op-amps are built from MOS transistors, the input currents are extremely small because the input transistors present a capacitor to the input, not a resistor, and capacitors are open circuits for DC. However, because of the capacitor inputs, currents are required to charge and discharge the input capacitances of the MOS transistors, so there are non-zero currents at the input of an MOS op-amp whenever v_+ or v_- is changing. In applications where these charging and discharging currents are important, we shall bring the input capacitances of the first-stage of amplification outside the op-amp symbol and show them explicitly. The rest of the the op-amp is then assumed to have zero input current.

14.10 Dynamic Effects

In our examination of the two-transistor MOS inverting amplifier, we observed two important dynamic effects: a finite *slew rate* when attempting to swing the output through large amplitudes, and a frequency-dependent transfer

---

[14]Decibels are 20 times the log of the amplitude ratio. See page 435.

function in the linear-gain region. Op-amps also have these characteristics. Typical slew-rate values are on the order of 1 V/μsec. And since a typical op-amp has three gain stages and, therefore, a large number of internal capacitances, we can expect, in general, a very complex transfer function. However, it is highly desirable for the op-amp to behave as a single-pole element over its useful frequency range. We have already seen that capacitive loading of the output node of the high-gain stage tends to produce a dominant low-frequency pole. In modern op-amps, what is called a *compensation capacitance* is intentionally added to the circuit, often at the high-gain output node, in order to control the location of this dominant pole. Thus, instead of the DC gain A_0, it is more correct to represent each of the gains as being s-dependent, so that we would write for the difference-mode transfer characteristic

$$v_o(s) = A(s)(v_+ - v_- - v_{off}) \tag{14.59}$$

where

$$A(s) = \frac{A_0 s_o}{s_o + s} \tag{14.60}$$

or, equivalently

$$A(s) = \frac{A_0}{1 + s\tau} \tag{14.61}$$

where τ is $1/s_o$. The quantity $A_0 s_o$ is called the *gain-bandwidth product* for the op-amp, and is an important figure of merit. A transfer function of this form suggests that there is an internal state variable which leads to a differential equation describing $v_o(t)$:

$$\frac{dv_o}{dt} + \frac{1}{\tau} v_o = A_0 [v_+(t) - v_-(t) - v_{off}] \tag{14.62}$$

Real op-amps have a more complex frequency behavior than this simple single-pole response. Poles at higher frequencies can lead to important effects when op-amps are used as elements in feedback-control circuits. However, for the rest of this chapter, we will treat the single-pole approximation as sufficient.

14.11   Basic Op-Amp Circuits

There are six basic linear op-amp circuits that are quite common. In order to learn the principles of analyzing op-amp circuits, we shall study the inverting amplifier in detail, after which we shall summarize the remaining examples quickly.

14.11.1   Inverting Amplifier

The *inverting amplifier* is shown in Fig. 14.25. It contains an op-amp and two resistors, one of which is connected as a *feedback* resistor between the output and the inverting input. As we shall discover in more detail in the following chapter, this is one example of a feedback system. However, for the moment, we will analyze it purely as a circuit, and not attempt to force it into the canonical feedback form.

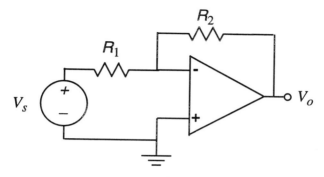

Figure 14.25.   The inverting amplifier.

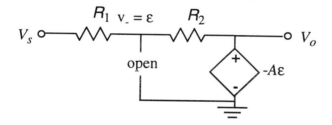

Figure 14.26.   Linearized model of the inverting amplifier.

The linearized equivalent circuit is shown in Fig. 14.26. Here, the op-amp is replaced by a linear dependent voltage source with gain A (which might be s-dependent), and the value of v_- is identified as ε, which will connect later to our description of this circuit as a feedback system. Note that, consistent with our assumption that op-amps draw no current at their inputs, the input is represented as an open circuit. This circuit is readily analyzed by writing KCL at the node joining R_1 and R_2.

$$\frac{V_s - \varepsilon}{R_1} = \frac{\varepsilon + A\varepsilon}{R_2} \tag{14.63}$$

which yields for ε

$$\frac{\varepsilon}{V_s} = \frac{R_2}{AR_1} \left[\frac{1}{1 + \dfrac{1}{A}\left(1 + \dfrac{R_2}{R_1}\right)} \right] \tag{14.64}$$

and for V_o

$$\frac{V_o}{V_s} = -\frac{R_2}{R_1}\left[\frac{1}{1 + \frac{1}{A}\left(1 + \frac{R_2}{R_1}\right)}\right] \tag{14.65}$$

This is an interesting result. In the limit $A \to \infty$, we have the simultaneous limits $\varepsilon \to 0$ and $V_o \to -R_2/R_1$. Thus for a sufficiently high-gain op-amp, the voltage gain is simply given by the ratio of two resistors. This ratio can be chosen quite accurately, especially in integrated circuits where resistance ratio is determined by layout. The result is that the transfer function does not depend on the value of A.

We now include the op-amp's a single-pole response. That is, we identify

$$A = A(s) = \frac{A_0 s_0}{s + s_0} \tag{14.66}$$

When substituted into the transfer function, we obtain

$$\frac{V_o}{V_s} = -\frac{R_2}{R_1}\left[\frac{A_0 s_0}{A_0 s_0 + \left(1 + \frac{R_2}{R_1}\right)(s + s_0)}\right] \tag{14.67}$$

This circuit has a single pole at

$$s = -\frac{\left(A_0 + 1 + \frac{R_2}{R_1}\right)s_0}{1 + \frac{R_2}{R_1}} \tag{14.68}$$

We now examine these results under the assumption that the DC *open-loop gain* A_0 is much larger than the ratio R_2/R_1, which is itself larger than unity. With these assumptions, the ratio R_2/R_1 is the magnitude of the circuit gain, which we now refer to as the DC *closed-loop gain*.

The magnitude of the sinusoidal-steady-state gain is plotted as a function of frequency in Fig. 14.27. We observe the single-pole roll-off of the open-loop op-amp transfer function with its pole at s_0, and the shifted and scaled single-pole roll-off of the closed-loop transfer function, with its pole shifted to higher frequency by the factor $A_0 R_1/R_2$. At the same time, the gain magnitude is scaled down by the same factor, from A_0 to R_2/R_1. The product of the gain and the bandwidth is preserved at $A_0 s_0$, the *gain-bandwidth product* for the op-amp, regardless of the specific values of R_2 and R_1 as long as their ratio remains larger than unity. Thus the gain-bandwidth product is an important figure of merit for the op-amp.

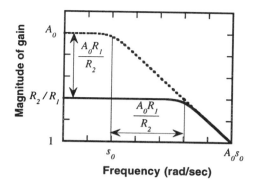

Figure 14.27. Frequency response for an inverting amplifier with $A_0 = 10^5$, $s_0 = 1$, $R_2/R_1 = 100$. The dashed line is the open-loop op-amp transfer function.

We note that the scaling factor $A_0 R_1/R_2$ is the ratio of the DC open-loop gain to the DC closed-loop gain. This ratio governs the decrease of the gain due to the feedback in the circuit and the corresponding increase in the bandwidth.

14.11.2 Short Method for Analyzing Op-Amp Circuits

We can extract from the previous analysis a short method for analyzing op-amp circuits. It is based on the fact that for a high-gain op-amp, the range of input voltages that correspond to the linear region is quite small. For example, if $A_0 = 10^5$ and the supply voltage equals 12 V, then in the linear-gain region, $|v_+ - v_-|$ is less than about 10^{-4} V, or 100 μV. This is typically much smaller than the input or output voltages and can be ignored algebraically compared to V_s or V_o. The feedback resistor connected to the negative terminal tends to stabilize the circuit in the linear region because if the output increases, the resistive connection tends to increase v_- which, in turn, tends to lower the output. Hence, we can assume that circuits with feedback to the inverting input are operating in the linear region. This leads to a very simple method for first-order analysis of op-amp circuits:

1. Assume the op-amp input current is zero.

2. Assume that $v_+ \approx v_-$. (When v_+ is connected to circuit ground, this is called the *virtual ground* approximation.)

3. Analyze the external circuit with these constraints to find the transfer function.

This method will be used in analysis of the remaining examples.

14.11.3 Noninverting Amplifier

The *noninverting amplifier* configuration is shown in Fig. 14.28.

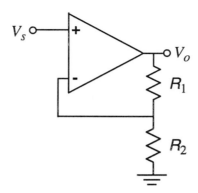

Figure 14.28.   The noninverting amplifier. In the limit of infinite R_2, this circuit becomes the unity-gain voltage follower, often used as a buffer circuit.

If we assume that the input currents are zero and that $v_+ \approx v_-$, we obtain the result that

$$V_s = \frac{R_2}{R_1 + R_2} V_o \qquad (14.69)$$

which results in the transfer function

$$\frac{V_o}{V_s} = 1 + \frac{R_1}{R_2} \qquad (14.70)$$

This circuit has a positive, hence, *noninverting* gain with a value that depends on the resistance ratio. Unlike the inverting amplifier, there is a non-negligible common-mode voltage equal to V_s at both inputs. Therefore, in this configuration, the common-mode rejection characteristic of the amplifier is extremely important.

In the limit that R_2 becomes infinite so that there is only a feedback connection between the output and the inverting terminal, the gain becomes unity, independent of the value of R_1. It is common, in this case, simply to connect the output directly to the inverting input without any resistor. This configuration is called the *unity-gain buffer* or *voltage follower*. It has the merit of drawing no current from the input circuit while having the ability to supply substantial current to whatever load is connected to the output.

14.11.4   Transimpedance Amplifier

A circuit that is closely related to the inverting amplifier is shown in Fig. 14.29. It is called the *transimpedance* amplifier. It is used as a current-to-voltage converter.

If we assume some current I_s into the circuit (whether from an ideal current source or any other circuit), and we employ the zero-input-current approxima-

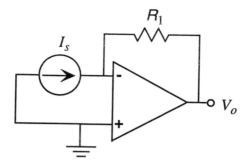

Figure 14.29.   The transimpedance amplifier.

tion for the op-amp and the virtual-ground approximation at the input, then we immediately conclude that

$$V_o = -R_1 I_s \qquad (14.71)$$

The incoming current is converted to a voltage at the output with a scale factor determined by the feedback resistor.

14.11.5   Integrator

A circuit that is related to the transimpedance amplifier is the *integrator*, shown in Fig. 14.30.

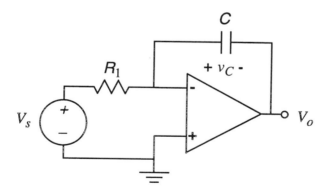

Figure 14.30.   The integrator circuit, in its ideal form.

Using the virtual-ground approximation, the current through R_1 is approximately V_s/R_1. This current must flow into the capacitor which integrates the current into a charge. That is, if v_C is the voltage across the capacitor,

$$C\frac{dv_C}{dt} = \frac{V_s}{R_1} \qquad (14.72)$$

But v_C is equal to $-V_o$. Thus we find

$$V_o = -\frac{1}{R_1 C} \int V_s(t) dt \tag{14.73}$$

Unlike the other circuits we have encountered, the integrator runs a serious risk of reaching the saturation limits of the linear region. For example, if a DC voltage is applied to its input, the output of the integrator is a linear ramp that eventually will hit the saturation level. Thus integrators are extremely sensitive to any DC currents, including the minute DC currents required at the inputs of non-MOS op-amps. In practice, therefore, it is common to provide some kind of DC path between the output and the inverting input so as to limit the DC transfer function. One way to achieve this is to place a very large resistor in parallel with the capacitor. Another method, which works well for circuits that are used for integrating transients, is to place a MOSFET in parallel with the capacitor, and switch the MOSFET from cutoff to its on state in order to discharge the capacitor voltage to zero just before measuring the next transient.

14.11.6   Differentiator

If the resistor and capacitor in the integrator are exchanged, the result is the *differentiator*, shown in Fig. 14.31.

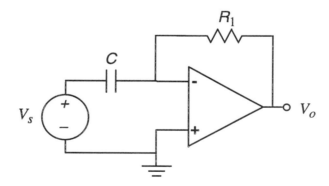

Figure 14.31.   The differentiator.

Using the short op-amp analysis method, we conclude that

$$V_o = -R_1 C \frac{dV_s}{dt} \tag{14.74}$$

This circuit computes the derivative of the input voltage. Its primary limitations are the maximum output voltage and the slew rate of the op-amp. If a step is applied to the input, the resulting output is not a perfect impulse, but rather a waveform whose shape depends on the internal amplifier dynamics. We can illustrate this to first order by going back to the full analysis with a linear but s-dependent transfer function. The result is

$$\frac{V_o}{V_s} = -sR_1C\frac{A_0 s_0}{s + (A_0 + 1)s_0} \qquad (14.75)$$

This is a response function that takes the scaled derivative of the input sR_1C and applies it to a single-pole transfer function with a pole at

$$s = -(A_0 + 1)s_0 \approx -A_0 s_0 \qquad (14.76)$$

Thus the response to a step input with this ideal model would be a waveform that is zero before $t = 0$, and starting at $t = 0$ has the form

$$V_s(t) = (A_0 s_0)R_1 Ce^{-(A_0 s_0)t} \qquad (14.77)$$

There would be a sudden rise to a large amplitude, given by $(A_0 s_0)R_1 C$, followed by a rapid exponential decay, since the gain-bandwidth product $A_0 s_0$ is large. However, the initial rise would be limited by the slew rate of the amplifier and possibly by the saturation level of the output, causing significant distortion. As a result, differentiator circuits must be used with care when working with rapidly switched signals.

14.12   Charge-Measuring Circuits

MOS transistors, because of their capacitive inputs, are extremely useful for measuring charge. There are two basic configurations of charge-measuring circuits, one that is purely analog, the other incorporating MOSFET switches.

14.12.1   Differential Charge Measurement

Figure 14.32 shows a differential charge-measurement circuit that uses two depletion-mode MOSFETS in combination with three op-amps in a feedback configuration.

The depletion-mode FETs have a conducting channel in the absence of any DC bias. Therefore we can assume non-zero drain currents I_1 and I_2 even if the source voltage V_s is zero. The elements R_s and C_s represent some general impedance that couples the source voltage to the gate of the first FET. The two FETs are presumed to be identical, and they have the same voltage V_{DD} applied to their drains. For clarity, their input capacitances are shown explicitly in the circuit. The transimpedance amplifiers convert the two currents into voltages which are, in turn, applied to the open-loop op-amp that drives the gate of the right-hand FET. This is a very-high-gain loop with negative feedback (an increase in I_2 causes a decrease in V_o). Therefore, we can use the virtual ground approximation at the input of this op-amp, and conclude that $I_1 = I_2$. That is, this circuit forces the two transistors to carry the same current. As a result, the

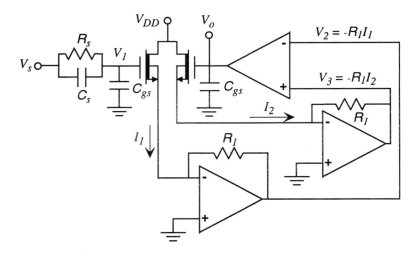

Figure 14.32.   Differential charge-measurement analog circuit.

output voltage V_o must be identical to the voltage V_1 which appears at the gate of the left-hand transistor, and this voltage is directly related to the charge on its input capacitance C_{gs}. Specifically, the relation between V_o and V_s is

$$\frac{V_o}{V_s} = \frac{1 + sR_sC_s}{1 + sR_s(C_s + C_{gs})} \tag{14.78}$$

The charging of C_{gs} through the parallel combination of R_s and C_s has unity transfer function at sufficiently low frequencies. At higher frequencies, well above the reciprocal of this charging time, the charge on C_{gs} is determined by a capacitive divider:

$$\frac{V_o}{V_s} = \frac{C_s}{C_s + C_{gs}} \tag{14.79}$$

Circuits of this type are widely used for measuring charges at high-impedance nodes such as V_1 without requiring any DC current drain from that node. It is the basis of the microdielectrometer, a commercial device that measures the complex dielectric constant of resins and fluids as they undergo chemical reactions such as curing, or epoxy-resin hardening.[15] The typical microdielectrometer impedance levels of R_s and C_s correspond to measuring a 1 pF capacitance at a frequency of 1 Hz, or about 10^{11} Ohms.

---

[15]Originally developed at the Massachusetts Institute of Technology [85], the microdielectrometer is now marketed by Holometrix-Micromet. Information may be found at www.micromet.com.

14.12.2 Switched-Capacitor Circuits

A second category of charge-measurement circuit involves a combination of FET switches, capacitors, and MOFET op-amps (the capacitive input is critical for this type of circuit). In this case, we use enhancement-mode FET's which are off in the absence of a gate voltage but develop a conducting channel when a voltage that exceeds the source and/or drain voltage by more than the threshold voltage is applied to the gate. The simplest example circuit is shown in Fig. 14.33.

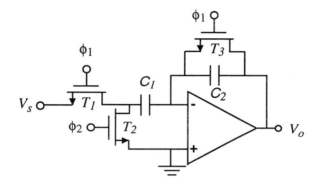

Figure 14.33. A switched-capacitor inverter.

The operation of this circuit depends on two non-overlapping clock signals ϕ_1 and ϕ_2. The waveforms are illustrated in Fig. 14.34.

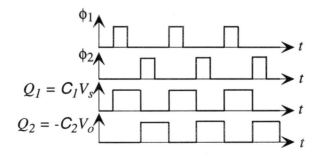

Figure 14.34. Schematic waveforms in the switched-capacitor inverter. Charging and discharging transients are assumed to be very fast compared to the clock intervals.

When ϕ_1 is high, transistors T_1 and T_3 are turned on, and transistor T_2 is off. Because of the conduction path through T_3, the op-amp behaves as a non-inverting voltage follower with its non-inverting input connected to ground.

Therefore, V_o is zero.[16] At the same time, capacitor C_1 charges to a value Q_1 = $C_1 V_s$ because it's right-hand terminal is connected to a virtual ground. This charge remains on C_1 even when ϕ_1 returns to a low value, because T_2 is still off, and there is no conduction path through T_1, T_3, or the op-amp input. When ϕ_2 goes high, turning on T_2, the voltage on the left-hand node of C_1 is pulled to zero. However, the charge on the capacitor cannot change instantaneously. Thus, the action of turning on T_2 pulls the inverting input voltage negative, which swings the output strongly positive. The capacitive coupling back to the inverting input through C_2 tends to pull the inverting input back toward zero. The circuit settles when the inverting node is once again at a virtual ground, which means that C_1 is now discharged. The positive charge $C_1 V_s$ which was on the left-hand plate of C_1 has flowed to ground while the negative charge $-C_1 V_s$ on the right hand plate of C_1 has shifted to the left-hand plate of C_2. This requires that the right-hand plate of C_2 have a charge $+C_1 V_s$. However, this charge must equal $C_2 V_0$. Hence, we conclude that the output voltage swings to a value given by

$$V_o = \frac{C_1}{C_2} V_s \qquad (14.80)$$

On an average basis, the rate of charge transport through this circuit is $C_1 V_s$ Coulombs of charge per clock period. It operates like an old-fashioned bucket brigade. This amount of charge is placed on C_1 when ϕ_1 is high, and it is then transferred to C_2, hence to the output, when ϕ_2 is high. If we were to average the output over one clock cycle, we would obtain, in effect, an inverting amplifier, with gain $-C_1/2C_2$. However, the primary use of circuits like these in MEMS systems is not to implement inverting amplifiers, but in capacitance measurement. We shall return to this subject in Chapter 19.

Related Reading

P. R. Gray and R. G. Meyer, *Analysis and Design of Analog Integrated Circuits*, Second Edition, New York: Wiley, 1984.

J. K. Roberge, *Operational Amplifiers: Theory and Practice*, New York: Wiley, 1975.

R. T. Howe and C. G. Sodini, *Microelectronics: An Integrated Approach*, Upper Saddle River, NJ: Prentice Hall, 1997.

S. D. Senturia and B. D. Wedlock, *Electronic Circuits and Applications*, New York: Wiley, 1975; reprinted by Krieger Publishing.

---

[16]This assumes zero voltage offset in the op-amp. The effect of offset voltage on circuits of this type is considered in Chapter 19.

S. M. Sze, *Physics of Semiconductor Devices*, Second Edition, New York: Wiley, 1981.

Problems

14.1 A junction diode with a 100 μm square area is formed in an n-type substrate with doping 10^{15} cm^{-3} by implantation and diffusion of boron. Assume that the junction depth is 2 μm and that for purposes of analysis, the p-region has a uniform concentration of 10^{17} cm^{-3}. Determine the potential drop across the space charge layer in equilibrium and the equilibrium space-charge-layer width. Find the reverse bias voltage at which this width increases by 50%. Assuming a diffusion length of 2 μm, estimate the value of I_0 for this diode.

14.2 The diode of Problem 14.1 is forward biased. Determine the voltage at which the minority carrier concentration at the edge of the space-charge-layer has doubled. This voltage is called the onset of *high-level injection*. On which side of the diode is high-level injection encountered first, the more heavily doped side or the more lightly doped side?

14.3 Using the same process for forming the diode in Problem 14.1, a semiconductor resistor is formed with a serpentine layout totalling 100 squares (W = 10 μm, L = 1 mm). Estimate the resistance of this resistor, and calculate the sensitivity of this resistance to the average reverse-bias voltage between the p-type resistor and the n-type substrate. How would this result change if the junction depth were shallower?

14.4 Estimate the temperature dependence of the reverse saturation current of the resistor of Problem 14.1. Compare its value at room temperature to that at 250°C. If the resistor is designed for use with a 10 μA DC current, how large a perturbation is this leakage current? (Reverse leakage currents ultimately limit the usefulness of junction-isolated semiconductor resistors at high temperatures.)

14.5 For the diode of Problem 14.1, what value of photocurrent is needed for the diode to have an open-circuit voltage of 0.4 V? Assuming that this current arises from a uniform volume hole-electron-pair generation rate in the space-charge layer, what is the value of this generation rate? Assuming one hole-electron pair per photon, and assuming each photon has an energy of 2 eV, what is the radiant energy (W/cm^2) that is required in the space-charge layer region?

14.6 The lightly shaded lines in Figure 14.17 illustrate a simplified way to estimate the transfer function of an amplifier. The horizontal lines correspond to the state of the circuit with the input set either to zero or to the supply voltage while the sloped line passes through the point where v_{in} equals v_{out} and has a slope given by the gain calculated from the incremental circuit. Use this method to estimate the transfer function of a CMOS circuit identical to the example of Fig. 14.17 except with $V_{T_n} = 1$ V and $V_{T_p} = -1.5$ V.

14.7 The transient calculation of Fig. 14.21 shows a different voltage level in the 0 to 0.5 μs interval for the two cases. Explain this difference.

14.8 The integrator circuit of Fig. 14.30 will saturate if a DC voltage is applied to its input. However, for AC input signals, the operation is that of an ideal integrator. If a second resistor R_2 is placed in parallel with the integrating capacitor so as to provide a negative DC feedback path for this circuit, determine how the transfer function is modified. What restriction on the possible value of R_2 will minimize the disturbance of the operation of the circuit as an integrator when AC signals are applied to the input?

Chapter 15

FEEDBACK SYSTEMS

Give me a fruitful error any time, full of seeds, bursting with its own corrections.
— Vilfredo Pareto, writing on Johann Kepler

15.1   Introduction

We have already encountered several examples of feedback systems and circuits that use feedback. In this chapter, we examine this subject in more systematic fashion, beginning with basic feedback concepts, feedback in linear systems, and, finally, some important aspects of feedback in nonlinear systems.

15.2   Basic Feedback Concepts

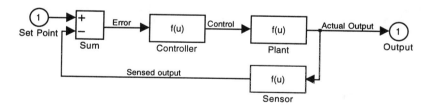

Figure 15.1.   A general feedback system, using SIMULINK blocks to represent generic transfer functions for each block.

A feedback system is one which is configured so as to make the output an implicit function of itself, as in Fig. 15.1. Conventionally, we think of feedback systems as controlling some object (called the "plant"), and on this basis, we associate this particular kind of block diagram with feedback systems. We saw an example such a system in the position controller introduced in Chapter 2.

Ideally, the sensor that measures the output is perfectly accurate so that the sensed output equals the actual output. In this case, the system reduces to a two-block system: the plant, and a controller that is driven by the difference between the desired output (the *set point*) and the actual output. This difference is called the *error*.

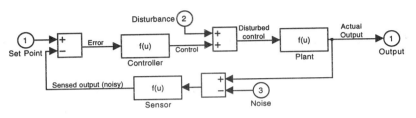

Figure 15.2.   A general feedback system with noise and disturbance added.

Life being what it is, sensors are not perfect, so it is important to make provision for both calibration errors in the sensor and errors due to random disturbances, called noise (see Fig. 15.2).[1]

In addition, there may be forces applied to the plant that do not come from the controller. An example would be a force-feedback accelerometer in which the proof mass is the plant, the set point is the desired position of the proof mass, and the controller is the actuator that moves the proof mass. When the proof mass is subjected to an acceleration, it is displaced from its set point by the inertial force. Such an additional force is called a *disturbance* even though, in the case of an accelerometer, it is what we want to measure.

The two diagrams make no restriction on the function blocks that serve as controller, plant, or sensor. They can be linear or non-linear, and can even have explicit time dependences. Our approach will be to restrict our attention to time-invariant systems, with no time dependence inside the blocks, and to start with the simpler case of linear function blocks.

15.3   Feedback in Linear Systems

We consider first the linear feedback system, written in transfer-function form, shown in Fig. 15.3. The error, denoted here by ε, is equal to $X_{in}(s) - X_{out}(s)$ and the output is $H(s)K(s)\varepsilon$. Thus we can readily write the overall transfer function for this system:

$$X_{out}(s) = H(s) \cdot K(s) \cdot [X_{in}(s) - X_{out}(s)] \qquad (15.1)$$

which yields

---

[1]Noise is discussed in detail in Chapter 16.

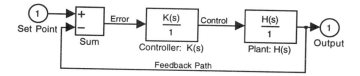

Figure 15.3. A linear feedback system with an ideal sensor, written with SIMULINK transfer-function blocks. (The strange-looking denominator of unity beneath the $K(s)$ and $H(s)$ transfer functions is required by the SIMULINK labeling syntax for a transfer-function block.)

$$X_{out}(s) = \frac{H(s)K(s)}{1 + H(s)K(s)} X_{in}(s) \qquad (15.2)$$

This result, known as *Black's formula*,[2] is the fundamental transfer characteristic of basic linear feedback systems. It is clear that if we want $X_{out}(s)$ to equal $X_{in}(s)$, we want to make the product $H(s)K(s)$ very large compared to unity. Since the plant has a finite $H(s)$ determined by its construction, the only way to get a large product is to have $K(s)$ be large. Thus we expect to find high-gain amplifiers as a critical part of feedback controllers. We might also be able to tailor the s-dependence of $K(s)$ to compensate for or modify undesirable aspects of the plant behavior. Thus, design of feedback controllers is a rich topic.

15.3.1 Feedback Amplifiers

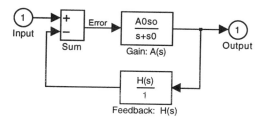

Figure 15.4. Feedback diagram representation of an amplifier with feedback. The system output, in this case, is taken from the output of the "controller" block, and the "plant" block is what provides the feedback to the input. This model applies to many op-amp circuits.

A variant on the two-block system of Fig 15.3 is to identify the controller as an amplifier with gain $A(s)$ and the plant as a feedback network with transfer

---

[2]Some historical notes on the development of feedback concepts, including the contributions of Black, Bode, and Nyquist, can be found in Chapter 5 of Siebert [43].

function $H(s)$, but take the amplifier output as the system output. This is illustrated in Fig. 15.4.

To illustrate the use of a feedback model for an amplifier, we consider a noninverting amplifier circuit of Fig. 14.28, for which

$$H(s) = \beta = \frac{R_2}{R_1 + R_2} \tag{15.3}$$

leading to the overall transfer function

$$\frac{V_{out}(s)}{V_{in}(s)} = \frac{K(s)}{1 + \beta K(s)} \tag{15.4}$$

Clearly, in the limit of $\beta K(s) \gg 1$, the transfer function reduces to $1/\beta$, or $1 + R_1/R_2$. However, as we have already seen, it is not possible for $\beta K(s)$ to be much greater than unity at all frequencies because the amplifier has a finite gain-bandwidth product. For a single-pole amplifier and a resistive (independent of s) feedback block, the result as derived in Section 14.11.1 is a single-pole transfer function which preserves the gain-bandwidth product of the amplifier.

15.3.2   Example: The Position Controller

We now return to the position-control example introduced in Chapter 2. The plant in this case is a spring-mass-dashpot system with the force-displacement characteristic given by

$$\frac{X(s)}{F(s)} = H(s) = \frac{1}{ms^2 + bs + k} \tag{15.5}$$

It will be convenient to express the frequency normalized to the undamped resonant frequency of the spring-mass system. Thus, we can write

$$H(s) = \left(\frac{1}{k}\right) \frac{1}{\hat{s}^2 + \frac{1}{Q}\hat{s} + 1} \tag{15.6}$$

where

$$\hat{s} = \frac{s}{\omega_o} \tag{15.7}$$

$$\omega_o = \sqrt{\frac{k}{m}} \tag{15.8}$$

$$Q = \frac{m\omega_o}{b} \tag{15.9}$$

To simplify the rest of the discussion, we will further assume that $k = 1$. In effect, we are measuring frequency in units of ω_o, and scaling the unit of force to make k unity.

15.3.2.1   Ideal Amplifier

If the controller is an ideal amplifier, with DC gain K_0 and no high-frequency roll-off, the overall transfer function is

$$\frac{X_{out}(s)}{X_{in}(s)} = \frac{K_0}{\hat{s}^2 + \frac{1}{Q}\hat{s} + K_0 + 1} \tag{15.10}$$

To make the static ($\hat{s} = 0$) transfer function approach unity, we simply make K_0 large. However, we note several important features of this result:

First, for any finite gain K_0, there is still a static error given by

$$\varepsilon = \frac{1}{1 + K_0} \tag{15.11}$$

Second, as the gain K_0 is increased, the system becomes increasingly under-damped, subject to long ringing oscillations in response to a set point change (or to a disturbance). This is seen by finding the poles of the closed-loop system:

$$\hat{s}_{1,2} = -\frac{1}{2Q} \pm \sqrt{\left(\frac{1}{2Q}\right)^2 - (1 + K_0)} \tag{15.12}$$

As K_0 increases, the poles approach the limit

$$\hat{s}_{1,2} = -\frac{1}{2Q} \pm j\sqrt{K_0} \tag{15.13}$$

and the closed-loop quality factor approaches $Q\sqrt{K_0}$. Thus a spring-mass-dashpot system with an initially critically damped response ($Q = 1/2$) develops a highly underdamped closed-loop response with quality factor $(\sqrt{K_0})/2$. Any small disturbance creates a long ringing transient response. This behavior is highly undesirable, but it is also somewhat unrealistic, because we know that all amplifiers have gains that roll off at high frequency. Before considering a more realistic amplifier model, we introduce a very useful way of examining the effect of feedback on frequency response, the *root-locus method*.

15.3.2.2   Root-Locus Plots

We have seen that the DC amplifier gain affects the frequency response of a closed-loop system. A *root-locus plot* is a parametric graph of the pole locations of a feedback system as a function of the DC gain.[3] The root-locus

---

[3]The MATLAB command `rlocus` is useful for constructing such plots.

plot for the example of the previous section is shown in Fig. 15.5. It is clearly seen in the root-locus plot that as K_0 increases, the poles move away from the real axis. There is no problem with stability. The poles remain in the left-hand-plane for all K_0. When we add amplifier roll-off, however, this is no longer true, as we shall discover in the next section.

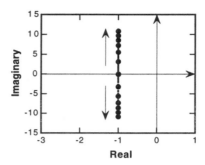

Figure 15.5.   The root-locus plot for a position controller with $Q = 1/2$ and an ideal amplifier with no roll-off. The short arrows indicate the variation of pole location with increasing K_0.

15.3.2.3   Single-Pole Controller

We now consider the case where the controller has a single-pole transfer function of the form

$$K(s) = \frac{K_0}{1 + \hat{s}\hat{\tau}} \tag{15.14}$$

Since $\hat{s}$ is measured in units of ω_o, the normalized time-constant $\hat{\tau}$ is equal to $\omega_o\tau$, where τ is the actual time constant of the single-pole circuit. It should be noted that the controller in this case could either be an op-amp alone, in which case K_0 would be the open-loop DC gain of the op-amp and τ would be the op-amp internal time constant, or the controller could itself be a feedback amplifier with a single-pole transfer characteristic.

Now there are three interesting questions. (1) Is there still a static error at DC? The answer is "yes," still equal to $1/(K_0 + 1)$. (2) What is the effect on the closed-loop dynamics of increasing K_0? (3) How important is the value of $\hat{\tau}$?

We will continue to use $Q = 1/2$ for the spring-mass-dashpot system. With this assumption, the closed-loop transfer function is

$$\frac{X_{out}(s)}{X_{in}(s)} = \frac{K_0}{\hat{\tau}\hat{s}^3 + (1 + 2\hat{\tau})\hat{s}^2 + (2 + \hat{\tau})\hat{s} + K_0 + 1} \tag{15.15}$$

The root-locus plot for the case $\hat{\tau} = 1$ is shown in Fig. 15.6. With this choice of $\hat{\tau}$, the transfer function can be written

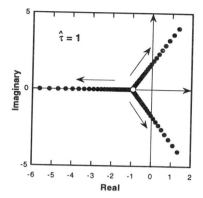

Figure 15.6. Root-locus plot for a single-pole controller with $\hat{\tau} = 1$. The open circle identifies the system triple pole when K_0 is vanishingly small.

$$\frac{X_{out}(s)}{X_{in}(s)} = \frac{K_0}{(\hat{s} + 1)^3 + K_0} \tag{15.16}$$

The poles are clearly at

$$\hat{s}_{1,2,3} = (-1)^{1/3}(K_0^{1/3} - 1) \tag{15.17}$$

The three cube roots of -1 are -1, $e^{j\pi/6}$, and $e^{j\pi/6}$. As K_0 increases in magnitude, the three pole locations follow vectors that start at the original pole location at -1, and move off in the complex plane at angles of $180°$, $60°$, and $-60°$. At a critical value of K_0, two of the poles reach the imaginary axis, and the system changes from a stable system to an unstable system.

The value of K_0 at which the system goes unstable can be calculated from the *Routh test*[4] which, for a cubic system, states that in order for all of the roots of a polynomial to be in the left-half plane, all terms must have the same sign, and the coefficient of the linear term must exceed the ratio of the constant term to the quadratic term. Thus, for stability, we require

$$Q\hat{\tau} + 1 > \frac{K_0 + 1}{1 + \hat{\tau}/Q} \tag{15.18}$$

and for $Q = 1/2$ and $\hat{\tau} = 1$, this reduces to the requirement that $K_0 < 7/2$.

What happens for different values of $\hat{\tau}$? Fig. 15.7 shows two root-locus plots, one for $\hat{\tau}$ much less than unity, the other for $\hat{\tau}$ much greater than unity.

With $\hat{\tau} = 0.1$, the controller has a much wider bandwidth than the resonant frequency of the plant, as indicated by the pole corresponding to the open circle

---

[4] See Siebert, p. 174 [43].

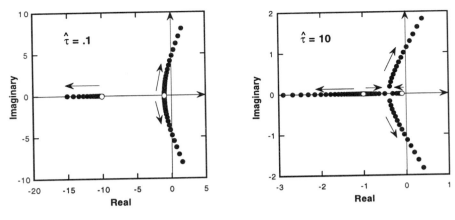

Figure 15.7. Root-locus plots for two different values of $\hat{\tau}$. The open circles show the closed-loop pole locations when K_0 is vanishingly small.

at -10. Hence, we would expect that the system would be stable to a higher value of DC gain compared to the $\hat{\tau} = 1$ example. Indeed, the Routh test predicts stability for $K_0 < 24.2$, and the achievable bandwidth, as measured by the frequency at which the instability occurs, is somewhat larger than the original plant resonance frequency. In contrast, for the case with $\hat{\tau} = 10$, the controller's bandwidth is smaller than the resonant frequency of the plant, as indicated by the open circle at -0.1. Now, with the plant poles located at higher frequency than the controller pole, the root-locus behavior is different. One of the plant poles moves to higher frequency with increasing K_0 while the other moves toward the origin, meets the low-frequency pole, at which point the two poles become complex and move inexorably toward those asymptotic values of the cube roots of (-1). Thus the system eventually exhibits an instability as in the other examples, but now at a frequency which is lower than the plant resonant frequency. The dynamics are dominated by the controller.

Our conclusions from these examples are several:

- The single-pole controller transfer function still results in a DC error.

- Because the overall system is now third order, there is a maximum DC gain for which the system remains stable.

- Better dynamic behavior is obtained when the bandwidth of the controller is greater than the bandwidth of the plant as measured, in this example, by its resonant frequency.

So what can we do to improve this situation? The answer is to introduce additional features into the controller transfer function, as explained below.

15.3.3   PID Control

In our previous examples, we have focused on a single-pole transfer function for the controller. A somewhat different approach, called *proportional-integral-differential control*, or PID for short, is to consider the controller transfer function to be a sum of three terms:

$$K(s) = K_0 \left(1 + \frac{\beta}{s} + \gamma s\right) \qquad (15.19)$$

where β and γ are constants, and K_0 is an overall gain factor. K_0 by itself corresponds to the constant-gain case already considered. It is called *proportional control* because the controller output is proportional to the error. The β term is called *integral control* because the controller output is proportional to the integral of the error, which corresponds to $1/s$ in the Laplace-transform domain. The γ term is called *differential control*, or *rate control* because it produces controller output proportional to the rate of change of the error, corresponding to the factor s in the Laplace-transform domain.

If we use this controller function in combination with a plant with a transfer function that we write in the form

$$H(s) = \frac{N(s)}{D(s)} \qquad (15.20)$$

we obtain the closed-loop transfer function

$$\frac{X_{out}(s)}{X_{in}(s)} = \frac{N(s)K_0(\gamma s^2 + s + \beta)}{sD(s) + N(s)K_0(\gamma s^2 + s + \beta)} \qquad (15.21)$$

Several features are readily apparent. First, in the DC limit, with s going to zero, this transfer function goes to unity exactly. There is no DC error! The cause of this improvement in DC response is the addition of integral control, which provides feedback proportional to the integral of the error. Whenever the error is non-zero, the integral-control term grows with time. Thus the only stable DC operation is to have zero error. Next, the combination of differential and integral control adds two zeros to the transfer function whose locations can be adjusted by the relative sizes of γ and β. These zeros have a significant effect on the overall transfer function and on its stability. We examine several examples.

15.3.3.1   The Effect of Integral Feedback

If we set γ to zero initially, we have what is called *proportional-integral*, or *PI* control. To examine the behavior of a PI system, we examine the root-locus behavior of a PI controller using our critically-damped normalized spring-mass-dashpot system as the plant. The normalized value for β is $\hat{\beta} = \beta/\omega_o$. With these definitions, the closed-loop transfer function becomes

$$\frac{X_{out}(\hat{s})}{X_{in}(\hat{s})} = \frac{K_0(\hat{s} + \hat{\beta})}{\hat{s}^3 + (1/Q)\hat{s}^2 + (1 + K_0)\hat{s} + K_0\hat{\beta}} \qquad (15.22)$$

This now a third-order system with one zero. If we examine the critically damped case, with $Q = 1/2$, and use $\hat{\beta} = 1$, the factor $\hat{s}+1$ in the numerator cancels a similar factor in the denominator, and the transfer function reduces to

$$\frac{X_{out}(\hat{s})}{X_{in}(\hat{s})} = \frac{K_0}{\hat{s}^2 + \hat{s} + K_0} \qquad (15.23)$$

This looks very similar to the constant-gain case with one important difference. The constant term in the denominator is now K_0 instead of $K_0 + 1$, so the DC ($\hat{s} = 0$) transfer function is unity, as we expect for integral control. The root locus plot, however, is very similar to the constant-gain case, in that the asymptotic behavior of the poles at large K_0 is

$$\hat{s}_{1,2} \rightarrow -\frac{1}{2} \pm j\sqrt{K_0} \qquad (15.24)$$

For other choices of β, more complex root-locus behavior is observed.

15.3.3.2   The Effect of Differential Feedback

If we keep $\hat{\beta} = 1$ and vary $\hat{\gamma}$, we find that the effect of the differential feedback on the root-locus plot is to shift the asymptotic pole locations away from the imaginary axis. In other words, differential feedback (also called *feed-forward compensation*) helps stabilize the system at high gain. The reason for this can be understood with a mechanical example. If one is trying to steer a car to follow a winding curve, one doesn't use only the present position of the car relative to the road. One looks ahead to see which way the road is turning, and one then adjusts the wheel in anticipation of a change of road direction. Differential feedback is determined by the rate of change of the error, not just by its magnitude. Thus, in a practical sense, differential feedback is like looking ahead to see how the error is changing and providing feedback based on that information.

Figure 15.8 shows the root-locus plot for a PID system with our critically damped normalized spring-mass-dashpot system as the plant, and with $\hat{\beta} = 1$ and $\hat{\gamma} = .01$. While it is not evident in the plot, one of the poles initially at -1 remains essentially unmoved as K_0 is increased. The other pole originally at -1 moves toward the integrator pole at $\hat{s} = 0$. They meet at -1/2 and, for higher gains, form a complex pair that follow an arc away from the imaginary axis. (If $\hat{\gamma}$ were zero, this asymptote would be perfectly vertical, as described in the previous section.) Depending on the precise values of $\hat{\beta}$ and $\hat{\gamma}$, variations in this behavior are observed, but the basic point is that the differential feedback helps

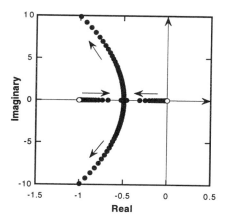

Figure 15.8.   Root-locus plot for the PID control of an initially critically damped spring-mass-dashpot system.

improve the quality of the high-frequency response by reducing the tendency toward ringing.

15.3.4   The Effect of Amplifier Bandwidth

We now need to consider the effect of finite amplifier bandwidth on the system. If we replace K_0 by $A_0/(1 + \hat{s}\hat{\tau})$, the transfer function becomes

$$\frac{X_{out}(\hat{s})}{X_{in}(\hat{s})} = \frac{A_0(\hat{\gamma}\hat{s}^2 + \hat{s} + \hat{\beta})}{\hat{\tau}\hat{s}^4 + (1 + \hat{\tau}/Q)s^3 + (A_0\hat{\gamma} + 1/Q)\hat{s}^2 + (1 + A_0)\hat{s} + A_0\hat{\beta}}$$

$$(15.25)$$

The amplifier time constant adds an additional pole to the transfer function which makes it more difficult to stabilize at high loop gains, and the root-locus behavior becomes complex, as shown in Fig. 15.9.

The first effect of the finite amplifier bandwidth, which in this example is chosen at twice the resonant frequency of the plant, is to modify the dynamics at low loop gain. In the limit of vanishing A_0, the poles of the system are the open circles in Fig. 15.9. The integrator pole at $\hat{s}=0$ is still at the origin, but the two critically damped plant poles are now split into a complex pair, and the high-frequency pole of the amplifier is shifted out from -2 to -2.7. Then, as A_0 is increased, the poles move in somewhat dizzying fashion. The two complex poles move toward the imaginary axis, and then move away and follow a vertical asymptote. The integrator pole originally at $\hat{s} =0$ moves along the negative real axis, passing between the two complex poles, eventually meeting up with the high-frequency amplifier pole, which moves toward the origin. Obviously,

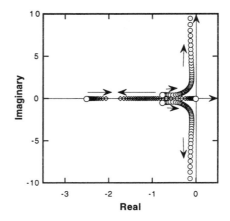

Figure 15.9.   Root-locus plot for the PID control of an initially critically damped plant using an amplifier with a bandwidth equal to twice the resonant frequency of the plant ($\hat{\tau}=0.5$, $\hat{\beta}=1$, $\hat{\gamma}=.27$).

depending on the gain factor, this system can have wildly different frequency responses. Our goal here is not to explore this subject to completion, but rather to illustrate the key ideas of how feedback can affect system dynamics.

15.3.5   Phase Margin

When systems have more than three poles, our analytical insights break down, and we need alternative ways of thinking about feedback system stability and frequency response. Going back to Black's formula, if the quantity $H(s)K(s)$ is ever equal to -1, then the closed-loop transfer function blows up, much like the fuse example of Section 11.6.2. Therefore, a useful way to examine system behavior is to plot the frequency response of the *loop transmission function* $H(s)K(s)$, and test whether it has a phase of -180 when it's magnitude is unity. There is a rich subject here generally referred to as the *Nyquist criterion* for system stability. We examine only one feature, namely, the *phase margin*. The phase margin is defined as the difference between the phase of the loop transmission and 180° at the point where the magnitude of the loop transmission is unity. If the phase margin is zero, then the loop transmission is exactly -1, and the system is clearly unstable.

Figure 15.10 shows the magnitude and phase angle of the loop transmission plotted against frequency. At the frequency where the magnitude goes through unity, the phase angle is at -155°, so the phase margin is 25°. This example shows a rather small margin. One typically prefers to operate feedback loops with phase margins above 45°.

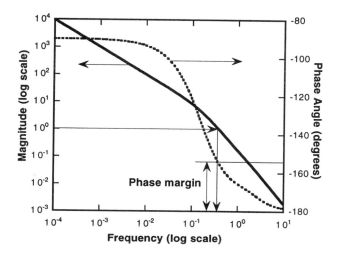

Figure 15.10. Magnitude and phase angle of the loop transmission function plotted *vs.* frequency. $A_0 = .35$, $\hat{\tau} = 0.5$, $\hat{\beta} = 1$, $\hat{\gamma} = .27$. The phase margin is the difference between the phase angle and -180° at the frequency where the magnitude is unity.

15.3.6 Noise and Disturbances

We now consider the added effects of noise $N(s)$ and disturbance $D(s)$, referring back to Fig. 15.2, with the plant represented by $H(s)$, the controller by $K(s)$, and the sensor by $M(s)$. The closed-loop output can be found from

$$X_{out}(s) = H(s)\left\{D(s) + K(s)\left[X_{in}(s) - M(s)\left(X_{out}(s) - N(s)\right)\right]\right\}$$

(15.26)

from which we obtain the following results for the output, the error ε, and the control signal denoted by F.

$$X_{out} = \frac{HK}{1+HKM}X_{in} + \frac{HKM}{1+HKM}N + \frac{H}{1+HKM}D$$

$$\varepsilon = \frac{1}{1+HKM}X_{in} + \frac{M}{1+HKM}N + \frac{HM}{1+HKM}D \qquad (15.27)$$

$$F = \frac{K}{1+KM}X_{in} + \frac{KM}{1+HKM}N + \frac{HKM}{1+HKM}D$$

where the s-dependence of the various quantities is suppressed to make the formulae more compact. We notice that now the loop transmission is the product $H(s)K(s)M(s)$. If the sensor is ideal, then $M(s) = 1$. If the sensor has an offset error, that can be represented as a constant term within $N(s)$. If the sensor has a scale error, then the magnitude of M at the frequencies of interest will differ from unity, but we can reasonably assume that this difference is not significant for purposes of determining loop behavior. Finally, the finite sensor bandwidth will contribute at least one additional pole to the loop transmission. This must be taken into account in designing the controller for good loop stability and dynamic behavior.

If the purpose of this feedback system is to make X_{out} equal X_{in} with small error, then we wish to make HKM large at the frequencies at which X_{in} might be changing. This makes the transfer function $HK/(1 + HKM)$ close to unity (again, assuming M is close to unity at all frequencies of interest). However, when we do that, we notice that the contribution of N to X_{out} is $HKM/(1+HKM)$, which is also close to unity. In other words, sensor noise in the range of frequencies where X_{in} is changing appears undiminished at the output, and can ultimately limit the accuracy of the control function.

If the purpose of the feedback system is to measure the disturbance D by finding the control F under closed-loop conditions with X_{in} held constant, then F becomes the system output, and we note that it equals $-HKM/(1+HKM)$ times D. Thus, controlling the loop so that $X_{out} = X_{in}$ over the range of frequencies where D is changing also establishes that F will equal $-D$. The noise contribution to F is similar to its contribution to X_{out}, scaled by H to convert set-point or position noise to an equivalent control or force noise.

Suppose, instead, it is desired to design the feedback loop so as to make X_{out} insensitive to disturbances. Then one wishes to make the quantity $H/(1 + HKM)$ small at the frequencies where the disturbance might be important. This is achieved by making KM large compared to H, in which case the noise term in the control becomes large. Thus there is a tradeoff that must be made. If one wishes to suppress disturbances over some range of frequencies, then over that same range of frequencies, the effect of noise is scaled up. Conversely, if one wishes to suppress the effect of noise on the output, by making HK small over the range of frequencies where the noise is important, than one suffers loss of loop accuracy, in that the error becomes large.

Our conclusion is that in a feedback loop with set-point inputs that may have one range of important frequencies, disturbances which have their own range of frequencies, possibly overlapping the set-point range, and noise which can be expected to be relatively broad-band in its frequency content, it is necessary to tailor the various transfer function to achieve an acceptable compromise between loop accuracy, disturbance rejection, and noise transmission.

15.3.7   Stabilization of Unstable Systems

We have seen that feedback can convert a stable plant into an unstable system at high loop gain. However, it is also possible to use feedback to stabilize systems that would be unstable under open loop conditions. A good example is a seal balancing a ball on its nose. Somehow, the seal is using feedback to keep the ball aloft.

We will look at only one simple example to demonstrate the principle. Suppose the plant has a right-half-plane pole, so that

$$H(s) = \frac{s_o}{s - s_o} \tag{15.28}$$

with s_o positive. If we use a single-pole controller with gain $K_0/(1 + s\tau)$, Black's formula yields

$$\frac{X_{out}}{X_{in}} = \frac{K_0 s_o}{\tau s^2 + s(1 - s_o \tau) + s_o(K_0 - 1)} \tag{15.29}$$

The closed-loop response has two poles, and in order to be sure that they are in the left-half plane, we apply the Routh test. For a second-order system, it is sufficient to have all three terms nonzero and of the same sign. Therefore, to be assured of stability, we require $K_0 > 1$ and $s_o \tau < 1$. The first requirement states that if the gain is high enough, the unstable pole can be stabilized. The second requirement states that the bandwidth of the controller must be greater than the bandwidth of the system with the unstable pole. When these two conditions are met, the closed-loop system is stable.

15.3.8 Controllability and Observability Revisited

In our introduction of eigenfunction methods in Section 7.2.5, we introduced the notions of *controllability* and *observability* of the modes of a system. A controllable mode was one whose amplitude could be affected by the system inputs, while an observable mode is one whose value affects the output (and hence can be sensed). Now that we understand how feedback can affect systems with multiple degrees of freedom, modifying the natural frequencies and even producing unstable modes from stable modes, we recognize the importance of controllability and observability. For robust control of a system, we need to be able to observe and control all of the modes that might affect either system performance or system stability. A high-frequency heavily damped mode is not a problem, but a mode with a frequency low-enough to interact with the system modes we care about under feedback conditions can be a problem. The way to make a mode observable is with sensors. The way to make a mode controllable is with actuators. Hence the type and placement of sensors and actuators within a microsystem may be dictated by the requirements of stability within a feedback system.

15.4 Feedback in Nonlinear Systems

When there is a nonlinear element present in a system, we are no longer able to use Laplace-transform methods for analysis. It is necessary to return to the state equations and carry out analysis in the time domain. Alternatively, if one knows that a system is operating in the neighborhood of a fixed operating point, it is possible to linearize the system about that operating point, thereby converting it back to a linear system for which the analysis tools of the previous section are readily available. Our attention will focus on the large-signal behavior of feedback systems with nonlinear elements. We consider two cases: (1) a quasi-static nonlinear system (with no state variables) for which we seek the quasi-static transfer characteristic; and (2) a dynamic second-order nonlinear system that can exhibit a variety of interesting behaviors.

15.4.1   Quasi-static Nonlinear Feedback Systems

Let us consider a system in which the plant has the following force-displacement characteristic:

$$X_{out} = X_0 \tan^{-1}\left(\frac{F}{F_0}\right) \tag{15.30}$$

where X_0 and F_0 are scale factors. This transfer function is illustrated in Fig. 15.11.

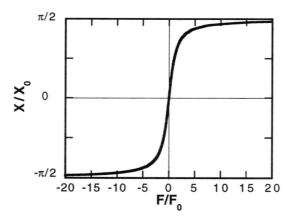

Figure 15.11.   Nonlinear plant transfer function.

If $|X_{out}/X_0|$ is kept much less than $\pi/2$, then the arctangent function is quite linear. However, as $|X_{out}/X_0|$ gets close to $\pi/2$, there is a saturating nonlinearity. The question is how well can the feedback loop maintain accurate set-point control in the presence of this nonlinearity.

If we use a constant-gain controller, with gain K_0, we obtain

$$X_{out} = X_0 \tan^{-1}\left(\frac{K_0(X_{in} - X_{out})}{F_0}\right) \tag{15.31}$$

If we take $F_0 = 1$ for convenience, this can be manipulated to

$$X_{out} = X_{in} + \frac{1}{K_0} \tan \left(\frac{X_{out}}{X_0} \right) \tag{15.32}$$

We note that the error in the output, which is the nonlinear tan term, is scaled down by a factor $1/K_0$. Thus as long as $\tan(X_{out}/X_{in})$ is much less than K_0, the nonlinearity is significantly suppressed. This is demonstrated in the graph of Fig. 15.12, which plots X_{out} and the error as functions of X_{in}, where X is normalized to units of $\pi/2$.

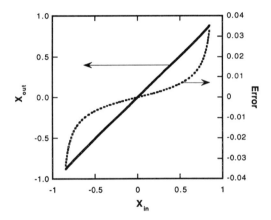

Figure 15.12. Closed-loop Input-output characteristics and error using an arctangent function for the plant.

We see that there is excellent linearity between X_{out} and X_{in} up to about $X_{in} = 0.88(\pi/2)$. Beyond this point, the error begins to grow significantly.

Since the value of F is proportional to the error, we see that the effect of the feedback loop is to apply a much larger control signal as saturation is approached, essentially pre-distorting the control signal to compensate for the saturating nonlinearity in the response function. This is the typical behavior for a nonlinear element in a quasi-static feedback loop.

15.5 Resonators and Oscillators

A *resonator* is a mechanical or electronic element that, at least in a linearized sense, has a pair of complex poles relatively near the imaginary axis so that its sinusoidal-steady-state behavior shows a strong peak in the vicinity of its undamped resonant frequency. An *oscillator* is an electronic circuit that, in combination with a suitable resonator or frequency-selective element, produces continuous oscillations. Oscillators are required in devices such as the Coriolis rate gyroscope discussed in Chapter 21, and in various types of resonant strain gauges. Our goal in this section is to understand how to turn a resonator into an oscillator.

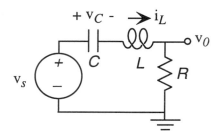

Figure 15.13.   A resonant *RLC* circuit.

The characteristic behavior of an oscillator is that its state-space trajectory forms a stable closed path called a *limit cycle*.[5] We begin with a purely electronic example, an *RLC* circuit with an underdamped response (see Fig. 15.13). The transfer function is

$$\frac{v_0(s)}{v_s(s)} = \frac{sRC}{s^2LC + sRC + 1} \tag{15.33}$$

The state equations for the two state variables i_L and v_C are

$$\frac{dv_C}{dt} = \frac{1}{C}i_L \tag{15.34}$$

$$\frac{di_L}{dt} = \frac{1}{L}(v_s - v_C - i_LR) \tag{15.35}$$

For any value of *R* that results in underdamped behavior, the zero-input response of this system is a damped sinusoid corresponding to a set of state trajectories that spiral toward the origin. The origin is a global fixed point for this system. In the absence of some additional features, this circuit does not exhibit sustained oscillations.

One way to get sustained oscillatory behavior is to reduce *R* to zero, creating an undamped linear system. This approach has two problems. First, it is physically unrealizable, except possibly for planetary motion in deep space. Second, and more importantly, the undamped zero-input-response state trajectories are ellipses whose size depends on the initial conditions. Such behavior is *NOT* a limit cycle. The test is to perturb the ellipse by changing initial conditions. The result for an undamped system is an ellipse with a different size. These

---

[5] A good discussion of limit cycles and the associated vocabulary can be found in Chapter 7 of Strogatz [45].

undamped oscillatory state trajectories are not stable, in that they do not return to their original form when perturbed.

A stable limit cycle, in contrast to this undamped behavior, occurs in a dissipative (damped) system with suitable gain elements. The exact same limit-cycle state trajectory will result from a range of varied initial conditions (called the *basin* of the limit-cycle *attractor*). The system may execute a temporary transient behavior, but settles in to the stable limit cycle at long times.

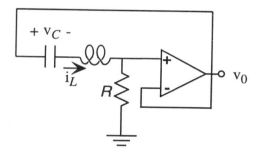

Figure 15.14.   A unity-gain buffer added to the linear RLC circuit.

We wish to add a suitable amplifier to this circuit in order to create a stable limit cycle. The most obvious choice is to use a linear amplifier, as illustrated in Fig 15.14. A unity-gain buffer amplifier takes the output voltage, which equals $i_L R$ and applies it as the input. The state equations for this system are found by replacing v_s by $i_L R$.

$$\frac{dv_C}{dt} = \frac{1}{C} i_L \tag{15.36}$$

$$\frac{di_L}{dt} = \frac{1}{L}(i_L R - v_C - i_L R) \tag{15.37}$$

The two terms involving R cancel, leaving the same state equations as for the $R = 0$ case. Thus a unity-gain linear amplifier cannot produce the conditions necessary to create a stable limit cycle.[6] Furthermore, if a linear gain greater than unity is used, the system is simply unstable, hence useless.

What is needed is *a gain greater than unity in combination with a suitable saturating nonlinearity*. Figure 15.15 shows a linear amplifier with gain A

---

[6]Unfortunately, some carelessness on this subject has crept into the literature. The reader is cautioned to be careful when reading papers about oscillators.

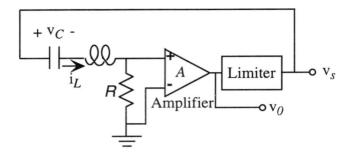

Figure 15.15.   An RLC oscillator, with a linear amplifier and a nonlinear limiter with saturating transfer characteristics.

in combination with a block called a *limiter*. The exact shape of the limiter transfer function is an issue of detail; for illustration, we will use the arctangent saturating nonlinearity already encountered in the previous section. We assume

$$v_s = V_1 \tan^{-1}\left(\frac{v_0}{V_2}\right) \tag{15.38}$$

where V_1 and V_2 are scale factors that determine the magnitude and sharpness of the saturation behavior. With this addition, the state equations now become

$$\frac{dv_C}{dt} = \frac{1}{C}i_L \tag{15.39}$$

$$\frac{di_L}{dt} = \frac{1}{L}\left[V_1 \tan^{-1}\left(\frac{Ai_L R}{V_2}\right) - v_C - i_L R\right] \tag{15.40}$$

We now have an interesting situation. The arctangent term is positive, and grows with increasing i_L. The damping term is negative, and also grows with increasing i_L. If we select the parameters such that the linearization of the arctangent term around the origin yields a more positive slope with respect to i_L than R, the global fixed point at the origin disappears. Specifically, if the following condition is met,

$$\frac{AV_1}{V_2} > 1 \tag{15.41}$$

then the global stability of the origin as a fixed point disappears. The linearized coefficient of R in the second state equation changes from negative to positive; the system is unstable for small values of i_L. However, as i_L grows, the arctangent term saturates, but the $-i_L R$ terms grows linearly. Therefore, at sufficiently large i_L, the resistive damping overcomes the arctangent term, and the system becomes damped and incrementally stable. Therefore i_L must tend to decrease.

If this were a static system, we would expect to find a stable DC fixed point, as in the previous section. However this system has dynamic state equations that tend to drive an oscillatory response, exchanging stored energy between the capacitor and the inductor. Therefore, a discussion of growth and reduction of i_L must be viewed in the context that there is simultaneously an oscillation going on. A useful way to think about it is in terms of energy. The total stored energy (expressed with co-energies) is

$$W' = \frac{1}{2} C v_C^2 + \frac{1}{2} L i_L^2 \tag{15.42}$$

The rate of change of stored energy is then found to be

$$\frac{dW'}{dt} = i_L \left[V_1 \tan^{-1} \left(\frac{A i_L R}{V_2} \right) - i_L R \right] \tag{15.43}$$

Because of the saturation behavior of the arctangent term, it is clear that dW'/dt is negative for large i_L. Similarly, if AV_1/V_2 is greater than unity, dW'/dt is positive for small i_L. The oscillatory nature of the zero-input-response state equations tends to drive i_L in an oscillatory fashion. Thus, during the part of the cycle when i_L is small, the stored energy in the system is growing, but during the part of the cycle when i_L is large, the system is losing energy. A *stable limit cycle* is one for which the stored energy is periodic, with the increase in stored energy during the small amplitude part of the cycle exactly compensated by the decrease in stored energy during the large-amplitude part of the cycle. We now illustrate this with some examples.

15.5.1 Simulink Model

It is straightforward to construct a SIMULINK model of this system as shown in Fig. 15.16.

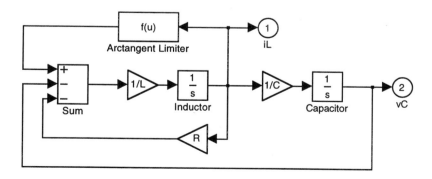

Figure 15.16. SIMULINK implementation of the state equations of an *RLC* oscillator with an arctangent limiter. The amplifier gain of *A* is included in the argument of the arctangent function.

This model has seven parameters: the values of R, L, and C, the arctangent scale factors V_1 and V_2, the amplifier gain A, and the initial voltage on the capacitor. This cannot be zero, even though the origin is no longer a stable fixed point, but it is a fixed point (try substituting $i_L = 0$ and $v_C = 0$ into the state equations!). Therefore, it is necessary to perturb the system at least slightly away from the origin in order for the system to oscillate and reach its limit cycle. We now explore two types of oscillation behavior.

15.5.2   The (Almost) Sinusoidal Oscillator

When the gain and scale parameters are set such that the net gain near the origin is small, the resulting limit cycle is almost perfectly sinusoidal. This regime of operation is called a *marginal oscillator*, a system that is just barely oscillating. To keep things simple, simulations are performed with C and L equal to unity, so that the undamped resonant frequency ω_o equals unity. The scale parameters V_1 and V_2 are also equal set to unity. This reduces the problem to the choice of A and R, and the selection of a non-zero initial condition. For illustration, we have selected a value of $R = 0.1$, so that the undriven Q of the circuit is 10.

To get marginal-oscillator behavior, any value of A somewhat greater than unity suffices. A value of 1.5 was selected for this example. By experimenting with the SIMULINK model, it was determined that the limit-cycle amplitude for this choice of A is about 11.5. Therefore, to speed up the settling time of the simulation, an initial condition of v_C equal to 11.5 was selected. However, any initial condition could be used; the only cost is the time required for the system to settle into its stable limit cycle.

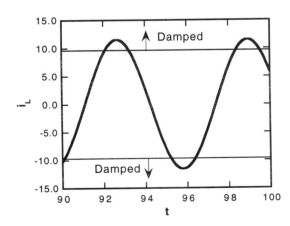

Figure 15.17.   Current waveform for the *RLC* oscillator operating as a marginal oscillator with $A = 1.5$.

The waveform for i_L is plotted in Fig. 15.17. The time-interval is taken from the end of the SIMULINK simulation to be certain that the final limit cycle behavior was reached. The waveform looks like a sinusoid, but it is *NOT* exactly a sinusoid. For $i_L < 9.674$, the arctangent term exceeds the resistive damping term, and the system is gaining energy.[7] For $I_L > 9.764$, the system is damped and is losing energy. The net result is a slightly distorted sinusoid, but the distortion cannot be readily observed in this plot. Characteristically, marginal oscillators exhibit this slightly distorted sinusoidal behavior. The period of the sinusoid is 6.07, slightly less than the value of 2π that would be expected if the system were oscillating at its undamped resonant frequency. Thus, the oscillation frequency is slightly larger than ω_o, an effect that arises from the tendency for waveforms to grow exponentially when i_L is small.

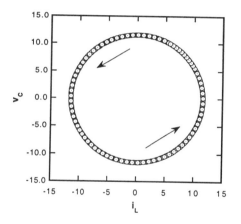

Figure 15.18. Limit cycle for the *RLC* oscillator operating as a marginal oscillator with $R = 0.1$ and $A = 1.5$.

The limit cycle is shown in Fig. 15.18, plotted as v_C against i_L, with time as a parameter. Because of symmetric choice of values for L and C, the limit cycle appear as circular (although it is a slightly distorted circle, as we know from the preceding discussion). The arrows indicate the direction of rotation.

Figure 15.19 shows the time-dependence of the stored energy superimposed on the i_L waveform. Note that each decrease in stored energy starts exactly when i_L enters the damped regions, and each increase in stored energy starts when i_L leaves the damped region. However, the magnitude of the oscillation in stored energy is small. This is characteristic of nearly sinuosidal oscillators.

---

[7]We recall that this energy is coming from the power supplies of the amplifier.

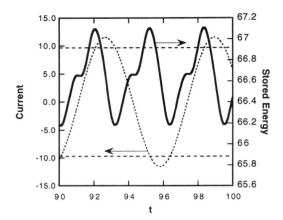

Figure 15.19.   Time dependence of the stored energy superimposed on the current waveform for the sinusoidal oscillator.

15.5.3   Relaxation Oscillation

A different kind of oscillation results if we make A very large. The tops of the sine waves then become noticeably flattened during the damping part of the cycle. An even more dramatic variation occurs if we make the initial circuit overdamped, and then make A large enough to create the conditions for oscillation. The waveforms and limit cycle are illustrated in Figs. 15.20 and 15.21.

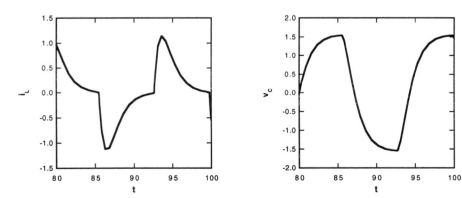

Figure 15.20.   Waveforms for the *RLC* oscillator operating as a relaxation oscillator with $R = 2$ and $A = 1000$.

Note the square-pulse-like capacitor voltage and the corresponding sequence of sort-pulse transients for the inductor current and the markedly distorted limit cycle.

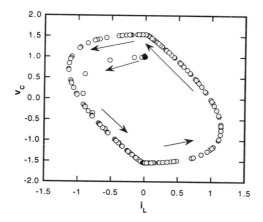

Figure 15.21. Startup transient and limit cycle for the *RLC* oscillator operating as a relaxation oscillator with $R = 2$ and $A = 1000$. The black circle shows the initial state with $v_C = 1$ and $i_L = 0$.

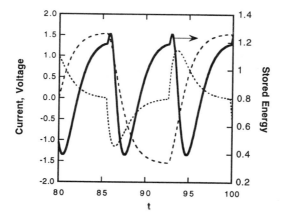

Figure 15.22. Time dependence of the stored energy superimposed on the current and voltage waveforms for the relaxation oscillator.

The stored energy for the relaxation oscillator is shown in Fig. 15.22 superimposed on the current and voltage waveforms. The behavior is complex, but each decrease in stored energy occurs when the current enters the damped region, and each increase occurs when the current leaves the damped region.

The variety of oscillator circuits and resulting waveforms is enormous. We have illustrated two types of oscillators using just one circuit configuration. The overarching message is that the behavior of an *oscillator*, as distinct from the resonator at its core, depends in detail on the nonlinearities of the driving circuit.

Related Reading

P. R. Gray and R. G. Meyer, *Analysis and Design of Analog Integrated Circuits*, Second Edition, New York: Wiley, 1984.

J. K. Roberge, *Operational Amplifiers: Theory and Practice*, New York: Wiley, 1975.

W. McC. Siebert, *Circuits, Signals, and Systems*, New York: McGraw Hill, 1986.

S. H. Strogatz, *Nonlinear Dynamics and Chaos*, Reading, MA: Addison Wesley, 1994.

Problems

15.1 For the feedback system of Fig. 15.2, assume the plant is an accelerometer modeled by a spring-mass-dashpot system with a mass of 0.1 milligram, a resonance frequency of 5 kHz, and a Q of 2, and a position sensor that has a unity-gain transfer function at DC (no error) and a single-pole frequency response with a 20 kHz roll-off. Is there any way a proportional controller can create an overall system response to an accelerator disturbance such that there is no overshoot? If "yes" what is the maximum gain that can be used. If "no," present a convincing argument based either on analysis or numerical modeling with MATLAB or SIMULINK.

15.2 For the system of Problem 15.1, design a PID controller that achieves an overall critically damped system response. Demonstrate with results from MATLAB or SIMULINK simulation.

15.3 For the system of Problem 15.1, now assume that the sensor has a 5% error in its DC position measurement. Can you design a feedback loop to compensate for this error? If "yes," show the design. If "no," explain.

15.4 For the system of Problem 15.1, determine the maximum proportional gain for which the system is stable. Repeat for a controller that is a proportional controller, but with a 10 kHz bandwidth.

15.5 An Oscillator Treasure Hunt: Build the SIMULINK model of the arctangent limited oscillator of Fig. 15.16 and study the variation of limit cycle with the parameters of the circuit, especially Q, and the parameters V_1 and V_2 of the limiter. Show that the shape of the limit cycle is independent of initial state (as long as the initial state is not the origin of state space).

15.6 Instead of using an arctangent nonlinearity, it is possible to use a milder nonlinearity of the form

$$v_s = a_1 v_0 - a_3 v_0^3$$

where a_1 and a_3 are constants. Assume that a_1 is set such that the system is just able to oscillate. By assuming a purely sinusoidal oscillation, substituting into the state equations, and collecting terms, it is possible to find first-order expressions for both the frequency of oscillation and the amplitude. Furthermore, if one assumes that one of the circuit parameters is perturbed, for example R goes to $R + \delta R$ in step-wise fashion, the result will be a new limit cycle, with different amplitude and frequency, but that there will be a characteristic time associated with making the transition from the first limit cycle to the second. Using linearization of the state equations with a substituted sinusoid trial solution, find the first-order dynamic response of the amplitude of the sinusoid to a step change in R. (A solution to a similar problem can be found in [86].)

Chapter 16

NOISE

How is't with me, when every noise appals me?

—William Shakespeare, in *Macbeth*

16.1   Introduction

Any unwanted signal can be considered to be "noise," in that it potentially interferes with the clarity and accuracy of the desired "signal." There is a wide variety of noise sources. They can be classified into two groups: *interfering signals*, and *random noise*.

Interfering signals are things such as inductive or capacitive pickup of the 60 Hz (or 50 Hz) power-line signal, direct parasitic coupling of excitation signals to the output of a system, pickup and accidental rectification of broadcast radio or television signals, and parasitic DC voltages at amplifier inputs created by thermoelectric voltages in response to non-uniform temperature in a circuit.

Random noise includes: random fluctuations of effort variables associated with the properties of a thermal reservoir at constant temperature; random fluctuations of flow variables associated with the discreteness of electronic charges and individual atoms or ions; mechanism-specific fluctuations due to the physics of device operation, such as the current fluctuations created by the capture and release of electrons by localized trap states at the surface of a semiconductor; and, spatially distributed temperature fluctuations in systems driven with fluctuating dissipative heat sources.

The approaches to dealing with the two types of noise sources are quite different. Interfering signals have a systematic or characteristic behavior that can be studied, isolated, and, with proper design or shielding, minimized. Random noise, on the other hand, is an intrinsic property of the components that comprise the system. Its effect on system performance must be analyzed statistically.

In this chapter, we examine first the problem of isolation of circuits from power-line interference (or other equivalent sources of disturbance) and introduce the notions of guards and shields. We then discuss the statistical characterization of random noise, examine several types of random-noise sources, and develop methods for calculating the signal-to-noise ratio. We conclude with a discussion of "drifts," a catch-all term for low-frequency random behavior not otherwise accounted for.

16.2   The Interference Problem

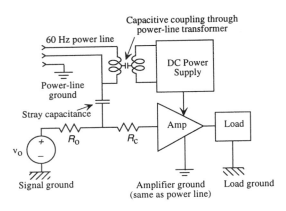

Figure 16.1.   Illustrating two paths for coupling of the power-line signal into a sensor and amplifier circuit. R_o is the source resistance, and R_c is the cable resistance for the cable connecting the sensor to the amplifier.

Figure 16.1 illustrates the problem of power-line interference, showing capacitive-coupling between power-line sources and a simple amplifier circuit. The source network in this case might be a sensor measuring some physical parameter, and the load might be the input to a data-processing system. Two capacitive couplings are shown between the power line and the circuit, a stray capacitance to the signal path, and a capacitance between the primary and secondary windings of the transformer that is an integral part of the DC power supply.

The transformer coupling capacitance can be quite large, on the order of 1000 pF. It is typically modeled as a single lumped element between the centers of the primary and secondary windings. The effect of this coupling capacitor is to create a common-mode AC signal at the center of the secondary winding of the power supply. To suppress the effects of this common-mode signal, it is common practice to use a center-tapped secondary winding, use the center tap as the power-supply ground, and then connect that ground back to power-line ground. This works well, but creates other problems, as explained below.

Three separate grounds are shown in the diagram: the power-line ground shared with the amplifier power supply, the signal ground, and the load ground. Unless steps are taken to ensure that these grounds are at the same potential, they can actually be at different AC potentials, as much as 1 to 10 V being common. The potential difference between the signal ground and the amplifier ground appears as a common-mode signal at the input of the amplifier and can get through the amplifier unless it has excellent common-mode-rejection characteristics.

The stray electrostatic coupling between the power line and the conductors in the circuit can be modeled as a capacitor. Its value can be on the order of 1 pF for unshielded circuits. If the source-plus-cable resistance is 1 kΩ, then there can be 40 μV of 60 Hz AC present on the signal line in addition to the desired signal v_o. To minimize this effect, the current due to this electrostatic coupling must be shunted around the source resistance.

16.2.1   Shields

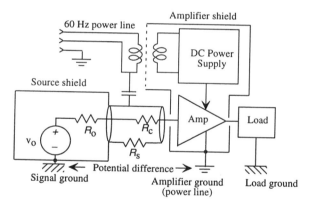

Figure 16.2.   Illustrating shielding of signal sources (sensors) and amplifiers. A shield connected to signal ground shunts capacitively-coupled currents away from the signal source. The cable shield surrounds but does not contact the inner conductor of the cable.

A *shield* is a conductor that nearly surrounds the space being shielded. Its purpose is to interrupt the capacitive current from the power line and shunt these currents to ground without passing through any internal component. Figure 16.2 illustrates one shield around the source network and its associated cable, and a second shield around the amplifier and its power supply. This second shield actually continues into and through the power transformer and greatly reduces the capacitive coupling between primary and secondary. Care must be taken in building shielded transformers to prevent eddy-current losses in the shielding material, but this can be done successfully, reducing the primary-to-secondary coupling capacitance to the order of 10 pF.

The shield around the source network intercepts the roughly 40 nA of capacitive current coupled from the power line through the 1 pF stray capacitance, and shunts it harmlessly to the signal ground. There is some capacitance between the shield and the circuit, but as long as the shield resistance R_s is small, the AC potential developed on the shield is also small, and hence the capacitive coupling to the signal line is negligible. The shield around the amplifier and its power supply has the same effect of reducing power-line common-mode coupling into the amplifier.

Electrostatic shields also provide some modest reduction in the inductive pickup of interfering magnetic fields. The mechanism is the attenuation due to magnetic diffusion of the field through the conductor. Robust magnetic shielding, however, requires high-permeability enclosures.

16.2.2   Ground Loops

In order to reduce the common-mode potential due to differences in local ground potentials, the most logical thing to do is to connect the two grounds together. What is interesting is that the way this connection is made is quite important. Figure 16.3 illustrates two methods.

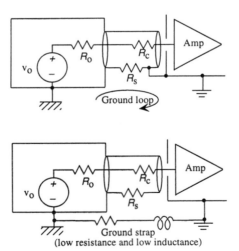

Figure 16.3.   Two methods to reduce the common-mode potential due to different local grounds. Using the cable shield produces a ground loop, with the possibility of significant inductively-coupled pickup. The ground strap avoids this problem.

One method is to connect the shield of the signal cable to the amplifier shield. This produces what is called a *ground loop*. A ground loop is a circuit path connected to nominal ground at two places. The rest of the loop is the unseen connection between grounds, possibly via the the power-line ground. The problem with a ground loop is that it can enclose a changing AC flux, and by Faraday's law of induction, currents can be induced which give rise to potentials at various point in the loop. To minimize these potentials, it is wise to use low-inductance and low-resistance connections. The cable shield might be low resistance, but it might have substantial inductance, and hence can produce common-mode signals between the signal and amplifier grounds from induced ground-loop currents. The second method, using a low-resistance low-inductance ground strap between signal and amplifier ground, and connecting each shield to its own local ground, is preferred. There is still a ground loop, but its effect can be reduced by the design of a ground strap, using, for example, a flat braided cable.

16.2.3   Guards

A *guard* is a shield that is connected to the common-mode potential of the signal being processed. Combinations of guards and shields can be very effective at isolating low-signal-level sources and their amplifiers from external interfering signals. A detailed example using a triple-shielded transformer and a separate amplifier ground is discussed in Senturia and Wedlock.[1] In that example, the guards are connected passively to their common-mode-potential reference points.

Another way to use guards is to connect them to the output of a unity-gain buffer amplifier that is driven by the common-mode signal. This can be used with cable shields, circuit shields, or, as in the example shown in Fig. 16.4, to reduce the effect of line-to-substrate capacitances in integrated circuit structures.

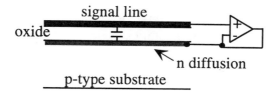

Figure 16.4.   Illustrating a driven guard. The effective parasitic capacitance between the signal line and the substrate is reduced by driving the lower electrode of the capacitance with a unity-gain buffer.

A signal line is shown over an oxide insulator on a semiconductor substrate. There can be substantial capacitance between the signal line and the substrate. The need to charge and discharge this capacitance when the signal changes can

---

[1]Senturia and Wedlock, *Electronic Circuits and Applications*, pp. 555-557 [39].

modify circuit behavior in an undesirable way. One method of preventing this problem is to use an inverting amplifier, which places the signal line at a virtual ground, hence restricting the voltage across this capacitor to be very small. But if one needs to use a noninverting amplifier, then the signal itself is a common mode voltage. If a unity-gain buffer is used at the end of the signal line and the output of that buffer is used to drive a diffusion that is placed beneath the signal line, then the signal-to-substrate capacitance is forced to have zero volts across it, hence is uncharged.

Another type of guard is called a *guard ring*, which is a conductor placed around the perimeter of a sensitive portion of a circuit, such as a sense electrode, to intercept any stray surface currents that might interfere with device operation. Functionally, though, a "guard" ring works as a "shield" in that it intercepts currents and shunts them to ground (one more example of confusing nomenclature).

16.3   Characterization of Signals

Before we can characterize random noise, it is important to specify what a signal is. The key feature of a signal is that it carries *information*. One might like to think that a sinusoidal signal, such as $A \cos(\omega t + \theta)$, carries information in that it has an amplitude, a frequency, and a phase. But technically, if all three of these quantities are constant, the waveform carries no information because, if one has a record of its behavior up to some time t_o, its behavior is completely predictable for all future time. Information is intrinsically linked to some degree of unpredictability of an outcome. If there are two possible outcomes (like a one-bit number), one can convey lots of information. This entire book, which is alleged to be information-bearing, is represented in the author's computer as a sequence of bits. But if there is only one possible outcome, there can be no information transfer.

A sinusoidally-encoded information-carrying waveform might be of the form

$$x(t) = A(t) \cos[\omega(t)t + \phi(t)] \qquad (16.1)$$

where now all three quantities, the amplitude, the frequency, and the phase, are time-dependent. This form is quite general. For example, a slowly varying DC pressure could be represented with ω and ϕ set to zero, while a frequency-modulated radio station could be represented with a constant A and ϕ, but a time-varying ω. When A is time-dependent, we call the waveform an *amplitude-modulated signal*, or, when ω and ϕ are zero, a *baseband signal*. When ω is time-dependent, we call the waveform a *frequency-modulated signal*, and when ϕ is time-dependent, we call it a *phase-modulated signal*.

Let us consider a baseband signal, presumed to be information carrying. By definition, it cannot be completely predictable. If we observe this signal for some time interval $\hat{t}$, and then compute its Fourier transform, we will get some spectral indication of its frequency content. If we repeat the observation for a second such interval, and again compute the Fourier transform, the result will not be identical (since the waveform is to some degree unpredictable), but *it will occupy the same part of the frequency spectrum*. After multiple repetitions using an ensemble of observations, we can characterize the region of the frequency spectrum where signal information can appear. This is called the *bandwidth* of the baseband signal. We must design our signal-processing system to accept the full bandwidth of the baseband signal or risk losing information. Generally, the faster a baseband signal can change, the wider its bandwidth.

16.3.1   Amplitude-Modulated Signals

The mathematics of frequency- and phase-modulated signals can get complex, but for amplitude-modulated signals, abbreviated AM, the math is straightforward and illustrates important fundamental concepts. Consider first a sinusoidally modulated sinusoid:

$$x(t) = A_0(1 + m \cos \omega_m t) \cos \omega_c t \qquad (16.2)$$

The frequency ω_c is called the *carrier frequency*, ω_m is called the *modulation frequency*, and m is called the modulation index. In keeping with the characteristics of a typical amplitude-modulated wave, we will assume that ω_m is small compared to ω_c and m is less than unity.

We can expand this waveform to discover that it consists of three sinuosoids at different frequencies:

$$x(t) = A_0 \cos \omega_c t + \frac{mA_0}{2} [\cos(\omega_c + \omega_m)t + \cos(\omega_c - \omega_m)t] \qquad (16.3)$$

Since this waveform consists of three sinusoids, each with constant parameters, it can carry no information. But we can learn a lot from examining it. We note that the AM wave consists of a carrier sinusoid plus two *sidebands*. These sidebands are located symmetrically about the carrier frequency, shifted by the modulation frequency ω_m. The amplitude of the sidebands depends on both the carrier amplitude A_0 and on the modulation index m. This is characteristic of all AM waves.

Now, if an information-carrying baseband signal $y(t)$ is used for the modulation instead of $\cos \omega_m t$, then the modulated signal carries information. The spectrum of the modulated waveform $x(t)$ consists of a spread of frequencies determined by the Fourier transform of $y(t)$. To illustrate this, we compute the Fourier transform of $x(t)$:

$$X(\omega) = \int_{-\infty}^{\infty} x(t)e^{-j\omega t}\, dt \tag{16.4}$$

The *spectrum* of a waveform is a plot of the amplitude of its Fourier transform. The spectrum of the sinusoidally-modulated $x(t)$ (Eq. 16.3), is given by

$$
\begin{aligned}
X(\omega) = {} & \pi A_0[\delta(\omega - \omega_c) + \delta(\omega + \omega_c)] \\
& + m\pi A_0[\delta(\omega - \omega_c - \omega_m) + \delta(\omega - \omega_c + \omega_m) \\
& + \delta(\omega + \omega_c - \omega_m) + \delta(\omega + \omega_c + \omega_m)]
\end{aligned}
\tag{16.5}
$$

This spectrum is plotted in Fig. 16.5.

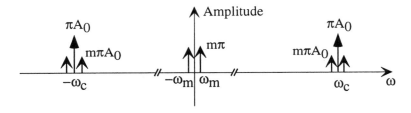

Figure 16.5. The spectrum of a sinusoidally modulated amplitude-modulated wave shown together with the spectrum of the baseband sinusoid. The relative amplitude of the baseband spectrum compared to the two parts of the amplitude-modulated spectrum is not to scale.

If we compute the spectrum of an AM wave which uses $y(t)$ as a modulation signal, we obtain

$$X(\omega) = \pi A_0(\delta(\omega - \omega_c) + \delta(\omega + \omega_c) + mA_0 \int_{-\infty}^{\infty} [y(t)(\cos\omega_c t)]e^{-j\omega t}\, dt \tag{16.6}$$

The second integral is the Fourier transform of the *product* of $y(t)$ and the cos carrier. The result is the *convolution* of the Fourier transform of the cos with that of $y(t)$. Convolution in the frequency domain of two signals $F(\omega)$ and $G(\omega)$ is written

$$F(\omega) \otimes G(\omega) = \int_{-\infty}^{\infty} F(\omega - \omega')G(\omega') \frac{d\omega'}{2\pi} \tag{16.7}$$

Thus, using the Fourier transform of the cos, given by $\pi[\delta(\omega - \omega_c) + \delta(\omega + \omega_c)]$, and denoting a typical Fourier transform of an observation of $y(t)$ as $Y(\omega)$, we find

$$X(\omega) = \pi A_0 (\delta(\omega - \omega_c) + \delta(\omega + \omega_c) + \frac{mA_0}{2} [Y(\omega - \omega_c) + Y(\omega + \omega_c)]$$

(16.8)

Now we have a true information-carrying amplitude-modulated waveform.

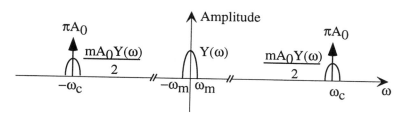

Figure 16.6. The spectrum of an amplitude-modulated wave shown together with the baseband spectrum $Y(\omega)$. The relative amplitude of the baseband spectrum compared to the two parts of the amplitude-modulated spectrum is not to scale.

Figure 16.6 shows the spectrum of the baseband signal $Y(\omega)$ as having an arbitrary amplitude within a confined band of frequencies bounded by $\pm\omega_m$. In the same figure, we show the AM spectrum consisting of two impulses at $\pm\omega_c$ with area πA_o and a copy of $Y(\omega)$, shifted and scaled by $mA_0/2$. The discrete sidebands around each carrier impulse are now smeared out into the complete spectrum of $y(t)$.

This discussion is sufficient to permit us to think constructively about signal characterization. Signals, if they are to carry information, must occupy finite bandwidth. The more rapidly the baseband signal changes, the wider the bandwidth needed to represent its frequency content. When dealing with baseband signals directly, we need only be concerned with the baseband spectrum. However, if the baseband signal is used to amplitude-modulate a carrier, then the resulting spectrum is the convolution of the baseband spectrum with the Fourier transform of the carrier. For frequency- and phase-modulated waveforms, the spectrum that results from the modulation is much more complex, and will not be covered here. However, the basic point is the same: the spectrum of the information-carrying waveform is determined by a combination of the carrier frequency, the manner of modulation, and the baseband spectrum.

16.4 Characterization of Random Noise

It would be nice if we could discuss noise sources using the same Fourier methods that were just used to discuss signals. However, random noise exhibits statistical characteristics that make such a characterization impossible. For one thing, the bandwidth required to represent noise signals is very large, covering virtually the entire spectrum of interest. Instead, we use modified Fourier methods based on noise power instead of noise amplitude. This works well.[2]

We assume that a noise waveform has zero average value. That is, for a noise waveform $v_n(t)$, it must be true that

$$\bar{v}_n = \lim_{\hat{t} \to \infty} \frac{1}{\hat{t}} \int_{-\hat{t}/2}^{\hat{t}/2} v_n(t) dt = 0 \qquad (16.9)$$

16.4.1   Mean-Square and Root-Mean-Square Noise

If we compute the average of the square of v_n, we get a non-zero result:

$$\overline{v_n^2} = \lim_{\hat{t} \to \infty} \frac{1}{\hat{t}} \int_{-\hat{t}/2}^{\hat{t}/2} [v_n(t)]^2 dt \qquad (16.10)$$

The quantity $\overline{v_n^2}$ is called the *mean-square noise*, and its square root is called the *root-mean-square noise*, and is denoted $v_{n,rms}$. The root-mean-square (RMS) noise is an amplitude that characterizes the noise, and thus can be compared with signal amplitudes (see below).

16.4.2   Addition of Uncorrelated Sources

If we have a signal waveform $v_s(t)$ to which is added a random noise waveform $v_n(t)$, we can compute the average and mean square of the total:

$$v = v_s + v_n \qquad (16.11)$$
$$\bar{v} = \bar{v}_s \qquad (16.12)$$
$$\overline{v^2} = \overline{v_s^2} + \overline{v_n^2} + \overline{v_s v_n} \qquad (16.13)$$

The quantity $\overline{v_s v_n}$ measures the correlation between the time-variation of v_s and that of v_n. With the exception of a few very specific examples, noise sources are not correlated with other waveforms, so this quantity is zero. The net result is that

$$\overline{v^2} = \overline{v_s^2} + \overline{v_n^2} \qquad (16.14)$$

This can be summarized as follows: *The mean-square amplitude of the sum of uncorrelated sources is the sum of the mean-square amplitude from each source.*

---

[2]Our treatment of random noise is at an elementary and intuitive level. More mathematical treatments can be found in the Related Reading at the end of the chapter.

16.4.3   Signal-to-Noise Ratio

The *signal-to-noise ratio* is defined as the ratio of the mean-square signal to the mean-square noise.

$$S/N = \frac{\overline{v_s^2}}{\overline{v_n^2}}$$ (16.15)

This ratio is often expressed in decibels (dB):

$$S/N = 10 \log \left(\frac{\overline{v_s^2}}{\overline{v_n^2}} \right) \quad \text{in dB}$$ (16.16)

This can be further expressed in terms of the RMS values of each quantity:

$$S/N = 20 \log \left(\frac{v_{s,rms}}{v_{n,rms}} \right) \quad \text{in dB}$$ (16.17)

16.4.4   Spectral Density Function

In spite of the fact that we cannot use normal Fourier methods to describe noise, there is a way to deal with the frequency content of noise. Let us assume that we have a narrow-band filter that passes all frequencies within a bandwidth δf around a center frequency f_o, where we now express frequency in Hertz to be compatible with standard texts on noise. A broadband noise source is applied to the input of the filter, and we measure the mean square noise at the output of the filter, denoted by $v_o^2(f_o, \delta f)$. It is a characteristic property of the noise sources we shall deal with that for narrow enough δf,

$$\overline{v_o^2(f_o, \delta f)} = S_n(f_o) \delta f$$ (16.18)

That is, the mean-square output of the narrow-band filter is proportional to its bandwidth. The proportionality factor, $S_n(f_o)$, is called the *spectral density function* of the noise source. The spectral density is related to the total mean-square noise for the broadband source, since if we tune the filter's center frequency f_o over a broad enough frequency range and add up the resulting mean-square noise, we account for all of the noise power.[3] Hence, it is true that

$$\overline{v_n^2} = \int_0^\infty S_n(f) df$$ (16.19)

---

[3] This assumes that noise components in different frequency ranges are not correlated with one another, a correct assumption for the noise sources considered here.

16.4.5   Noise in Linear Systems

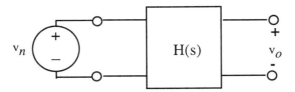

Figure 16.7.   A noise source connected to a linear system with transfer function $H(s)$.

Figure 16.7 shows a noise source v_n with spectral density function $S_n(f)$ connected at the input port of a linear system with system transfer function $H(s)$. If we measure the spectral density function at the output, we will obtain[4]

$$S_o(f) = |H(j2\pi f)|^2 S_n(f) \tag{16.20}$$

The spectral density at the output is equal to the spectral density at the input multiplied by the square of the transfer-function magnitude evaluated at $s = j2\pi f$. Therefore, the mean square noise at the output is given by

$$\overline{v_o^2} = \int_0^\infty |H(j2\pi f)|^2 S_n(f)\, df \tag{16.21}$$

With this result, we are now equipped to calculate signal-to-noise ratios in any linear system, provided we know the spectral density functions of the various noise sources in a system.

16.5   Noise Sources

There are three basic sources of random noise: fluctuations associated with dissipative processes that connect to a thermal reservoir; fluctuations associated with the discreteness of particles that constitute a flow; and fluctuations associated with the capture and release of particles participating in a flow. All lead to measurable noise.[5]

16.5.1   Thermal Noise

Any dissipative process that is coupled to a thermal reservoir results in fluctuations, even in equilibrium. Thus, an electrical resistor, even when carrying

---

[4]See, for example, Siebert, Chapter 19 [43].
[5]Extensive discussion of noise sources can be found in van der Ziel [87] and in Chapter 11 of Gray and Meyer [88].

no current, has a fluctuating voltage (with zero average) across its terminals. The mean-square amplitude of the fluctuations, as measured by the spectral density function of the voltage, is proportional to the temperature of the resistor, that is, to the temperature of the thermal reservoir with which the electrical element is in equilibrium. And, as we shall see below, these fluctuations provide a mechanism for exchanging energy between the thermal reservoir and the electrical energy domain.

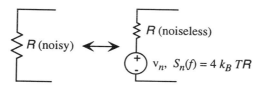

Figure 16.8.   Equivalent circuit for the thermal noise in a resistor.

Thermal noise in an electrical resistor is called *Johnson noise*. The electrical model of a real (hence, noisy) resistor is a noise voltage source in series with an ideal (noiseless) resistor, as illustrated in Fig. 16.8. The spectral density function of the voltage noise is

$$S_n(f) = 4k_B T R \qquad (16.22)$$

where k_B is Boltzmann's constant, T is the temperature of the resistor, and R is the value of the resistance. Because this spectral density function is constant over all frequencies, it is called *white noise*, analogous to white light which has a uniform mix of all visible wavelengths. If one looks more carefully at the physics underlying thermal noise, there is an upper cutoff frequency for this noise source, but it is at very high frequencies, corresponding to far-infrared wavelengths. Thus, for all practical purposes, we can treat this noise source as white at all frequencies of interest.

Thermal noise is associated with all dissipative processes that couple to the thermal energy domain. Therefore, not only do electrical resistors have thermal noise, but every dissipative element that we have modeled as a resistor also has thermal noise, and the spectral density is $4k_B T R$ provided that thermal noise is the only important noise source (we will discuss other noise sources below). This means that the fluidic damping element in an accelerometer, when modeled as a resistor, has a noise source associated with it. The fluctuation is in the effort variable, hence the force on the proof mass. This type of noise is called *Brownian motion noise* because it arises from the interaction of the proof-mass with the fluctuations in gas pressure arising from the random thermal motions of individual atoms in the gas. In practice, it is difficult to build an accelerometer that can observe this Brownian-motion noise because electrical noise in the measurement circuit is typically larger.

The mean-square noise from a thermal noise source depends on the bandwidth of the circuit in which it is placed. We examine this issue in the following section.

16.5.2   Noise Bandwidth

If thermal noise is passed through an ideal filter with unity amplitude over a bandwidth Δf, and zero elsewhere, then the mean-square noise at the output of the filter is

$$\overline{v_n^2} = 4k_B T R \Delta f \tag{16.23}$$

This leads to the concept of a *noise bandwidth*, the effective bandwidth that determines the mean-square noise due to a white noise source. In general, to find the noise bandwidth for a white noise source, one computes the transfer function between the noise source and the desired output, then integrates the magnitude squared of the transfer function over all frequencies. The result is the noise bandwidth. An example will illustrate the method and also will bring out an important and fundamental property of thermal noise.

Consider a noisy resistor in parallel with a capacitor, the simplest lumped model of a dissipative system. Assume that the resistor is at temperature T. Since the resistor has a fluctuating voltage across it, the capacitor must also have a fluctuating voltage across it, meaning that there is, on average, some stored energy in the capacitor. Thus, *the thermal fluctuations in the resistor bring the capacitor into thermodynamic equilibrium with the thermal reservoir.*

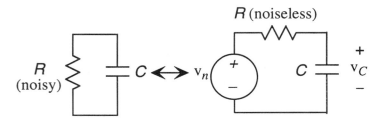

Figure 16.9.   A noisy resistor in parallel with a capacitor. The equivalent circuit is on the right.

Figure 16.9 shows the equivalent circuit for this case, with the thermal noise source in series with a noiseless resistor. This is a linear system, and we can calculate the transfer function.

$$H(s) = \frac{V_C(s)}{V_n(s)} = \frac{1}{1 + sRC} \tag{16.24}$$

Thus, the noise bandwidth is

$$\Delta f = \int_0^\infty \frac{1}{1 + (2\pi f RC)^2} \, df \tag{16.25}$$

This integral is readily evaluated to yield

$$\Delta f = \frac{1}{4RC} \tag{16.26}$$

The mean-square voltage on the capacitor is

$$\overline{v_C^2} = 4k_B T R \Delta f = \frac{k_B T}{C} \tag{16.27}$$

This is a remarkable statement. It says that when a capacitor is allowed to come into equilibrium with a thermal reservoir, it develops a non-zero mean-square voltage proportional to $k_B T$. In fact, the average energy stored in the capacitor is $(1/2)k_B T$ and this value does not depend on the size of the resistor! As long as there is some finite resistance coupling a capacitor to a thermal reservoir, regardless of how large or small, the capacitor comes into thermal equilibrium with the reservoir and develops an average stored energy of $(1/2)k_B T$, and an associated mean-square voltage $k_B T / C$.

We can extend this to the mechanical domain. In an accelerometer, the damping mechanism couples the spring to the thermal reservoir. Hence, the mechanical spring has an average thermal energy of $(1/2)k_B T$, and a mean-square force fluctuation of $k k_B T$, where k is the spring constant.

The noise bandwidth of more complex systems can be calculated similarly, as long as the spectral density corresponds to white noise. For non-white noise sources, such as flicker noise discussed below, the frequency dependence of the spectral density function must be included inside the integral. The concept of a noise bandwidth is less useful in such cases.

16.5.3 Shot Noise

When an average DC current is created by discrete particles that must cross a potential barrier, such as in a pn junction, the random arrival rate of charge carriers crossing the barrier creates a fluctuating current, called *shot noise*. If $i(t)$ is such a DC current which, because of shot noise, also has a noise component, we can write

$$i(t) = I_{DC} + i_n(t) \tag{16.28}$$

where I_{DC} is the DC current, and $i_n(t)$ is a noise current with zero average value.

Shot noise models based on random arrival of totally independent electrons results in a white-noise spectral density function for white noise of the form

$$S_n(f) = 2q_e I_{DC} \tag{16.29}$$

where q_e is the electronic charge. More detailed models that include the finite transit time of the electron result in an upper-frequency limit for shot noise, but this frequency is much higher than frequencies of interest in practical circuits. Therefore, we can treat shot noise as a white noise source.

As a rule of thumb, shot-noise sources should be associated with DC currents crossing potential barriers, as in pn diodes, while thermal noise sources should be associated with resistors. The full details require statistical analysis of the individual transport mechanisms, and are somewhat complex [87]. A circuit model that includes shot noise will be presented in the following section.

16.5.4   Flicker Noise

It is found experimentally that the noise currents in diodes and FETs have an additional low-frequency component that varies with the magnitude of the DC current. It is called *flicker noise*, or "$1/f$" noise, based on the shape of the spectral density function. Flicker noise arises from the capture and release of charge carriers in localized "trap" states in the semiconductor. Because there is typically a distribution of the binding energies of such traps, there is a corresponding distribution of capture and release times τ_t. Because these times depend exponentially on the binding energy, the quantity $\log \tau_t$ is distributed similar to the binding energies, and for a reasonably uniform distribution, this results in a $1/f$ spectral density function.

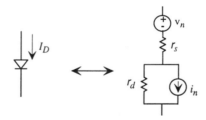

Figure 16.10. An ideal exponential diode carrying a DC current together with its associated incremental model including shot noise, flicker noise, and thermal noise.

Figure 16.10 shows a pn junction diode carrying a DC current I_D. On the right is the incremental model of the diode. The resistor r_d is the small-signal diode resistance, given by

$$\frac{1}{r_d} = \frac{dI_D}{dV_D} = \frac{q_e I_D}{k_B T} \tag{16.30}$$

and r_s is any additional series resistance that comes from the neutral regions of the semiconductor and the diode contacts. A thermal noise source v_n, with spectral density $4k_B T r_s$ is associated with r_s, and a combined shot-noise and flicker-source i_n has the spectral density function[6]

$$S_i(f) = 2q_e I_D + K\frac{(I_D)^a}{f} \tag{16.31}$$

The first term is the shot noise due to charge carriers crossing the depletion region, and the second term is the flicker noise. The exponent a varies in the range 0.5 to 2, depending on device details. The $1/f$ dependence can also have an exponent that differs slightly from unity. The relative importance of the flicker noise compared to the shot noise depends on the value of K, and that can differ widely from device to device. The frequency at which the shot noise equals the flicker noise is called the *flicker-noise corner frequency*. Above this frequency, the noise spectrum is white; below it, the spectrum has $1/f$ behavior. As we shall see in the following section, the corner frequency in MOSFET devices can be rather high.

16.5.5   Amplifier Noise

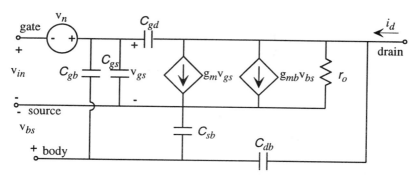

Figure 16.11.   Small-signal model for a MOSFET with an input-referred noise source added.

Amplifiers add noise to the signals they amplify. Because of the gain of successive stages, noise from the input stage is most critical. Thus, we can get a good estimate of the noise behavior of amplifiers by creating a noise model for the input transistors, and projecting all of the noise produced by the amplifier to an equivalent, or *input-referred* noise source. Figure 16.11 shows an equivalent small-signal (linearized) circuit model for a noisy MOSFET. It is identical to the small-signal model introduced in Chapter 14, but it has a voltage-noise source v_n added between the input and the reference-point for the transconductance dependent source.

---

[6]See Gray and Meyer, p. 644 [88].

The spectral density function for the v_n noise source in Fig. 16.11 takes into account all of the major noise sources in the transistor, scaled back to a value that they would have if they appeared at the position of this voltage source. The result depends on the experimental observation that the $1/f$ noise in a typical MOSFET is independent of I_D and is inversely proportional to the gate-channel-capacitance. Thus, the spectral density function has the form

$$S_n(f) = 4k_B T \left(\frac{2}{3g_m} \right) (1 + F_n) + \frac{K_f}{WL\hat{C}_{ox}f} \tag{16.32}$$

where g_m is the transistor transconductance at the operating-point, I_D is the operating-point drain current, K_f is a scale factor for the $1/f$ noise in the transistor channel, and $WL\hat{C}_{ox}$ is the gate-to-channel capacitance. The factor of 2/3 comes from a more detailed model of the channel as a noise source.[7] A typical value for K_f is on the order of 10^{-24} V^{2}F, or 10^{-12} V^{2}pF. The factor F_n is called the *noise factor*,[8] and is an empirical positive scale factor with a maximum of about 5. It can account for a myriad of otherwise unmodeled effects in the first and later stages of amplification. It is included here as a "reality check" when computing signal-to-noise ratios. That is, the "reality" could be a factor of 2-5 worse than what is calculated with $F_n = 0$.

The important conclusion is that $1/f$ noise dominates at low frequencies. Failure to include this term in noise models leads to gross overestimates of achievable signal-to-noise ratio at low frequencies. We now consider an example.

16.6   Example: A Resistance Thermometer

Fig. 16.12 shows a resistance-thermometer circuit, selected to illustrate the method of making practical noise estimates. In practice, one would use more elaborate circuits for the measurement, such as a Wheatstone bridge, but the noise calculations then become too cumbersome for a first example.

The resistor R_T is presumed to be a resistance thermometer, a long narrow thin-film metal resistor located on a portion of a microstructure whose temperature is required. The resistance-temperature characteristic for R_T is assumed to be

$$R_T = R_0(1 + \alpha_R \Delta T) \tag{16.33}$$

where R_0 is the nominal room-temperature resistance, assumed to be 1 kΩ, and α_R is the temperature coefficient of resistance, which we will take to be 3

---

[7]For additional discussion, see Chapter 5 of van der Ziel [87] or Chapter 11 of Gray and Meyer [88].

[8]This is not the same as the *noise figure* for an amplifier, which is the ratio of the signal-to-noise ratio at the output to the signal-to-noise ratio of a resistive source network at the input of the amplifier.

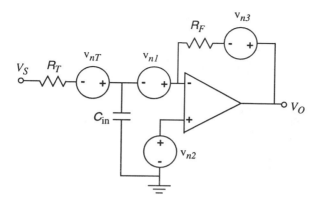

Figure 16.12. A temperature-dependent resistor is connected into an inverting amplifier with a nominal gain of 100. The circuit is shown with all noise sources, together with an input capacitance at the inverting input.

$\times 10^{-3}$ K^{-1}, a value typical of metals like aluminum. ΔT is the temperature variation that we wish to measure. V_S is a source voltage applied to the resistance thermometer. We recall from Section 11.6.4 that we cannot set V_S too large, or self-heating will perturb the accuracy of the thermometer. This limits the amount of current we can supply to the op-amp circuit, and hence the magnitude of the output signal.

There are four noise sources in this circuit: the thermal noise associated with the resistance thermometer R_T, input-referred noise sources at both op-amp inputs, and thermal noise associated with the feedback resistor R_F. The output of this circuit can be readily found using the ideal op-amp assumption:

$$V_0 = -\frac{R_F}{R_T}(V_S + v_{nT} + v_{n1}) + \left(1 + \frac{R_F}{R_T}\right) v_{n2} + v_{n3} \qquad (16.34)$$

Note that the capacitor does not appear in the transfer function because it is connected to a virtual ground. The term in this expression that is "signal" is

$$V_{0,s} = -\frac{R_F}{R_T} V_S \qquad (16.35)$$

We wish to determine the minimum variation in this quantity that can be detected in the presence of the noise. In order to clarify this variation, we substitute for R_T and use Taylor's theorem to obtain the incremental output signal $v_{0,s}$

$$v_{0,s} = -\frac{R_F V_S \alpha_R}{R_0} \Delta T \qquad (16.36)$$

To calculate the spectral density function of the noise at the output, we recognize that the four noise sources are not correlated with each other, since each noise source is associated with a physically distinct device. Therefore, the spectral density of the fluctuations in V_0 is the suitably weighted spectral densities of the various sources:

$$S_{n,0}(f) = \left(\frac{R_F}{R_T}\right)^2 [S_{n,T}(f) + S_{n,1}(f)] + \left(1 + \frac{R_F}{R_T}\right)^2 S_{n,2}(f) + S_{n,3}(f)$$
(16.37)

The two resistor terms are $4k_BTR_T$ and $4k_BTR_F$, respectively. For purposes of calculating the noise, we shall use R_0 in place of R_T, since the resistor value is a weak function of temperature. The amplifier input-noise spectral densities are (assuming the ideal case, $F_n = 1$):

$$S_{n,1} = S_{n,2} = 4k_BT \left(\frac{2}{3g_m}\right) + \frac{K_f}{WL\hat{C}_{ox}f}$$
(16.38)

For this example, we shall assume that the op-amp input transistors have channel lengths $L = 2$ μm, channel widths $W = 30$ μm, and a gate-oxide thickness of 15 nm. We further assume that the operating point is set to provide a g_m of 300 μSiemens, corresponding to a drain current of about 100 μA in each input transistor. This is a fairly large current for the input of an op-amp. For smaller currents, the g_m would decrease, so the input-referred amplifier noise would be increased.

We assume R_F is selected to be A_0R_0, with $A_0 = 100$. Neglecting 1 compared to A_0 yields

$$S_{n,0}(f) = A_0^2 \left[4k_BT \left(R_0 + \frac{4}{3g_m} + \frac{R_0}{A_0}\right) + \frac{2K_f}{WL\hat{C}_{ox}f}\right]$$
(16.39)

We see immediately that the feedback resistor noise (the third term in parentheses) is smaller than the input resistor noise by a factor of A_0, and hence can be neglected. The white-noise part of the spectrum thus depends on the relative sizes of R_0 and g_m. For the values selected here, the term $4/3g_m$ has a value of 4.4 kΩ, and thus is 4.4 times as large as the intrinsic thermal noise of the source resistor. In order to make these values comparable, it would be necessary to increase the input-transistor drain currents by a factor of 4.4, to almost 0.5 mA. Alternatively, op-amps with bipolar transistors at their input typically have lower noise in this impedance range, and might be preferred for this application.[9]

---

[9]Chapter 11 of Gray and Meyer discusses noise in bipolar transistor circuits [88].

This example demonstrates the tight coupling between system performance and the details of the technology used to build a MEMS device. If lower noise op-amps are required and one chooses to build a monolithic integrated microsystem, the noise requirement could become the driving force for selecting a bipolar technology rather than an MOS technology to fabricate the microsystem. In fact, the integrated piezoresistive pressure sensor Case Study of Chapter 18 uses bipolar rather than MOS op-amps.

We now use numerical values for all quantities assuming a nominal value of 300 K for the temperature of the resistive noise sources and a value of $10^{-12}V^2pF$ for K_f, a value which results in a $1/f$ corner frequency of about 100 kHz. The result is

$$S_{n,0}(f) = 8.9 \times 10^{-13} + \frac{7.2 \times 10^{-8}}{f} \tag{16.40}$$

We note that not only does the amplifier contribute a factor of 4.4 times as much thermal noise as the source resistor, but the $1/f$ noise is huge by comparison. At a frequency of 1 Hz, the $1/f$ noise is almost 10^5 times the thermal noise. To demonstrate how important this is, let us calculate the mean-square noise for various assumptions about the source waveform V_S and the characteristics of the voltmeter used to measure the output.

16.6.1   Using a DC source

Let us assume V_S is a DC source, and we measure V_0 with a "DC" Voltmeter that makes a reading every second, using a single-pole averaging circuit with a 50 msec time constant. Therefore, for purposes of calculating white noise, we can use a noise bandwidth of $1/4\tau_b$, where $\tau_b = 50ms$. (Alternatively, the amplifier circuit itself might be designed to be what limits the noise bandwidth.) To calculate the $1/f$ noise, we must also take into account the lower-cutoff frequency determined by the 1-second observation time per measurement, denoted by $\hat{t}_m$.

$$\overline{v_{n,0}^2} = \frac{8.9 \times 10^{-13}}{4\tau_b} + \int_{1/\hat{t}_m}^{\infty} \frac{7.2 \times 10^{-8}}{f[1 + (2\pi f\tau_b)^2]} \, df \tag{16.41}$$

where a lower limit of $1/\hat{t}_m$ is an estimate for the lower-cutoff frequency (the result is not highly sensitive to the precise value). The integral can be evaluated.

$$\int_{1/\hat{t}_m}^{\infty} \frac{df}{f[1 + (2\pi f\tau_b)^2]} = \frac{1}{2} \ln\left[1 + \left(\frac{\hat{t}_m}{2\pi\tau_b}\right)^2\right] \approx \ln\left(\frac{\hat{t}_m}{2\pi\tau_b}\right) \tag{16.42}$$

Thus we conclude that for this example, $\overline{v_{n,0}^2}$ equals 8.3×10^{-8} V^2, corresponding to an RMS value of 290 μV. The signal-to-noise ratio is

$$S/N = \frac{|v_{0,s}|}{\sqrt{\overline{v_{n,0}^2}}} = \frac{A_0 V_S \alpha_R}{\sqrt{8.3 \times 10^{-8}}} \Delta T \qquad (16.43)$$

Therefore, if we assume $S/N = 1$ at the threshold level of detectability, the minimum detectable temperature change is calculated to be

$$\Delta T_{min} = \frac{2.88 \times 10^{-6}}{V_S} \qquad (16.44)$$

or about 3 μK for a 1-Volt excitation signal. This looks very attractive. However, to detect a temperature change this small, it is necessary to detect a resistance change of 1 part in 10^8. When we examined resistor self-heating in Section 11.6.4, we were interested in a maximum allowed temperature rise of 40 mK, which is 10^4 times as large as this calculated minimum, and even for that example, it was necessary to use currents of less than 300 μA. Therefore, resistor self-heating forces us to reduce V_S. But as we reduce V_S, the minimum detectable ΔT due to noise sources goes up. There is clearly an optimum value of V_S to use, and at this optimum, the true minimum detectable ΔT is found. Determining that optimum is a good design problem.

16.6.2   Modulation of an AC Carrier

Since $1/f$ noise dominates, it is useful to try a more elaborate scheme to detect the temperature change. Suppose we use an AC signal for V_S, and use a rectifier at the output of the amplifier followed by a DC meter with a low-pass filter, again with noise bandwidth $1/4\tau_b$. Things now change quite a lot. As an example, let use use a 1 kHz sinusoid for V_S. The "signal" now appears as sidebands on the 1 kHz carrier, confined to a bandwidth of $1/4\tau_b$ on either side. Thus, the thermal-noise bandwidth has doubled, to $1/2\tau_b$, but the $1/f$ integration limits have now been reduced to the same limits as the thermal noise, namely, a very narrow bandwidth about the 1 kHz carrier. Instead of integrating the $1/f$ noise from $1/\hat{t}_m$ to infinity, we can treat the $1/f$ part of the spectrum as effectively constant over this narrow noise bandwidth, and evaluate the total mean-square noise as:

$$\overline{v_{0,n}^2} = \frac{S_n(f)_{f=1kHz}}{2\tau_b} = 7.2 \times 10^{-10} \ V^2 \qquad (16.45)$$

The RMS noise is now only 2.7 μV, as compared with 290 μV for the DC source example. Therefore, it is possible to reduce V_S by a factor of 100, thereby reducing the self-heating driving force by a factor of about 10^4, and still maintain the same signal-to-noise ratio. Clearly, the optimum detection level will be better for this case.

16.6.3   CAUTION: Modulation Does Not Always Work

Modulation of an AC carrier only works when the $1/f$ noise is in the amplifier, not in the sensing resistor. Piezoresistors, which we will study in Chapter 18, have an intrinsic $1/f$ noise which is a *resistance fluctuation*. This is such an important issue, we treat it here with care. Assume that the resistor $R(T)$ has the following behavior:

$$R_T = R_0[1 + \eta(t)](1 + \alpha\Delta T) \qquad (16.46)$$

where $\eta(t)$ is random noise with zero mean and a $1/f$ frequency characteristic. Using the virtual-ground approximation, the current through R_T is V_S/R_T, and the contribution of V_S to the output is $-R_F V_S/R_T$. Thus the output is

$$V_0(t) = -\left(\frac{R_F V_S}{R_0}\right)\frac{1}{[1 + \eta(t)](1 + \alpha\Delta T)} \qquad (16.47)$$

Assuming the two terms in the denominator do not depart significantly from unity, we can use Taylor's theorem. Thus, to lowest order, the output is

$$V_0(t) = -\left(\frac{R_F V_S}{R_0}\right)(1 - \eta(t) - \alpha\Delta T) \qquad (16.48)$$

Note that the $\eta(t)$ term, which represents the $1/f$ noise in the resistor value, propagates through the signal path in *exactly* the same way as a temperature signal. It is part of the baseband signal. Therefore, after rectification and low-pass filtering, this component of $1/f$ noise will still be present, and we are right back to the DC-driven case.

16.7   Drifts

We now combine several ideas from very different sources:

1. It is observed that many measurements are ultimately limited by $1/f$ noise or, at least, by slow fluctuations in output. These slow fluctuations are usually called *drifts*, whether they appear as systematic and monotonic trends or as a form of very-low-frequency noise. Because the effects of $1/f$ noise get worse at low frequencies just like the observed drifts, it is common to think of drifts as having their origin in a "$1/f$-like" mechanism.

2. We have just noticed that when component values fluctuate with a $1/f$ frequency characteristic, the $1/f$ noise becomes part of the baseband signal, and cannot be eliminated with clever modulation techniques.

3. The eigenfunctions of the heat-flow equation have a dispersion relation that varies as $\sqrt{f}$, potentially leading to spatially varying thermal fluctuations

that could be responsible for component values having $1/f$ spectral density functions.

We explore this third point further. The heat-flow equation is of the form

$$\frac{\partial T(x,t)}{\partial t} = D\nabla^2 T(x,t) \tag{16.49}$$

where $T(x,t)$ is the temperature distribution and D is the thermal diffusivity. If we take the space and time Fourier transforms of this equation, using plane waves of the form $exp[j(\mathbf{k} \cdot \mathbf{r} - \omega t)]$, we obtain the following dispersion relation:

$$j\omega = Dk^2 \tag{16.50}$$

where $k = |\mathbf{k}|$, and where the allowed values of $\mathbf{k}$ are determined by boundary conditions. Thus each spatial mode allowed by the boundary conditions has a time dependence of the form

$$e^{\pm\sqrt{j\omega}t} = e^{\pm\sqrt{\omega/2}t}e^{\pm j\sqrt{\omega/2}t} \tag{16.51}$$

Every spatial mode is damped at a rate determined by ω, and has an oscillatory response with a Q of $1/\sqrt{2}$. Waveforms with this Q do not exhibit overshoot or ringing, but a disturbance that must be represented by a superposition of such spatial modes will exhibit a frequency dependence that could have $1/f$ character. As an example, we recall from the example of Section 12.10 that when the self-heated resistor was modeled with a full superposition of terms, the heat transfer function had a $1/\sqrt{\omega}$ behavior above the corner frequency of the lowest spatial mode (ω was the angular frequency of the current heating the resistor). In a full MEMS device the size of a silicon chip, the lowest spatial mode will have a very low corner frequency, so the response to thermal signals can be expected to have $1/\sqrt{\omega}$ behavior over a wide frequency range. Further, because the resistor value depends on temperature, the measured total resistance can also develop a $1/\sqrt{\omega}$ dependence in response to an external thermal excitation. Thus, any random temperature signal (for example some radiant energy) could create resistance fluctuations, and the spectral density of these fluctuations would go as $1/\omega$, since the spectral density function goes as the square of the fluctuating quantity.

This is not a proof of anything. It is a plausibility argument as to why so many phenomena appear to exhibit $1/f$ noise. Dissipative processes that obey the heat-flow or diffusion equation can be thought of as "$1/f$ responses waiting to happen." Any random external excitation, for example, a small radiant thermal signal, could manifest itself as fluctuations in component values having $1/f$ character, and, as we have seen, these fluctuations appear in the baseband and, hence, will ultimately limit achievable signal-to-noise ratio.

Related Reading

P. R. Gray and R. G. Meyer, *Analog Integrated Circuits*, Second Edition, New York: Wiley, 1984.

M. Schwartz, *Information Transmission, Modulation, and Noise*,Second Edition, New York: McGraw Hill, 1970.

S. D. Senturia and B. D. Wedlock, *Electronic Circuits and Applications*, New York: Wiley, 1975; reprinted by Krieger Publishing.

W. McC. Siebert, *Circuits, Signals, and Systems*, Cambridge, MA: MIT Press, 1986.

A. van der Ziel, *Noise: Sources, Characterization, Measurement*, Englewood Cliffs, NJ: Prentice-Hall, 1970.

Problems

16.1 When using resistors for sensing applications, the Wheatstone bridge circuit below is often used. R_1 and R_2 have the same nominal value of R_0, but have different variations with respect to the variable being measured. For example, for temperature measurement, the R_1 resistors might be located on the object whose temperature is being measured, while R_2 would be located on the substrate at a reference temperature. For strain measurement (discussed in Section 18.2.4), the different piezoresistors would be positioned so that they responded differently to the structural deformation. In either case, the R_1 and R_2 values will differ slightly.

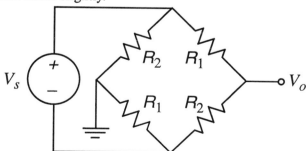

Assuming that the bridge output V_0 is connected to the input of a high-impedance amplifier, effectively an open circuit, show that the equivalent circuit for this bridge is a suitably sized voltage source that depends on R_2 - R_1 in series with a resistance of value R_0 and a single Johnson noise source with spectral density $4k_BTR_0$.

16.2 A microaccelerometer with a proof mass of 300 ng and a resonant frequency of 25 kHz is critically damped and is in thermal equilibrium at room temperature. What is the root mean-square amplitude of the fluctuation of the position of the proof mass?

16.3 A flux of particles is diffusing in a linear concentration gradient. By analogy to shot noise, develop an expression for the spectral density of the noise component of the diffusion current.

16.4 Combine the self-heated resistor example of Section 11.6.4 with the noise calculation from the DC temperature measurement example of Section 16.6.1 and find the optimum value of V_s and the corresponding minimal detectable temperature difference.

16.5 Repeat Problem 16.4 for the AC modulated temperature measurement example of Section 16.6.2.

V
CASE STUDIES

Chapter 17

PACKAGING

In MEMS devices, especially, the package and the die are inseparable.
—D. J. Monk and M. K. Shah

17.1   Introduction to the Case Studies

This part of the book presents a series of Case Studies that build on the material presented so far. Along the way, some new domain-specific material is presented: piezoresistance in Chapter 18, capacitance-measuring circuits in Chapter 19, optical diffraction in Chapter 20, piezoelectricity in Chapter 21, and some DNA fundamentals in Chapter 22. But our main purpose is to integrate the many ideas already introduced by focusing on specific examples.

Each Case-Study Chapter examines only one or two major examples. The risk of selecting so few is that the reader may feel the author is implicitly "endorsing" one particular approach as the "right" one. But that is not the case. The criteria for selecting the particular mix of examples were (1) to cover a variety of sensing and actuation methods, (2) to explore different realms of device application, and (3) to draw, as much as possible, on *real commercial products*. In each case, there are viable competitive approaches, but there is not enough space in a book like this to explore multiple examples in detail. The reader is urged to consult more general survey references, such as Kovacs [9], to see a more complete spectrum of ways in which different problems have been approached.

We begin the discussion of Case Studies with three "back-end" problems that affect every design: packaging, test, and calibration.

17.2   Packaging, Test, and Calibration

In Chapter 2, we suggested that packaging and package design must be closely coupled with system and device design. In this chapter, we present some "common-sense" guidelines for package design, and then illustrate some of the problems with the packaging of an automotive pressure sensor.

Packaging of microsensors and microactuators differs significantly from its microelectronic counterpart. In microelectronics, it is possible for a single package type to be used for many different types of chips as long as the chip has the appropriate size and the right number of wire bonds. The goal of the package is to protect the chip from all outside influences, provide for electrical connection at pins or solder bumps, and provide a heat-flow path for heat generated by on-chip power dissipation. The detailed electronic function of the chip is not important. Exchange of information on die size and pin-out requirements, power dissipation by the chip, allowed chip temperature during operation, and, possibly, allowable levels of stress at the chip level produced by the package is sufficient to permit package designers to design packages. Package standards have been developed to support this communication between package designers and device designers.

In MEMS, the situation is very different. The detailed function of the MEMS chip is critical to the design of the package. For example, accelerometers do not require direct coupling to the outside media, but pressure sensors do require such coupling. And while accelerometers might work better at close to atmospheric pressure, resonant devices such as rate gyros might require vacuum packaging. Furthermore, instead of one package design serving many different chip designs, it is quite common for a single chip design, such as a bulk micromachined pressure sensor, to be used with many different package types to support various customer needs, fragmenting the packaging market into many tiny markets. This makes the development of package standards almost impossible. Furthermore, this market fragmentation tends to increase the cost of the package relative to the device, in some cases resulting in a package that is noticeably more expensive than the device it is enclosing. For example, the cost of a packaged and calibrated integrated silicon pressure sensor, a relatively mature device technology, breaks down as follows: 35% for the silicon chip containing the mechanical element and the bipolar or CMOS circuitry, 45% for the package, and 20% for calibration and test [89].

The problems of device calibration and test are closely linked with packaging. Obviously, any sensor or actuator whose *quantitative* performance is important must be tested and possibly adjusted or trimmed so that its performance meets specifications on accuracy and sensitivity over the operating range of temperatures. A microelectronic circuit that only performs an electronic function can be tested at the wafer level, *prior to die separation and packaging*, using wafer-probe stations and automated electronic test equipment. Of course it must be tested again, after packaging, but at least bad die identified at the wafer level can be removed from the packaging flow. The situation is different with MEMS devices, whether fabricated as monolithic or hybrid devices. While the electrical functionality of MEMS devices can be tested at the wafer level, it is much more difficult to arrange for application of stimuli such as pressure or acceleration at the wafer level. Therefore, unless the packaging functions can be brought to the wafer level, testing and calibration must be performed at the die level, after die separation, and after at least some of the packaging has been completed.

The difficulty of testing prior to packaging has serious implications for the organization of the MEMS industry. Microelectronic foundries, which can do their testing prior to packaging, can deliver what are called *known good die*, ready for packaging, to their customers. The customers can commit a package to each die with confidence that the die works. In contrast, and by the very nature of the problem, MEMS foundries cannot do this unless the foundry operation includes enough packaging to permit testing of mechanical as well as electrical functionality. Without access to known good die, one must commit expensive packaging steps to devices that may not be good, with the risk of increasing costs for scrapped material.

A question: *If the package, test, and calibration are so important and so expensive, why aren't they a more prominent part of the MEMS conferences and publications?* The answer is interesting: It is *not* because there are not significant intellectual and engineering challenges. Quite the contrary. Rather, it is this author's experience that packaging, test, and calibration are intrinsically industrial and commercial fields; it is difficult for university researchers, the group that publishes a lot, to engage the real intellectual core of these subjects because they depend on having access to large quantities of manufactured devices and to the physical facilities with which to do manufacturing-level packaging and trimming operations. And since successful execution of packaging, test, and calibration can determine the commercial success or failure of an entire product line, detailed information critical to these operations is typically kept extremely private by manufacturers. Publications from industry tend to be descriptive rather than detailed, and the minute details that make the difference between success and failure tend to be treated as trade secrets.

Can we still make progress with this subject? The answer is "yes." There are some good guidelines for thinking about package design, and there are a few well-documented discussions of packaging and calibration procedures for a few devices [90, 91, 92].

17.3   An Approach to Packaging

This section must begin with a disclaimer. It is extremely difficult to prove that the procedures and guidelines we present for designing MEMS devices and their packages will guarantee success. But in providing a way of thinking about the subject, we create a structure within which to argue out the many choices that must be made, and we raise some warning flags about issues that can be overlooked until one is too far down the development path to permit easy fixes. Figure 17.1 summarizes the steps.

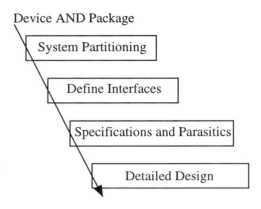

Figure 17.1.   Schematic flow-chart for device and package design. In practice, each level is visited iteratively until the designs converge.

The figure suggests a one-way flow, but in practice, thinking simultaneously about the design problem at each of the various levels is typically necessary, requiring in a highly iterative design process. At the very outset, alternative device concepts must be assessed using some combination of modeling and prototyping. In order to do either, one must move rather far down the design path. Results from early work either guide the path to refinement of the design, or may force a major change in device or package concept. The goal is that the device/package design iteration not become an endless loop. Development time is expensive, and there is an opportunity cost due to lost sales from products that are late to market. The most important guideline toward avoiding such loops is stated below:

Design the MEMS Device AND the Package at the Same Time

This is easier said than done. In many organizations, the persons who do device and process design are different from those who are responsible for package design. In some cases, they are in completely different parts of the organization in different locations, possibly even different companies. Nevertheless, because the MEMS device function directly affects how it is packaged, this cooperative and concurrent design is essential.

Partition the System Wisely

System partitioning refers to the decision as to how much of the system electronics is built as part of the MEMS device. The ultimate driving force for this partitioning decision is the customer's total system cost. It might be true that a MEMS device with no integrated electronics plus an ordinary printed circuit board containing the electronics is the lowest-cost and best solution for one customer. A different customer might have technical requirements that make a fully integrated monolithic device the lowest-cost solution. But cost data are hard to acquire before one does some amount of manufacturing, at least at the prototype level. Before one does manufacturing, one must make a design, and this design requires a partitioning decision. So the partitioning decision, which may have a huge impact on cost, must be made quite early in the design cycle.

In the absence of a customer requirement for a specific level of integration, *this author recommends that the amount of electronic functionality integrated monolithically with the MEMS device be strictly minimized.* This recommendation is not an absolute, in the sense that no monolithically integrated device could ever be expected to succeed. Indeed, the Case-Study example to be presented later in this chapter is a monolithically integrated pressure sensor. However, this "minimalist" recommendation forces the discussion of integration into the correct context, which is determining the relative costs and customer benefits of various partitionings.

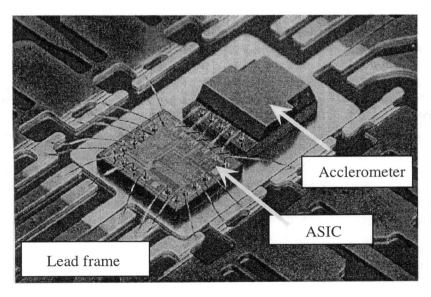

Figure 17.2. Illustrating a two-chip hybrid approach to system partitioning. The accelerometer contains no electronics. All of the electronics is on the ASIC chip. (Source: Motorola; reprinted with permission).

Figure 17.2 shows one approach to the partitioning problem: use two chips. One is the MEMS device, in this case, an accelerometer. The other is a custom integrated circuit (an "application-specific integrated circuit," or ASIC) that contains the required signal-conditioning, calibration, and output circuitry. Both chips are attached to the same lead frame. The accelerometer is wire bonded to the ASIC, and the ASIC is wire bonded to the lead frame. Then the entire assembly is molded into a plastic case for protection.

In contrast, the monolithic pressure sensor presented later in this chapter and in Chapter 18 integrates the electronics with the mechanical diaphragm. Such integration adds complexity to the basic device process and decreases the flexibility of the designers of both the MEMS components and the electronics. Integration also significantly raises the front-end cost to get through a development cycle because both the MEMS device and the circuitry must be built in a common process and a change in either requires starting both over again. However, integration does potentially reduce overall system size and can improve interconnect reliability. Because companies tend not to publish their detailed cost data, the question of whether integration ultimately reduces cost must ultimately be decided in the marketplace. Generally, though, one would expect fully integrated products only in very large markets because the manufacturer is better able to amortize the higher development cost over the life of the product.

Obviously, the decision to build a system as a monolithically integrated device or as a multi-device hybrid has a huge effect on packaging. The partitioning decision reached by any particular organization is, ultimately, a *judgment* based on a combination of technological and business experience and on some kind of business model that suggests a higher likelihood of commercial success for the path selected. Questions of customer requirements, manufacturing yields, and costs for calibration and test enter into this discussion. There are no general answers, but there are specific examples in the marketplace.

Define System Interfaces

Once the partitioning decision is made, it is possible to specify the various interfaces that the system architecture must support. For the MEMS component, one can identify how the transducer is be appropriately connected to the transduced quantity and identify the environmental issues that the device will face: the chemical constituents to be encountered and the temperature, pressure and stress ranges. With these in mind, one can assess possible materials for the device construction and the associated package, their chemical and thermal stability, mechanical properties (such as fracture strength and thermal expansion), and, where appropriate, biocompatibility. For the part of the system that is external to the MEMS device, one can create a functional description of the circuit requirements, including the number of electrical interconnects needed and their electrical-impedance requirements, in case it is necessary to drive long connecting cables between the MEMS device and the host electronics. At this phase, it is also useful to begin constructing a list of parasitic effects that can arise: manufacturing tolerances creating variations in device dimensions or material properties; parasitic resistances, capacitances, and inductances created by interconnect; effects of stress from packaging on device performance; effects of temperature on device and system performance; risk of corrosion and its effects.

Design Specifications

The next step is to create formal specifications for the device, both performance specifications and specifications that constrain the device and package design. Examples of performance specifications include the required performance over temperature, the operating environment, and reliability issues such as electromagnetic interference or shock survivability resulting from dropping the device. Examples of the design specifications include an overall device and package concept, complete with choices of materials, a plausible fabrication and assembly process for both the device and the package, and details of the package-device interface that are critical, such as the number and location of electrical connections and the way that the device must couple to the external environment for its operation as a transducer. At this point, it is also essential to provide for test and calibration. This includes defining a test procedure for sensor acceptance prior to packaging, a procedure for accepting a package prior to using it for packaging, checkpoints during assembly to assure satisfactory performance, and provision for calibration and final test. If these are not thought through prior to committing to a detailed design, very costly revisions may be required later.

Detailed Design

The final step is the detailed device design, consisting of a fabrication process and the artwork for masks. Developing such a design requires modeling and may also require some short-loop fabrication experiments to verify that a proposed structural element can be built with acceptable tolerances. For the package, the design consists of a fabrication process and drawings for the required tooling and for the package itself. The third element in the detailed design is a process flow for assembly, including the acceptance procedure for the device and package, test procedures, and calibration procedures.

17.4   A Commercial Pressure-Sensor Case Study

We have outlined a model process of design conception and refinement. It would be wonderful if one could document exactly what actual design process resulted in a particular commercial device. Unfortunately, because the details of the iterations needed to reach commercial success are highly proprietary, we can only view the end result, often with critical details quietly covered. Nevertheless, one can learn a lot by examining examples. The Motorola Manifold-Absolute-Pressure (MAP) Sensor, designed for the automotive market, provides a good Case Study [92, 93].[1] The purpose of measuring the absolute pressure in the intake manifold of an automobile engine is to calculate the mass airflow into the engine, one of the variables needed by the engine control system to control the air-fuel ratio. In this chapter, we examine the packaging of the MAP sensor. In Chapter 18, we will examine the mechanics of the sensor itself and also the details of the signal-conditioning and calibration circuit.

Table 17.1.   Manifold-Absolute-Pressure-sensor requirements. Entries show range of specifications across several different products. (Source: Motorola Sensor Device Data, 1998).

Property	Specification
Transfer function	$V_{out} = (K_1 P - K_2)V_S$ $\pm$ error (K_1 and K_2 are constants)
Pressure range	15 - 105 kPa (some devices to 250 kPa)
Ratiometricity	1% $\pm$ 0.5% output change for a 1% change in V_S
Power supply	5 V $\pm$ regulation spec.
Output current drive	Sink 0.08 - 1 mA; Source .2 to 5 mA, depending on device
Thermal cycle (unpowered)	200 to 1000 cycles, -40/125°C, 60 min/cycle
Hot storage (powered)	125°C, 100-1000 hours
Hot storage (unpowered)	125°C, 500-1000 hours
Cold storage (unpowered)	-40°C, 96-1000 hours
Humidity (powered)	96-1000 hours, 60 to 85°C, 85 to 90% RH
Humidity (unpowered)	96 hours, 121°C, 100% RH (autoclave)
Drop	1-5 drops of 1 meter
Mechanical shock	$5g$-$100g$ pulses of 10 msec duration
Electromagnetic compatibility	50-200 V/m, 1 - 1000 MHz

Table 17.1 illustrates the kinds of specifications that a commercial MAP sensor must meet. The specification entries show ranges that cover a variety of individual MAP products and are intended to illustrate the types of specifications that must be addressed by the device and package designer. Each individual product has specific requirements for each entry in this table and also has additional specifications describing the "error" allowed in the transfer function and other performance details. Not included in the table are external exposure to automotive fluids, such as gasoline, brake fluid and motor oil, and field-use hazards such as dust.

---

[1]This Case Study is based on materials and commentary provided by David Monk of the Motorola Sensor Products Division.

17.4.1 Device Concept

The Motorola pressure-sensor family consists of bulk micromachined silicon diaphragms sensed with piezoresistance. Figure 17.3 illustrates the device concept.

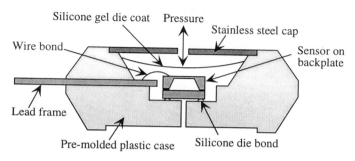

Figure 17.3. The device concept: a schematic cross-section of the MAP sensor and first-level package. (Redrawn after [92].)

An absolute pressure sensor requires a backplate to seal the space beneath the diaphragm under vacuum conditions. This backplate also strengthens the sensor structure, for ease of handling. Figure 17.3 shows a silicon die mounted on its backplate which is die-bonded with a low-stress silicone gel into a pre-molded plastic package, wire-bonded to a lead frame (with up to 8 pins), covered with a silicone gel for protection against the ambient, and finally covered with a stainless-steel cover with a pressure access port. Figure 17.4 shows a device photograph.

17.4.2 System Partitioning

Motorola's system partitioning decision, in this case, was to include signal-conditioning circuitry and trimmable calibration resistors on the same silicon die as the mechanical diaphragm. Compared to industry practice, this was an unusual decision. Most pressure-sensor manufacturers provide uncompensated sensors that require external circuitry to achieve temperature compensation, signal conditioning, and calibration. The advantages of the integrated approach are a smaller overall device and improved interconnect reliability as a result of the integration of the electronics. It is also possible that the integration improves the overall electromagnetic compatibility of the device by placing all of the signal-conditioning circuitry in a small space close to a grounded substrate. Finally, moving required amplifiers from the customer's circuit board to the MEMS device may lower the customer's system costs. In choosing a fully integrated sensor, Motorola made a business decision that the high development cost of the integrated sensor would ultimately be recovered from the commercial success of the product. It is interesting that Motorola did

Figure 17.4. Photograph of a six-pin version of a MAP sensor prior to final gel encapsulation. (Source: Motorola; reprinted with permission.)

limit the amount of electronics placed on the chip. Only the temperature-compensation and calibration circuits are placed on the chip; EMI filters and buffer amplifiers needed to drive low-impedance loads are added externally when needed.

17.4.3   Interfaces

Once the partitioning decision has been made, the electrical interface can be defined. The on-chip circuitry requires only three leads when the device is used: power, ground, and output. However, in order to calibrate the device, up to eight different contacts must be made to the on-chip circuitry. A critical decision concerns when to calibrate: before or after packaging? Because packaging can introduce stresses that might shift the calibration, Motorola decided to perform the calibration after the die has been mounted into the package and wire bonded. This means that the package must have extra pins to support the calibration. This is a good example of how system-level issues, such as calibration, affect both the device and package design.

The media interface (the pressure port) provides coupling between the diaphragm on the sensor die and the outside world. The use of a metal cap which is inserted after die attach, wire bonding and calibration serves both to protect the die and to provide pressure access.[2] However, corrosive chemicals in the external medium might destroy the wire bonds or compromise the on-chip cir-

---

[2]The stainless steel cover also provides a good surface for labelling each part [89].

cuitry. Therefore, after calibration, the die is covered with a protective silicone gel. The gel is a low-modulus rubber, with a Poisson ratio of very close to 1/2. It transfers the external pressure to the surface of the diaphragm as if it were an incompressible liquid. Because this gel must be applied *after calibration*, the effect of the gel on the calibration must be determined in advance, and the calibration targets must be pre-compensated so that devices meet the specifications after gel application.

17.4.4 Details

We describe here some of the details of the packaging as described by Monk and Shah [93] and in Motorola's publications [92].

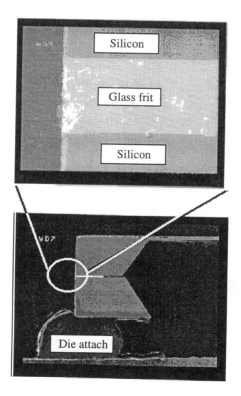

Figure 17.5. Cross-section and close-up of glass frit used to bond sensor wafer to support wafer. The support, in this case, is etched through, creating a differential pressure sensor. The die-attach is visible in the lower photograph. (Source: Motorola; edited and reprinted with permission.)

Wafer-level packaging

After the diaphragm sensor chip has been fabricated, it is bonded to its backplate at the wafer level (prior to die separation). To reduce thermal-expansion-mismatch stresses, a silicon backplate was selected. But because the sensor contains integrated circuits, high-temperature silicon fusion bonding could not be used. Anodic bonding to glass, which is used successfully by many pressure-sensor manufacturers (including Motorola) offers an alternative. However, in this particular case, a glass-frit bond was used (see Fig. 17.5) because it was more forgiving of deviations from perfect wafer flatness than anodic bonding. The challenge was to achieve a glass-frit seal, fired at between 450 and 500°C, that is void free and does not introduce unacceptable stress into the sensor. Further, because the sensor die includes electronic circuits, the glass frit must meet higher standards of purity than would be required if only the passive sensor were involved.

Plastic molding of the housing

The earliest packages used for this application were injection molded thermoplastics such as polyesters. However, difficulties were encountered both with the resistance of these materials to hot-wet conditions and with the development of leaks at the interface between the lead frame and the plastic housing. Epoxy materials can be tailored so that their thermal expansion is reasonably well matched to that of the lead frame, and epoxies can form a chemical bond with suitably treated metal surfaces. However, there is a large tooling cost associated with the transfer molding of epoxy cases, and because of the high pressures needed during molding, additional pins are required to hold the lead frame in position prior to hardening of the epoxy. The cavity created by these pins must then be backfilled with epoxy in a separate step. The interface between the original epoxy and the backfill epoxy represents a potential leak hazard. While transfer-molded epoxy packages have been used extensively, more recently Motorola has moved back to lower-cost injection-molded packages made from PPS (poly(phenylene sulfide)), a high-temperature thermoplastic [89].

Die attach

The separated die are placed on a square pad of silicone gel which attaches the die to the package (see Figs. 17.4 and 17.5). While RTV ("room-temperature-Vulcanized", or room-temperature "cured") silicones were used extensively, Motorola now uses a gel-like silicone or flourosilicone that requires a 150°C oven cure [89]. Being a soft, rubbery material, this die-attach adhesive does not transfer large stresses from the package to the die. A significant problem in the use of these materials in a manufacturing environment is the lot-to-lot variation in the materials themselves. (This problem is also encountered in the materials used for the gel encapsulant.)

Wire bond

The soft platform on which the die sits poses a challenge for wire bonding, which typically requires coupling of ultrasonic energy into the bonding region. Furthermore, the heat transfer through the die attach is relatively poor, making thermosonic bonding difficult to control. Motorola has developed fixturing that uses the lead frame to assist in conducting heat into the package during wire bonding. Optimization of the ultrasonic bonding operation is done empirically [89].

Trim technology

The on-chip resistors whose values are trimmed to achieve calibration and temperature compensation are made of CrSi alloy, a material with a very low temperature coefficient of resistance. These resistors can be cut by a laser to modify their resistance by as much as 300%, providing a wide range of adjustment. However, laser trimming is an expensive step, and it is a "bottleneck" type of manufacturing operation, requiring serial access to each resistor. And because laser trimming generates heat, trimming must take place before final device encapsulation, which makes the device vulnerable to small shifts in calibration produced by post-trimming stresses.

The fact that on-chip laser trimming is required is one of the consequences of the original partitioning decision to place the circuitry on the same die as the sensor. Manufacturers who use a hybrid approach, separating the sensor from the electronics, have greater flexibility in how the calibration is achieved. Until recently, though, laser trimming was the industry standard, at least for automotive pressure sensors. More recently, manufacturers have been experimenting with the use of electrically-programmable read-only memories (EPROM's) that can be built into an ASIC chip containing the signal-conditioning and calibration circuitry. The circuitry is designed to use digital information stored in the EPROM to implement the calibration. The chip is mounted on the same lead frame as the sensor, and the two are encapsulated together. Calibration is performed after encapsulation, with the calibration information being written into the EPROM by signals provided at the terminals of the fully-packaged device.

Silicone gel

The silicone gel used to provide sensor protection has the same problems as the die attach: material selection and lot-to-lot variation. The gel must adhere to a wide range of materials, have a low modulus, and provide passivation against corrosion. As stated earlier, this gel coat can create calibration shifts. When using laser trimming for calibration, the calibration targets must be compensated to account for this shift. One specific advantage of using MOS circuits that contain EPROM's instead of bipolar circuits that require laser trimming is that calibration can performed after all packaging operations, including the application of the gel coat, have been completed. However, as we learned in Chapter 16, MOS amplifiers have higher noise levels at low frequencies when amplifying signals from low-impedance sources. Thus we see that even minute details of assembly, such as whether one can calibrate before or after the application of a protective gel, interact with global choices about device technology.

Electrical connector

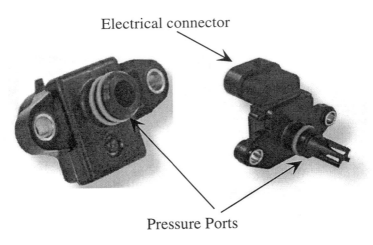

Pressure Ports

Figure 17.6.   Two different next-level packages for MAP sensors. (Source: Motorola; edited and reprinted with permission.)

Next-level assembly

In order for the MAP sensor to go into a car, a next-level assembly is required, typically a circuit board that holds the chip and a molded plastic housing to allow coupling to the automobile intake manifold or the attachment of a vacuum hose. The housing also provides ease of handling by mechanics and general protection against damage. Two examples are shown in Fig. 17.6. The pressure-access ports are very different in the two packages. The electrical connector can be seen only in the right-hand example, but a similar connector is present on the left-hand example. This illustrates the market fragmentation that occurs when customer requirements create needs for multiple package types for a single product. This fragmentation has driven Motorola toward an assembly method referred to as "mass customization," the ability to operate a standard assembly process for several products, even though the package outline changes from customer to customer [89].

The circuit board is not shown in Fig. 17.6. The sensor is mounted on the circuit board with solder. Details include what solder composition to use and how much gold to pre-flash onto the lead frame to promote soldering. Furthermore, after soldering and cleaning, the circuit board is itself sprayed with a protective coating. The pressure interface must be designed so that the mounting, soldering, cleaning, and passivating operations can be done in a highly automated fashion without introducing contamination.

Electrical and Reliability Testing

Each device must be subjected to electrical test, pressure test and temperature tests. One interesting challenge is to verify that the electrical test equipment is itself properly calibrated. So-called "golden" units, whose characteristics are carefully documented, are used to cross-check various electrical-test stations.

Reliability testing (hot storage, cold storage, hot-wet conditions, shock tests, drop tests, exposure to various media) is not performed on every unit, but it can still be a bottleneck because some of the tests take a long time. A standard distribution of characteristics from manufactured devices is developed, after which each lot of devices is sampled to assure that it meets specifications.

17.4.5   A Final Comment

We close this discussion, and this Chapter, with a quote from Monk and Shah: "The last point that we would like to emphasize is that in MEMS devices, especially, the package and the die are inseparable. That is to say, the package affects the electrical output of the device, and, in many cases, the die affects the needed packaging. Therefore, a collaboration between "front-end" and "back-end" engineering is necessary to build products that are optimized for high volume manufacturing." It is hoped that the example presented here underscores this message.

Related Reading

ASM International, Handbook Committee, M. L. Minges (technical chairman), *Electronic Materials Handbook: Volume 1, Packaging*, Materials Park, OH: ASM International, 1989.

N. Maluf, *An Introduction to Microelectromechanical Systems Engineering*, Boston: Artech House, 2000, Chapter 6.

D. J. Monk and M. K. Shah, Packaging and testing considerations for commercialization of bulk micromachined piezoresistive pressure sensors, *Proc. Commercialization of Microsystems '96*, Kona, HA, October 1996, pp. 136-149.

L. Ristic (ed.), *Sensor Technology and Devices*, Boston: Artech House, 1994, Chapter 6.

R. R. Tummala and E. J. Rymaszewski (eds.), *Microelectronics Packaging Handbook*, New York: Van Nostrand Reinhold, 1989.

Problems

17.1 A Packaging Treasure Hunt: Go to the literature (either journals or commercial web sites) and find an example of a pressure sensor other than the Motorola MAP sensor. Identify the system partitioning used, the device concept, and key details that affect either the package design or the choice of device fabrication technology.

17.2 Another Treasure Hunt: Repeat Problem 17.1 for either an accelerometer, an optical MEMS device, a rate gyroscope, a microfluidic component or bioMEMS device, or a chemical sensor, selecting devices other than those covered in Chapters 18-23.

Chapter 18

A PIEZORESISTIVE PRESSURE SENSOR

I've measured it from side to side:
'Tis three feet long and two feet wide.

—Wordsworth, in *The Thorn*

18.1   Sensing Pressure

Pressure measurement is a key part of many systems, both commercial and industrial. Silicon has proved to be an astonishingly good material from which to build small pressure sensors. They are millimeter-sized, somewhat smaller than Wordsworth's objects described above. Pressure sensors presently constitute the largest market segment of mechanical MEMS devices.

Since pressure is a normal stress (force per unit area), one could imagine sensing pressure directly by using a piezoelectric material which can transduce normal stress into voltage.[1] Alternatively, one could apply the pressure to one side of a deformable diaphragm, a reference pressure to the other side, and determine how much the diaphragm deforms. This latter approach is by far the dominant one, in both macrofabricated and microfabricated pressure sensors.

There are several ways of sensing the deformation of a diaphragm across which a differential pressure has been applied. The most obvious is to determine the displacement of the diaphragm using either capacitance change, some optical signature, or even the change in current in a tunneling tip. These *position-measuring methods* are examined in Chapter 19 in the context of acceleration measurement, but they apply equally to pressure sensing.

An indirect but very powerful way to sense the deformation is to measure the bending strain in the diaphragm. Silicon has the property of *piezoresistance*, a

---

[1]Piezoelectric materials are discussed in Chapter 21.

change in resistance with stress (or strain). It is a material that is ideally suited to this type of device. The Case-Study example in this chapter is a silicon piezoresistive pressure sensor.

Another indirect method is to create a resonant structure, either of silicon or quartz, and couple the resonator to the diaphragm in such a way that the diaphragm displacement creates a changing stress in the resonant structure, thereby shifting its resonant frequency. Resonant measurement devices are discussed in Chapter 21 in the context of a rate-gyroscope, but resonant methods can also be applied to other mechanical measurements such as pressure and acceleration.

The specific device selected for the Case Study is the same Motorola manifold-absolute-pressure (MAP) sensor already discussed as an example of packaging in Chapter 17. In this chapter, we examine how the device is fabricated, how it works, and how the signal-conditioning and calibration electronics are organized and used. At the end of the chapter, we examine a recently announced set of device improvements which provide excellent insight into the many interactions among device fabrication, calibration and trim, and packaging. As a first step, though, we must learn about piezoresistance.

18.2   Piezoresistance

Piezoresistivity is the dependence of electrical resistivity on strain. The resistivity of a material depends on the internal atom positions and their motions. Strains change these arrangements and, hence, the resistivity. Historically, the quantitative formulation of the piezoresistive effect has been in terms of stress rather than strain, which is the origin of the *piezo* prefix (from the Greek *peizin*, to press [94]).

The electronic states of a material depend on the atomic constituents and on their positions. In a crystalline material, these states form quasi-continua in energy called *energy bands* and are filled according to the requirements of the Pauli exclusion principle to a highest filled level. In metals, this highest filled level occurs in the middle of a band, resulting in a large number of empty states lying adjacent in energy to the highest filled states. Application of an electric field slightly shifts the occupancy of these levels, favoring carriers moving in the direction of the field, resulting in a current.[2] Changing the internal atomic positions by applying stresses to the metal distorts the energy bands slightly, resulting in small changes in the amount of conduction that results from an applied field. This is the piezoresistive effect at its simplest. Metal-film strain

---

[2]Whether carriers move "along" or "opposite" to the applied field direction depends in detail on the energy bands. Generally, the electrons behave as negatively charged particles, hence move on average in a direction opposite to the electric field.

gauges that are bonded onto components in which it is desired to measure strain are based on this principle.

In a semiconductor, the highest filled level nominally occurs at the edge of the valence band, and the next available band, called the conduction band is empty. However, as discussed in Chapter 14, a semiconductor in thermal equilibrium has some carriers on the conduction band, called "conduction electrons" which behave as negatively charged carriers, and also some vacancies in the valence band, which result in positively charged carriers called "holes." The numbers of each type of carrier depend sensitively on the doping and the band gap, which is the difference in energy between the highest valence-band energy and the lowest conduction-band energy. When the internal atomic positions in a semiconductor are changed by the application of a stress, these band-edge energies move by small amounts. But even small shifts can have enormous effects on the conductivity properties. A detailed discussion of the mechanism in terms of the detailed energy bands in semiconductors is well beyond the scope of this text; suffice it say that when the crystalline lattice is strained, conductivity is enhanced in some directions and reduced in others. When expressed as a percentage of the original conductivity, the effect can be considerably larger than in metals. It is not surprising, therefore, that piezoresistance in semiconductors provides an excellent strain-measurement method. However, it proves difficult to make transferable thin-film strain gauges that can be bonded to other surfaces. Instead, one builds the piezoresistive strain gage directly into the structure of interest. This was the motivation for much of the early work in the micromachining of microsensors. If the sense element cannot not easily be removed from the silicon wafer, then make the entire device structure, with sense element included, out of the same silicon wafer! Many very successful devices, especially silicon pressure sensors, exploit this approach.

18.2.1   Analytic Formulation in Cubic Materials

The general formulation of the piezoresistivity of a material can introduce a dizzying level of complexity. To begin with, the resistivity must be formulated as a second-rank tensor, coupling an electric field to a current density. Furthermore, stress is itself a second rank tensor, so the piezoresistive effect requires a fourth-rank tensor for its full description.

Assuming that the piezoresistive effect is linear (which is true for small strains but can fail at sufficiently large strains [95]), the relationship between electric field and current density is written

$$\mathcal{E} = [\rho_e + \Pi \cdot \sigma] \cdot \mathbf{J} \tag{18.1}$$

where ρ_e is the resistivity tensor, Π is the full fourth-rank piezoresistive tensor having dimensions of a resistivity divided by a stress, σ is the full second-rank stress tensor, and J is the current density.

In a cubic material, this can be considerably simplified. First, the resistivity tensor is diagonal, with all three diagonal entries equal. Thus, in a cubic material, the resistivity is a scalar ρ_e. Further, just as we did with the elastic constants,[3] we can write the components of the stress tensor as a six-element array, with the index pairs (ij) being reassigned according to

$$[11 \ , \ 22 \ , \ 33 \ , \ 23 \ , \ 31 \ , \ 12] \leftrightarrow [1 \ , \ 2 \ , \ 3 \ , \ 4 \ , \ 5 \ , \ 6] \tag{18.2}$$

With this notation, using cubic symmetry for the piezoresistive coefficients and using the intrinsic symmetry of the stress tensor, the resulting field-current relationship can be written

$$\frac{\mathcal{E}_1}{\rho_e} = [1 + \pi_{11}\sigma_1 + \pi_{12}(\sigma_2 + \sigma_3)] J_1 + \pi_{44}(\tau_{12}J_2 + \tau_{13}J_3) \tag{18.3}$$

$$\frac{\mathcal{E}_2}{\rho_e} = [1 + \pi_{11}\sigma_2 + \pi_{12}(\sigma_1 + \sigma_3)] J_2 + \pi_{44}(\tau_{12}J_1 + \tau_{23}J_3) \tag{18.4}$$

$$\frac{\mathcal{E}_3}{\rho_e} = [1 + \pi_{11}\sigma_3 + \pi_{12}(\sigma_1 + \sigma_2)] J_3 + \pi_{44}(\tau_{13}J_1 + \tau_{23}J_2) \tag{18.5}$$

where the three independent piezoresistive coefficients are

$$\begin{aligned} \rho_e\pi_{11} &= \Pi_{1111} \\ \rho_e\pi_{12} &= \Pi_{1122} \\ \rho_e\pi_{44} &= 2\Pi_{2323} \end{aligned} \tag{18.6}$$

Note that the π_{IJ} coefficients have the unstrained resistivity factored out, so the units of the π_{IJ} coefficients are inverse Pascals.

18.2.2   Longitudinal and Transverse Piezoresistance

If a relatively long, relatively narrow resistor is defined in a planar structure, for example, by ion implantation followed by diffusion, then the primary current density and electric field are both along the long axis of the resistor. This axis need not coincide with the cubic crystal axes. Therefore, it is necessary to know how to transform the piezoresistive equations to an arbitrary coordinate system. The structures are typically designed so that one of the axes of principal in-plane stress is also along the resistor axis. This permits a simplification of the piezoresistive formulation to the form

---

[3] See Section 8.2.7.

$$\frac{\Delta R}{R} = \pi_l \sigma_l + \pi_t \sigma_t \tag{18.7}$$

where R is the resistance of the resistor, and the subscripts l and t refer to longitudinal and transverse stresses with respect to the resistor axis.

The general expressions for π_l and π_t are obtained by applying coordinate transforms to the original full tensors [96]. The results are [97]

$$\pi_l = \pi_{11} - 2\left(\pi_{11} - \pi_{12} - \pi_{44}\right)\left(l_1^2 m_1^2 + l_1^2 n_1^2 + m_1^2 n_1^2\right) \tag{18.8}$$

and

$$\pi_t = \pi_{12} + \left(\pi_{11} - \pi_{12} - \pi_{44}\right)\left(l_1^2 l_2^2 + m_1^2 m_2^2 + n_1^2 n_2^2\right) \tag{18.9}$$

where (l_1, m_1, n_1) and (l_2, m_2, n_2) are the sets of direction cosines between the longitudinal resistor direction (subscript 1) and the crystal axis, and between the transverse resistor direction (subscript 2) and the crystal axes.

In many silicon micromachined devices, resistors are oriented along [110] directions in (100) wafers.[4] The longitudinal direction cosines are $(1/\sqrt{2}, 1/\sqrt{2}, 0)$ and the transverse direction cosines are $(-1/\sqrt{2}, 1/\sqrt{2}, 0)$. This results in

$$\pi_{l,110} = \frac{1}{2}\left(\pi_{11} + \pi_{12} + \pi_{44}\right) \tag{18.10}$$

and

$$\pi_{t,110} = \frac{1}{2}\left(\pi_{11} + \pi_{12} - \pi_{44}\right) \tag{18.11}$$

18.2.3 Piezoresistive Coefficients of Silicon

The piezoresistive coefficients have been measured for many materials. Of primary interest in MEMS are the coefficients for silicon. These coefficients depend strongly on the doping type, a reflection of the fact that the detailed valence-band and conduction-band structures in silicon are very different. Table 18.1 gives typical values for p-type and n-type silicon.

These coefficients are weak functions of doping level for doping below about 10^{19} cm^{-3} but then decrease markedly at high doping. The coefficients decrease with increasing temperature, dropping to about 0.7 of their room-temperature value at 150°C. The temperature dependence is somewhat nonlinear, which aggravates the problem of compensating for the temperature

---

[4]Note, however, that the pressure-sensor Case Study of this chapter uses a piezoresistor oriented along a [100] direction.

Table 18.1. Typical room-temperature piezoresistance coefficients for n- and p-type silicon [98].

Type	Resistivity	π_{11}	π_{12}	π_{44}
Units	Ω-cm	10^{-11} Pa^{-1}	10^{-11} Pa^{-1}	10^{-11} Pa^{-1}
n-type	11.7	-102.2	53.4	-13.6
p-type	7.8	6.6	-1.1	138.1

dependence. Details can be found in [97, 98]. An important fact is that at higher doping, the temperature dependence of the piezoresistive coefficients becomes small. Therefore, if it is desired to operate a piezoresistive sensor over a wide temperature range, there may be a design advantage in sacrificing piezoresistive sensitivity in exchange for small temperature dependences by using heavily doped piezoresistors.

18.2.4   Structural Examples

Figure 18.1 uses a bulk micromachined accelerometer structure to illustrate the use of piezoresistors to measure stress, and thereby, to infer the proof-mass motion. As the proof-mass moves in response to inertial forces, the flexure bends. If the elastic behavior of the substrate support can be ignored, then the maximum bending stress at the surface occurs right at the cantilever support.[5]

Two resistor orientations are illustrated in Fig. 18.1. Assuming that the resistor orientations are along a [110] direction, as would be the case for a bulk micromachined accelerometer, then the π_l and π_t coefficients for these resistors, using the coefficients in Table 18.1 and assuming a uniform stress throughout the piezoresistor, become

$$\text{n}-\text{type}: \quad \pi_l = -31.2 \qquad \pi_t = -17.6$$

$$\text{p}-\text{type}: \quad \pi_l = 71.8 \qquad \pi_t = -66.3$$

(18.12)

The units of these π coefficients are 10^{-11} Pa^{-1}. Note that the the p-type piezoresistors have a larger sensitivity and that the π_l and π_t coefficients have opposite signs and almost equal magnitudes. This makes p-type piezoresistors well-suited for full-bridge applications, which we shall examine later. It must be remembered that to use a p-type piezoresistor, one must diffuse it

---

[5]If the flexure thickness becomes comparable to the substrate thickness, as would be the case in very-high-g accelerometer or very-high-pressure sensor-diaphragm designs, the point of maximum stress actually shifts onto the substrate. Numerical modeling is required in such cases to determine where to place the piezoresistors [99].

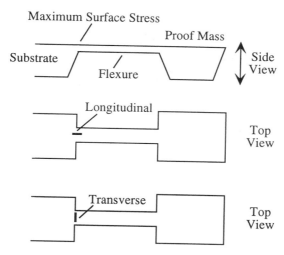

Figure 18.1.   Illustrating lateral and transverse piezoresistor placements using an accelerometer flexure as an example.

into an n-type substrate to achieve junction isolation.   Alternatively, using single-crystal silicon-on-insulator structures or polycrystalline silicon surface-micromachined structures, one can use the insulating layers to isolate piezore-sistors from each other and from the substrate.[6]

When the actual size of the resistor is taken into account, the transverse resistor orientation has the potential for the largest response because, if it can be placed at exactly the right point, the entire resistor will experience the maximum bending stress. However, this orientation is very susceptible to manufacturing variations that can arise from small photolithographic alignment errors. The longitudinal resistor, on the other hand, must extend over some finite length along the cantilever and, for alignment reasons, may also extend onto the support. Therefore, not every part of the resistor experiences the maximum stress, and some loss of sensitivity will result.[7] However, because some averaging is already taking place, small misalignments or placement errors will have a less severe effect than for the transverse orientation.

Which orientation is best? That is a system issue. The system require-ments dictate specifications for sensitivity, accuracy and precision. Because of manufacturing variations, some form of trim or calibration adjustment may be needed. The allowable manufacturing variations must be matched to the capa-bilities of the compensation circuit being used. If the ultimate in sensitivity is

---

[6]Because of the average effect of multiple grain orientations, polysilicon has noticeably smaller piezoresistive coefficients than single-crystal silicon [100].
[7]How to perform averaging over nonuniform stress is discussed in the next section.

required, then it may be necessary to use the transverse orientation, and provide a compensation circuit with enough range to null out manufacturing variations. On the other hand, a longitudinal orientation, because it is more forgiving, may allow a less robust compensation circuit and a less costly manufacturing and calibration procedure. We will examine the interaction between manufacturing tolerances, device design, and calibration in the Case Study (Section 18.3).

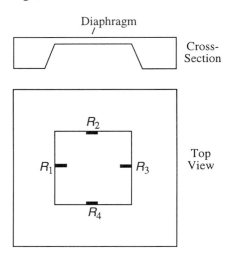

Figure 18.2.   Illustrating the placement of piezoresistors on a bulk micromachined diaphragm pressure sensor.

Figure 18.2 illustrates one possible placement of four piezoresistors on a diaphragm pressure sensor. Since this is a bulk micromachined structure, all resistor axes are along one of the <110> directions, and are aligned along an axis of principal stress at the edge of the plate. However, resistors R_1 and R_3 experience stresses that are rotated 90° compared with the stresses experienced by R_2 and R_4. That is, the longitudinal stress on R_1 and R_3 is the transverse stress at R_2 and R_4, and vice versa. Furthermore, because these resistors are on a plate rather than a cantilever, each resistor experiences both a longitudinal and transverse stress. For example, if resistor R_1 experiences a longitudinal stress σ_l, then it must simultaneously experience a transverse stress $\nu\sigma_l$, where ν is the Poisson ratio. Otherwise, it would shrink in the transverse direction, which is not allowed in a plate. As a result, the total change in resistance for R_1, again assuming uniform stress over the entire resistor, would be

$$\frac{\Delta R_1}{R_1} = (\pi_l + \nu\pi_l)\,\sigma_l \tag{18.13}$$

The Poisson ratio happens to have a minimum in the [110] direction of a (100) plane, with a value of 0.064 [101]. Thus, using the typical p-type π_{ij}

coefficients, we find

$$\frac{\Delta R_1}{R_1} = \left(67.6 \times 10^{-11}\right)\sigma_l \tag{18.14}$$

The corresponding piezoresistive coefficient for R_2, found by reversing the roles of longitudinal and transverse stresses, is

$$\frac{\Delta R_2}{R_2} = -\left(61.7 \times 10^{-11}\right)\sigma_l \tag{18.15}$$

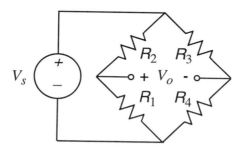

Figure 18.3.   A Wheatstone-bridge circuit constructed from the resistors in Fig. 18.2.

If these four resistors are connected in a Wheatstone bridge configuration, as shown in Fig. 18.3, then the fact that resistors in the same leg of the bridge move in opposite directions makes the bridge output larger. If we write

$$R_1 = R_3 = (1 + \alpha_1)R0 \tag{18.16}$$

and

$$R_2 = R_4 = (1 - \alpha_2)R0 \tag{18.17}$$

where α_1 and α_2 represent the product of the effective piezoresistive coefficient and the stress, then for the bridge

$$\frac{V_o}{V_s} = \frac{R_1 R_3 - R_2 R_4}{(R_1 + R_2)(R_3 + R_4)} \approx \frac{2(\alpha_1 + \alpha_2)}{1 + \alpha_1 - \alpha_2} \tag{18.18}$$

Since α_1 and α_2 are typically small (on the order of .02 or less), and differ from each other by only 10%, this bridge gives an optimally large output without a large nonlinearity.

18.2.5   Averaging over Stress and Doping Variations

The calculation of piezoresistive responses in real structures must take account of the fact that piezoresistors are typically formed by diffusion, hence, have nonuniform doping. They also span a finite area on the device, hence, have nonuniform stress. For a complete representation of all effects, one should solve the Poisson equation for the electrostatic potential throughout the piezore-sistor, subject to boundary conditions of applied potentials at its contacts and subject to stress variations created by deformation of the structural elements. From this potential, one determines the electric field, then the current density, then the total current. However, because the direction of current flow is along the resistor axis and, except at the contacts, parallel to the surface, considerable simplification is possible. Furthermore, the change in resistivity due to stress is small, typically 2% or less.

A good approximation is to begin with the the doping variation which occurs in a direction normal to the surface. The resistor can be thought of as a stack of slices, each slice having a slightly different doping. The slices are connected electrically in parallel because they share the same contacts. Thus to calculate the unstrained resistance R_0, we sum the conductances of each slice, leading us to evaluate the integral

$$\frac{1}{R_0} = G_0 = \int_0^{z_j} \frac{W}{L \rho_{e,o}(z)} \, dz \qquad (18.19)$$

where $\rho_{e,o}(z)$ is the unstrained doping-dependent electrical resistivity, L is the length of the resistor, and z_j is depth of the resistor, equal to the edge of the space-charge layer at the isolating junction.

The value of z_j relative to the structural thickness is quite important. Recall that in pure bending, the stress goes from tensile at one surface to compressive at the other. If the piezoresistor goes all the way through the structure, the stress response will average to zero. At the other extreme, if x_j is very small compared to the structural thickness, one can use the surface stress to estimate the response. In practical structures, x_j may not negligibly small, so stress-averaging may be necessary.

We consider first a case of a cantilever in pure bending, with no variation of bending stress along the axial direction. Equivalently, we assume a constant radius of curvature ρ.[8] We assume the top surface is at $z = 0$, and that the neutral axis is at $H/2$, where H is the thickness of the cantilever. The longitudinal stress due to bending is then

$$\sigma_l(z) = \frac{E(H/2 - z)}{\rho} \qquad (18.20)$$

where E is the Young's modulus, and where the signs have been chosen such that the top surface is presumed to be bent into tension. Since this is a cantilever,

---

[8]*Notation alert*: This is one of the places where ρ as radius of curvature and ρ_e as electrical resistivity must coexist and must be distinguished by the reader.

we can ignore transverse stresses. The conductance $g(z)$ of a single differential slice of thickness dz is

$$g(z) = \frac{W\,dz}{L\rho_e(z)} \tag{18.21}$$

where $\rho_e(z)$ is the strained resistivity which now varies with depth due both to doping and to stress. The stress-dependence is expressed as

$$\rho_e(z) = \rho_{e,o}(z)\left[1 + \pi_l \sigma_l(z)\right] \tag{18.22}$$

We take advantage of the fact that the $\pi_l \sigma_l(z)$ term is small to write

$$g(z) = \frac{W\,dz}{L\rho_{e,o}(z)}\left[1 - \pi_l \sigma_l(z)\right] \tag{18.23}$$

The total conductance is found by integrating over all the slices.

$$\frac{1}{R} = \int_0^{z_j} \frac{W}{L\rho_{e,o}(z)}\left[1 - \pi_l \sigma_l(z)\right]dz \tag{18.24}$$

We recognize the first term inside the integral as giving rise to $1/R_0$. Thus we can write

$$\frac{1}{R} = \frac{1}{R_0}\left[1 - R_0\int_0^{z_j} \frac{W}{L\rho_{e,o}(z)}\pi_l \sigma_l(z)\,dz\right] \tag{18.25}$$

This can be transformed, again using the fact that the piezoresistive term is small, to

$$R = R_0\left[1 + R_0\int_0^{z_j} \frac{W}{L\rho_{e,o}(z)}\pi_l \sigma_l(z)\,dz\right] \tag{18.26}$$

Simplifications beyond this point depend on approximations. If we assume that x_j is much less than $H/2$, we can replace σ_l by $EH/2\rho$, but there still may be some z-dependence in the π_l coefficient because of doping dependence. For general variations due to doping, numerical integration may be required.

Of perhaps equivalent importance is the variation of stress along the length of a resistor that is in the longitudinal orientation. To model that effect, we can think of cutting the resistor along its length into segments, and using Eq. 18.26 to evaluate the resistance of each segment and sum the result, since the segments are electrically in series. In this way, the piezoresistive responses from regions of different stress are correctly captured, as illustrated in the following example.

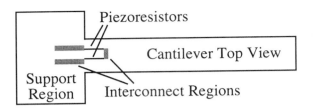

Figure 18.4.   An example using piezoresistance to measure the deflection of a cantilever.

18.2.6   A Numerical Example

Assume that an n-type cantilever of length 200 μm, width 20 μm, and thickness 5 μm is bent by a point load at its free end. Assume that a longitudinal p-type piezoresistor is created as shown in Fig. 18.4. Because it is necessary to make contact to the piezoresistor, and it would be undesirable to have a competing transverse-oriented resistor creating a canceling effect, the piezoresistor is formed into two parallel narrow longitudinal strips. Assumed dimensions for this example are 20 μm in length, 2 μm in width, and a depth into the substrate of $z_j = 0.2$ μm. The shaded regions labelled "interconnect" are wide diffusions, preferably much more heavily doped than the piezoresistor portions such that the total resistance is dominated by the series combination of the two long thin piezoresistors. We assume that the cantilever tip is displaced downward (into the page in Fig. 18.4) by an amount w_{max}. The cantilever equation for a point load, expressed in terms of w_{max} is

$$w = \frac{3}{2}w_{max}\left(\frac{x}{L_c}\right)^2\left(1 - \frac{x}{3L_c}\right) \tag{18.27}$$

where x is measured from the support, and L_c is the cantilever length. The magnitude of the radius of curvature is

$$\frac{1}{\rho} = \left|\frac{d^2w}{dx^2}\right| = \frac{3w_{max}(L_c - x)}{L_c^3} \tag{18.28}$$

For this example, the depth of the piezoresistor is sufficiently small compared to the beam thickness that we can use the surface stress to estimate the response. This stress is

$$\sigma_l = \frac{EH}{2\rho} = \frac{3Ew_{max}(L_c - x)}{2L_c^3} \tag{18.29}$$

Using the given values, and assuming $w_{max} = 1$ μm, $E = 160$ GPa, we obtain a linearly varying stress from 30 MPa at $x = 0$ to 27 MPa at $x = 20$ μm. Since the piezoresistance is a linear function of stress, the average stress value of 28.5 MPa can be used to estimate the response. Using $\pi_l = 71.8 \times 10^{-11}$ Pa, we find

$$\frac{\Delta R}{R} = \pi_l \sigma_l = 0.02 \tag{18.30}$$

Thus we obtain a 2% change in resistance due to bending.

18.3 The Motorola MAP Sensor

The Motorola manifold-absolute-pressure (MAP) sensor was introduced in Chapter 17. It uses piezoresistance to measure diaphragm bending and integrates the signal-conditioning and calibration circuitry onto the same chip as the diaphragm. In the discussion to follow, we examine a simplified view of the process flow, the specific piezoresistor configuration used, and details of the associated circuitry and trim procedures, and we discuss the impact of several recent improvements.[9]

18.3.1 Process Flow

The Motorola MAP sensor is unusual in several respects. First, it is the only high-volume fully-integrated silicon pressure sensor in the automotive market. Second, it uses *bipolar transistors* instead of MOS transistors in building the on-chip circuitry. Since conventional bipolar technologies are built on (111) silicon wafers, and bulk micromachining of silicon is preferably done on (100) wafers, the decision to create a new bipolar process optimized for (100) wafers was a large step. Third, it uses only one piezoresistor, oriented in a [100] direction and located near the edge of the diaphragm, to sense the bending stress [92]. Motorola has given the name "Xducer" to this single-resistor approach.[10]

Figure 18.5 illustrates a typical bipolar process. The transistors are built in an epitaxial n-silicon layer that is grown on a (100) p-type substrate. Individual transistors must be laterally isolated from one another. This is accomplished by a deep p-type diffusion through the n-epi layer, creating a p-n junction all the way around the device. (There are other ways of accomplishing this isolation, for example, deep trench etching with back-filled oxide.) Prior to the growth of the epitaxial layer, an $n+$ diffusion is placed beneath each transistor region. This *buried layer* helps provide low-resistance connection to the collector of each transistor. When the epitaxial layer is grown, this buried layer diffuses part way into the epilayer, as shown in the final diagram.

---

[9]This Case Study was developed based on materials and commentary provided by David Monk, William Newton, and Andrew McNeil of the Motorola Sensor Products Division.

[10]Xducer is a registered Trade Mark of Motorola, Inc. We shall see later in Section 18.3.6 that Motorola has now replaced the Xducer with a new design. However, because the documentation of the circuitry and trim procedures is more complete for the Xducer, we use the Xducer as the subject for the bulk of this Case Study.

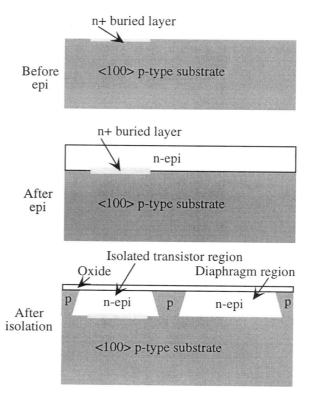

Figure 18.5.   Illustrating junction isolation of transistor regions in a bipolar-transistor technology.

Figure 18.6 shows how a basic bipolar process can be adapted for an integrated pressure sensor. The bipolar transistors require a p+diffusion to form the base followed by an n+ diffusion to form the emitter. The lightly p-doped piezoresistors require their own process step. However, the p+ base diffusion can also be used as the highly-doped interconnect for the piezoresistors and, as in all standard bipolar processes, the n+ emitter diffusion can be used to make the collector contact. The n-epi layer itself, supported by the n+ buried layer, serves as the collector. The transistors are used to build the three op-amps needed for the signal-conditioning and calibration circuitry. Not shown in the figure is provision for the deposition of CrSi alloy thin-film resistors that are trimmed during the calibration step. Also not shown is a passivation overcoat, such as a deposited oxide, over the electronics portion of the chip.

The n-epitaxial layer is used for the diaphragm. This affords an opportunity for good thickness control using a combination of epitaxial growth to set the

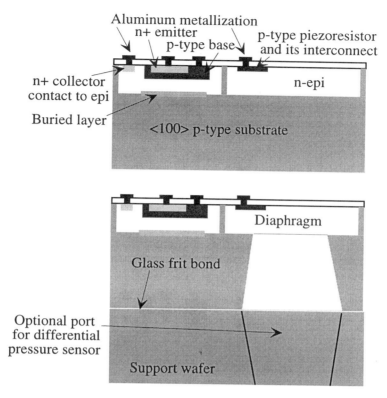

Figure 18.6. Adaptation of a junction-isolated bipolar process to create the diaphragm out of the epitaxial layer. The width of the isolation between devices has been reduced for clarity.

thickness plus a p-n junction etch stop during the diaphragm formation.[11] After diaphragm etching, the wafer is glass-frit bonded to a support wafer, as discussed in Chapter 17. For differential pressures sensors, this support wafer is pre-etched with openings to allow backside access to the diaphragm.

18.3.2 Details of the Diaphragm and Piezoresistor

Typical diaphragm dimensions are 1000×1000 μm square, with a thickness of 20 μm. The piezoresistor is located near the edge center, where stress is highest. The single resistor is oriented at a 45° angle to the side of the square diaphragm, as shown in Fig. 18.7.

---

[11]The original Xducer pressure transducers used a timed etch to set the diaphragm thickness. One of the recent improvements has been the adoption of the electrochemical etch stop for improved thickness control [102].

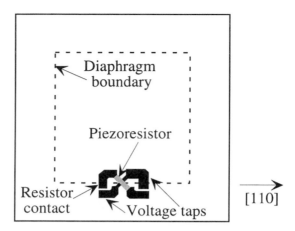

Figure 18.7. Illustrating the position and orientation of the Motorola Xducer piezoresistor. There are two resistor contacts which provide the current in the piezoresistor and two voltage taps which measure a transverse voltage set up by by the stresses in the resistor.

The p-type piezoresistor has four contacts, two of which provide the current along the resistor axis. The other two are transverse voltage taps which are connected to a high-impedance op-amp input so as to draw no current. The dark-shaded contact regions are much wider than the piezoresistor and are doped p+; hence these regions have lower resistance than the piezoresistor itself.

Given that the diaphragm edge is along a [110] direction in a (100) plane, it means that the resistor axis is actually along a [100] direction. We select an axis system in which the "1" axis is along the resistor length, the "2" axis is in the plane of the diaphragm, and the "3" axis is normal to the diaphragm. The only non-zero current density is J_1. However, the electric field in the transverse direction is not zero. We can see this from the piezoresistive equations for the case $J_2 = J_3 = 0$, and σ_3 and $\tau_{13} = 0$, which would be the case for small-amplitude bending of the diaphragm under a pressure load.

$$\mathcal{E}_1 = \rho_e \left(1 + \pi_{11}\sigma_1 + \pi_{12}\sigma_2\right) J_1 \tag{18.31}$$

$$\mathcal{E}_2 = \rho_e \pi_{44}\tau_{12} J_1 \tag{18.32}$$

$$\mathcal{E}_3 = 0 \tag{18.33}$$

To make a first-order analysis, we shall assume that the current density J_1 does not vary with depth, so no averaging in the depth direction is required. The voltage along the length of the piezoresistor, denoted by V_1, is then the integral of the $\mathcal{E}_1$ equation:

$$V_1 = \rho_e L_R J_1 \left[1 + \frac{1}{L_R} \int_0^{L_R} (\pi_{11}\sigma_1 + \pi_{12}\sigma_2) \, dx_1 \right] \qquad (18.34)$$

where L_R is the length of the piezoresistor and x_1 is the spatial coordinate for an axis aligned along the resistor length. The prefactor is the voltage due to the unstrained resistivity of the piezoresistor. The integral term is the change in voltage due to a length averaging of the piezoresistance effect. Because part of the resistor is off the diaphragm, where stress is low, and part of the resistor is away from the edge of the diaphragm, it is reasonable to assume that the average stresses giving rise to axial piezoresistivity are substantially less than the maximum stresses at the diaphragm edge. Since diaphragm-edge stresses tend to produce resistance changes of only a few percent, it is reasonable, at least to first order, to neglect this integral term. Obviously, a more detailed analysis would have to include both the variation of piezoresistor doping with depth and a quantitative calculation of this average axial-piezoresistive effect.

The transverse voltage V_2 is $\mathcal{E}_2 W_R$, where W_R is the width of the piezoresistor, yielding

$$V_2 = \pi_{44}\tau_{12} \left(\frac{W_R}{L_R} \right) V_1 \qquad (18.35)$$

The unstressed resistivity cancels out, just as it does in the more conventional Wheatstone-bridge methods. The transverse voltage is proportional to the longitudinal voltage and, provided that the axial piezoresistive effect due to the average stresses is sufficiently small, the transverse voltage depends only on the shear stress τ_{12} *in the region between the voltage taps.* The placement of the voltage taps determines where the sensitive region is. Figure 18.7 shows the taps right at the diaphragm edge. In practice, this location would be shifted slightly onto the diaphragm to allow for some misalignment error. In our analysis to follow, we shall assume that the placement of the taps is exactly at the diaphragm edge.

18.3.3   Stress Analysis

In Section 10.4.6, we made an approximate model of the bending of a plate under the effects of a pressure load. We used a trial solution of the form

$$\hat{w} = \frac{c_1}{4} \left[1 + \cos\left(\frac{2\pi x}{L} \right) \right] \left[1 + \cos\left(\frac{2\pi y}{L} \right) \right] \qquad (18.36)$$

where $\hat{w}$ is the displacement function, L is the edge-length of the diaphragm, c_1 is the displacement at the center of the diaphragm, and x and y are the in-plane coordinates, now expressed in a coordinate system in which the origin is at the center of the diaphragm and the axes are parallel to the diaphragm edges.

The energy-method analysis with this trial function led to a load-deflection equation of the form

$$P = \left\{ C_r \left[\frac{\sigma_0 H}{L^2} \right] + C_b \left[\frac{EH^3}{(1-\nu^2)L^4} \right] \right\} c_1 + C_s f_s(\nu) \left[\frac{EH}{(1-\nu)L^4} \right] c_1^3 \tag{18.37}$$

where C_r is the coefficient of the residual-stress term, C_b is the coefficient of the plate-bending term, and C_s is the coefficient of the large-amplitude in-plane stretching term. For the present analysis, C_r is zero because the diaphragm is bulk micromachined. Further, we shall neglect packaging-induced stresses (although, as discussed in Chapter 17, such stresses may create small shifts in calibration). Further, we shall assume a small-amplitude loading, in which case we can ignore the C_s term. The value of the C_b from the variational analysis is $\pi^4/6$. Thus the energy-method solution has the pressure-deflection relation

$$P = \frac{\pi^4 E H^3}{6(1-\nu^2)L^4} c_1 \tag{18.38}$$

To calculate the shear stress at the diaphragm edge, we proceed in three steps. First, we find the x-directed radius of curvature due to bending at the center of the edge ($y = 0$). Then, assuming that the piezoresistor depth is small compared to the diaphragm thickness (a reasonable assumption for a 20 μm thick diaphragm), we use the radius of curvature to find the stress at the surface. Because this is a plate, the y-directed stress is equal to the Poisson ratio times the x-directed stress. Finally, we resolve these two principal-axis stress components into the shear stress in a coordinate system rotated by 45°, with one axis along the piezoresistor axis and the second transverse to that axis.

The x-directed radius of curvature at the middle of the diaphragm edge is

$$\frac{1}{\rho_x} = \frac{\partial^2 \hat{w}}{\partial x^2}\bigg|_{x=L/2,y=0} = \left(\frac{2\pi}{L} \right)^2 \frac{c_1}{2} \tag{18.39}$$

The magnitude of the x-directed surface stress is

$$\sigma_x = \frac{EH}{2\rho_x} \tag{18.40}$$

If we substitute for the various quantities, we can express σ_x as

$$\sigma_x = \frac{1}{\pi^2}(1-\nu^2)\left(\frac{L}{H} \right)^2 P \tag{18.41}$$

and because we are dealing with a plate, the y-directed stress at the center of the edge is

$$\sigma_y = \nu\sigma_x \tag{18.42}$$

The numerical factor in front of the σ_x expression can be evaluated, with $\nu =$ .06 for a [110] direction in a (100) plane, to yield

$$\sigma_x = 0.606 \left(\frac{L}{H}\right)^2 P \tag{18.43}$$

This is a reasonably good value, considering the rather crude energy-method solution employed. Clark and Wise [103] did a full numerical simulation of the bending of a square diaphragm using finite-difference methods, and obtained the result

$$\sigma_x = 0.294 \left(\frac{L}{H}\right)^2 P \tag{18.44}$$

Our variational solution has all of the correct dimensionality, and gives numerical agreement with the exact stress solution to within a factor of 2. This degree of agreement is not unusual between single-parameter energy models and exact solutions, as discussed in Chapter 10. If a trial solution with two adjustable parameters were used, better agreement could be expected. We shall use the Clark and Wise result going forward from here.

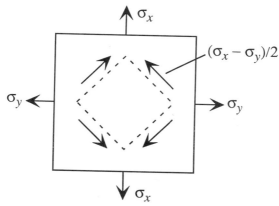

Figure 18.8. The x- and y-directed in-plane axial stresses are equivalent to a shear stress in the rotated coordinate system.

Using the methods of Section 9.2.3 for resolving stresses on inclined sections, we find, referring to Fig. 18.8, that the two principal-axis stresses, σ_x and σ_y are equivalent to a single net shear stress in the rotated coordinate system given by

$$\tau_{12} = \frac{\sigma_x - \sigma_y}{2} = 0.141 \left(\frac{L}{H}\right)^2 P \tag{18.45}$$

where the numerical factor now includes the additional $(1-\nu)/2$ factor.

With these results, we can now calculate the output of the Xducer, using a typical value of π_{44} of 138×10^{-11} Pa^{-1}.

$$\frac{V_2}{V_1} = (0.195 \times 10^{-9}) \left(\frac{L}{H}\right)^2 \left(\frac{W_R}{L_R}\right) P \tag{18.46}$$

For the given plate dimensions, $L/H = 50$. We do not know the exact dimensions of the piezoresistor, but a reasonable guess is $W_R/L_R = 1/5$. With these values, the transfer function becomes

$$\frac{V_2}{V_1} = K_P P \tag{18.47}$$

where $K_P = 9.6 \times 10^{-8}$ Pa^{-1}. It is customary to take out a factor of 10^3 so that the unit of pressure is kPa, and to take out a second factor of 10^3 to convert V_2 from V to mV. With these adjustments,

$$K_P = .096 \text{ mV/(V-kPa)} \tag{18.48}$$

This turns out to be a very accurate result. The 100 kPa 5V pressure sensors typically have K_P values of 0.1 mV/(V-kPa) [89]. We shall use this 0.096 value as we move on to examine the signal-conditioning and calibration circuitry.

18.3.4   Signal-Conditioning and Calibration

The piezoresistor output V_2 requires considerable amplification and modification before a calibrated pressure output is achieved that remains correct over the full operating temperature range of the device. Figure 18.9 illustrates the problem.

The data of Fig. 18.9 are based on published Motorola data [104] but have been scaled to a 5 V supply voltage applied to a sensor with a K_P of 0.096 mV/V-kPa. The definitions of *offset* and *span* are illustrated in the Figure for the data at 25°C. There is some offset at zero pressure attributable to such details as the precise transverse location of the voltage taps. This offset increases with increasing temperature, consistent with the variation of resistivity in the device. The slope of the characteristics decreases with increasing temperature, indicative of the reduction of piezoresistive sensitivity with increasing temperature. Because of the offset and sensitivity have opposite temperature dependences, the curves all intersect within a small region, referred to as the "pivot point."

The goal of the calibration and trim operation is to embed the raw sensor into a circuit such that the overall circuit output has a constant offset, independent of

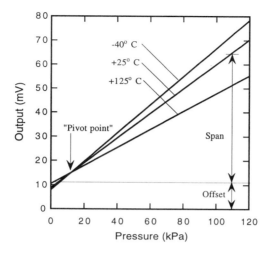

Figure 18.9. Typical uncompensated output data for a Motorola pressure sensor. (Source: Gragg *et al.* [104] scaled to match the applied voltage and device sensitivity of the text example.)

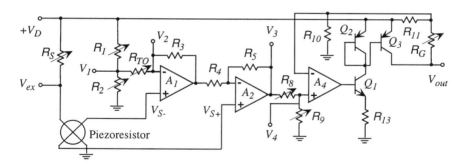

Figure 18.10. Signal-conditioning circuit for the Motorola pressure sensor. Resistors with arrows are laser trimmed. (Source: Motorola Sensor Device Data [92], p. 3-175, redrawn for clarity.)

temperature, and a span over pressure that is also independent of temperature.[12] Along the way, the circuit should provide amplification so that the output signal is measured in volts rather than millivolts. Motorola [92] publishes a circuit to

---

[12] A good discussion of signal conditioning, temperature compensation, and trim of piezoresistors can be found in [105].

accomplish these functions, as shown in Fig. 18.10. All of the resistors shown in this circuit are thin-film resistors with nominally zero temperature coefficient. Therefore, the only temperature dependence comes from the piezoresistance itself.

There are five steps to the calibration: temperature compensation of the span; offset adjust, which consists of two trimming steps; temperature compensation of the offset (TCO); and overall gain to set full-scale output. Each step requires the laser trimming of one resistor. However, because the target goal can be missed to either side, two of the steps require the presence of two trimmable resistors, one of which is trimmed if the target is missed on the high side, the other if missed on the low side. Thus there are seven trimmable resistors in the circuit. In order to be able to access each of these resistors during trim and calibration, a total of five test points are added to the three terminals needed to operate the device once the calibration is complete. Because the device is placed in its package and wire-bonded prior to calibration, the package must have additional leads for these test points, a maximum of eight leads. Outputs V_1 - V_4 and V_{ex} are used in calibration, but not during device operation. The three leads needed for operation are the supply voltage, V_D the output V_{out} and ground.

Because Motorola does not publish resistor values, this circuit can only be discussed in somewhat general terms. Nevertheless, the procedure is enlightening, and worth careful examination.

18.3.4.1   Temperature Compensation of Span

The resistor R_S and the piezoresistor are connected in series. As temperature is increased, the value of the piezoresistor, denoted below by R_P, increases. However, the piezoresistive response decreases with increasing temperature, both because of the intrinsic temperature dependence of piezoresistance and the stress stiffening of the diaphragm caused by thermal-expansion mismatch between the silicon and the passivation layers. Without knowing the doping and the constitutive properties of these passivation layers, it is not possible to provide numbers for the temperature coefficients. But the way this circuit is set up, one can conclude that the temperature coefficient of resistance for R_P is larger than the temperature coefficient of the piezoresistive response. This is consistent with the choice of a lightly doped p-type piezoresistor [105, 106]. The value of R_S is trimmed to provide first-order cancellation of these two effects. The analysis is as follows:

$$V_{ex} = \frac{R_P}{R_S + R_P} V_D \qquad (18.49)$$

The piezoresistor output, $(V_{S+} - V_{S-})$, is (except for the offset) proportional to the product of V_{ex}, the pressure P, and the piezoresistive sensitivity K_P.

Thus, ignoring the offset for the moment, if α_R is the temperature coefficient of R_P, and α_P is the temperature coefficient of K_P, the overall output depends on temperature as

$$\frac{V_{S+} - V_{S-}}{PV_D} = \frac{K_{P0}(1 - \alpha_P \Delta T)R_{P0}(1 + \alpha_R \Delta T)}{R_S + R_{P0}(1 + \alpha_R \Delta T)} \tag{18.50}$$

where R_{P0} and K_{P0} are the room-temperature nominal values of R_P and K_P. The goal is to adjust R_S so that this expression has a zero temperature coefficient to first order. This is accomplished by adjusting R_S until

$$R_S = \frac{\alpha_P}{\alpha_R - \alpha_P} R_{P0} \tag{18.51}$$

or, equivalently, until

$$V_{ex} = \frac{\alpha_R - \alpha_P}{\alpha_R} V_D \tag{18.52}$$

With experimental data for α_R and α_P determined from test devices, the circuit can be designed such that the trimmable range of R_S is large enough to encompass the expected variations.

18.3.4.2 Offset Adjust

In the complete circuit, there are two potential sources of offset in the output: One is the intrinsic device offset that comes from imperfect matching of the two voltage taps. However, with the circuit now referenced to a common ground, there is a much larger potential offset from the common-mode voltage at the sensor taps. If we assume that the voltage taps are approximately half-way down the length of the piezoresistor, then even without a stress present, both voltage taps are at voltage $V_{ex}/2$. This is the common-mode voltage. When a pressure is applied, a difference appears between V_{S+} and V_{S-}. However, the op-amp circuits operate on the total voltages. Therefore, the common-mode signal, denoted by V_{CM} must be included:

$$V_{S+} = V_{CM} + K_P P V_{ex}/2 \tag{18.53}$$
$$V_{S-} = V_{CM} - K_P P V_{ex}/2 \tag{18.54}$$

Referring gain to the circuit of Fig. 18.10, with the pressure set to its lowest value (nominally zero, but actually 20 kPa), R_1 or R_2 is trimmed such that V_1 is made equal to V_2. Under these circumstances, there is no current through R_{TO}, and amplifier A_1 behaves as a unity-gain voltage follower. This means that its output is equal to V_{S-}. Amplifier A_2 accepts this as an input to R_4, but also has V_{S+} applied to its non-inverting input. Therefore, the output V_3 is given by

$$V_3 = -\frac{R_5}{R_4}V_{S-} + \left(1 + \frac{R_5}{R_4}\right)V_{S+} \qquad (18.55)$$

or,

$$V_3 = V_{CM} + \frac{R_5}{R_4}K_P PV_{ex} \qquad (18.56)$$

That is, the A_2 amplifier increases the difference-mode signal by a gain R_5/R_4, but passes the common-mode signal through at unity gain. This common-mode signal is attenuated by the voltage divider formed by R_8 and R_9, as sensed at voltage tap V_4. The voltage divider ratio is trimmed until the desired offset voltage is reached.

18.3.4.3   Temperature Compensation of Offset

The device is then heated, and, with minimum pressure applied, the offset is once-again trimmed. The most recent devices include an on-chip heater that supports this calibration step [102]. Because of the temperature increase in V_{CM}, V_2 is now above V_1. The resistor R_{TO} is adjusted until the offset at the output agrees with the room temperature offset.

18.3.4.4   Span Adjust

The operation of the combination of A_4, the three bipolar transistors Q_1-Q_3, and the resistors R_{10}, R_{11}, R_{13} and R_G is, in effect, a large non-inverting amplifier with an input of V_4 and an overall gain set by R_G. Detailed analysis of the output circuit requires a plunge into bipolar circuits, and will not be done here. For those with some familiarity with such circuits, Q_1 is an emitter follower, setting the current in R_{13} according to the voltage at the output of $A4$. Q_2 and Q_3 form a *current mirror*, in effect reflecting that current so that it is available to drive the output. Feedback from the output through R_G and through the divider network made up of R_{10} and R_{11} together with the supply voltage set the final output voltage. Trim of R_G with pressure applied sets the overall circuit gain. This completes the calibration sequence.

18.3.5   Device Noise

It is interesting that Motorola does not publish a noise specification for these sensors. System noise comes from several sources: Johnson noise in the piezoresistors and fixed resistors; $1/f$ noise in the piezoresistors, possibly driven by slow thermal effects; and front-end amplifier noise, presumed to have both a white and $1/f$ component. The specification that comes closest to a noise parameter is the offset drift, typically specified as 0.5% of the full-scale span [92]. If the device has a 100 kPa span, the equivalent drift in pressure units is 0.5 kPa, and for a 50 mV full-scale part, the equivalent bridge output is 250

μV. The Johnson noise for any reasonable resistor value and noise bandwidth is much smaller than this amount, indicating that, as with so many measurements, drifts and parametric stability rather than random noise limit the threshold for minimum detectable signal.

18.3.6   Recent Design Changes

Just before this book went to press, Motorola released information on a number of design changes [102]. We have already referred to the replacement of the Xducer by a new piezoresistor design, the conversion to electrochemical etch stop for improved diaphragm thickness control, and the inclusion of an on-chip heater to support TCO adjust. These design changes are based on the desire to improve performance and lower cost. One way to lower cost is to make the chip smaller, giving more functional die per processed wafer. Motorola has done this in two ways: first, shrinking the size of the diaphragm and, second, reducing the op-amp count from three to two.[13]

The reason that improved thickness control leads to reduced chip size is that acceptable manufacturing tolerances can be achieved on thinner diaphragms. But with a thinner diaphragm, it is also possible to scale the edge length of the diaphragm without loss of sensitivity. Thus, an improvement in control of one dimension (thickness) allows a dramatic reduction in chip size, from 3.05 mm square to 2.76 mm square, a 22% area reduction.

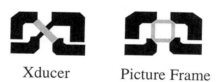

Xducer   Picture Frame

Figure 18.11.   Schematic comparison between the Xducer on the left and the "picture frame" design on the right. The black regions are p+ doped interconnect while the shaded regions are lightly p-doped piezoresistors. (After [102].)

If the diaphragm is made smaller, then it is necessary to improve the dimensional control and placement accuracy of the piezoresistors. Also, if possible, it is desirable to achieve higher sensitivity. Motorola has reported a new "picture frame" piezoresistor design that replaces the Xducer (see Fig. 18.11). This design is identical in function to the Wheatstone bridge of Fig. 18.2 except that, like the Xducer, all four resistors are located at the center of same edge of the diaphragm. The two resistors parallel to the edge experience almost the same stress and, by symmetry, the two resistors perpendicular to the edge experience

---

[13]Because the details of the trim associated with the new circuit are not yet published, we cannot comment in detail on this aspect.

the same stress. Therefore, the analysis of the bridge in Section 18.2.4 applies directly.

The picture frame transducer has several advantages over the design of Fig. 18.2. By placing all four piezoresistors in one location, the area dedicated to interconnect is reduced to that of the Xducer. However, the full-bridge implementation provides about a 40% increase in span compared to the Xducer. Further, by placing all four resistors in the same location, resistor matching can be improved. With the picture frame layout, however, it is not possible to locate all four piezoresistors at the maximum stress location. The amount of signal lost by nonoptimum placement can be reduced by shrinking the piezoresistors themselves. The resistor size is 25 μm long by 8 μm wide [89]. This is small enough that the piezoresistors have stress levels nearly as high as the bridge with optimally placed resistors as in Fig. 18.2, yet produces acceptable variations in transducer output.

18.3.7   Higher-Order Effects

The preceding discussion has carried us through a first-order calculation of device sensitivity, and a step-by-step calibration procedure. There are, however, many issues that we have stepped around. These include: the longitudinal piezoresistance, which affects the total device resistance; detailed modeling of the stress distribution in the diaphragm, including the effects of the elasticity of the tapered walls at the diaphragm edges; issues of resistor placement and stress averaging, both in the longitudinal and transverse piezoresistivity responses; nonlinearities in the temperature coefficient of resistance over the wide range from -40°C to 125°C; effects of doping variation and lithographic errors on device characteristics. Most of these higher-order issues are amenable to numerical modeling.

Finally, while this chapter has emphasized the use of piezoresistance to measure pressure, the principle can be equally applied to any other mechanical measurement where there is motion of a part, resulting in stress in its supports.

Related Reading

G. Bitko, A. McNeil, and R. Frank, Improving the MEMS Pressure Sensor, *Sensors Magazine*, July, 2000, pp. 62-67.

G. T. A. Kovacs, *Micromachined Transducers Sourcebook.* New York: WCB McGraw-Hill, 1998.

Motorola, Inc. *Motorola Sensor Device Data/Handbook* Rev. 4, 1998.

A. Nathan and H. Baltes, *Microtransducer CAD: Physical and Computational Aspects*, Vienna: Springer, 1999.

NovaSensor, *Silicon Sensors and Microstructures*, Fremont CA: NovaSensor, 1990.

L. Ristic (ed.), *Sensor Technology and Devices*, Boston: Artech House, 1994.

Problems

18.1 The Wheatstone bridge configuration of Fig. 18.2 has an intrinsic nonlinearity due to the small mismatch between the magnitudes of π_l and π_t. Determine the variation of π_l and π_t with resistor orientation in the plane of a (100) wafer. Is there an orientation for which π_l exactly equals the negative of π_t, so that this noninearity could be eliminated by design?

18.2 Determine how the sensitivity of the example of Section 18.2.6 is affected if the depth of the junction, instead of being 0.2 μm, is half the thickness of the cantilever.

18.3 The method used for the temperature compensation of span relied on the fact that α_R was greater than α_P. Suggest a circuit that could be used for compensation if the reverse were true, and show that the first-order compensation is effective.

18.4 Analyze the sensitivity of the picture frame design of Fig. 18.11, and compare it to that of the Xducer and to Motorola's published statement that the picture frame achieves a 40% increase in span.

18.5 Estimate the self heating of the Xducer assuming that the Xducer has the same dimensions as one of the picture-frame resistors (8 × 25 microns) and has a doping on the order of 10^{17} Ohm-cm. How does this error compare with the published offset drift?

18.6 A Piezoresistive Treasure Hunt: Go to the literature, either in journals or in published commercial device specifications available on the web, and locate a device other than a MAP sensor that uses piezoresistive sensing. Identify the piezoresistors, estimate the mechanical loads these piezoresistors will experience and, on the basis of these estimates, determine a first-order value for the full-scale signal available prior to amplification.

Chapter 19

A CAPACITIVE ACCELEROMETER

Shake, Rattle, ...
> —First part of song title, Big Joe Turner, 1952 (see also Chapter 21)

19.1   Introduction

The measurement of acceleration, in addition to being a central element of inertial guidance systems, has application to a wide variety of industrial and commercial problems including crash detection for air-bag deployment in cars, vibration analysis of industrial machinery, and provision of feedback signals to steady the image in a video recorder against hand-held vibration. In Section 2.3, we introduced a position-control system, shown again in Fig. 19.1. The system consists of an object which moves in response to an applied force, a sensor which detects the object's position, a difference amplifier that compares the actual position with the desired position (the set point), and a controller which applies a force to the object tending to return it toward the desired position. We now rethink this system as applied to the problem of measuring accelerations.

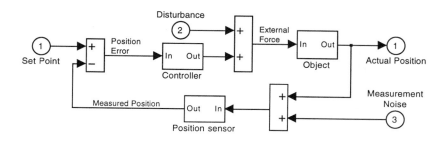

Figure 19.1.   The position-control system of Section 2.3.

We consider first the object, now to be called the *proof mass*, in combination with the sensor. In the absence of the force feedback, this combination functions as an *open-loop* force sensor. When the force feedback is added, creating a *closed-loop* force sensor, the disturbance becomes the measurand, and the controller output which counters the disturbance serves as the system output.

Most accelerometers on the market are of the open-loop type, but there are examples of closed-loop accelerometers as well. In either case, a proof mass is held by some kind of elastic support attached to a rigid frame. Acceleration of the frame causes the mass to move relative to the frame, bending or stretching the support. Detection of the acceleration is accomplished either by direct observation of the changed position of the proof mass relative to the frame, using methods to be discussed in this chapter, or indirectly, by detection of the deformation of the support. We have already examined piezoresistance as a method of measuring strains in microstructures. In Chapter 21 we shall examine another method based on piezoelectricity. There are commercial examples of accelerometers that use both types of indirect position measurement. However, in this chapter, in order to round out our discussion of measurement methods, we choose to focus on *direct position measurement.*

19.2   Fundamentals of Quasi-Static Accelerometers

In Chapter 2 we introduced the spring-mash-dashpot lumped-element system, and in Chapter 7 we studied its dynamic response. Here we revisit this system, viewing it as the key element in an accelerometer.

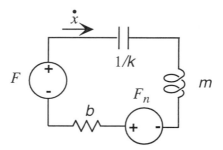

Figure 19.2.   Spring-mass-dashpot model for an accelerometer including the intrinsic noise due to damping.

Figure 19.2 shows a lumped model for a spring-mass-dashpot system with two force sources. F represents the external force and F_n represents the equivalent force noise analogous to Johnson noise in a resistor. The velocity response of this system using s-plane notation is

$$\dot{x} = sx = \frac{F + F_n}{ms + b + k/s} \tag{19.1}$$

The resulting force-displacement characteristic is

$$x = \frac{F + F_n}{ms^2 + bs + k} \tag{19.2}$$

This system has an undamped resonant frequency w_o equal to $\sqrt{k/m}$ and a quality factor Q equal to mw_o/b.

When being used as an accelerometer, m is the proof mass and k is the spring constant of the support. The damping b typically comes from squeezed-film air damping. It is typically adjusted to achieve approximately critical damping so as to achieve a fast response without risk of overshoot or ringing in the response. A *quasi-static* accelerometer is one in which the motion of the proof mass follows the time-evolution of the applied inertial force without significant retardation or attenuation. Therefore, one designs the accelerometer to have a resonant frequency much larger than the expected maximum frequency component of the acceleration signal. In all of the following discussion, we shall assume that the frequencies of interest are well below w_o. In that case, we can use the quasi-static response:

$$x = \frac{F + F_n}{k} \tag{19.3}$$

We examine, first, the signal term, then the noise term. The inertial force F equals the proof mass m times the acceleration to be measured, denoted here by a. This, together with the definition of w_o, leads to a fundamental characteristic of quasi-static accelerometers. The displacement and acceleration are scaled by the square of the resonance frequency:

$$x = \frac{a}{w_o^2} \tag{19.4}$$

This equation has several important implications. The most obvious one is that the scale factor depends only on the resonant frequency and is not affected by the choice of a large mass and stiff spring or a small mass and a compliant spring. Only the ratio enters into the response. The second implication is that if one needs to make an accelerometer that responds quickly, hence has a high resonant frequency, then the amplitude of the position signal to be sensed will be small. For example, for 50 g Analog Devices accelerometer (where g is the acceleration of gravity),[1] having a resonant frequency of 24.7 kHz, the maximum quasi-static displacement of the proof mass is 20 nm. On the

---

[1] Serious notation alert! This chapter includes g for grams as part of units and g representing the acceleration of gravity (9.8 m/sec^2), both as an input and in units such as 1 mg = $10^{-3}g$. In an attempt to maintain the reader's sanity, an upper-case G rather than a lower-case g will be used to denote gaps in this chapter (which, of course, must not be confused with the standard symbol for conductance).

other hand, if a 1 kHz resonant frequency is acceptable for the application, the maximum displacement becomes 1.2 μm.

We now consider the intrinsic noise due to damping. The spectral density function of the force noise is $4k_BTb$, just like Johnson noise in a resistor. This noise typically arises from fluid damping. It is called *Brownian motion noise*, although in fact, the form of the force-noise spectral density function does not depend on the specific mechanism producing the damping.

It is common to convert this to an equivalent acceleration. We do that in two steps. First, we note that in a 1 Hz noise bandwidth, the mean-square force noise is $\sqrt{4k_BTb}$. To convert that to displacement, we divide by the spring constant k, and to convert the displacement to equivalent acceleration, we multiply by ω_o^2. The net result is that the mean-square equivalent acceleration noise is

$$a_{n,\mathrm{rms}} = \sqrt{\frac{4k_BT\omega_o}{mQ}} \qquad (19.5)$$

For a device with a 24.7 kHz resonant frequency, a proof mass of 2.2×10^{-10} kg, and a Q of 5, the rms acceleration noise is 4.83×10^{-3} m/(sec^2-$\sqrt{\mathrm{Hz}}$), or about 0.5 mg/$\sqrt{\mathrm{Hz}}$. This suggests that it should be possible to achieve huge signal-to-noise ratios with microaccelerometers.

In practice, the signal-to-noise ratio, and, hence, the achievable accelerometer sensitivity, is dominated by other effects: the noise contributed by the position-measuring circuit, especially its first stage of amplification; the need to build additional stiffness into the structure (raising its resonant frequency) to prevent either sticking of parts during fabrication or excess fragility; or residual calibration errors and drift problems. We now consider position measurement methods.

19.3   Position Measurement With Capacitance

There are many methods of direct position measurement, for example, capacitance change, inductance change, optical methods, and scanning-probe tips. Of these, capacitance change is the most widely used in microaccelerometers. We shall analyze a commercial capacitive accelerometer from Analog Devices as the major Case Study in this chapter. Inductance change is widely used in macro-sensors, but has not yet entered the market in commercial microsensors. Optical position sensing in microstructures is a new subject which may become important in the future. Position measurement with scanning-probe tips, specifically electron-tunneling tips, has some very interesting features. We shall examine this subject briefly at the end of the chapter, and compare it with the capacitance-measurement approach.

Capacitance measurement is one of the most versatile methods of position measurement. Figure 19.3 illustrates several types of capacitors that are in common use.

Figure 19.3. Illustrating a variety of capacitor structures that can be used for position sensing.

The parallel-plate capacitor can vary either with vertical motion of a movable plate, modifying the gap, or by transverse motion of one plate relative to another, modifying the effective area of the capacitor. Interdigital capacitors vary with the degree of engagement of the fingers. Also, displacement of one of the electrodes out of the plane of the figure would modify the capacitance, but this is not a configuration in common use. The fringing capacitance deploys an interdigital set of electrodes on one substrate and detects the change in interdigital capacitance as the electrodes are brought into proximity with a third electrode. If this electrode is grounded, as shown in the figure, proximity reduces the interdigital capacitance. If this electrode is floating, proximity increases the interdigital capacitance.

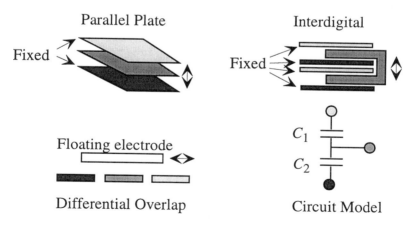

Figure 19.4. Examples of differential capacitors.

Figure 19.4 illustrates *differential capacitors* that can be used for position sensing. In all three examples, there are three electrodes used for the measurement, with two capacitors that are nominally of equal size when the moveable component is centered. Shading of the terminals in the circuit model matches the shading of the capacitor components in each example. Motion of the moveable component in the indicated direction increases one capacitance and decreases the other. A variant of the parallel-plate differential capacitor would have the middle and lower plate fixed and only the upper plate moveable. In this configuration, motion of the upper plate modifies one capacitor while the other remains constant.

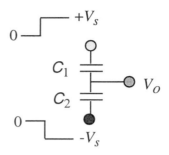

Figure 19.5.   A typical circuit use of a differential capacitor

Differential capacitors have the virtue of cancelling many effects to first order, providing a signal that is zero at the balance point and carries a sign that indicates the direction of motion. From a system point of view, a differential capacitor accomplishes linearization about the balance point. Consider the parallel-plate example, with the gap of the upper capacitor G_1 and that of the lower capacitor G_2. We assume equal area of both capacitors. A voltage $+V_s$ is applied to the upper plate. Simultaneously, a voltage $-V_s$ is applied to the lower plate. The voltage that appears at the output is

$$V_o = -V_s + \frac{C_1}{C_1 + C_2}(2V_s) = \frac{C_1 - C_2}{C_1 + C_2}V_s \qquad (19.6)$$

Since the areas are equal, this can be rearranged to yield

$$V_o = \frac{G_2 - G_1}{G_1 + G_2}V_s \qquad (19.7)$$

If the two gaps are equal, the output voltage is zero. However, if the middle plate moves so that one gap is larger than the other, the output voltage is a *linear* function of this change.

19.3.1   Circuits for Capacitance Measurement

There are many circuit configurations that are used for capacitance measurement. We examine a few examples. The starting point in all cases is the charge-voltage relationship for the capacitor. We can assume this relationship is linear (that is, there is no nonlinear dielectric medium involved), but the capacitance will, in general, depend on displacement. Thus we can write

$$Q = C(x)V \tag{19.8}$$

where Q is the capacitor charge, V is the voltage across the capacitor, and $C(x)$ is the capacitance that depends on one or more displacement coordinates. The current in the capacitor is the time-derivative of the charge:

$$i_C = C(x)\frac{dV}{dt} + V\frac{\partial C}{\partial x}\frac{dx}{dt} \tag{19.9}$$

where the partial derivative with respect to x is used to remind us that if the capacitance depends on more than one displacement coordinate, there will be a term proportional to the time rate of change of each of those coordinates.

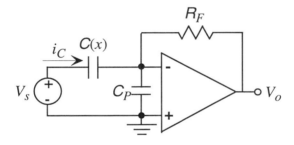

Figure 19.6.   Use of a transimpedance amplifier to capture the capacitor current.

The simplest circuit for measuring capacitance is shown in Fig. 19.6. Both the sensor capacitance $C(x)$ and a parasitic capacitance to ground C_P must be included because the interconnect between the sensor and the amplifier always adds some amount of parasitic capacitance. Here, a transimpedance amplifier is used to capture the current through the capacitor $C(x)$. The advantage of this configuration is that, because of the virtual ground at the op-amp input, there is negligible charge on the parasitic capacitance and it does not affect the measurement. The output is $V_o = -R_F i_C$.

If V_s is a DC source, V_o is proportional to the velocity dx/dt. While an output proportional to velocity is very useful when constructing oscillators or force-feedback systems, a measurement of velocity is not equivalent to a measurement of position. To obtain position, one must either use a properly initialized integrator to reconstruct the position from the velocity output, or use a time-varying source waveform, as discussed below.

If we use a sinusoid as the source V_s in Fig. 19.6, we can determine capacitance directly. For example, if the position, hence the capacitance, is constant, and $V_s = V_{so} \cos \omega t$, then the output of the amplifier is $-\omega V_{so} C(x) \sin \omega t$. The value of $C(x)$ can be determined from the amplitude of the output sine wave. However, if x is time varying, there is a second term in the output proportional to dx/dt. If these two terms were of comparable size, the output would be a mix of direct position information via the $C(x)$ term and velocity information, via the dx/dt term. Therefore, this approach is most useful when the dx/dt term can be presumed to be negligible, a situation that can usually be achieved by making the frequency of the V_s sufficiently large. The precise criterion is the subject of Problem 19.1.

In either case, whether using DC or AC source waveforms, the act of measuring the capacitance creates an electrostatic force which disturbs the position of the moveable element. The magnitude of the force is $(1/2)(\partial C/\partial x)V_s^2$. For a DC voltage, the force is constant. For a sinusoidal voltage, presumed to be at a frequency far above the resonant frequency of the structure so that the high-frequency component at 2ω produces negligible motion, the effective force is the time average of V_s^2, which is $V_{so}^2/2$. The magnitude of the position disturbance depends on the compliance of the structure. Accurate measurement requires either the use of a sufficiently small voltage so that the position disturbance is negligible, the use of very short pulses so that the measurement is completed before the moveable element can change position, or the introduction of some method of correcting for the disturbance as part of the sensor calibration.

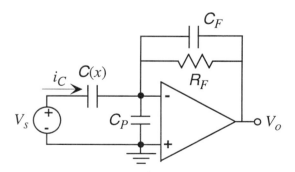

Figure 19.7.   Adding a feedback capacitor to the circuit of Fig. 19.6.

When using a high-frequency AC source, so that the velocity-dependent component of the current can be ignored, the circuit of Fig. 19.7 is useful. We assume that the value of R_F is chosen such that at the measurement frequency, $\omega R_F C_F$ is large compared to unity. The output is then

$$V_0 \approx -\frac{i_C}{sC_F}V_s \approx -\frac{C(x)}{C_F}V_s \tag{19.10}$$

The function of the resistor is to provide DC feedback to the op-amp input so that the DC value at the inverting input is clamped at zero. Alternatively, this resistor could be connected between the inverting input and ground. Without the resistor, the potential at the input node could drift away from zero, and the amplifier output could saturate.

It is also possible to use the op-amp circuits introduced in Section 14.12 to measure capacitance. The first of these, a differential-amplifier approach (Section 14.12.1), requires very close matching of the input capacitances to ground at the two transistor inputs, including both the device input capacitance and any parasitic capacitances arising from interconnect. Hence, it is most suitable for applications in which the transistors are integrated with the capacitive position sensing element. Furthermore, it requires quantitative determination of the total parasitic capacitance as part of the calibration.

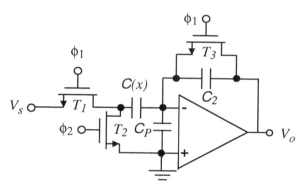

Figure 19.8.   A switched-capacitor inverter.

The second approach is based on the circuit of Fig.19.8. We recall from Section 14.12.2 that this circuit uses two non-overlapping clock pulses ϕ_1 and ϕ_2 to switch the transistors from open circuits to closed connections. As with the previous sinusoidal example, we will assume that the switching frequency for the clocks is much higher than any characteristic frequency of the motion so that within one clock cycle, we can assume $C(x)$ is constant. A DC source is used for V_s. When ϕ_1 turns on transistors T_1 and T_3, the amplifier operates as a unity-gain buffer with a virtual ground at the inverting input, and capacitor $C(x)$ acquires a charge $C(x)V_s$. Then ϕ_1 is turned off, isolating $C(x)$ and also turning the op-amp into an integrator, ready to collect charge from $C(x)$. A short time later, ϕ_2 is turned on, grounding the left-terminal of $C(x)$, which creates a negative-going signal at the inverting input which drives V_o strongly positive, pulling the inverting node back toward zero. The circuit settles when

$V_0 = [C(x)/C_2]V_s$. Then the clock cycle is repeated, charging $C(x)$ to V_s and discharging C_2 back to zero. The output alternates between zero[2] and $[C(x)/C_2]V_s$. Putting this signal through a low-pass filter provides an averaged output equal to $f[C(x)/C_2]V_s$, where f is the fractional duty cycle (equal to 0.5 for symmetric clock timing). As with the inverting amplifier examples, this circuit suppresses the effect of the parasitic capacitance because when the circuit is settled, the inverting input is at virtual ground potential.

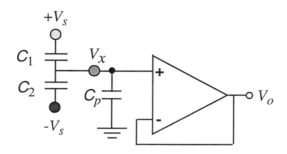

Figure 19.9.   Use of a follower to measure the output voltage of a differential capacitor.

When a differential capacitor is used, the voltage on the shared terminal is measured. Figure 19.9 illustrates the most direct approach using a unity-gain buffer to sense the output voltage, labeled here as V_x. However, now the parasitic capacitance contributes to the output. Assuming symmetric positive and negative sinusoidal or pulse signals ($\pm V_s$) applied to the outer terminals, $V(x)$ is given by

$$V_x = \frac{C_1 - C_2}{C_1 + C_2 + C_P} V_s \tag{19.11}$$

The parasitic capacitance reduces the signal, and also affects the calibration of the measurement. One way to mitigate this problem in fully-integrated designs is to fabricate a guard electrode beneath the interconnect that is driven by the output V_0. Since V_0 is almost exactly equal to V_x, the net voltage across the parasitic interconnect capacitance is very small, just like in the virtual-ground examples. However, this does add fabrication complexity, and it is difficult to cancel all of the parasitic capacitance this way. An alternate method is to use oppositely phased sinusoidal sources for $+V_s$ and $-V_s$, and to replace the $C(x)$ connection of Fig. 19.6 or 19.7 with a connection to the shared terminal of the

---

[2]We assume here that the op-amp has no input offset. Circuits that deal with input offset errors are discussed in Sections 19.3.3 and 19.3.4 below.

differential capacitor. It can be readily shown that for the case of Fig. 19.7, the output is

$$V_0 = -\frac{C_1 - C_2}{C_F}V_s \qquad (19.12)$$

19.3.2 Demodulation Methods

When using sinusoidal sources to measure capacitance, the output of the primary measuring circuit is a sinusoid whose amplitude is proportional to the capacitance of interest. To extract this amplitude in a way that can follow the relatively slow variations associated with the changing of $C(x)$, one can use a peak detector or a synchronous demodulator.

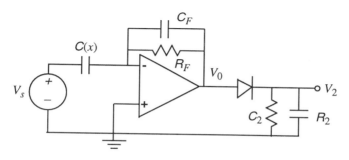

Figure 19.10.   Use of a peak detector to demodulate the capacitance signal of Fig. 19.7.

The *peak detector* is illustrated in Fig. 19.10, with waveforms in Fig. 19.11. The op-amp output V_0 is a relatively high-frequency sinusoid proportional to $C(x)$. If the $R_2 C_2$ time constant is selected to be long compared to the period of the sinusoid, yet short compared with the characteristic time for a change in $C(x)$, the detected output V_2 is a slowly varying signal which follows the amplitude of the sinusoid, albeit with some high-frequency ripple artifacts still present.[3]

The *synchronous demodulator* is a circuit for demodulating a periodic waveform, whether sinusoidal or pulsed. One way to approach synchronous demodulation is with an analog multiplier, with one input being the modulated carrier of interest and the other being a sinusoid or square wave at the same frequency as the carrier. As an example, assume that the waveform of interest is $A(t)\cos\omega t$, where $A(t)$ is a slowly varying amplitude corresponding

---

[3]Details of the operation of this circuit, including methods for estimating the ripple, can be found in Senturia and Wedlock, pp. 489-492 [39].

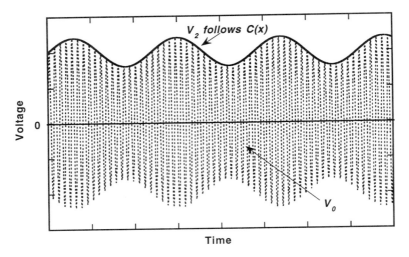

Figure 19.11. Example of the use of a peak detector to demodulate a modulated sinusoid for the case where $C(x)$ is changing slowly with time in a sinusoidal fashion. The dashed line is the output V_0 while the solid line is the demodulated output V_2. Ripple effects are not shown.

to changes in $C(x)$. If we multiply this by a reference sinusoid at the same frequency, $B \cos(\omega t + \theta)$, where θ is a phase shift, the result is

$$[A(t) \cos \omega t][B \cos(\omega t + \theta)] = \frac{A(t)B}{2}[\cos \theta + \cos(2\omega t + \theta)] \quad (19.13)$$

If this is put through a low-pass filter with a corner frequency that rejects the component at 2ω but does not filter the slow variations in $A(t)$, the result[4] is $A(t)B \cos \theta$. Note that the output is a faithful copy of $A(t)$, but its scale factor depends on the phase-difference between the carrier and the reference signal with which it is multiplied. For this reason, a detector of this type is also called a *phase-sensitive detector*. Because of this phase sensitivity, it is essential that the reference signal have the correct phase. If the transimpedance amplifier of Fig. 19.6 is used, the reference signal must be shifted by $\pi/2$ with respect to the phase of the V_s sinusoid. On the other hand, if the integrating version of Fig. 19.7 is used, the reference signal must be in phase with V_s.

An alternate approach to synchronous demodulation is with a track-and-hold circuit in combination with a switch, as illustrated in Fig. 19.12. For purposes of illustration, we shall assume that the two clocks, while strictly non-overlapping, have an extremely small time interval when neither clock is

---

[4]A similar result would be obtained if we used a square wave instead of a sinusoid for the reference waveform.

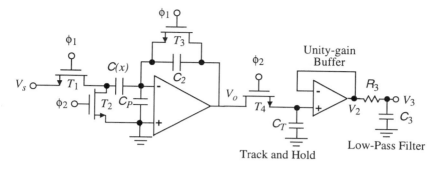

Figure 19.12. Addition of an elementary synchronous demodulator and low-pass filter to the switched capacitor circuit of Fig. 19.8.

on. Thus, when ϕ_2 is on, and the V_0 output equals $C(x)V_s/C_F$, the track-and-hold capacitor C_T is charged[5] to this value through transistor T_4, so the output V_2 equals V_0. When ϕ_1 is on, C_T is disconnected, and hence holds the previous value of V_0. Once ϕ_2 turns on again, the V_2 output is updated to a new value of $C(x)V_s/C_F$. Thus the output is a stair-step waveform that follows samples of $C(x)$, one sample per clock cycle. In this example, the phase of the signal that controls T_4 is automatically correct. However, in a more general version of this type of circuit, it is necessary to pay attention to the phase of the reference waveform, just as in the multiplier example discussed above. The $R_3 C_3$ section of the circuit is a low-pass filter with a time constant long compared to the switching period but short compared to the expected variations. The filter smoothes out the steps in V_2 between samples so that V_3 is a signal that accurately follows $C(x)$.

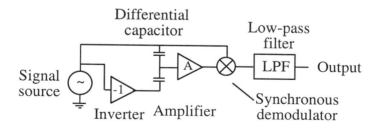

Figure 19.13. A schematic system diagram for differential capacitance measurement using a synchronous demodulator. The notation is a hybrid, with circuit blocks on the left and system-style blocks on the right. This kind of mixed pictorial representation is common.

---

[5]The size of transistor T_4 must be large enough to charge C_T in a time short compared to the clock period when T_4 switched to its nonsaturation region. One way to do this is to make C_T small, but if C_T is too small, leakage currents back through the transistor during the hold interval will compromise the result.

Putting all the pieces together, we have in Fig. 19.13 a complete capacitance measurement system. The system consists of a signal source, a unity-gain inverter to create the antiphase signal needed for the differential measurement, the differential capacitor, an amplifier, a synchronous demodulator, and a low-pass filter. While the system looks complex, if the differential capacitor is built in a CMOS-compatible process, then the entire system can be built on a chip.

19.3.3   Chopper-Stabilized Amplifiers

As discussed in Section 14.9, op-amps have input offset voltages. These offsets can create errors in the inferred capacitance value, and because the offsets can be temperature dependent, it is difficult to remove their effects with calibration. There are many techniques for dealing with offsets, most of which involve some form of high-frequency switching. For example, a *chopper-stabilized* op-amp uses transistor switches to alternate the input of a non-inverting amplifier between the input signal and ground. We analyze this approach below.

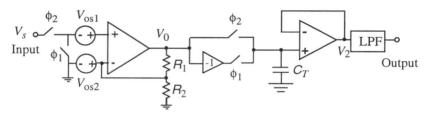

Figure 19.14.   A chopper-stabilized amplifier, with voltage sources added to show the input off-set voltages. Switching transistors are represented as simple switches driven by non-overlapping clocks ϕ_1 and ϕ_2.

Figure 19.14 shows an example of a chopper-stabilized amplifier, with one input and one output. The components of the chopper-stabilized amplifier are an op-amp (shown here connected to implement a non-inverting gain), a synchronous demodulator, a low-pass filter, and transistor switches driven with alternating non-overlapping clocks. The circuit shows two voltage sources representing the input-offset voltages at each of the terminals. The voltage at V_0 is given by

$$V_0 = \frac{A(R_1 + R_2)}{AR_1 + R_1 + R_2}(v_+ - V_{os2}) \qquad (19.14)$$

where A is the open-loop gain of the op-amp and v_+ is the voltage at the non-inverting input of the op-amp. In the limit of large A, this reduces to the familiar non-inverting gain $(R_1 + R_2)/R_1$. During the ϕ_1 phase, v_+ is V_{os1}, while during the ϕ_2 phase, v_+ is $V_s + V_{os1}$. Thus the V_0 signal is a square wave that alternates between the amplified difference in offset voltages, and the amplified signal plus the difference in offset voltages. Because the amplified offset voltages appear in both phases, they contribute to the DC average value of the V_0 signal, but not to the height of the square wave. In order to demodulate this signal successfully, a more complex synchronous demodulator than the example of Fig. 19.8 is needed. The synchronous demodulator must use both clock phases and an inverter to subtract alternative op-amp outputs. The output of the unity-gain buffer amplifier alternates between the negative of the amplified offset and the positive amplified signal plus offset. When this goes through the low-pass filter, only the amplified signal remains.

Chopper-stabilized amplifiers have the additional benefit of cancelling out the low-frequency amplifier noise, specifically and most importantly, those components of $1/f$ noise that appears in the difference between v_+ and v_- and are slow enough not change value during one clock cycle. However, any $1/f$ noise that may be present on the incoming signal is passed through the system unchanged. Furthermore, this version of the chopper amplifier is affected by parasitic capacitance at the input, since the input node is switched between ground and V_s.

19.3.4   Correlated Double Sampling

A second method for offset cancellation (and, along with it, cancellation of low-frequency $1/f$ amplifier noise) is called *correlated double sampling*. A circuit example is shown in Fig. 19.15. In this example, inverting amplifiers are used which do not suffer from interference from parasitic capacitance at the input.

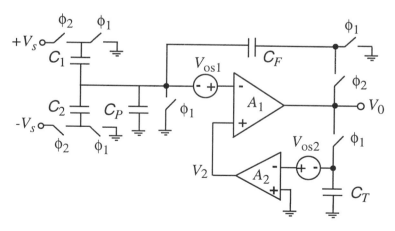

Figure 19.15.   A correlated double sampling circuit for use with a differential capacitor. V_{os1} and V_{os2} represent the net differential offset voltages of the two op-amps.

When ϕ_1 is on, the sensor capacitances C_1 and C_2 and the feedback capacitance C_F have all their terminals grounded. The output voltage for the ϕ_1 phase, denoted $V_{0,1}$ is given by

$$V_{0,1} = -\frac{A_1}{1 + A_1 A_2} V_{os1} - \frac{A_1 A_2}{1 + A_1 A_2} V_{os2} \tag{19.15}$$

This value is stored on capacitor C_T, and remains there during the ϕ_2 part of the cycle. It is readily shown that when ϕ_2 is on, the output has the value

$$V_{0,2} = -\frac{A_1(C_1 - C_2)}{C_1 + C_2 + A_1 C_F} V_s$$
$$-\left[\frac{C_1 + C_2 + C_F}{C_1 + C_2 + (1 + A_1)C_F}\right]\left[\frac{A_1 V_{os1} + A_1 A_2 V_{os2}}{1 + A_1 A_2}\right] \tag{19.16}$$

The first term is the desired sensor signal, scaled by a factor that approaches $-(C_1 - C_2)V_s/C_F$ for large A_1. The second term is the residual amplified offset. Note that the this term is identical to $V_{0,1}$, except scaled by a factor that approaches $(1/A_1)[1 + (C_1 + C_2)/C_F]$ for large A_1. The net result is a reduction in the effect of offset V_{os1} by a factor of order $A_1 A_2$ and offset V_{os2} by a factor of order A_1.

Both the chopper stabilized amplifier and correlated double sampling circuits are readily integrated into CMOS processes designed for analog circuit implementation.

19.3.5   Signal-to-Noise Issues

Clearly, the larger the voltage that is used in any capacitance measurement, the larger the signal. However, along with this voltage goes an actuating force that can disturb the position of the moveable element of the sensor. This force is proportional to the square of the applied voltage (or the time average of the square if the voltage is a high-frequency AC waveform) and also to the gradient term $\partial C/\partial x$. This gradient term determines the sensitivity of the capacitive sensor. The more sensitive the device, the lower the voltage that can be used without what may prove to be an unacceptable applied force.

In circuits that use switched-capacitor configurations, such as in the previous examples, the capacitors are connected directly to voltage sources. Therefore, if these capacitors also determine the bandwidth of the system, the mean-square noise will be at least kT/C. What is more typical is that the system bandwidth is set by the low-pass filter, and that front-end amplifier noise dominates. Another potential noise source is the non-ideality of the transistor switches. The internal transistor capacitances must be charged and discharged every switching cycle. To the extent that the amount of charge on these internal capacitances varies with the signal level being switched, a switch-dependent error can be introduced into the transfer function which must be corrected with calibration.

19.4   A Capacitive Accelerometer Case Study

Analog Devices (ADI) manufactures a family of surface-micromachined capacitive accelerometers.[6] The devices are monolithic, having a single silicon chip that contains the micromachined proof-mass and spring supports and the circuitry that implements all of the electronic functions required to give an analog output signal proportional to acceleration. A photograph of a complete chip (with 8 wire bonds attached) is shown in Fig. 19.16. The mechanical sensor region is in the center; the surrounding regions contain the electronics.

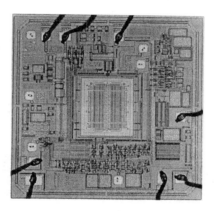

Figure 19.16.   Die photo of ADXL150 chip, with eight wire bonds attached. The die is 1.94 mm on a side. The sensor region is 753 × 657 μm. (Source: Analog Devices; reprinted with permission.)

Figure 19.17 shows a schematic of the sensor section of the accelerometer. It is fabricated with surface micromachining of polysilicon. The shuttle forms the proof mass. It is suspended on folded springs that are attached to the substrate only at the anchor points. A number of cantilevered electrodes are attached to the shuttle. In the sense region, each cantilever is positioned between two fixed electrodes, forming a lateral differential capacitor of the type shown in Fig. 19.4. There are 42 repeat units of this type on the device. There is also a self-test region with similar electrode arrangements, but these electrodes are connected to a drive circuit that can apply an electrostatic force on the shuttle, displacing the shuttle for self-test purposes.

---

[6]This Case Study is based in part on materials and commentary provided by Michael Judy of Analog Devices and Stephen Bart of Microcosm Technologies.

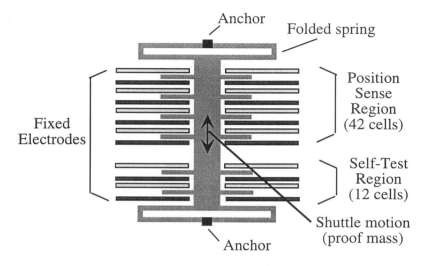

Figure 19.17.   Schematic illustration of the design of the sensor portion of the ADXL150 accelerometer. Similarly shaded fixed electrodes in the sense region are connected together forming a lateral differential capacitor with a moveable proof-mass shuttle. Similarly shaded fixed electrodes in the self-test region are also connected together, forming a second differential capacitor that is used to apply electrostatic forces to displace the shuttle.

Figure 19.18.   Die photo of ADXL150 sensor region. (Source: Analog Devices; reprinted with permission.)

Figure 19.18 shows a photograph of the sensor region of the chip. The 12 self-test cells (each cell having two fixed and one moveable electrode) are visible, three each on each side of the two ends of the shuttle, and the 42 sense cells are visible in the central portion. The device is 500 μm from anchor to anchor. Each sensor and self-test cell has 104 μm of overlap between the

Figure 19.19.   Enlarged view of electrodes on an Analog Devices accelerometer.   (Source: Analog Devices; reprinted with permission.)

moveable and fixed electrodes.   An expanded view of the electrodes from a similar accelerometer product is shown in Fig. 19.19.

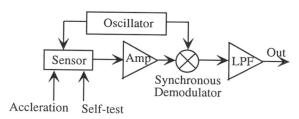

Figure 19.20.   Basic block diagram of the ADXL150 measurement system.

The block diagram of the ADXL150 measurement system is shown in Fig. 19.20. The first monolithic accelerometer of this type, the ADXL50 (the device shown in the photograph on page 5), used a force-feedback system. While the performance of an open-loop system depends on the linearity of the spring suspension and the accuracy of the position-measurement, the performance of a closed-loop force-feedback system depends on the sensitivity of the position-measuring system in combination with the linearity and accuracy of the forcing actuator.   More recent devices, including the ADXL150, use an open-loop system as shown in Fig. 19.20, suggesting that sufficient linearity and accuracy can be achieved without force-feedback. (At the end of this chapter, we shall examine a tunneling accelerometer which, because of its highly non-linear response to position changes, requires a force-feedback implementation.)

Figure 19.21 illustrates, in highly exaggerated fashion, the motion of the proof mass shuttle in response either to an inertial acceleration or to an elec-

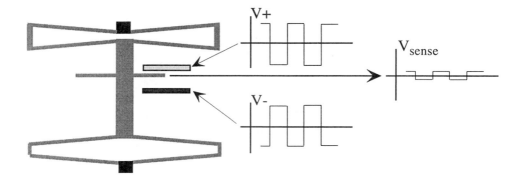

Figure 19.21. Highly exaggerated sketch of a displaced shuttle, due either to an external acceleration or to voltages applied to the self-test electrodes. The drive waveforms and resulting sense waveform are also shown.

trostatic force applied from the self-test electrodes. We note that the effect of a displacement is to unbalance the differential capacitor. The two fixed electrodes are driven with oppositely polarized square waves, and the resulting sensor output is a square wave which measures the unbalance in the differential capacitor. This output is amplified, synchronously demodulated, and then low-pass filtered to give the output signal. Not shown in the diagram is an additional circuit that can add an offset to the output signal. The offset is discussed in the next section.

The magnitude of the square-wave sensor output depends both on the mechanical displacement due to acceleration (or self-test) and on the amplitude of the applied square waves. A low-pass filter is recommended at the supply input to remove high-frequency switching transients and clock signals.

We now examine, in turn, the performance specifications for the ADLX76, the sensor element and its dynamic model, fabrication and packaging issues, and, finally, noise and accuracy characteristics.

19.4.1   Specifications

A selected set of specifications for the ADXL150 is shown in Table 19.1. As with the pressure-sensor example of the previous two chapters, the list of specifications is lengthy, and addresses many of the subtle effects that must be confronted when manufacturing a sensor product.

The form of the transfer function for this system is

$$V_{out} = \frac{V_s}{2} \pm \alpha + \beta a V_s \tag{19.17}$$

where V_s is the actual supply voltage, a is the acceleration being measured, and α and β are parameters. The zero-g ($a = 0$) output is the bias. It is nominally 2.5

Table 19.1.   Selected Specifications of ADXL150 Capacitive Accelerometer. (Source: Analog Devices Data Sheet.)

Property	Specification
Sensitivity	38mV/g
Full-scale range	$\pm$ 50 g
Transfer function form	see text
Package type	14-pin cerpak
Temperature range	-40 to +85°C
Supply voltage	4 - 6 V
Nonlinearity	0.2 %
Package alignment error	$\pm$ 1°
Transverse sensitivity	$\pm$ 2%
Zero-g output voltage (Bias)	$V_s/2 \pm 0.35$ V
Temperature drift (from 25°C to T_{min} or T_{max})	0.2 g
Noise from 10 Hz to nominal bandwidth	1 m$g/\sqrt{\text{Hz}}$
Clock noise	5 mV peak-to-peak
Bandwidth	400 or 1000 Hz, customer choice
Temperature drift of bandwidth	50 Hz
Sensor resonant frequency	24 kHz
Self test output change	400 mV
Absolute maximum acceleration	2000 g (unpowered)
	500 g (powered)
Drop test	1.2 meters
Min/max storage temperature	-65 to 150 °C
Max lead temperature (10 seconds)	245 °C

V (half of the nominal value of V_s), but can vary by the amount of device offset α, wich is in the range. $\pm$ 350 mV. If DC measurements of acceleration are required, then this offset must be nulled out. The chip provides a compensation pin to which an external voltage can be applied to reduce the zero-g offset at fixed temperature. The 0.2 g bias drift with temperature then becomes the limitation on offset. The value of β depends on the sensor design. For a device with 38 mV/g sensitivity, $\beta = 7.6 \times 10^{-3}$ g^{-1}. An important feature of the transfer function is that, except for the small nonlinearity, both the offset and acceleration output are proportional to the supply voltage. The device is said to be *ratiometric* to the supply. When designing off-chip compensation circuits, for example, with resistive voltage dividers, this ratiometric feature is very useful, since both the on-chip and off-chip circuits will vary similarly with small variations in supply voltage.

There are four features that are laser trimmed at the wafer level: offset, sensitivity, self-test, and clock frequency. We note that the device has eight leads. But only five of them are used during device operation: power supply, ground, sensor output, self-test, and offset compensation. The remaining leads are used for various testing functions. We recall that the pressure-sensor example of Chapter 18 also had extra leads that were used for calibration and test.

19.4.2 Sensor Design and Modeling

The shuttle, electrodes, and folded spring are polysilicon, 2 μm thick. The nominal gap between the fingers is 1.3 μm, and the length of the overlap region is 104 μm. If we use parallel-plate estimates to get the correct order of magnitude of the capacitances, we write

$$C_{sense} \approx 42 \frac{\varepsilon_0 H L_o}{G_0 \pm y} \qquad (19.18)$$

where H is the polysilicon thickness, L_o is the length of the overlap region, G_0 is the nominal gap, 42 is the number of identical sensor cells (connected in parallel), and y is the displacement of the shuttle with respect to the nominal rest position halfway between the pair of fixed electrodes. One capacitor increases in value; the other decreases. Assuming small motions, the result is

$$C_{sense} \approx 60 \text{ fF} \left[1 \pm \frac{y}{G_0}\right] \qquad (19.19)$$

The actual capacitance is larger than this 60 fF estimate, because for H and G_0 comparable, fringing fields are important. A better estimate for the at-rest capacitance is about 100 fF.

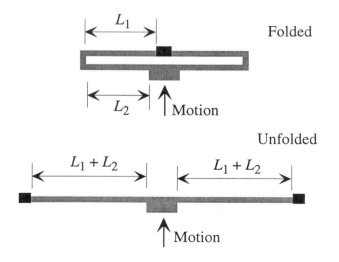

Figure 19.22.   Illustration of a folded spring, and its unfolded equivalent.

Figure 19.22 illustrates how one of the folded springs, with two segments each of length L_1 and L_2 can be "unfolded" to create a doubly-clamped beam of length $2L$, where L is equal to $L_1 + L_2$. The beam has a point load at its midpoint corresponding to the inertial load of the proof mass. The spring constant can be estimated from our analysis of the point-loaded beam in Section 10.4.1, in which we used energy methods to estimate the spring constant k of a beam loaded with a point force at its center (see Eq. 10.61).

$$k = \frac{F}{c} = \left(\frac{\pi^4}{6}\right)\left[\frac{EWH^3}{(2L)^3}\right] \tag{19.20}$$

where F is the applied load, c is the central beam displacement, E is the Young's modulus, W is the beam width, H is the beam thickness, and $2L$ is the total length of the combined unfolded springs.[7] In this case, W corresponds to the polysilicon thickness, since the beam is bending in the plane of the wafer, and H corresponds to the lithographically defined in-plane width of the beam element, estimated to be 2.5 μm. If we assume that L_1 and L_2 are equal to one another and are on the order of 75 μm so that $2L = 300$ μm, we obtain an estimate for the stiffness of the folded spring at each end of the proof mass to be 2.8 N/m. If we combine the effects of the springs at the two anchors, the net stiffness would be doubled, to 5.6 N/m. This is very close to the actual value of 5.4 N/m supplied by ADI based on actual devices. This estimate does not include the spring-softening effect of the applied voltage (see Section 7.3.2, especially Fig. 7.5). An estimate of this effect[8] using a nominal gap of 1.3 μm and a nominal at-rest capacitance of 100 fF with 2.5 V across the capacitor is a net reduction in stiffness of 0.4 N/m.

The mass of the shuttle with attached cantilever electrodes can be estimated from device dimensions. The ADI-supplied value is $m = 2.2 \times 10^{-10}$ kg. Combining this mass with the stiffness, the undamped resonant frequency ω_o is 1.55×10^5 rad/sec, corresponding to 24.7 kHz. Thus the operational bandwidth of 1000 Hz for the accelerometer is much less than the resonant frequency.

The stated Q-factor of the mechanical resonance is 5. This is actually a difficult quantity to calculate because the effect of squeezed film damping between the fingers must be added to the Couette flow beneath the proof mass as the shuttle displaces. Further, because of the low aspect ratio of the gaps between the fingers, even squeezed-film damping estimates are inaccurate. Nevertheless, it is useful to see how close we come with a Couette flow model

[7]The coefficient in front of this expression has value 16.2, while the exact solution for a doubly-clamped beam with a point load in the center is 16. This can be readily found by solving the beam equation with a central point load using two cubic polynomials, one to the left of the load, the other to the right. The two solutions must be symmetric, and the discontinuity in the fourth derivative at the center of the beam must match the point load.
[8]See Problem 19.4.

(see Sec. 13.3.1), assuming a shuttle area A of 500×50 μm, and an air gap h beneath the shuttle of 1.6 μm, corresponding to the thickness of the sacrificial oxide. The damping constant b, obtained from Eq. 13.39 is

$$b = \frac{\eta A}{h} \qquad (19.21)$$

where η is the viscosity of the gas surrounding the device (we use 18 μPa-sec for air) and A is the area of the plates. The value of b is 2.8×10^{-7} (N-sec)/m, corresponding to a quality factor

$$Q = \frac{m\omega_o}{b} \approx 120 \qquad (19.22)$$

This estimate is much higher than the actual Q, indicating that the Couette flow significantly underestimates the total damping force.

We can also estimate the squeezed-film damping between the capacitor fingers. If we use the fact that there are 54 moving fingers, hence 108 gaps, and use Eq. 13.81 to estimate the squeezed film damping between the capacitor fingers (even knowing that it will be a poor approximation because the geometric requirements for squeezed film damping are not met in this structure), we obtain an estimated squeezed film damping b value of 7.3×10^{-7} (N-sec)/m, which is about three times what we estimated for the Couette flow term. The total b value combining these two estimates 1.01×10^{-6} (N-sec)/m. The corresponding Q value is 34, still much larger than the actual Q of 5. Microcosm Technologies [107], using numerical fluid modeling simulations for the full device geometry, was able to calculate a much better estimate of the total fluid damping for the ADXL76 accelerometer, a device close in design to the ADXL150 considered here. Their result agrees with experiment to within 30%.

This example illustrates how, with complex geometries, it can be very difficult to predict the Q of a mechanical resonance that depends on fluid flow. The geometric and fluid-flow difficulty is compounded by the fact that for many of these surface-micromachined structures, there are perforations added to the proof mass to speed up the release etch.

19.4.3   Fabrication and Packaging

The ADXL150 accelerometer is fabricated in an Analog Devices process named iMEMS. The process combines MOS transistors, bipolar transistors, and polysilicon micromachined structures into a single process flow. A 1994 version of the process that is publicly documented [108] used 24 masks, with 13 used for the electronics and 11 for the mechanical structure and interconnect to the electronics. It is interesting to examine the interface between the circuit and sensor areas in this process as an example of the intricacy of a fully merged process designed to combine freely moving mechanical parts with fully protected electronic components.

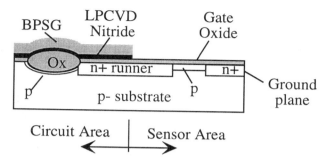

Figure 19.23. The interconnect region between the circuit area and sensor area in the iMEMS process after completion of the transistors, but before construction of the mechanical element. The gate polysilicon for the transistors has already been deposited, but is not shown in this figure.

The process starts with a lightly doped p-type wafer $(p-)$ in which are fabricated MOS and bipolar transistors – an n-well for the PMOS devices, sources, drains, and polysilicon gates for the MOS transistors and implanted and diffused bases and emitters for the bipolar transistors. Along the way, the interconnects between the circuit region and the sensor region are doped $n+$ to create *runners* as well as a *ground plane* beneath what will become the capacitive sensor. The rest of the region beneath the sensor receives a boron implant to help isolate the various $n+$ diffused interconnect and ground-plane regions. After the transistors are completed and overcoated with an LPCVD nitride layer and a deposited oxide (BPSG), a central sensor region, called the moat, is cleared all the way down to gate oxide. The resulting structure is illustrated schematically in Fig. 19.23.

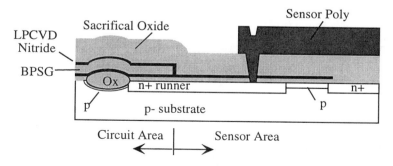

Figure 19.24. Illustrating the iMEMS process after deposition and patterning of the sensor polysilicon.

The device is then overcoated with LPCVD nitride to serve as an etch stop for the release etch. This nitride is removed above what will become the sensor ground plane. The sacrificial oxide layer (1.6 μm thick) is deposited and a contact opening is etched to the $n+$ interconnect. The sensor polysilicon is deposited, doped by phosphorus implant and patterned. This brings the process to the drawing in Fig. 19.24. There is also a contact between the sensor polysilicon and the $n+$ ground plane, but this is not shown in the figure.

The sensor polysilicon is deposited as a partially amorphous film to assure tensile stress after anneal. It is 2 μm thick, and is doped by phosphorous implant to a sheet resistance of 150 Ohms/square. Grain size in the polysilicon ranges from 0.3 to 0.5 μm. The final stress in the polysilicon (after all thermal processing) is targeted in the 40-75 MPa range. This tensile stress does not affect the spring stiffness. The folded spring design permits relaxation of the axial stress in the spring beams on release. However, when the shuttle is released, it can bend the folded springs or even buckle out of plane; therefore, the polysilicon stress is designed to be slightly tensile to prevent any warping or buckling of the spring or shuttle that could be created by compressive stress.

It is also important to control the vertical stress gradient in the polysilicon in order to prevent unacceptable bending of the cantilevered electrodes. In the 1994 iMEMS documentation, the recommended maximum length for a cantilevered structure was 150-180 μm. More recent process improvements have greatly reduced the warpage due to stress gradients so that it is no longer a major design limitation.

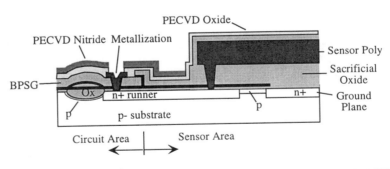

Figure 19.25. Illustrating the iMEMS process after deposition and patterning of PECVD oxide and nitride, opening of contact holes, and metallization.

An additional thin oxide is then deposited over the wafer, and both this oxide and the sacrificial oxide are removed from the circuit area and the immediately adjacent sensor area. Contact openings are patterned in the circuit area to connect to the $n+$ runners. Pt silicide is formed in the contact openings, followed by TiW/Al metallization. Also deposited in this part of the process, but not shown, are the trimmable SiCr resistors. A PECVD oxide followed by a PECVD nitride film is then deposited, each being patterned as shown in Fig. 19.25.

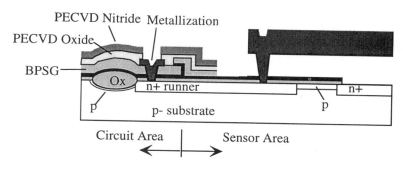

Figure 19.26. Schematic of the final released structure in the iMEMS process.

Fig. 19.26 illustrates the final released structure. Note that the *p*-type region beneath the suspended polysilicon structure is covered with LPCVD nitride and gate oxide. However, the *n*+ ground plane in the region beneath the capacitive structure is exposed. The circuit area, except for the metallization, is completely protected with a combination of PECVD oxide and PECVD nitride as well as a layer of LPCVD nitride beneath the BPSG.

An important issue is when to do die separation, before or after the release etch. Because die separation involves sawing the wafer, fragile structures can be broken. In addition, the sawing slurry produces particles that can contaminate the mechanical structure and render it nonfunctional. The Analog Devices solution to this problem [109] is to perform the release etch at the wafer level (clearly an economical approach) and then to saw the wafer from the back of the wafer instead of the usual process of sawing from the front. During sawing, the released mechanical elements are protected by two layers of tape on the front side of the wafer. The first layer of tape has holes punched into it corresponding to the sensor moats on a wafer. The thickness of this tape is greater than the height of the mechanical elements on the wafer. Thus each hole creates a little well around a sensor element. These wells are then covered by a second layer of tape, which is continuous, "wrapping" each sensor element in a cavity during die sawing. There is a coating applied to the mechanical elements to prevent stiction, but the details of that process are proprietary. After die separation, the devices are mounted in a standard 14-pin cerpak package, wire bonded, and lid sealed.

19.4.4 Noise and Accuracy

The sensitivity of the accelerometer is determined by the noise, which specified as 1 mg/$\sqrt{\text{Hz}}$ in a bandwidth from 10 Hz to 1000 Hz, about twice the

Brownian noise estimate of 0.5 m$g/\sqrt{\text{Hz}}$. If we assume that the stated noise is white over the 1000 Hz bandwidth, the root-mean-square acceleration noise δa would be 32 mg. Recalling that displacement and acceleration for a quasi-static accelerometer are related by the square of the resonant frequency, we infer that the root-mean-square effective position error δx would then be

$$\delta x = \frac{\delta a}{\omega_o^2} = \frac{(32 \times 10^{-3})(9.8)}{(1.55 \times 10^5)^2} = 0.013 \text{ nm} \tag{19.23}$$

The corresponding capacitance change can be found from Eq. 19.19, but using 100 fF as the undisturbed capacitance. The nominal gap is 1.3 μm, so the minimum detectable capacitance change is 10^{-18} F, or one aF. The published noise spectrum shows no $1/f$ noise, at least above 10 Hz, indicating not only that the use of modulation and synchronous detection suppresses $1/f$ amplifier noise, but also that the intrinsic $1/f$-noise component of the capacitive sensor is quite low.

The offset of the sensor has an interesting origin. First of all, the device must live in a one-g world (unless it is launched into space). Therefore, the orientation of the device relative to the earth's gravitational field affects its apparent offset. Further, if the gaps between fixed and moveable electrodes are not perfectly vertical and identical to each other, the as-formed differential capacitance element is not perfectly balanced at the rest position. One of the two capacitors is slightly larger than the other, even in the absence of any applied inertial load. When the drive voltage is applied, there is a net force on the moveable element tending to shrink the smaller of the two gaps. This force is balanced by deformation of the spring. Hence, the act of turning on the voltage to measure the capacitance actually increases the positional offset of the device! As an example, if the gaps are mismatched by 1%, then the unbalance force with an applied voltage of 2.5 V across each capacitor is 0.01 μN, which is enough to move the shuttle by 0.002 μm, a substantial shift. A displacement of this size corresponds to an acceleration signal of almost 5 g. Since this effect depends on the drive voltage, as does the spring-softening effect discussed earlier, the stability of the drive-voltage amplitude is critical to the accuracy of the sensor. Drifts in this voltage due to temperature change can produce a corresponding change in offset and sensitivity, but for a ratiometric device, these effects are somewhat mitigated.

To assure that the etching is accurate, ADI studied the pattern dependence of the etching tool used to cut the mechanical structure and found that the etch rate and uniformity depended on the size of the gap [110]. Therefore, the design is constrained such that all critical features must be etched with gaps that are the same size. This includes the beams for the springs. They are surrounded by dummy polysilicon features so that the etched gap defining the springs is the same as that defining the electrodes.

Cross-axis sensitivity is another important accelerometer characteristic. The device should be sensitive in one axis only and should be packaged so that the sensitive axis can be clearly established when the packaged device is mounted for its end-use application. Misalignment of the device relative to the fiducial markings on the package can result in an apparent cross-axis sensitivity. But the device itself can also have an intrinsic cross-axis response. The risk of a cross-axis response can be estimated by comparing either the stiffnesses or resonant frequencies of the relevant deformation modes. There are two cross-axis accelerations of concern, one in the plane of the shuttle, and one normal to the plane of the shuttle. The in-plane cross-axis stiffness is very large compared to that of the sensing-mode stiffness due to the folded-spring design. However, the beams of the folded springs have nearly a square cross section. This means that vertical vibration modes will have nearly the same resonant frequency as the in-plane sense mode. But the damping of these two modes is very different. We have already seen that the damping of the sense mode produces a Q of about 5. In contrast, the vertical mode will be heavily damped by the squeezed-film damping beneath the shuttle. Hence, transient accelerations applied in the vertical direction will not produce large displacements. Further, the differential capacitor in combination with the sense circuitry will not respond, at least to first order, to vertical motion of the proof mass. Hence, in spite of the fact that this sensor has nearby cross-axis modes, the cross-axis sensitivity is acceptably low.

Additional contributions to accuracy come from the gain of the various amplifier stages, which are laser trimmed to achieve the correct full-scale output. As is evident from this discussion, there are many detailed issues that contribute to the overall performance of a device of this type. We now examine briefly a different way to make an accelerometer, and compare some of its features to the capacitive approach.

19.5 Position Measurement With Tunneling Tips

Perhaps the most sensitive method for measuring position is to exploit the exponential dependence of current on the distance between an atomically sharp tunneling tip and an electrode. Scanning tunneling microscopes which use this principle are well-established commercially. Microfabrication of structures that incorporate tunneling tips permit the implementation of a wide variety of sensors. In this section,[9] we examine briefly an accelerometer based on tunneling [111], with particular attention to three features: first, the very high sensitivity to displacement characteristic of tunneling; second, the fact that

---

[9]This section is based in part on materials and commentary provided by Thomas Kenny and Cheng-Hsien Liu of Stanford University.

successful deployment of tunneling in a sensor device requires force feedback; and, third, the unusual noise characteristics of the tunneling sensor.

Figure 19.27. The basic tunneling-tip transducer. The current between electrodes is an exponential function of the displacement d_t.

Figure 19.27 shows a sharp conducting tip near a flat electrode. When the distance between the tip and the electrode approaches nanometer dimensions, a conducting path is created by direct tunneling of electrons across the gap. Tunneling is a quantum-mechanical phenomenon, and the details fall well outside the scope of this book. From an engineering point of view, the important fact is that if a bias voltage V_B is applied between the electrodes, a tunneling current I_t flows. The dependence of I_t on the gap d_t is of the form:

$$I_t \propto V_B e^{-\alpha_I \sqrt{\Phi} d_t} \qquad (19.24)$$

where α_I is a constant equal to 1.025 Å^{-1} $\text{eV}^{-0.5}$, and Φ is the effective height of the quantum mechanical barrier through which the electrons must tunnel, which is determined to be about 0.2 eV. The net result is an exponential dependence of current on displacement with a decay length of .046 nm. Typical current levels are nanoamps when the displacement is of order 1 nm at voltages of order 0.2 V.

A distance change of only 0.01 nm results in a 4.5% change in tunnel current, and changes well below this level can be measured. However, because of the strong exponential dependence of current on distance, it is necessary to use force feedback with these tips to keep the distance and corresponding current within a useful operating range. Furthermore, because it is virtually impossible to fabricate accurate 1 nm gaps, the usual practice is to fabricate the tip and counter electrode well spaced from one another, and use an actuator to bring the tip and counter electrode into proximity. Then, during operation, the small changes in the actuator force required to maintain a fixed tunneling current can be interpreted in terms of the acceleration applied to the proof mass.

The specific device we shall examine is shown schematically in Figs. 19.28 and 19.29. It is fabricated from two wafers that are bonded together after each wafer is fully processed. The tunnel tip is formed from silicon by undercutting an oxide mask with KOH etching. The tip is then coated with a thin oxide, a thin nitride, and an evaporated Cr/Pt/Au metal sandwich. The underlying silicon is then removed beneath the tip, creating a tip on a nitride cantilever. The advantage of this compliant support for the tip is to prevent crushing the

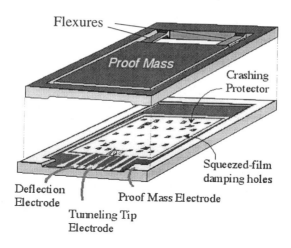

Figure 19.28. Exploded view of the tunneling accelerometer. (Source: T. Kenny, Stanford U.; edited and reprinted with permission.)

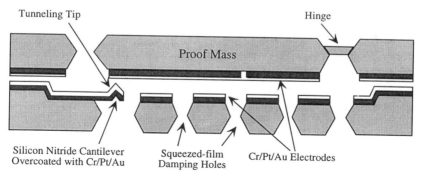

Figure 19.29. Cross-section of the tunneling accelerometer. There is a thin oxide beneath the nitride layer that is not shown in the Figure. (Source: T. Kenny, Stanford U.; edited and reprinted with permission.)

tip if there is accidental mechanical contact with the counter electrode. The resonance frequency of the cantilever is high, about 40 kHz, which is far above the resonant frequency for the proof mass. Additional protection is provided by the "crashing protectors," which are unmetalized protrusions that prevent the proof mass from crushing the delicate tip during normal device handling. The tips fabricated with this process are not particularly sharp by the standards used in scanning tunneling microscopes. But since these tips are used only for displacement measurement, their lateral dimensions are not as critical. There will always be some atomic feature that protrudes out the most and, hence, is closest to the counter electrode, and that feature will dominate the tunneling current. A photograph of a completed tip is shown in Fig. 19.30.

Nitride cantilever with tunneling tip

Figure 19.30.   Photograph of a fabricated tunnel tip. (Source: T. Kenny, Stanford U.; reprinted with permission.)

The lower wafer contains gold electrodes, one of which contacts the electrode on the lower face of the proof mass, a second contacts the tunnel tip, and a third extends beneath the proof mass to provide actuation. As fabricated, the space between the proof mass and the tip is relatively large. A large DC voltage V_{DC} (of order 25 V) is applied between the deflection electrode and the proof mass. To this large voltage is added a small additional time-varying voltage v_{ac} whose magnitude is controlled by feedback from the tunneling current. By using feedback to keep the tunnel current constant, the proof mass is forced to maintain a fixed distance from the tunnel tip.[10] The total actuation force acting on the proof mass is proportional to the square of the applied voltage, that is, to $(V_{DC} + v_{ac})^2$, which expands to $V_{DC}^2 + 2V_{DC}v_{ac}$ plus a small term that goes as v_{ac}^2. Thus the total force has a component that is linear in v_{ac}. Under closed-loop circumstances, with v_{ac} adjusted to maintain a constant tunneling current, measurement of v_{ac} is equivalent to measuring the inertial force that is disturbing the feedback loop.

The noise characteristics of this accelerometer are interesting. First, because the device itself has a strong exponential nonlinearity, one can think of the tunneling current as having a large gauge factor. We recall that the gauge factor of a piezoresistor was the ratio of the fractional change in resistance to

---

[10]Details of the control circuit can be found in [112].

the strain, which is the fractional change in dimension. In this case, we can also define a gauge factor as the ratio fractional change in tunnel current to the fractional change in gap.

$$\text{G.F.} = \left| \frac{\delta I_t / I_t}{\delta d_t / d_t} \right| = \frac{d_t}{.046 \text{ nm}} \tag{19.25}$$

where, in normal operation, d_t is about 1 nm. Therefore, the gauge factor for this sensor is about 22. As a direct position sensor, the tunneling sensor is about 20 times as sensitive as a capacitance sensor, which typically has a gauge factor of unity.[11] This means that the tunneling sensor acts as if it has the first stage of amplification built right into the sensor, and its noise performance, rather than being dominated by electronic amplifier noise, is directly related to the intrinsic noise of the device itself.

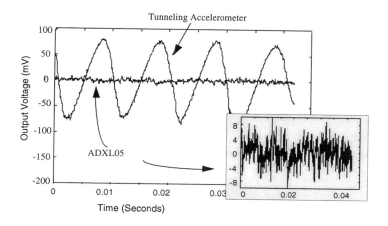

Figure 19.31.   Comparison of the response of a tunneling accelerometer to that of an Analog Devices ADXL05 when driven with identical 1.5 mg accelerations at 100 Hz. (Source: T. Kenny, Stanford U.; edited and reprinted with permission.)

Experience with the tunneling sensors has shown that sub milli-g sensitivity is readily achieved. Figure 19.31 shows a direct comparison between a tunneling accelerometer and a 5-g version of the Analog Devices capacitive accelerometer (ADXL05). Both devices are subjected to a sinusoidal acceleration of 1.5 mg at 100 Hz. The devices do have $1/f$ noise, much of which has been attributed to thermal-drift-driven bimetal effects in the hinges. The newest device designs [113] have reduced the $1/f$ noise by a factor of three,

---

[11] This is readily seen from the fact that capacitances typically vary only with the first power of a displacement coordinate, rather than as an exponential.

achieving equivalent 20 nano-$g/\sqrt{\text{Hz}}$ noise levels in devices with a 5 Hz - 1.5 kHz bandwidth.

The tunneling devices are not as repeatable, device to device, as the capacitive sensors, both in their basic sensitivities and in their noise characteristics. Thus, it may be a while before these highly sensitive devices find their way into mass-market applications. However, the contrast between the use of capacitance *vs.* tunneling to measure position is both interesting and instructive.

Related Reading

N. Yazdi, F. Ayazi, and K. Najafi, Micromachined inertial sensors, *Proc. IEEE*, Vol. 86, pp. 1640-1659, 1998.

Problems

19.1 For the circuit of Fig. 19.6, assume that $x = x_o \cos \omega_m t$ and $V_s = V_{so} \cos \omega_c t$. Determine the spectrum of the output signal, and determine a criterion for neglect of the contribution of the velocity to the output.

19.2 Derive the switched-capacitor amplifier result in Equation 19.12, and find the corresponding result when using the circuit of Fig. 19.6.

19.3 Derive the correlated double sampling result in Equation 19.16.

19.4 Estimate the effect of electrostatic spring softening on the Analog Devices accelerometer, as discussed in Section 19.4.2.

19.5 An Accelerometer Treasure Hunt: Go to the literature or to a commercial web site and locate a piezoresistive accelerometer or a capacitive accelerometer other than one from Analog Devices. Create an approximate model for the accelerometer sensitivity and dynamic response, and compare the stated noise threshold with the Brownian motion noise floor.

Chapter 20

ELECTROSTATIC PROJECTION DISPLAYS

. . . he beholds the light, and whence it flows,
He sees it in his joy;

—Wordsworth, *Ode to Duty*

20.1   Introduction

MEMS devices are well-suited for interacting with light. The structural dimensions are on the same order as the wavelength of infrared or visible light. Control of light with either reflection or diffraction can be achieved with relatively small motions. Further, microfabrication starting materials have mirror-smooth surfaces and, with suitable care in fabrication, this smoothness can be retained in deposited layers. Finally, actuators for the control of light only need to manipulate the structure itself; large amounts of external work are not required.

There has been a huge expansion of work devoted to optical MEMS devices and systems. Much of it is summarized in the excellent review by Walker and Nagel [114]. Much of the rest of it can be found on the business pages of the newspaper: high-profile high-dollar corporate buyouts of small MEMS companies by the optical telecommunications industry.[1] This chapter explores only one aspect of this new and exciting area, the use of MEMS devices to create projection display systems. Along the way, we shall study diffraction theory and the electrostatic actuation of beams, including the use of their quasi-static

---

[1]To name a few, in the first half of calendar year 2000 Nortel purchased Xros and CoreTek, JDS Uniphase purchased Cronos, Corning purchased IntelliSense, and Cypress Semiconductor purchased Silicon Light Machines. The average price: more than one billion dollars.

and resonant characteristics to infer both material properties such as residual stress and the effect of small gap sizes on squeezed-film damping.

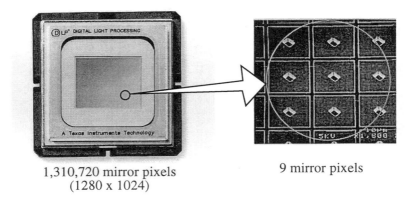

1,310,720 mirror pixels   9 mirror pixels
(1280 x 1024)

Figure 20.1.   The Texas Instruments deformable mirror device (DMD). A two-dimensional array of tiltable micromirrors is built above a silicon chip containing drive circuitry. (Source: Texas Instruments Digital Imaging Division; enhanced and reprinted with permission.)

There are two basic approaches now in use for MEMS-based projection displays: reflective displays, pioneered by Texas Instruments, and diffractive displays, pioneered by Silicon Light Machines. The Texas Instruments approach [23], called the DMD,[2] is illustrated in Fig. 20.1. A large two-dimensional array of aluminum micro-mirrors is fabricated over a silicon integrated circuit that contains the drive electronics along with electrodes for electrostatically tipping the mirrors to various positions. Each mirror controls one pixel of the image.

A display system using the DMD chip is illustrated in Fig. 20.2. The DMD chip is mounted on a circuit board that contains the circuitry needed to convert a digital representation of an incoming image into the appropriate sequence of control signals to actuate the mirror array. Incident light from a collimated source is either reflected into the image direction, when the mirror is in one position, or deflected out of the image direction when the mirror is tipped to a different position. Gray scale is achieved by time division, controlling the length of time within the frame interval that the mirror is tipped to reflect light in the image direction. This technology is being commercialized in hand-carried projectors that can be used for projection of digital images directly from a

---

[2]DMD is a registered Trade Mark of Texas Instruments, Inc.

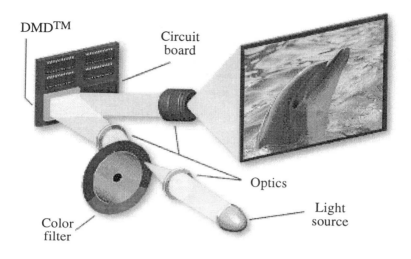

DMD™

Circuit board

Optics

Light source

Color filter

Figure 20.2. Schematic illustration of a display system using one DMD chip in combination with a color wheel. Three-chip systems, one each for red, green, and blue, are also being manufactured. (Source: Texas Instruments Digital Imaging Division; reprinted with permission.)

computer file, and is also being developed for large-auditorium projection systems.

The Silicon Light Machines approach, called a *grating light valve* display or GLV display,[3] is different (see Fig. 20.3). Instead of using a mirror to reflect the light from each pixel of an image, an array of small electrostatically actuated diffraction gratings is used. When unactuated, the array reflects incident light back to the source, but when actuated, the array diffracts light at a specific angle which can be collected by the optical system. A small diffraction grating of six beams forms one pixel of the image. Maximum diffraction intensity is achieved when the beams are displaced by $\lambda/4$, where λ is the wavelength of the light.

Figure 20.4 illustrates how a single GLV pixel is used. Incoming light is directed onto the pixel by a centrally located mirror. When the pixel is actuated, diffracted light is collected between the solid aperture and the mirror, focused by a lens, and directed onto the screen. Without any actuation, the screen is dark because the incident light is reflected back along its own path.

The original GLV display concept was based on a two-dimensional array of diffractive pixels [115], just as in the two-dimensional Texas Instruments micro-mirror array. More recently, the Silicon Light Machines GLV products use a *linear array* of pixels, which is a much smaller microfabricated device

---

[3]GLV and grating light valve are registered Trade Marks of Silicon Light Machines.

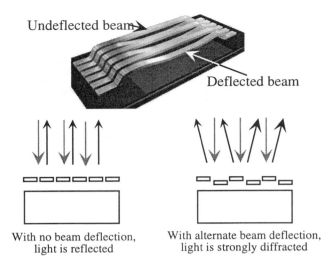

With no beam deflection,   With alternate beam deflection,
light is reflected   light is strongly diffracted

Figure 20.3.   Schematic illustration of the pixel design for a grating-light-valve (GLV) display. With no deflected beams, the pixel reflects normal-incident light. When alternate beams in the pixel are deflected, the pixel diffracts light at an angle. (Source: Silicon Light Machines; enhanced and reprinted with permission.)

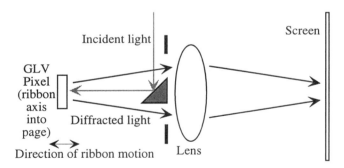

Figure 20.4.   Illustrating how the diffractive projection from a single pixel is achieved.

with far fewer moving parts than the corresponding two-dimensional array. A wafer with multiple linear arrays is shown in Fig. 20.5. Figure 20.6 illustrates how such a linear array can be used, together with a horizontal scanning mirror, to create a full two-dimensional display. This approach depends on having extremely rapid actuation speed so that each GLV pixel can project a complete horizontal line of image pixels within one frame time.

Gray scale can be controlled by adjusting the amount of time during which the pixel is actuated, just as with the DMD. However, the diffractive approach offers an alternate method of achieving gray scale. The diffracted intensity at

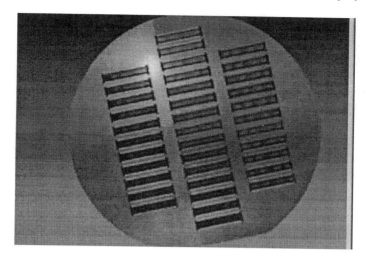

Figure 20.5. A wafer of GLV linear arrays. (Source: Silicon Light Machines; reprinted with permission.)

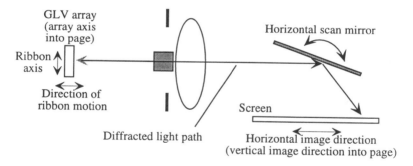

Figure 20.6. Illustrating the combination of a linear grating-light-valve array, forming the vertical components of an image, with a horizontal scanning mirror to create a full two-dimensional image. Note that this figure is rotated 90° with respect to Fig. 20.4. The direction of the diffracted light is now in and out of the paper and therefore appears to pass through the position of the prism.

a fixed viewing angle depends on the amplitude of the grating displacement within a pixel. Therefore, by using varied analog voltages to actuate each pixel, gray scale can be controlled directly.

Both approaches depend on the electrostatic actuation of suspended microstructures. The DMD device uses the pull-in instability discussed in Section 6.4.3 to tip each mirror until it hits a limiting position that fixes the reflection angle. Each mirror element is supported by an elastically linear torsional spring. As one electrode or the other is actuated, electrostatic attraction tips the mirror toward the active electrode. With sufficient voltage, pull-in is exceeded and the mirror tips until it contacts the landing pad.

The original GLV device also used pull-in [115], in this case, vertical pull-in of beams until they contacted the electrode, resulting in a central region of each beam that is displaced yet flat. This is an effective geometry for a diffraction grating, but it introduces potential problems with charging of the dielectric materials that insulate the electrodes and contact charge transfer effects when surfaces touch. It also requires use of time division within a frame interval to control gray scale. The Silicon Light Machines version of this device uses analog control of gray scale instead of fully pulled in beams and, since pull-in is not used, contact charging is not an issue.

We now examine the basic electromechanics of the DMD device, which can be analyzed with rigid-body pull-in methods. Then we examine the electrostatic actuation of elastic beams, and the dependence of their behavior on residual stress. Finally, we examine the GLV display system in more detail, including device fabrication and packaging.

20.2   Electromechanics of the DMD Device
20.2.1   Electrode Structure

The tipping motion of the DMD device is based on the change of capacitance between two initially parallel plates when one plate is free to rotate about a support.

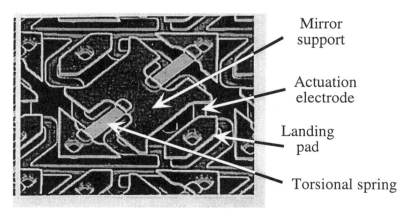

Mirror
support

Actuation
electrode

Landing
pad

Torsional spring

Figure 20.7. One pixel of the DMD device without the mirror and its central post. The H-shaped mirror support is free to rotate about the axis of the torsional spring in response to voltage applied to the actuation electrode. (Source: Texas Instruments Digital Imaging Division; enhanced and reprinted with permission.)

Figure 20.7 shows a single pixel without the mirror element. There is an H-shaped mirror support suspended above the substrate by a thin torsional spring.

The mirror is built over the middle of this support with a central post elevating the mirror above the plane of the support. A pair of actuation electrodes beneath the mirror support allows for tipping of the mirror support about the axis of the torsional spring, and since the mirror is attached to the center of this support, the mirror tips when the support does. To prevent short circuits, and to provide for an accurately controlled motion, there are landing pads for the corner of the mirror element. These pads, as well as the mirror element and its support, are all grounded so there is no electrical event when the mirror strikes the landing pad. The electrodes can be seen more clearly in Fig. 20.8. The mirror, its post and support have been broken away to reveal the actuation electrodes and landing pads.

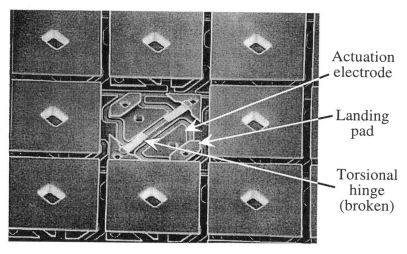

Actuation electrode

Landing pad

Torsional hinge (broken)

Figure 20.8. A DMD pixel array with one mirror, its post and mirror support broken away to reveal the actuation electrode beneath. (Source: Texas Instruments Digital Imaging Division; enhanced and reprinted with permission.)

20.2.2   Torsional Pull-in

We shall analyze the pull-in of the torsional mirror with two methods. We begin with the energy-based method of calculating the capacitance as a function of angle, and demonstrating that the torque which results is a nonlinear and increasing function of the angle. Hence, we can argue that there will be an angle at which the equilibrium between the torque and a linear restoring force will become unstable. The second method, that used by Hornbeck in his original analysis of the DMD structure [116], is to calculate the torque directly from a parallel-plate approximation to the tilted capacitor.

Figure 20.9 shows a simplified schematic of a single pixel, with the torsional spring, the mirror support, and one actuation electrode. This geometry is

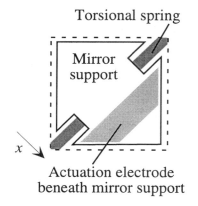

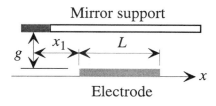

Figure 20.9.   Geometry of the DMD mirror.

similar to the DMD design analyzed by Hornbeck. Assuming that the mirror support is grounded, we can find the torque on the mirror support by calculating the capacitance C between the actuation electrode and the mirror support as a function of tilt angle θ_0 (see Fig. 20.10). Assuming voltage-controlled actuation, we then formulate the co-energy:

$$W^*(\theta_0) = \frac{1}{2}CV^2 \tag{20.1}$$

and find the torque as the negative gradient of the co-energy with respect to θ_0. Because the width of the electrode (in the direction perpendicular to the x-axis) is not uniform, this calculation is best done numerically. For illustration though, we shall carry it through assuming a uniform electrode width in order to get an analytical result we can work with.

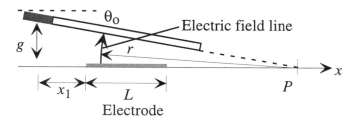

Figure 20.10.   Illustrating the field and capacitance calculation for the titled DMD.

Figure 20.10 illustrates the field calculation. The point P is where the projection of the tilted mirror hits the axis. Assuming a voltage V applied to the electrode, then the solution to the Laplace equation in the space between the electrode and the support (neglecting fringing) is

$$\phi(\theta) = V\left(1 - \frac{\theta}{\theta_o}\right) \tag{20.2}$$

where θ_o is the tilt angle, and the positive orientation of θ is taken as clockwise for convenience. The electric field is the negative cylindrical-coordinate gradient of ϕ:

$$\mathcal{E} = \frac{V}{r\theta_o}\hat{\imath}_\theta \tag{20.3}$$

The electric field follows circular arcs so that it can be normal to both surfaces, a requirement at the surface of a conductor under static conditions. The charge density on the surface of the electrode is equal to $\varepsilon_o\mathcal{E}$ evaluated at the electrode surface. Therefore, the total charge on the electrode is

$$Q = \frac{\varepsilon_0 V}{\theta_0} \int_{P-(x_1+L)}^{P-x_1} \frac{W(r)dr}{r} \tag{20.4}$$

where $W(r)$ is the variable width of the electrode. We shall assume a constant width W to illustrate the method. The capacitance is the ratio of charge to voltage. The tilted capacitor has the value

$$C = \frac{\varepsilon_0 W}{\theta_0} \ln\left[\frac{P - x_1}{P - (x_1 + L)}\right] \tag{20.5}$$

This can be related to the device dimensions by noting that

$$P = g \cot\theta_o \tag{20.6}$$

Thus

$$C = \frac{\varepsilon_0 W}{\theta_0} \ln\left[\frac{1 - \left(\frac{x_1}{g}\right)\tan\theta_o}{1 - \left(\frac{x_1 + L}{g}\right)\tan\theta_o}\right] \tag{20.7}$$

In the limit of small θ_0, this becomes

$$C = \frac{\varepsilon_0 WL}{g}\frac{\tan\theta_0}{\theta_0} \tag{20.8}$$

This result reduces to the parallel-plate result for θ_0 approaching zero. However, this small-angle approximation is not suitable for torque calculations because it neglects the very important nonlinearities that create the pull-in. Figure 20.11 plots the capacitance versus angle calculated from the full equation, normalized to the parallel-plate value C_0. The dimensions used for illustration purposes were $g = 1$ μm, $x_1 = 5$ μm, and $L = 10$ μm. The width cancels out of the normalized capacitance.

The results in Fig. 20.11 are extremely well-represented by a linear-cubic fit:

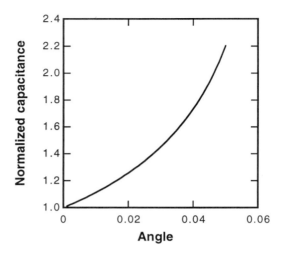

Figure 20.11.   Capacitance of a DMD example versus angle with $g = 5$ μm, $x_1 = 5$ μm, $L =$ 10 μm.

$$\frac{C(\theta_0)}{C(0)} = 1 + a_1\theta_0 + a_3\theta_0^3 \tag{20.9}$$

where for this particular geometry, $a_1 = 10.36$ and $a_3 = 5211$. This leads to an electrostatic co-energy

$$W^*(\theta_0) = \frac{1}{2}C(0)V^2(1 + a_1\theta_0 + a_3\theta_0^3) \tag{20.10}$$

and, therefore, to a torque τ of

$$\tau = -\frac{\partial W^*(\theta_0)}{\partial\theta_0} = -\frac{1}{2}C(0)V^2\left[a_1 + 3a_3\theta_0^2\right] \tag{20.11}$$

Static equilibrium occurs when the magnitude of this torque equals the restoring torque of the torsional spring, denoted here by $k_\theta\theta_0$. The result is a tilt angle given by

$$\theta_0 = -\frac{k_\theta}{3a_3C_0V^2} \pm \sqrt{\left(\frac{k_\theta}{3a_3C_0V^2}\right)^2 - \frac{a_1}{3a_3}} \tag{20.12}$$

This has a real solution provided that

$$\left(\frac{k_\theta}{3a_3C_0V^2}\right)^2 \geq \frac{a_1}{3a_3} \tag{20.13}$$

Thus, there is a real solution provided that V is less than the pull-in voltage, given by

$$V_{PI} = \left(\frac{k_\theta^2}{3a_1 a_3 C_0^2} \right)^{1/4} \tag{20.14}$$

Hornbeck's analysis of this same problem proceeds somewhat differently [116]. He treats the tilted plate as if it were still parallel enough to use parallel-plate capacitance relations to calculate the capacitance, the charge density, and the attractive force between the plates as a function of position along the plate. He weights the attractive force by its distance from the torsional spring, computing the torque directly. His expression for the magnitude of the torque is

$$\tau = \int_A \left(\frac{\varepsilon_0 V^2}{2[g_0 - w(x)]^2} \right) x \, dx \tag{20.15}$$

where A is the area of the electrode, g_0 is the original gap, and $w(x)$ is the position-dependent displacement of the tilted beam from its original location, equal to $x \tan(\theta_0)$. We recognize this as the parallel-plate force weighted by the distance from the torsional support. This equation is integrated numerically to yield a relation very similar to our analytical example. The result is a well-defined pull-in voltage. In operation as a digital display device, voltages that exceed the pull-in voltage are used to switch the position of the mirror. After it is switched, much lower voltages can be used to hold it in position, taking advantage of the hysteresis that is inherent in systems that exhibit pull-in.

20.3 Electromechanics of Electrostatically Actuated Beams

We now examine the electromechanics of the GLV device. The DMD analysis of the previous section treated the moving element as a rigid body. In this section, we extend the analysis to deformable microstructures such as thin beams (see Fig. 20.12). The problem is that the actuation force is a nonlinear function of the displacement, making exact analytical solution of the governing beam equation impossible. Hence, we must either use approximations or we must develop accurate macro-models that capture the nonlinearity correctly.

In Section 9.6.2, we examined the effect of residual axial stress in modifying the stiffness of a beam with a distributed load. The result was the Euler beam equation

$$EI \frac{d^4 w}{dx^4} - (\sigma_0 W H) \frac{d^2 w}{dx^2} = q \tag{20.16}$$

where w is the beam displacement, W and H are the beam width and height, respectively, E is the Young's modulus, I is the moment of inertia, and q is the distributed load in Newtons per unit length. If the load is uniform, then the solution for clamped boundary conditions is, from Section 9.6.2,

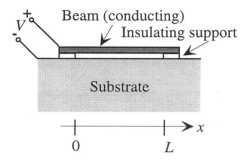

Figure 20.12.   Illustrating an electrostatically actuated beam

$$w = \frac{q}{2N} \left[x(L - x) + \frac{L\{\cosh[k_o(x - L/2)] - \cosh(k_oL/2)\}}{k_o \sinh(k_oL/2)} \right] \quad (20.17)$$

where N is the tension in the beam, equal to $\sigma_0 W H$, σ_0 is the residual stress, and

$$k_o = \sqrt{\frac{12N}{EWH^3}} \quad (20.18)$$

The maximum deflection, at the center of the beam, is

$$w_{\max} = \frac{qL}{4N} \left(\frac{L}{2} - 2\frac{\cosh(k_oL/2) - 1}{k_o \sinh(k_oL/2)} \right) \quad (20.19)$$

From this result, we can calculate an effective spring constant for the load-deflection behavior. The distributed load q is the force per unit beam length, and hence is equivalent to an effective pressure P_e times the beam width W. For electrostatic actuation with voltage V across a gap g_0, the pressure is

$$P_e = \frac{\varepsilon V^2}{2g_0^2} \quad (20.20)$$

The total force acting on the beam is WLP_e. In keeping with our notion that a lumped-parameter spring is a force divided by a distance, we would define the spring constant as WLP_e/w_{max} as was done in Section 9.6.2. However, several authors have chosen to express the effective spring constant as P_e/w_{max}, with units of Pa/m instead of N/m. In order to alert readers to this possibility, we adopt here this modified spring constant definition, but denote it as K'_{eff}

$$K'_{\text{eff}} = \frac{P_e}{w_{max}} = \frac{8N}{WL^2} \left[\frac{1}{1 + 2\dfrac{1 - \cosh u)}{u \sinh u}} \right] \quad (20.21)$$

where $u = k_o L/2$. We can then estimate the pull-in voltage using the lumped-element parallel-plate pull-in expression:

$$V_{PI} = \sqrt{\frac{8 k'_{\text{eff}} g_0^3}{27 \varepsilon_0}} \tag{20.22}$$

where g_0 is the original gap, and where the area is omitted from the denominator because we used a pressure rather than a total force to calculate k'_{eff}. This expression yields a pull-in voltage which, to first order, captures the dependence of pull-in voltage on beam dimensions, elastic modulus, and residual stress. However, because the gap under the beam is non-uniform, the parallel-plate approximation is off by about 20%. Since the *analytic form* of the pull-in voltage expression is basically correct, we can employ a sophisticated macro-modeling tool. We use this expression, but with some adjustable constants selected so that the pull-in voltage expression matches the calculated behavior based on full non-linear numerical simulations. This has been done, with a factor added to account for electrostatic fringing at the edges of the beam. The result is [117]

$$V_{PI} = \sqrt{\frac{4 \gamma_1 S}{\varepsilon_0 L^2 D(\gamma_2, k_o, L) \left[1 + \gamma_3 \frac{g_0}{W}\right]}} \tag{20.23}$$

where

$$D(\gamma_2, k_o, L) = 1 + \frac{2 \left[1 - \cosh\left(\gamma_2 k_o L/2\right)\right]}{(\gamma_2 k_o L/2) \sinh(\gamma_2 k_o L/2)} \tag{20.24}$$

and where the k_o is rewritten in terms of a *stress parameter* $S = \sigma_o H g_0^3$ and a *bending parameter* $B = E H^3 g_0^3$, such that

$$k_o = \sqrt{\frac{12 S}{B}} \tag{20.25}$$

The values of the constants for a fixed-fixed beam with fully clamped boundary conditions are $\gamma_1 = 2.79$, $\gamma_2 = 0.97$, and the fringing-field correction term γ_3 is 0.42.

The maximum beam deflection at pull-in for beams thicker than about half the gap, interestingly enough, is about 1/3 of the gap, just as with the lumped-element parallel-plate pull-in example. However, it has been demonstrated that for a thick enough beam, segmenting the actuation electrode so that actuation occurs only in regions nearer the supports permits full-gap travel without pull-in. This method is called *leveraged bending* [118].

Thinner beams, which require deflection by more than about one beam thickness to reach pull-in, become *stress-stiffened* and exhibit nonlinear restoring-force behavior. This has three effects. First, the pull-in voltage goes up compared to the non-stress-stiffened case. Second, the deformed shape changes somewhat. Third, the maximum travel at pull-in increases to almost half the gap. Details can be found in [118].

20.3.1   M-Test

What is interesting about the beam pull-in result is that while it has the appearance of an analytic, hence potentially approximate solution, it is actually quantitatively correct for a range of beam dimensions and constitutive properties. Hence it can be used, in combination with accurate geometrical data, to *measure* the residual stress and elastic modulus of the beams. Used in this manner, the use of pull-in voltage measurements for material property measurement is referred to as *M-Test*.

The M-Test method relies on an array of beams of different lengths. From measured values of the pull-in voltage *vs.* length, one extracts values of the stress parameter S and the bending parameter B by fitting to Equation 20.23. Then, when combined with accurate thickness and gap measurements, values for the residual axial stress and elastic modulus can be determined to within about 3%. Details of the method, including the error analyses, can be found in [117, 119, 120].

20.4   The Grating-Light-Valve Display

We now examine the GLV display as a Case Study.[4] We begin with the theory of using diffraction to control pixel intensity in a projected image.

20.4.1   Diffraction Theory

Scalar waves are time- and space-dependent disturbances that obey the scalar wave equation. In one dimension

$$\frac{\partial^2 \Psi}{\partial x^2} = \frac{1}{c^2} \frac{\partial^2 \Psi}{\partial t^2} \tag{20.26}$$

where $\Psi(x, t)$ is the wave amplitude and c is the wave velocity. The intensity of a wave is given by $|\Psi(x, t)|^2$. In a homogeneous medium, there are two basic types of solutions to this equation: *traveling waves* of the form $exp[ik(x \pm ct)]$, or *standing waves* which are the *sin* and *cos* functions built from linear combinations of the two traveling waves. In either case, the quantity k is called

---

[4]This Case Study was based in part on materials and commentary provided by Josef Berger, David Amm, and Chris Gudeman of Silicon Light Machines.

the *wavenumber*. It equals $2\pi/\lambda$, where λ is the *wavelength* of the wave. An important quantity is the *total phase* ϕ, equal to $k(x \pm ct)$ for the traveling wave. Points of constant phase must be equivalent. Hence to find out how the points of constant phase travel, we can set ϕ equal to a constant ϕ_0, and differentiate with respect to time

$$\frac{d\phi}{dt} = k\left(\frac{dx}{dt} \pm c\right) \tag{20.27}$$

For points of constant phase, $d\phi/dt = 0$. Hence points of constant phase move with a *phase velocity* v_p equal to c. The traveling wave with the minus sign moves in the positive x direction, the one with the plus sign in the negative x direction. Because all points on the wave move with the same velocity, it is possible to describe a wave by taking a snapshot of the wave at some convenient fixed time, such as $t = 0$.

A *plane wave* is a wave in which the points of constant phase form planes perpendicular to the direction of propagation and spaced apart by one wavelength. The intensity of a plane wave is independent of position. A *spherical wave* is a wave in which the points of constant phase lie on concentric spheres spaced apart by a wavelength. The intensity of a spherical wave decreases as the inverse square of the distance from the center.

Light is not a scalar wave. It is a vector wave field, combining both electric and magnetic fields, both of which are polarized perpendicular to the direction of wave propagation and to each other. However, for purposes of studying such fundamental optical behavior as diffraction, it is possible to use the scalar wave equation to find the wave amplitude far away from the source of the diffraction and use its squared magnitude for the intensity. This is the procedure we shall follow.

Huygen's Principle states that every point on a surface of constant phase can be considered as a source of a spherical propagating wave. At any observation point, one need only superpose all these propagating waves to find the resulting wave amplitude. The wave amplitude is then squared to find the intensity. We shall use Huygen's Principle to find the diffraction pattern from a 6-element GLV pixel.

Figure 20.13 illustrates the geometry for calculating the diffraction from a single pixel. Each beam is represented as ideally flat and displaced in a perfectly parallel manner.[5] According to Huygen's Principle, the wave amplitude at position (R, y) is given by the superposition integral:

---

[5] The effect of beam bending is considered in Section 20.4.3.4.

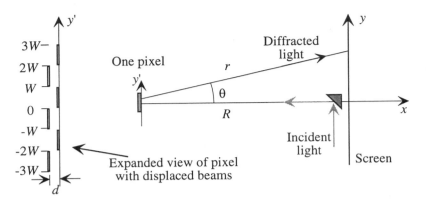

Figure 20.13.   The geometry for calculating the diffraction from a single six-element GLV pixel, with alternate beams displaced by d. The beam axis is into the paper.

$$\Psi(R, y) \propto \int_{-3W}^{3W} \frac{e^{i[kr + \delta\phi(y')]}}{r} dy' \tag{20.28}$$

where

$$r = \sqrt{R^2 + (y - y')^2} \tag{20.29}$$

and where $\delta\phi(y')$ is the change in the phase of the wave caused by the extra path length for the displaced mirrors. This path length increase, compared to reflection from the undisplaced mirrors, is $d(1 + \cos\theta)$ for normal-incident light. The phase change $\delta\phi$ is then $kd(1 + \cos\theta)$.

We can neglect the variation of r in the denominator of the superposition integral, but we must examine the details of the variation of r within the phase of the exponential. Since both R and r are much greater than y', the expression for r can be expanded to yield

$$r \approx r_0 \left(1 - \frac{yy'}{r_0^2}\right) \tag{20.30}$$

where r_0 is the value of r when y' is zero. The ratio y/r_0 equals $\sin\theta$. Hence, the wave amplitude can be approximated as

$$\Psi(R, y) \propto e^{ikr_0} \int_{-3W}^{3W} e^{-iky' \sin\theta} e^{i\delta\phi(y')} dy' \tag{20.31}$$

The pixel is made up of three identical pairs. With appropriate change of variables, the superposition integral becomes

$$\Psi(R, y) = exp^{ikr_0} \left(1 + e^{i2kW \sin\theta} + e^{-i2Wk \sin\theta}\right) \Psi_0(R, y) \tag{20.32}$$

where

$$\Psi_0(R,y) = \int_{-W}^{0} e^{-iky' \sin\theta} e^{ikd(1 + \cos\theta)} \, dy' + \int_{0}^{W} e^{-iky' \sin\theta} \, dy'$$

(20.33)

These expressions can be evaluated to yield

$$\Psi(R,y) \quad \propto \quad exp^{ikr_0}[1 + 2\cos(2kW \sin\theta)]$$

$$\times \quad \left[e^{ikW \sin\theta/2} \, e^{ikd(1 + \cos\theta)} + e^{-ikW \sin\theta/2} \right]$$

$$\times \quad \frac{\sin\left(\dfrac{kW \sin\theta}{2}\right)}{\left(\dfrac{kW \sin\theta}{2}\right)}$$

(20.34)

The third line of this equation is the famous Fraunhofer diffraction result from a single slit of width W. The remaining factors superpose the variously phased contributions from each of the six beams of width W. The peak diffraction intensity occurs in the first diffraction order, where

$$\frac{kW \sin\theta}{2} = \frac{\pi}{2}$$

(20.35)

This implies that the peak diffraction intensity occurs at an angle such that $kW \sin\theta = \pi$. Since the GLV display is designed to operate at the angle for maximum diffraction intensity, we can substitute this value into the expression for Ψ to yield

$$\Psi(R,y) \quad \propto \quad 3\frac{e^{ikr_0}}{(\pi/2)} \left(e^{i\pi/2} \, e^{ikd(1 + \cos\theta)} + e^{-i\pi/2} \right)$$

(20.36)

Since $exp(i\pi/2) = i$, we can evaluate the diffracted intensity I_d, which is proportional to the magnitude squared of Ψ, as

$$I_d \propto 1 - \cos[kd(1 + \cos\theta)]$$

(20.37)

For wavelengths of light in the visible, varying between about 0.4 and 0.6 μm, and for beam widths on the order of 2 μm or more, θ is a small angle, and $\cos\theta$ is close to unity. Hence we can estimate that the maximum diffracted intensity occurs when

$$\cos(2kd) = -1$$

(20.38)

which corresponds to $2kd = \pi$. Thus, the intensity of the diffracted signal goes from zero, when $d = 0$, to a maximum when

$$d = \frac{\lambda}{4} \qquad (20.39)$$

To achieve optimum diffraction for different colors, yet still maintain the same diffraction angle, one could build GLV pixels with different values of W. However, a far superior approach is to exploit the fact that the largest motion ever required is determined by the longest wavelength of interest, the red light. To make a GLV display for green or blue light, one can use the exact same device, but displace it less, still achieving analog control of intensity. This means that only one device type can be used for all three colors. Three arrays are used, aligned such that the diffracted light from each is combined at the entrance to the projection optics.

20.4.2   Device Fabrication and Packaging

The commercial process for fabricating GLV arrays is proprietary. However, a plausible process can be imagined based on what we have learned about surface micromachining. Each array contains 1088 pixels, each of which has six beam elements.

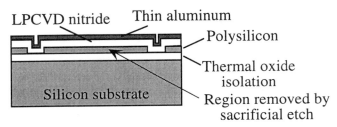

Figure 20.14.   Illustrating a plausible process for creating GLV beams. The polysilicon is the sacrificial layer. When the beams are patterned, access is provided for XeF$_2$ to undercut and release the beams.

Figure 20.14 illustrates a plausible sequence with which to make the beams. Each beam is a silicon nitride surface-micromachined beam coated with a thin layer of aluminum to provide reflectivity. The sacrificial layer is polysilicon. The release etch is the vapor XeF$_2$, which exhibits a high selectivity to etching silicon compared to the other common microelectronic materials (LPCVD silicon nitride, thermal SiO$_2$, PECVD oxide, aluminum, and titanium). Because a vapor etch is used, there is no release stiction, and because the device is operated in a non-contact mode, in-use stiction is also not a problem.

Figure 20.15 shows an electron-beam microscope photograph of an edge of a GLV array. Each pixel is six ribbons wide. Alternate beams of each pixel are driven with an analog voltage to control diffracted intensity. The other beams have a reference bias applied, for example, ground potential. Because of the large number of pixels, the total number of contacts to this chip is quite large.

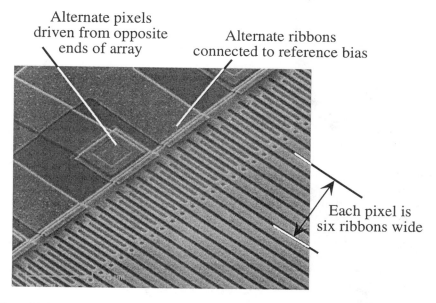

Alternate pixels
driven from opposite
ends of array

Alternate ribbons
connected to reference bias

Each pixel is
six ribbons wide

Figure 20.15. Photograph of the edge of a GLV array. Note that two levels of interconnect are used to provide actuation voltage to alternate beams within one pixel from one end of the array and a reference bias voltage to the fixed beams from the other end of the array. (Source: Silicon Light Machines; enhanced and reprinted with permission.)

To permit efficient packaging, control voltages for successive pixels are brought in at opposite edges of the array. One solution to the problem of connecting each beam to the right voltage is to use a two-level interconnect scheme with an interlevel dielectric. The cross-over is clearly visible in Fig. 20.15.

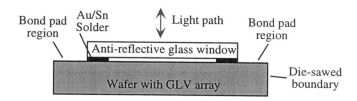

Bond pad
region

Au/Sn
Solder

Light path

Bond pad
region

Anti-reflective glass window

Wafer with GLV array

Die-sawed
boundary

Figure 20.16. Illustrating the sealing of the GLV array after release. A wafer is released and immediately placed in an apparatus which bonds individual windows over every GLV array with solder. Die sawing occurs after sealing.

Figure 20.16 shows a single GLV die after window sealing and die separation. The window attachment is done immediately after the array is released. A wafer of devices has individual windows soldered over each array, after which the wafer can be safely handled without fear of contamination of the arrays.

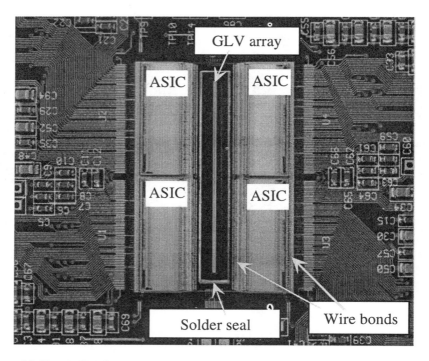

Figure 20.17. A GLV linear array wire-bonded directly to four ASIC chips mounted on the same circuit board. The ASIC's are, in turn, wire-bonded to the circuit board. (Source: Silicon Light Machines; enhanced and reprinted with permission.)

Figure 20.17 shows an encapsulated linear GLV array mounted on its circuit board and wire-bonded to four ASIC chips. Each ASIC handles 272 pixels and also provides the required reference bias for the non-actuated beams. A full display requires three of these boards, one for each color.

20.4.3   Quantitative Estimates of GLV Device Performance

We now combine the diffraction results and the electrostatic actuation results to produce a voltage-intensity model for the GLV pixel. The steps are as follows: (1) extract the effective spring constant for the voltage-deflection behavior from our analysis of the electrostatic actuation of beams; (2) determine the values of constitutive parameters, especially residual stress, that control beam stiffness; and (3) calculate the voltage-intensity characteristic. Along the way, we shall encounter the fact that the motion of each beam is strongly affected by squeezed-film damping. It turns out that this device is a wonderful example with which to illustrate the effects of small gaps on gas behavior.

20.4.3.1   Device Dimensions

Various device dimensions have been reported for the GLV pixel structure, depending on the type of study being performed. Beam widths are typically 3 μm, although a range of widths was used to study squeezed-film damping. Layer thicknesses are typically 80-90 nm for the silicon nitride and 50 nm for the aluminum. Beam lengths are on the order of 100 - 200 μm, but a great range of lengths was used for the damping study. The gap between the beam and the substrate must be at least three times the expected travel to avoid pull-in. Since the required motion is $\lambda/4$, a gap on the order of λ is sufficient, which for red light is 650 nm. Experimental devices with both larger and smaller gaps have also been used to study resonant and gas-damping effects.

20.4.3.2   Spring Constant

We can estimate the spring constant for the beam from Eq. 20.21. Because the silicon nitride ribbon has a very high tensile stress, on the order of 800 MPa (see below), one can expect that the elastic behavior of the ribbon is well into the stress-dominated regime, where u is very large. In this limit, the spring constant becomes

$$k'_{eff} = \frac{8\sigma_0 H}{L^2} \tag{20.40}$$

Therefore, the deflection d in response to an applied actuation voltage V is approximately[6]

$$d = \frac{L^2}{8\sigma_0 H} \frac{\varepsilon V^2}{2g_0^2} \tag{20.41}$$

Since we know the device dimensions, the only unknown quantity is the residual stress. Determination of that stress from resonant frequency measurements is the subject of the next section.

20.4.3.3   Determination of Residual Stress

The residual stress in both the silicon nitride and the aluminum films was determined from measurements of the resonance frequency of the pixel beams with varying degrees of aluminum coverage (see Fig. 20.18). This section draws on the paper by Payne *et al.* [121], which makes excellent use of the energy methods and Rayleigh-Ritz procedure presented in Section 10.5.

The basic approach is to excite a pixel into resonance with a suitable sinusoidal signal, then remove the excitation and watch the unforced damped

---

[6]For greater accuracy, we could capture the nonlinearity to first order by using $(g_0 - d)$ instead of g_0 in the denominator.

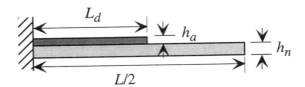

Figure 20.18. Illustrating the partial coverage of a GLV silicon nitride (lighter) beam with aluminum (darker). The coverage is symmetrical from both ends; only half the beam is shown.

oscillation of the pixel using the diffracted signal as a detector. The advantage of this approach is that one is certain that the materials are exactly those used in the real device. The disadvantage is that the geometry is not as ideal as one might like for rigorous quantitative measurements. Nevertheless, by making systematic variations of device dimensions, it is possible to extract reasonably accurate stress values.

Because silicon nitride is an insulator, it is not possible to actuate a beam with no aluminum on it. Payne's approach was to measure the *variation* of resonance frequency for a *sequence of beams* with different lengths of aluminum-covered region. This is just like the M-Test approach, in which a systematic variation of geometry is used to reveal the underlying constitutive properties. We outline the procedure and results here. Details are in the cited paper.

For a beam in the tension-dominated regime, the elastic energy due to a trial-function deformation $\hat{w}$ is dominated by the work done against the already-present residual stress σ_0. Thus, the elastic energy in a deformed beam can be written

$$W_e = \frac{N}{2} \int_0^L \left(\frac{d\hat{w}}{dx}\right)^2 dx \tag{20.42}$$

The maximum kinetic energy, assuming a sinusoidal time dependence for the shape of the trial function, is

$$W_k = \frac{W\omega^2}{2} \int_0^L \rho_m(x)h(x)\hat{w}(x)^2 \, dx \tag{20.43}$$

where ω is the frequency of the sinusoid, $\rho_m(x)$ is the x–dependent average mass density and $h(x)$ is the beam thickness, which now varies as a function of position along the beam. Using the Rayleigh-Ritz procedure, we estimate the resonant frequency as

$$\omega = \left[\frac{N \int_0^L \left(\frac{d\hat{w}}{dx}\right)^2 dx}{W \int_0^L \rho_m(x)h(x)\hat{w}(x)^2 \, dx}\right]^{1/2} \tag{20.44}$$

The numerator is, in effect, a spring constant, and the denominator is, in effect, a mass. To evaluate the denominator, we note that the x-dependent mass density times the x-dependent thickness is simply the mass per unit length:

$$\rho_m(x)h(x) = \rho_a h_a + \rho_n h_n \qquad (20.45)$$

for regions with aluminum, and

$$\rho_m(x)h(x) = \rho_n h_n \qquad (20.46)$$

for regions without aluminum, where where the subscripts a and n denote the density and thicknesses of aluminum and nitride, respectively.

For a *uniform* tension-dominated beam, the exact deformation is of the form $Ax(L-x)$, where A is a constant. We can use this form as our trial function for the more complex beam structure. We introduce a coverage parameter p, equal to $2L_d/L$, which gives the fraction of the beam length that is covered with aluminum. The denominator of the Rayleigh-Ritz expression can be evaluated to yield a total effective mass M

$$M = \left[p^3 \left(\frac{3}{8}p^2 - \frac{15}{8}p + \frac{5}{2} \right) \right] \rho_a W h_a + \rho_n W h_n \qquad (20.47)$$

The numerator of the Rayleigh-Ritz expression requires a determination of the tension in the beam after release. While the tension is uniform along the length of the beam, the relation of the value of that tension to the original intrinsic stresses in the nitride and aluminum must be developed. This is identical to the problem we solved in Section 9.6.1, in which we needed to find the extension of a coated cantilever after release.

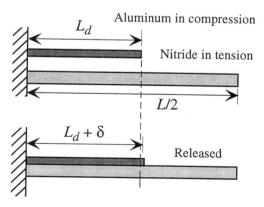

Figure 20.19. Illustrating the method of finding the axial stress after release. In the upper figure, the two films are presumed to have their respective stresses, and are unreleased. In the bottom figure, the films have been attached in their pre-stressed states, then allowed to relax. Because the aluminum is compressive, the boundary shifts to the right by an amount δ on release.

Figure 20.19 illustrates the procedure. If we assume a tensile (positive) stress σ_n in the nitride prior to release, and a compressive stress $-\sigma_a$ in the aluminum prior to release, then bond the two layers together and release them, the interface moves to the right by an amount δ. To find delta, we require that the tension be equal in the two parts of the beam. This leads to

$$\delta = \frac{L}{2} \frac{\sigma_a h_a p(1-p)}{E_a h_a(1-p) + E_n h_n} \tag{20.48}$$

From this, the stress in the uncoated nitride is

$$\sigma_{n,final} = \frac{E_n h_n \sigma_n + p E_n h_a \sigma_a + (1-p) E_a h_a \sigma_n}{E_n h_n + (1-p) E_a h_a} \tag{20.49}$$

and the final tension in the beam is then $W h_n \sigma_{n,final}$. Using this result, Payne calculates the Rayleigh-Ritz resonant frequency as

$$\omega =$$

$$\frac{\sqrt{10}}{L} \left[\frac{h_n \left[E_n h_n \sigma_n + p E_n h_a \sigma_a + (1-p) E_a h_a \sigma_n \right]}{\left[E_n h_n + (1-p) E_a h_a \right] \left[h_n \rho_n + p^3 \left(\frac{3p^2}{8} - \frac{15p}{8} + \frac{5}{2} \right) h_a \rho_a \right]} \right]^{1/2} \tag{20.50}$$

Measurements were made on beams of length 200 μm with varying values of L_d. The beam was excited strongly into resonance by a sinusoidal excitation which was tuned to get a large amplitude. The sinusoid was then turned off and the free oscillation was measured by sensing the diffracted signal. The resulting data were fitted to a damped sinusoid and the resonance frequency was extracted. As expected, the resonant frequency decreased with increasing L_d, consistent with a compressive stress on the aluminum. The fit of the frequency-*vs.*-L_d data to the above expression permitted extraction of values of 801 $\pm$ 2 MPa for the tensile nitride stress, and -92 $\pm$ 8 for the compressive aluminum stress.[7] With these values in hand, it is possible to design the pixel to achieve a desired voltage-displacement characteristic.

---

[7] These values are the uniaxial stress values after release. Prior to release, the biaxial stress would be larger by a factor $1/(1-\nu)$.

20.4.3.4 Beam Curvature

We previously analyzed the diffraction assuming that the displaced beams were perfectly flat. In fact, though, the actuated beams bend, as illustrated in Fig. 20.20.

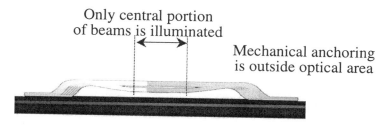

Only central portion of beams is illuminated

Mechanical anchoring is outside optical area

Figure 20.20. Illustrating the bending of GLV beams that occurs during actuation. Only the central portion of the beams is illuminated. [Source: Silicon Light Machines, reprinted with permission.]

In Eq. (20.19), we calculated the deflected shape of a GLV beam assuming perfectly clamped supports. In practice, the supports have finite compliance, but that is a second-order effect. Of more immediate interest is the effect of the beam curvature in the central illuminated portion of the deflected beam. For tension-dominated beams, the bending is nearly perfectly parabolic. We can calculate the radius of curvature from the expression

$$\hat{w} = Ax(L - x) \tag{20.51}$$

The maximum deflection is at the center, and equals $AL^2/4$. In use, this maximum deflection is $\lambda/4$. Thus, in use, the largest value of A is λ/L^2. The radius of curvature is $1/2A$. Hence, the most curvature and correspondingly smallest radius of curvature is $L^2/2\lambda$. If we use 650 nm as a typical wavelength, and 200 μm as a typical beam length, we estimate the smallest radius of curvature as 3 cm.

If all of the beams, including the reference beams, were similarly curved, then the effect of the curvature would be to focus the diffracted waves. However, the reference beams remain flat, and most of the time, most of the bent beams are deflected far less than the maximum and hence have much larger radii of curvature. This may explain the fact that the curvature of the deflected beams does not appear to create any significant artifact in the projected image [122].

20.4.3.5 Voltage-Intensity Characteristic

We can now assemble the pieces. Substituting $p = 1$ into Eq. (20.49), corresponding to a fully coated beam, and using the extracted stress values for the individual films, the material densities ($\rho_a = 2700$ kg/m^3, $\rho_n = 3440$ kg/m^3), elastic moduli ($E_a = 70$ GPa, $E_n = 250$ GPa), and film thicknesses (h_a

= 50 nm, h_n = 90 nm), we conclude that the axial tensile stress in the film is σ_0 = 482 MPa.

Combining the diffraction *vs.* displacement characteristic with the spring constant and the value of residual stress just obtained under the assumption that the diffraction angle is small, we find

$$I_d \propto 1 - \cos\left(\frac{\pi \varepsilon L^2 V^2}{4\lambda \sigma_0 H g_0^2}\right) \tag{20.52}$$

which can be simplified to

$$I_d \propto 1 - \cos\left[\pi \left(\frac{V}{V_{max}}\right)^2\right] \tag{20.53}$$

where V_{max} is the voltage at which the first-order diffraction is at its maximum value, corresponding to a beam displacement of $\lambda/4$.

$$V_{max} = \sqrt{\frac{4\lambda \sigma_0 H g_0^2}{\varepsilon L^2}} \tag{20.54}$$

Using the values $L = 300$ μm, $\lambda = 650$ nm, $\sigma_0 = 482$ MPa, and $H = 0.14$ μm, we obtain $V_{max} = 9.6$ V. Silicon Light Machines reports that their devices use voltages below 15 V, so our estimate is quite reasonable. Voltages in this range are readily compatible with commercially available ASIC circuits.

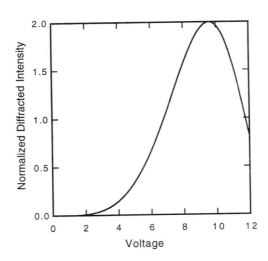

Figure 20.21.   Calculated voltage-diffraction response curve for a GLV example. The shape of the curve is identical to published experimental data.

Figure 20.21 shows the diffraction intensity *vs.* voltage characteristic for this case. In use, the voltage is restricted to be less than V_{max} so that the intensity is a monotonic function of actuation voltage.

20.4.3.6   Squeezed-Film Damping of the GLV Beams

Switching speed is essential to the successful commercial application of the GLV display. The beam resonance frequency is in the range of 1-2 MHz, which suggests that complete on-off switching should be achievable in about 1 μs. The dominant damping mechanism is squeezed-film damping. If there is too much damping, the switching speed will be sub-optimal, and if there is not enough damping, the beam will vibrate for a long time after being repositioned. The ideal is to achieve slightly underdamped behavior so that the switching time and settling time are both short.

In an extensive study of squeezed-film damping behavior of a variety of GLV pixel structures, Gudeman *et al.* [123] examined the resonant damping coefficient of pixels in various pressures of He, Ne, Ar, Kr and N_2. The gaps used in their tests were very small, from 150 to 750 nm. The method used was identical to that described earlier for the stress study. The pixel was excited with a sinusoid to achieve large resonant motion, then the excitation was turned off. The output intensity as a function of time was fit to a model in which the motion of the beam is a damped sinusoid. Both the resonance frequency and damping constant were obtained.

Table 20.1.   Summary of viscosity, mean free path, effective viscosity, and damping constant at 20 °C. All gases are at 1 atm pressure unless indicated otherwise. From [123].

Gas	Viscosity η (μPa-s)	Mean Free Path λ_m (nm)	Effective Viscosity η_{eff} (μPa-s)	Decay Time τ_d (μs)
He	19.41	194	6.96	1.3
Ne	31.11	138	13.73	0.71
N_2	17.49	67	10.91	0.82
Ar	22.17	70	13.51	0.69
Kr	25.04	53	16.80	0.55
N_2 (0.5 atm)	17.49	134	7.51	1.16
N_2 (1.5 atm)	17.49	44	11.82	0.75

Table 20.1 shows the data from Gudeman *et al.* It was found that the resonant frequency was independent of gas composition, but the damping constant varied significantly. Because of the small gaps used, it was necessary to take account of the Knudsen number correction to the gas viscosity, as described in Section 13.4. One would expect the damping rate to be proportional to the effective viscosity, given by

$$\eta_{\text{eff}} = \frac{\eta}{1 + 6K_n} \qquad (20.55)$$

where η is the bulk viscosity of the gas, and K_n is the Knudsen number, defined as the ratio of molecular mean free path λ_m to the gap. The decay time τ_d for the structure is the inverse of the damping rate. Hence, one would expect that the decay constant is proportional to $1/\eta_{\text{eff}}$. There is a dramatic difference in Table 20.1between the bulk viscosity and the effective viscosity in these small-gap structures due to the wide variation in mean free path. The decay time is plotted against the reciprocal of effective viscosity in Fig. 20.22. Note the nice linear response. Note also that by selecting different gases for the device environment, a significant change in switching speed occurs.

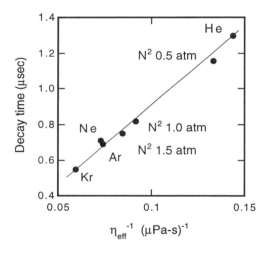

Figure 20.22.   Plot of the data from Table 20.1.

20.5   A Comparison

We conclude with a brief comparison between the DMD and the GLV device. The DMD display requires one mirror element per pixel in a full two-dimensional array, while the GLV display requires six beams per pixel, but only in a linear array. Therefore, the GLV display has far fewer moveable MEMS elements than the corresponding DMD display. The price of this device simplification is that the optical system is more complex, and must include a scanning mirror that must be accurately synchronized with the pixel drive voltages.

The DMD chip uses a two-level structure to achieve a high fill-factor for each pixel, but there remain dead zones around each mirror, and these could lead to visible artifacts at pixel boundaries. The GLV chip has no pixel boundaries in either direction.

Finally, if one mirror in a DMD array fails to function, the result is a dark spot on the screen. If one beam of a GLV pixel fails to function, the result is a decreased diffraction intensity, but not a dark spot.

Ultimately, of course, it is the cost, performance, and reliability of the final product that determines success in the marketplace. What is interesting here is that two very different optical MEMS approaches compete in the projection marketplace. In 1990 there were no commercial MEMS projection systems; by 2000 there were at least two, and possibly more under development elsewhere.

Related Reading

M. Born and E. Wolf, *Principles of Optics: Electromagnetic Theory of Propagation, Interference and Diffraction of Light*, Sixth Edition, Oxford: Pergamon Press, 1980.

E. Hecht, *Optics*, Second Edition, Reading, MA: Addison-Wesley, 1987.

S. J. Walker and D. J. Nagel, "Optics and MEMS," Technical Report NRL/MR/6336–99-7975, Naval Res. Lab., Washington DC, 1999. PDF version available at http://mstd.nrl.navy.mil/6330/6336/moems.htm.

Problems

20.1 Calculate the torsional stiffness k_θ that will result in a 5 Volt pull-in voltage for the DMD structure. Then, using a very rough approximation that the torsional stiffness of a beam of length L_b and moment of inertia I_b is GI_b/L_b, where G is the shear modulus of the beam, select the dimensions for an aluminum torsion spring that will achieve this pull-in voltage. (Details of basic torsional elements can be found in Chapter 3 of Gere and Timoshenko [50].)

20.2 The Silicon Light Machines pixel could be made smaller if fewer beams could be used per pixel. Could the Silicon Light Machines display be made with only four beams per pixel? Two beams per pixel? Explain.

20.3 Modify Eq. 20.41 to include the effect of gap change with actuation, and compare the resulting voltage-intensity plot up to the first intensity maximum with the more approximate calculation assuming a fixed g_0 in the denominator. How does the voltage at maximum intensity compare? Explain the difference, if any.

20.4 An Optics Treasure Hunt: Go to the literature or to a commercial web site and find a micromirror device or a Fabry-Perot optical filter. Determine the actuation method and, using the modeling methods you have learned, estimate the expected performance. Compare with the published behavior of the device.

Chapter 21

A PIEZOELECTRIC RATE GYROSCOPE

. . . and Roll

—Second part of song title, Big Joe Turner, 1952 (see also Chapter 19)

21.1   Introduction

A complete description of the motion of an object requires knowledge of all of its velocity components. We have already seen that it is possible to measure the linear acceleration of an inertial frame from which a proof mass is elastically suspended. But if the frame to which the proof mass is attached is rotating, additional complexities enter. Therefore, a complete *inertial measurement unit* requires that one measure both linear accelerations and rotation rates. The device that measures rotation rate is called a *rate gyroscope*. In this chapter, we shall examine the principles of rate gyroscope operation and then explore the use of *piezoelectricity* in quartz to build a micromachined tuning-fork gyroscope that has already successfully entered the automotive market in large volume.

Recall that in Chapter 18 we examined piezoresistivity as a method of strain measurement, and that in Chapter 19 we examined capacitive and tunneling methods for direct position measurement. In both chapters, we stressed that the sensing method could be applied to a much broader class of devices. The same is true of piezoelectricity. Thus, while we emphasize the rate-gyroscope application of piezoelectrics in this chapter, the reader should realize that there are many piezoelectric sensing devices, and we are touching only one corner of a very large field.

21.2   Kinematics of Rotation

To examine the effects of rotation on sensed motion, we imagine a particle constrained to move in a plane (see Fig. 21.1). The coordinate frame to

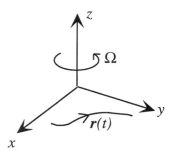

Figure 21.1.   A planar trajectory $\mathbf{r}(t)$ in a reference frame that is rotating with angular velocity Ω about an axis perpendicular to the plane.

which the mass is attached is rotated about an axis normal to the plane. As a result, the motion of the particle, when viewed in an *inertial frame*, one that is not accelerated or rotated, includes components from both the motion of the particle relative to the rotating frame, and the rotation of that frame relative to the inertial reference frame. If we denote the rotation vector Ω as

$$\Omega = \Omega \hat{\mathbf{i}}_z \tag{21.1}$$

then the particle velocity $\mathbf{v}_i$ as viewed in the inertial frame is related to the velocity in the rotating frame $\mathbf{v}_r$ by[1]

$$\mathbf{v}_i = \mathbf{v}_r + \Omega \times \mathbf{r} \tag{21.2}$$

where $\mathbf{r}$ is the time-dependent position vector in the rotating frame, that is, the position relative to the center of rotation. Abstracting from this result, the operator equivalent in the rotating frame to taking the time derivative in the inertial frame is

$$\left.\frac{d}{dt}\right|_i = \left.\frac{d}{dt}\right|_r + \Omega \times \tag{21.3}$$

Applying this operator to $\mathbf{v}_i$, we find the acceleration of the particle in the inertial frame:

$$\mathbf{a}_i = \mathbf{a}_r + 2\Omega \times \mathbf{v}_r + \Omega \times (\Omega \times \mathbf{r}) + \dot{\Omega} \times \mathbf{r} \tag{21.4}$$

where $\dot{\Omega}$ is the angular acceleration of the rotating frame, and $\mathbf{a}_r$ is the apparent acceleration in the rotating frame.

We can apply Newton's Second Law only in the inertial frame. Hence, the force $\mathbf{F}_i$ giving rise to the inertial acceleration $\mathbf{a}_i$ is $m\mathbf{a}_i$. Thus, the product

---

[1]See any text on classical mechanics, such as Chapter 4 of Goldstein [124].

$m\mathbf{a}_r$, which is the apparent result of applying Newton's Second Law in the rotating frame, differs from the applied force $\mathbf{F}_i$. The relation is

$$m\,\mathbf{a}_r = \mathbf{F}_i - 2m\,\boldsymbol{\Omega} \times \mathbf{v}_r - m\,\boldsymbol{\Omega} \times (\boldsymbol{\Omega} \times \mathbf{r}) - m\,\dot{\boldsymbol{\Omega}} \times \mathbf{r} \qquad (21.5)$$

If we apply this result to a particle constrained to move in the x-y plane, we obtain

$$m\ddot{x} = F_{i,x} + 2m\,\Omega\dot{y} + m\Omega^2 x + m\,\dot{\Omega}y \qquad (21.6)$$

$$m\ddot{y} = F_{i,y} - 2m\,\Omega\dot{x} + m\Omega^2 y - m\,\dot{\Omega}x \qquad (21.7)$$

These are the equations that govern the operation of the Coriolis Rate Gyroscope, which is examined in the next section.

21.3 The Coriolis Rate Gyroscope

The principle most-often exploited in building micromachined gyroscopes is the Coriolis force, expressed by the $\boldsymbol{\Omega} \times \mathbf{v}$ term in the equations of motion derived in the previous section. One effective way to use this force is to create a resonant motion of fixed amplitude in a direction perpendicular to the axis of rotation. The Coriolis force then induces motion in the third direction, perpendicular both to the direction of rotation and to the driven motion.

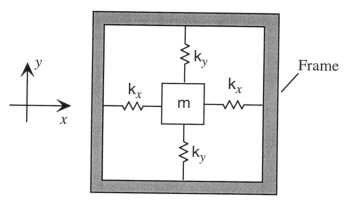

Figure 21.2. A simplified model for analyzing gyro behavior. The frame is presumed to rotate about the z-axis.

Figure 21.2 illustrates a simple model for analyzing gyro behavior. A single mass m is suspended by four springs. Assuming small-amplitude motions, which will be true in resonant gyroscopes, the x-motion and y-motions can be considered independently, coupled only through the Coriolis-force term and the term involving angular acceleration. Therefore, one can determine two undamped resonant frequencies for the system:

$$\omega_x = \sqrt{\frac{2k_x}{m}} \tag{21.8}$$

and

$$\omega_y = \sqrt{\frac{2k_y}{m}} \tag{21.9}$$

We also assume that associated with each of the two in-plane resonances is some kind of damping, giving rise to damping factors b_x and b_y, respectively.

The x-directed resonant mode is driven from an appropriate circuit to create a steady sinusoidal oscillation at a drive frequency, denoted by ω_d. To get a large driven-mode amplitude and hence a large x-directed velocity, one selects ω_d to match the resonant frequency ω_x, although as we observed when examining oscillators in Chapter 15, the actual frequency of a limit-cycle oscillation may differ slightly from the undamped resonant frequency, depending on the details of the circuit driving the resonance. If ω_d exactly matches the mechanical resonance ω_x, there is a 90° phase shift between the *driving force* and resulting *displacement*, and, therefore, the *velocity* of the drive mode has the same phase as the driving force. This will be important when determining the phase to use as a reference for demodulating the output signal. To proceed, we assume that the x-displacement is given by

$$x = x_o \cos \omega_d t \tag{21.10}$$

where ω_d is very close to ω_x. The resulting system has one degree of freedom, the y-directed motion, governed by

$$m\ddot{y} + b_y\dot{y} + m(\omega_y^2 - \Omega^2)y = -2m\Omega\frac{dx}{dt} - m\dot{\Omega}x \tag{21.11}$$

The resonant frequency ω_y is typically designed to be much greater than the rotational angular velocity Ω. Thus, the Ω^2 term can be safely neglected. Dividing through by the mass, and introducing the quality factor of the y-resonance as $Q_y = m\omega_y/b$, this becomes

$$\ddot{y} + \frac{\omega_y}{Q_y}\dot{y} + \omega_y^2 = -2\Omega\frac{dx}{dt} - \dot{\Omega}x \tag{21.12}$$

The right-hand side (RHS) is a mix of signals dependent on the rotation rate times the x velocity and on the angular acceleration $\dot{\Omega}$ times the position x. While the term proportional to angular acceleration is typically small, the variation of Ω with time does affect the gyro response in important ways, as will be discussed in detail below.

21.3.1 Sinusoidal Response Function

We assume a time-varying rotation rate. While we can use the Fourier transform to represent the time variation of $\Omega(t)$ as a superposition of sinusoids, for convenience, we shall assume a single sinusoid, keeping in mind that it is a placeholder for the complete $\Omega(t)$ waveform. That is, we assume

$$\Omega = \Omega_o \cos \omega_a t \tag{21.13}$$

where ω_a is a characteristic frequency for a change in the rotation rate. The RHS of Eq. 21.12 becomes

$$\text{RHS} = m\,\Omega_o \left[2\omega_d \sin \omega_d t \cos \omega_a t + \omega_a \sin \omega_a t \cos \omega_d t\right] \tag{21.14}$$

After some manipulations with trigonometric identities, we obtain

$$\text{RHS} = m\,\Omega_o \left\{ \left(\omega_d + \frac{\omega_a}{2}\right) \sin[(\omega_d + \omega_a)t] + \left(\omega_d - \frac{\omega_a}{2}\right) \sin[(\omega_d - \omega_a)t] \right\} \tag{21.15}$$

While it is generally true that ω_a will be much smaller than ω_d, the angular accelerations introduce sidebands spaced from ω_d by ω_a. The Coriolis force no longer occurs exactly at ω_d when there is angular acceleration. If we were to use a more general $\Omega(t)$ waveform, the RHS would be a complicated superposition of sideband pairs, one for each Fourier component of the original waveform, occupying a total band of $\omega_d \pm \omega_{a,max}$, with a bandwidth of $2\omega_{a,max}$.

In order to understand the response of the gyro to this complex waveform, we first find the response to a single complex sinusoid of arbitrary frequency, recalling that the final response is found by taking the real part of the complex response. We assume

$$\text{RHS} = Ae^{j\omega_t t} \tag{21.16}$$

The complex amplitude the y-motion $\hat{Y}$ is

$$\hat{Y} = \frac{1}{\omega_y^2 - \omega_t^2 + j\delta_y \omega_t} A \tag{21.17}$$

where we have introduced δ_y as the intrinsic bandwidth of the y-resonance, equal to ω_y/Q_y. The result depends on the difference between ω_y and ω_t. If we define

$$\delta_\omega = \omega_y - \omega_t \tag{21.18}$$

and assume that δ_ω is small compared to ω_y or ω_t, then

$$\hat{Y} \approx \left(\frac{1}{2\delta_\omega + j\delta_y}\right)\left(\frac{A}{\omega_t}\right) \tag{21.19}$$

We now consider the matching between the two resonant frequencies ω_x and ω_y. In some designs, these are tuned to be identical. In others, they are intentionally offset from one another by a shift we shall denote as Δ_ω. Given that ω_d will be very close to the value of ω_x, and given that ω_t lies in the band $\omega_d \pm \omega_{a,max}$, we can write

$$\delta_\omega = \Delta_\omega \pm \omega_a \tag{21.20}$$

where now Δ_ω is a constant. This leads to the transfer function

$$\hat{Y} \approx \left(\frac{1}{2(\Delta_\omega \pm \omega_a) + j\delta_y}\right)\left(\frac{A}{\omega_t}\right) \tag{21.21}$$

The sinusoidal response depends on three quantities: the intrinsic bandwidth of the y-resonance, the offset between ω_x and ω_y, and the characteristic frequency of angular acceleration. We now consider various cases.

21.3.2   Steady Rotation

If the reference frame is rotating at constant angular velocity, then ω_a is zero, and ω_t equals ω_d. The RHS of Eq. 21.14 is $2\omega_d x_o \Omega_o \sin(\omega_d t)$. This can be expressed with a complex exponential, as in Eq. 21.16,

$$\text{RHS} = -2j\omega_d x_o \Omega_o e^{j\omega_d t} \tag{21.22}$$

If the drive and sense resonance frequencies are very closely matched, so that $\Delta_\omega \ll \delta_y$, then the y-displacement is in phase with the x-displacement, and hence shifted by 90° from the x-direction drive force, and has amplitude

$$\hat{Y} \approx -\left(\frac{2}{j\delta_y}\right)\left(\frac{j\omega_d x_o \Omega_o}{\omega_d}\right) = -\frac{2}{\delta_y}x_o \Omega_o \tag{21.23}$$

On the other hand, if Δ is much larger δ_y, the y-displacement is 90 degrees out of phase with the x-displacement (represented by the factor j), and hence is in phase with the x-direction drive force, and has amplitude

$$\hat{Y} \approx -\frac{j}{\Delta_\omega}x_o \Omega_o \tag{21.24}$$

The second case has a smaller amplitude than the first case, and it has a different phase. Further, the second case has the advantage of a larger bandwidth for responding to angular accelerations, as discussed below. And in either case, the y-motion must be sensed either with direct position sensing or by sensing the stress in the elastic supports.

21.3.3 Response to Angular Accelerations

If there is angular acceleration present, then the RHS of Eq. 21.14 has two terms, one with a prefactor of ω_d, the other with ω_a. In practice, ω_a will be small compared to ω_d, so it is a good first approximation to neglect the term in the amplitude that depends directly on the angular acceleration. However, the frequency shift of the velocity-term sidebands from ω_d to $\omega_d \pm \omega_a$ cannot be ignored. Thus, we consider

$$\text{RHS} = 2\omega_d x_o \Omega_o \sin \omega_d t \cos \omega_a t \tag{21.25}$$

which can be thought of as a sum of two sinusoids, one with ω_t equal to $\omega_d + \omega_a$, the other with ω_t equal to $\omega_d - \omega_a$. In order to be able to make an accurate reconstruction of $\Omega(t)$, the y-motion responses to these two sidebands should have the same scale factor and phase shift. A good way to achieve this is to design the gyro such that Δ_ω is greater than $\omega_{a,max}$, and at the same time, much greater than δ_y. There is a cost in signal intensity by the ratio $2\delta_y/\Delta_\omega$, but all frequency components of the rotation signal experience the same scale factor and phase shift. That means that the y-motion quasi-statically follows the form of the ω_d term of the RHS, shifted in phase by 90 degrees. Recovery of the full $\Omega(t)$ waveform can be accomplished by synchronous demodulation with a reference signal whose phase matches to $\sin \omega_d t$.

21.3.4 Generalized Gyroscopic Modes

If we examine the equations of motion that govern rate-gyro behavior, we notice that there is no fundamental requirement that the gyro have separate springs and masses. Any vibratory mode can serve either as a drive or sense mode for a gyro. The only requirement is that they be configured so that the Coriolis force can couple the two modes with a term proportional to rotation rate. Here are several examples that are based on actual devices (see [125] for an extensive bibliography).

Figure 21.3 illustrates the basic tuning-fork gyroscope. The two tines of the tuning fork are driven to vibrate as shown, in the plane of the structure. A rotation about the axis of the tuning fork induces alternating out-of-plane motion of the tines which can be sensed. If the tuning-fork beams are square in cross section, then the two modes are closely matched in frequency. Intentional mismatch can be achieved by building the tuning fork with rectangular cross-section beams.

Figure 21.4 illustrates a different way of building a rate gyro. A massive disk is suspended on springs (not shown in the figure), and is driven into

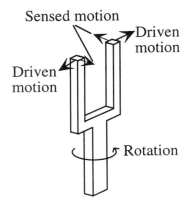

Figure 21.3.   Illustrating a single-ended tuning-fork gyroscope.

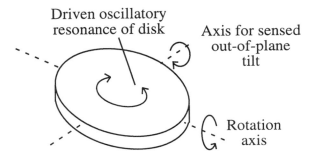

Figure 21.4.   Illustrating a disk gyroscope. The springs and frame supporting the disk are not shown.

an oscillatory movement in the plane of the disk. If the oscillating disk is rotated about an in-plane axis, the disk tilts out-of-plane around an axis that is orthogonal to the axis of rotation. In this case, because the drive and sense motions are so different, the support structure must be carefully engineered to establish a desired value of Δ_ω. Figure 21.5 shows an example of the disk gyro fabricated with a combination of surface micromachining and DRIE.

Figure 21.6 illustrates the so-called wine-glass gyro, based on the resonances of a circular ring. In practice, one does not build the whole wine glass, just the ring, shown as heavily shaded in the Figure. It is supported by springs (not shown) and driven into vibration in the plane of the ring. If the ring is perfectly round and uniformly suspended, there are two oscillatory in-plane modes that have the same frequency. Their motions are identical, except one mode has its primary axes of motion rotated by 45 degrees with respect to axes of the other mode. One of the modes is sinusoidally driven with a fixed amplitude.

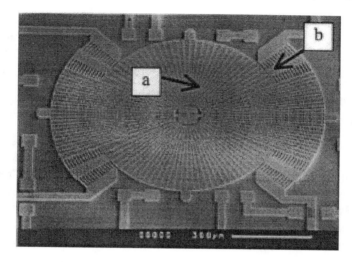

Figure 21.5. A resonant rate gyroscope, fabricated with a combination of surface microma-chining with thick polysilicon in combination with deep reactive ion etching. The oscillating disk is shown at (a) and the comb-drive electrostatic actuator for driving the in-plane oscillation is shown at (b). (Reprinted with permission from SAE paper 1999-01-0931 ©1999 Society of Automotive Engineers.)

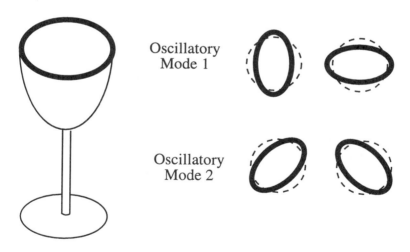

Figure 21.6. Illustrating the wine-glass gyro. The two resonant modes are illustrated at the side of the diagram. One mode is driven at constant amplitude , and the rotation-induced motion in the other mode is sensed by capacitive position sensors located around the perimeter of the ring.

Rotation about an axis normal to the plane of the ring induces motion in the second mode, which can be capacitively sensed.

This completes our preliminary discussion of gyro operation. Many devices in the literature can be successfully understood with the ideas presented above, including the choice of reference phase for the demodulation step. However, as we shall see, piezoelectric gyros have some additional complexity due to the intrinsic nature of piezoelectric coupling.

21.4   Piezoelectricity

In this section, we shall examine the coupling between internal dielectric polarization and strain, an effect called *piezoelectricity*. As was the case with piezoresistivity, the origin of piezoelectric effects lies in the details of the atomic arrangements within crystalline materials. However, electrical resistivity is intrinsically dissipative. Hence, the only use for piezoresistivity is the passive sensing of strain (or stress). There is no stored-energy component that depends on strain, hence, no opportunity for actuation.

In contrast, the internal dielectric polarization enters into the electrostatic stored energy through the relation

$$\mathcal{D} = \varepsilon_0 \mathcal{E} + \mathcal{P} \tag{21.26}$$

where $\mathcal{E}$ is the electric field, $\mathcal{D}$ is the electric displacement, $\mathcal{P}$ is the polarization, and ε_0 is the permittivity of free space. To illustrate using a linear medium, the stored electrostatic energy density is

$$\hat{\mathcal{W}}_e = \frac{1}{2} \mathcal{D} \cdot \mathcal{E} \tag{21.27}$$

Since strain can modify $\mathcal{P}$ and hence $\mathcal{D}$, it means that there is a strain-dependent component of the stored energy. Therefore, the gradient of the stored energy with respect to strain is nonzero. This can give rise to stresses, hence, to actuation forces. These actuation forces can be used to set a structure into vibration. Most piezoelectric sensing devices, such as thickness-shear-mode (TSM) bulk quartz resonant sensors, surface-acoustic-wave chemical sensors (SAW), and the rate gyroscope example discussed here, are based on mechanical vibratory motions that are electrically excited.[2]

21.4.1   The Origin of Piezoelectricity

When a crystal lacks a center of inversion symmetry, it means that the crystal can support an internal electric polarization in the absence of an applied external electric field, essentially an array of aligned electric dipoles within its structure. If a material has a center of inversion symmetry, then such an internal

---

[2]See Ballantine [126] for a discussion of a wide variety of vibrating acoustic sensing devices.

polarization is impossible.[3]   There are many materials that lack inversion symmetry. Examples include crystalline quartz, zinc oxide, lithium niobate, and poly-vinylidene fluoride (which is a partially ordered "semi-crystalline" polymer; within the crystalline regions there is a lack of inversion symmetry).

Most of the materials that lack inversion symmetry exhibit *piezoelectricity*, the creation of an internal polarization in response to stress. Some of these materials, called *ferroelectrics*, even have a spontaneous internal polarization at equilibrium, and this internal polarization can be modified by the application of a stress. The mechanism for creating the polarization is straightforward: stress causes internal atomic rearrangements, which shifts the charge balance among the atomic constituents. When a center of inversion is lacking, the rearrangements can result in a net polarization. Similarly, when a material that lacks inversion symmetry is subjected to an external electric field, which creates an internal polarization, it is possible for this polarization to be accompanied by a strain.

The fact that the internal polarization contributes to the stored energy means that piezoelectrics can be used both for sensing and for actuation. In practice, piezoelectrics are not very good at *quasi-static* sensing of strain because of the parasitic effects of small DC leakage currents, but they are highly effective for sensing of vibratory or resonant motions. As actuators, they can create large forces and hence can be used for exciting resonances and vibrations. They are also useful as quasi-static micropositioners, especially within a feedback loop that can correct for any hysteresis, drifts, or nonlinearities.

There is an important coupling to the thermal energy domain. Materials with spontaneous internal polarization exhibit a change in polarization with temperature change. And for materials that acquire polarization as a result of strain, temperature changes can modify that polarization through thermal-expansion effects. Materials which exhibit temperature-dependent polarization are called *pyroelectric*.[4]

What is common to all three classes of behavior, piezoelectric, pyroelectric, ferroelectric, is that the material must lack inversion symmetry. Which of the materials exhibit which of these behaviors, however, depends in great detail on the internal structure; each material must be considered as a separate case.

21.4.2   Analytical Formulation of Piezoelectricity

The piezoelectric effect links stress (a second-rank tensor) with polarization (a first-rank tensor). Therefore, the description of the piezoelectric effect

---

[3]Inverting the coordinates would reverse the polarization, and if the crystal is symmetric to coordinate inversion, the polarization must be unchanged. Hence, the only possible value for that polarization is zero.
[4]The pyroelectric effect will not be discussed here. The reader is referred to Ikeda [11] and to Section 8.6 of Nathan and Baltes [127].

requires a third-rank tensor. Because this tensor has sufficient symmetry with respect to its stress-dependence, we can use the same method that was used when collapsing the elastic constants (a fourth-rank tensor) to a 2×2 matrix.

In the description to follow, lower-case indices i, j run over the integers (1,2,3), and are used to index the components of the electric constituents (electric field, electric displacement, and polarization), while upper-case indices I, J run over the integers (1,2,3,4,5,6), and are used to index the reduced-notation components of the mechanical constituents (stress and strain). The stiffness and compliance coefficients C_{IJ} and S_{IJ} have their usual meaning. The dielectric permittivity is denoted by ε, and strain by ϵ, requiring either sharp eyes or a good cup of coffee when examining the equations. New coefficients are needed to represent the piezoelectric coupling between stress and electric displacement, and between applied electric field and strain.

There are several different formulations of the piezoelectric equations, any of them derivable from any other. It is most common to use the form in which electric field and stress are the independent variables:

$$\epsilon_I = \sum_J S_{IJ}^{\mathcal{E}} \sigma_J + \sum_j \mathcal{E}_j d_{jI} \tag{21.28}$$

$$D_i = \sum_J d_{iJ} \sigma_J + \sum_j \varepsilon_{ij}^{\sigma} \mathcal{E}_j \tag{21.29}$$

where the 3×6 array of piezoelectric stress coefficients d_{iJ} are the reduced-notation representation of a third-rank tensor d_{ijk}, and where it is understood that when the indices I and J have values of 4, 5, or 6 as subscripts to stress or strain, it is the corresponding shear stress τ or shear strain γ that is implied. The units for the d_{iJ} array are Coulombs/Newton. The superscript on the S_{IJ} means that the compliance constants are to be measured at constant electric field (presumably zero), and the superscript on the ε_{ij} means that the permittivity is to be measured at constant stress (also presumably zero). The possibility of pyroelectric interactions also requires specification of whether the parameters are measured at constant temperature or adiabatically (constant entropy); however, in practice, the difference between the d_{iJ} coefficients for the two cases is small [127], and will not be distinguished here.

While the formulation in terms of d_{iJ} coefficients is useful for analyzing actuators, an alternate formulation, presented below, is useful for analyzing vibration and waves:

$$\sigma_I = \sum_J C_{IJ}^{\mathcal{E}} \epsilon_J - \sum_j \mathcal{E}_j e_{jI} \tag{21.30}$$

$$D_i = \sum_J e_{iJ} \epsilon_J + \sum_j \varepsilon_{ij}^{\epsilon} \mathcal{E}_j \tag{21.31}$$

where the e_{iJ} piezoelectric coefficients have the dimensions of Coul/meter^2, and where the stiffness constants are to be measured at constant electric field, and the permittivity at constant strain. The e_{iJ} are related to the d_{iJ} through a tensor transformation involving the elastic constants measured at constant electric field. The relationships are (see [11])

$$d_{iJ} = \sum_K e_{iK} S^{\mathcal{E}}_{KJ} \tag{21.32}$$

$$e_{iJ} = \sum_K d_{iK} C^{\mathcal{E}}_{KJ} \tag{21.33}$$

21.4.3   Piezoelectric Materials

There are three basic classes of piezoelectric materials used in microfabrication: piezoelectric substrates such as quartz, lithium niobate, and gallium arsenide; thin-film piezoelectrics, such as zinc oxide, aluminum nitride and lead zirconate-titanate (PZT); and polymer-film piezoelectrics, such as polyvinylidene flouride (PVDF). Selected data for two of these materials are presented in Table 21.1.

Table 21.1.   Constitutive data for selected piezoelectric materials. The first index on the d- or e-coefficients refers to the electric-field variables, while the second index refers to the mechanical variable (stress or strain). Sources: [52, 126, 127, 128].

Quantity		α-Quartz	Zinc Oxide
Density (kg/m^3)		2648.5	5676
Relative permittivity	ε_{11}	4.5	8.3-9.3
	ε_{33}	4.6	8.9-11
Stiffness (GPa)	C_{11}	86.8	209.7
	C_{33}	105.75	210.9
	C_{12}	7.04	121.1
	C_{13}	119.1	105.1
	C_{44}	58.2	42.47
	C_{14}	-18.04	
Piezoelectric	d_{11}	-2.31	
d (pC/N)	d_{14}	-0.727	
e (C/m^2)	e_{13}		-.573
	e_{33}		1.32
	e_{15}		-0.48

The use of the data of Table 21.1 requires considerable care, because the non-zero elastic constants and piezoelectric coefficients appear in the various matrices in positions that depend in detail on the crystal class. For example, the stiffness matrix in quartz takes the form [53]

$$
\text{Quartz :} \quad
\begin{pmatrix}
C_{11} & C_{12} & C_{13} & C_{14} & 0 & 0 \\
C_{12} & C_{11} & C_{13} & -C_{14} & 0 & 0 \\
C_{13} & C_{13} & C_{33} & 0 & 0 & 0 \\
C_{14} & -C_{14} & 0 & C_{44} & 0 & 0 \\
0 & 0 & 0 & 0 & C_{44} & C_{14} \\
0 & 0 & 0 & 0 & C_{14} & (C_{11} - C_{12})/2
\end{pmatrix}
\tag{21.34}
$$

An additional detail: there is a sign-convention issue, because quartz crystals can grow in what is called "left-handed" or "right-handed" forms. In right-handed quartz, by convention, the positive end of the x_1 axis develops a negative charge on extension. The charge is positive in left-handed quartz. This affects the signs that are reported for the piezoelectric coefficients. The data in the table are for right-handed quartz. The array of d_{iJ} coefficients has the form

$$
\text{Quartz :} \quad
\begin{pmatrix}
d_{11} & -d_{11} & 0 & d_{14} & 0 & 0 \\
0 & 0 & 0 & 0 & -d_{14} & -2d_{11} \\
0 & 0 & 0 & 0 & 0 & 0
\end{pmatrix}
\tag{21.35}
$$

To find the e_{iJ} coefficients, one calculates

$$
e = d \cdot C
\tag{21.36}
$$

Performing this calculation for the right-handed quartz data yields e_{iJ} which are the negatives of those reported in [126], which does not specify the handedness of the quartz. These signs may be important in quasi-static situations, where one really cares whether the material is extending or shrinking, but in vibration and wave-propagation situations, this sign ambiguity corresponds at most to a 180 degree phase shift in the resulting vibration, which can be readily tracked through a system.

The corresponding elastic-constant and piezoelectric-coefficient arrays for zinc oxide are

$$
\text{Zinc Oxide :} \quad
\begin{pmatrix}
C_{11} & C_{12} & C_{13} & 0 & 0 & 0 \\
C_{12} & C_{11} & C_{13} & 0 & 0 & 0 \\
C_{13} & C_{13} & C_{33} & 0 & 0 & 0 \\
0 & 0 & 0 & C_{44} & 0 & 0 \\
0 & 0 & 0 & 0 & C_{44} & 0 \\
0 & 0 & 0 & 0 & 0 & (C_{11} - C_{12})/2
\end{pmatrix}
\tag{21.37}
$$

and

$$\text{Zinc Oxide}: \quad \begin{pmatrix} 0 & 0 & 0 & 0 & e_{15} & 0 \\ 0 & 0 & 0 & e_{15} & 0 & 0 \\ e_{31} & e_{31} & e_{33} & 0 & 0 & 0 \end{pmatrix} \quad (21.38)$$

The d_{iJ} coefficients can be found by inverting the C_{IJ} matrix to find S_{IJ}, and then using Eq. (21.33).

21.4.4   Piezoelectric Actuation

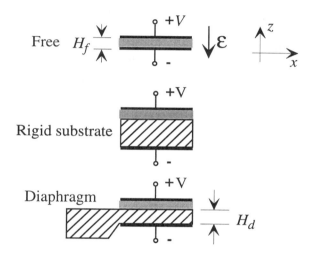

Figure 21.7.   A piezoelectric thin film, either as a free film, or deposited onto a rigid or bendable (diaphragm) substrate.

Consider a thin slab of ZnO, deposited (typically by sputtering) so that it's z axis is normal to the plane of the slab. Figure 21.7 shows three cases. A free film (which is not realizable in practice, but is a good starting point), a film deposited on a rigid substrate, and a film deposited on a diaphragm (or cantilever). For the free film, if a voltage is applied as shown, there is an electric field

$$\mathcal{E}_3 = -\frac{V}{H_f} \quad (21.39)$$

Because the film is free, all stresses are zero. Hence, we can use the d_{iJ} formulation to find the strains:

$$\epsilon_x = d_{31}\mathcal{E}_3 \quad (21.40)$$

$$\epsilon_y = d_{31}\mathcal{E}_3 \tag{21.41}$$

$$\epsilon_z = d_{33}\mathcal{E}_3 \tag{21.42}$$

The values of d_{31} and d_{33} can be obtained from the e-coefficients and elastic constants of Table 21.1. The results are d_{31} = -5.4 pC/N, and d_{33} = 11.7 pC/N. The film thickness increases, and there is a corresponding shrinkage strain in x and y equal to about half the z-axis strain. Note that the z-axis strain is inversely proportional to the film thickness H_f. This means that the absolute change in thickness depends on voltage but not on thickness, and has the magnitude $d_{33}V$. Furthermore, the volume strain, which is the sum of the three axial strains, is nearly zero. Finally, note that if the direction of the electric field were reversed, all the strains would change sign.

If this same film is placed on a rigid substrate, so that ϵ_x and ϵ_y are forced to be zero, then in-plane stresses develop. In that case, it is just as convenient to use the e-coefficient formulation, yielding the equations

$$\sigma_x = C_{13}\epsilon_z - e_{31}\mathcal{E}_3 \tag{21.43}$$

$$\sigma_y = C_{13}\epsilon_z - e_{13}\mathcal{E}_3 \tag{21.44}$$

$$0 = C_{33}\epsilon_z - e_{33}\mathcal{E}_3 \tag{21.45}$$

The solution is

$$\epsilon_z = \frac{e_{33}}{C_{33}}\mathcal{E}_3 \tag{21.46}$$

and

$$\sigma_x = \sigma_y = \left[\frac{C_{13}e_{33}}{C_{33}} - e_{31}\right]\mathcal{E}_3 \tag{21.47}$$

Using the values in Table 21.1, and assuming an electric field strength of 10^6 V/m, equivalent to 1 Volt applied across a 1 μm film, we obtain a z-axis strain of 6.3×10^{-6}, and in-plane biaxial stress of 1.23 MPa.

The in-plane stresses produce a bending moment. If the substrate can deform, for example, as a diaphragm, then bending will result from the application of voltage. If the piezoelectric film is thin compared to the diaphragm thickness, one can approximate the bending moment as

$$M \approx \frac{\sigma_x H_f H_d}{2} \tag{21.48}$$

Again, the piezoelectric film thickness cancels out, with the result that

$$M \approx \left[\frac{C_{13}e_{33}}{C_{33}} - e_{31}\right]\frac{V H_d}{2} \tag{21.49}$$

This is a biaxial moment, and will produce biaxial bending, as was examined in detail in Chapter 9.

The applications of this effect are numerous. Piezoelectric layers on deformable substrates can be used to initiate vibrations when excited in synchrony with the mechanical mode of the structure being excited, and can also be used quasistatically for micropositioning. However, because of charge-leakage and hysteresis effects, micropositioning applications are significantly improved when the piezoelectric actuator is within a feedback loop.

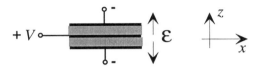

Figure 21.8.   A two-layer piezoelectric bimorph

Free-standing piezoelectric sheets can be made using ceramics such as PZT. These layers must be *poled*, a treatment in an electric field to establish the preferred piezoelectric direction within the film. The poled sheets can then be stacked, as illustrated in Fig. 21.8. If these sheets are stacked so that the poled directions are all along the same direction, then when the structure is driven with a voltage, the upper layer has a negative electric field, the lower layer a positive electric field. The in-plane stresses in the two layers oppose one another, producing a large bending moment. Structures of this type are called *piezoelectric bimorphs*. Alternatively, if layers are stacked with alternating poled directions, then the electric fields *relative to the poled direction* are the same in both layers, so the z-axis displacements add. *Piezoelectric stacks* built of many such layers can be used for actuators with many microns of motion range. Further, because of the relatively high elastic modulus of the piezoelectric materials, these actuators can create large forces.

21.4.5   Sensing with Piezoelectricity

While piezoelectrics can, in principle, be used for quasi-static sensing of strain or stress, they are better suited to time-varying sensing applications, where DC drifts and hysteresis are not important. The fact that piezoelectric actuation, when suitably synchronized with the mechanical modes of a structure, can be used to set up vibrations in structures creates a wide range of sensing applications. Quartz is a wonderful material for building mechanical resonators such as tuning forks for watches, and various types of sensors for mechanical loads, accelerations, and rotations. Quartz can be cut and polished into wafers, and can be bulk micromachined using selective etches and metallized to create electrodes. Because quartz is a good insulator, leakage currents tend to be small. Further, there is now a rich technology base for choosing various crystalline orientations of the slices cut from quartz crystals for different applications (see, for example, [10]).

Electrode arrays on piezoelectric substrates can also set up traveling acoustic waves. Applications include acoustic delay lines, and a wide range of chemical sensors obtained by applying chemically-selective coatings to the devices. This subject has been extensively examined by Ballantine *et al.* [126].

21.5   A Quartz Rate Gyroscope Case Study

The Systron Donner division of BEI Technologies manufactures quartz Coriolis gyroscopes for a wide variety of inertial applications, including the recent high-volume introduction into the automotive market for steering- and chassis-control systems.[5] The basic approach is the tuning fork. It is built by etching the structures from single-crystal quartz wafers that are cut so that the z-axis of the quartz, which is the hexagonal axis, is normal to the plane of the wafer. Such wafers are called "Z-cut" wafers.

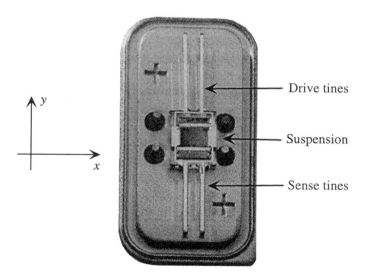

Figure 21.9. A quartz Coriolis gyro mounted in its package prior to lid sealing. (Source: Systron Donner; enhanced and reprinted with permission.)

Figure 21.9 shows one of the Systron Donner Quartz Rotation Sensors (QRS) after it has been fabricated, separated from the wafer, and mounted

---

[5]This Case Study is based in part on materials provided by Brad Sage of Systron Donner.

into its package prior to sealing. It has a double tuning-fork structure. One pair of tines are the driven tines. They have electrodes evaporated on them which create suitable electric fields in the quartz to drive in-plane bending motion (the details are explained later). If the tuning fork is rotated about its axis, the Coriolis force creates out-of-plane bending of the driven tines. This out-of-plane bending creates reaction moments in the supporting suspension, which transfers the moments to the sense tines. As a result, the sense tines also develop an out-of-plane motion. A differently shaped set of electrodes is evaporated onto the sense tines to detect this out-of-plane motion. In order to prevent unwanted coupling between the driven and sensed motions, the in-plane and out-of-plane resonant frequencies are intentionally offset from one another. We now examine the operation of this device, starting with a simpler case: the single-ended tuning fork.

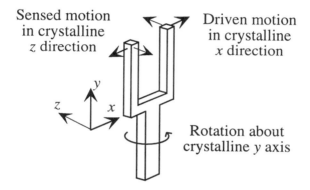

Figure 21.10.   The orientation of the tuning fork elements relative to the crystalline axes of the quartz wafer is critical.

21.5.1   Electrode Structures

The orientation of the tuning fork elements is critical to successful gyro operation. Figure 21.10 shows a schematic illustration of how the tuning fork elements must be oriented with respect to the crystalline axes. The driven motion must be in the crystalline x-direction and the rotation axis must coincide with the crystalline y-axis. The Coriolis-induced motion is then in the crystalline z-direction. The reasons for this choice lie in the details of the quartz piezoelectric coefficients.

Referring back to Eq. 21.35, we see that the piezoelectric coefficients that involve axial strains have only two non-zero elements: d_{11} which couples an x-directed electric field with an x-directed strain, and d_{12}, which is numerically equal to $-d_{11}$, and which couples an x-directed electric field with a y-directed axial strain. An electric field cannot set up a z-directed axial strain in Z-cut quartz; all three piezoelectric coefficients in the third column are zero.

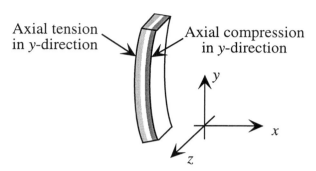

Axial tension
in y-direction

Axial compression
in y-direction

Figure 21.11. Illustrating a single bent tine, with the beam axis along y and the bending direction along x.

Since in order for the tines to vibrate, they must be made to bend, we examine a single bent tine, shown in Figure 21.11. The bending produces axial tension in the y-direction near one surface, and axial compression in the y-direction near the other surface. Of course, in pure bending, there is a smooth variation from maximum tension at one surface to maximum compression at the other. The shaded regions in the figure are simply for emphasis.

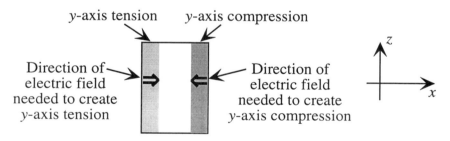

y-axis tension y-axis compression

Direction of
electric field
needed to create
y-axis tension

Direction of
electric field
needed to create
y-axis compression

Figure 21.12. Cross section of the bent tine from Fig. 21.11. Note that the y-axis is now into the plane of the drawing.

Figure 21.12 shows the cross section, in the x-z plane, of one of the bent tines from Fig. 21.11. Also shown is the direction of electric field necessary to achieve the y-directed tension and compression, taking advantage of the non-zero d_{12} piezoelectric coefficient in quartz.

Note that the electric field must change its direction from one face to the other. At first, it is not easy to imagine how to do this. However, because z-directed electric fields have no effect in quartz, the electrode configuration of Fig. 21.13 does the trick. We recall that the electric field is always normal to the surface of a conductor. Therefore, close to the electrodes, the fields are

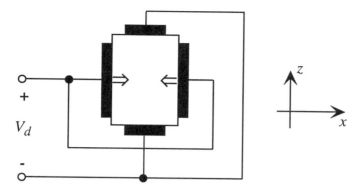

Figure 21.13.   Orientation and polarity of drive electrodes that will create the desired electric fields near the surface of the quartz beams.

x-directed. Toward the middle of the beam, the x component of field becomes small, but that is not a problem. Recall that the bending moment in a beam has its largest contributions from strains at its surfaces. Contributions to the moment from strains toward the middle of the beam are small. Therefore, even though this electrode configuration produces a primarily x-directed electric field only near the surface, it is precisely near the surface that the induced strains are most effective in creating a bending moment that can excite the beam into motion.

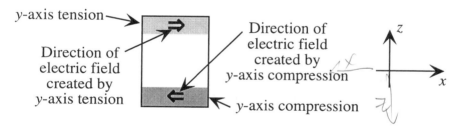

Figure 21.14.   Cross-section of a tine that is bent in the sense mode.

In the sense mode, the bending is in the z-direction, but the axial tension and compression due to bending is still in the y-direction. Figure 21.14 shows the cross section of a tine that has been bent in the z-direction. Note that the electric field set up by the bending is in the x direction, and that it has one direction in the tension region and the other direction in the compression region.

Figure 21.15 shows an electrode configuration that can sense the electric fields created by z-directed bending. Again, we recognize that there is actually a smooth variation of strain, and, hence, of x-directed electric field from the top surface to the bottom surface. The field is zero at the neutral axis. The placement of electrodes close to the top and bottom surfaces captures the region

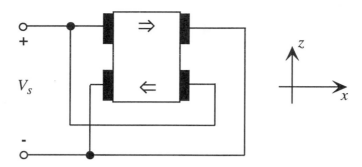

Figure 21.15. An electrode design that senses the z-directed bending without interacting with the drive mode.

of maximum electric field. While it would be possible to sense the bending with only a single pair of electrodes, having two sets both doubles the signal strength and makes the electrodes unresponsive to the drive-mode motion.

In principle, it would be possible to build both types of electrodes onto the same set of tines, using part of the length for driving and part for sensing. In practice, this produces severe interconnect problems, aggravates the problem of stray capacitive couplings between the drive and sense lines, and runs the risk that unless the sense electrodes are perfectly positioned, they might detect some of the drive motion. Therefore, the Systron Donner design uses a double tuning fork, coupled with an elastic suspension between the two forks.

21.5.2 Lumped-Element Modeling of Piezoelectric Devices

We shall use the Systron Donner QRS device as a vehicle for exploring the creation of lumped-element models for piezoelectric sensors and actuators. We do this in a series of steps, starting with the most elementary situation: quasi-static bending.

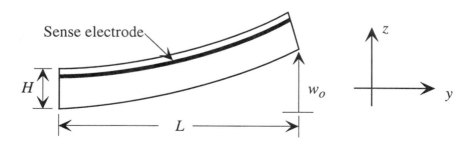

Figure 21.16. Highly exaggerated side view of a gyro tine bent in the sense mode. A single electrode is shown near the top of the tine. The tine support is at the left; the maximum z-displacement of the tine is w_o.

The easiest mode to analyze is the sense mode. We shall analyze it by finding the quasi-static bending response to an applied DC voltage. Figure 21.16 shown, in highly exaggerated form, the bending of a single tine in the sense mode. The support is at the left; the maximum z-displacement of the free end of the tine is w_o. A single electrode pair is located a distance z_e from the neutral axis of the tine (see Fig. 21.17). The electrode extends the full length of the tine in the y direction, and has height h in the z direction.

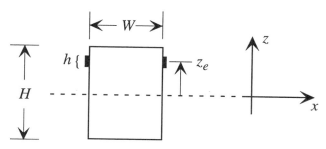

Figure 21.17.   Cross section of the tine of Fig. 21.16.

If a voltage is applied between the two electrodes, an electric field is set up in the x direction. This electric field creates a y directed axial strain and a corresponding axial stress. Because this axial stress is located away from the neutral axis of the beam, a bending moment is created which bends the beam to a specific radius of curvature and deflects the tip by distance w_o. Thus, we can find a relation between the applied voltage and the tip displacement. The creation of a y-directed axial stress also produces x-directed polarization through the d_{12} coefficient of the quartz. Therefore, there is a stress-dependent component to the dielectric response and, thus, a stress-dependent component of the electrical capacitance between the electrode pair. Our goal is to walk through the various steps of these relationships and construct a lumped mechanical model for the beam displacement and a corresponding lumped electrical model for the added component of electrical capacitance between the electrodes.

Because there are many steps to keep track of, the flow chart of Fig 21.18 is provided. We shall work our way through the various pathways and create a complete lumped model.

The x-directed electric field $\mathcal{E}_x$ is V/W (neglecting fringing for purposes of illustration). This sets up a y-directed axial strain $\epsilon_y = d_{12}\mathcal{E}_x$. The corresponding y-directed axial stress σ_y is $C_{22}d_{12}\mathcal{E}_x$, where the stiffness constant C_{22} is numerically equal to C_{11} for quartz. We use C_{11} hereafter.

There are, at this point, two branches in the flow chart. We consider first the branch leading to the charge on the electrode. It has two components. The first component results from the dielectric polarization of the medium in the absence of any piezoelectric effects. An x-directed electric field $\mathcal{E}_x$ creates an

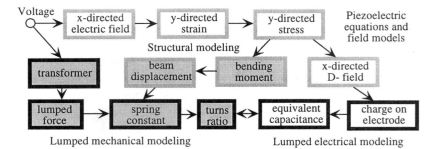

Figure 21.18. A flow chart for the creation of quasi-static lumped electrical and mechanical models of piezoelectric actuation and sensing. Boxes with solid black borders are lumped-element modeling steps; those with shaded borders are continuum or field-modeling steps. Boxes with clear backgrounds are electric or piezoelectric; those with shaded backgrounds are mechanical.

x-directed electric displacement $D_{x,\varepsilon}$ equal to $\varepsilon_{11}\mathcal{E}_x$. The magnitude of the electrode charge per unit area is equal to $D_{x,\varepsilon}$, so the total electrode charge due to the dielectric response is $hL\varepsilon_{11}\mathcal{E}_x$. The second component of electrode charge comes from the piezoelectric response. The y-directed stress creates an x-directed displacement field $\mathcal{D}_{x,e}$ equal to $d_{12}\sigma_y$, which equals $C_{11}d_{12}^2\mathcal{E}_x$. The corresponding charge on the electrode is then $hLC_{11}d_{12}^2\mathcal{E}_x$. Putting these terms together, we write

$$Q = Q_\varepsilon + Q_e \tag{21.50}$$

where

$$Q_\varepsilon = \left[\varepsilon_{11}\left(\frac{hL}{W}\right)\right]V \tag{21.51}$$

and

$$Q_e = \left[C_{11}d_{12}^2\left(\frac{hL}{W}\right)\right]V \tag{21.52}$$

Both of these relationships express quasi-static capacitances, since charge is proportional to voltage. The equivalent electrical capacitance at the electrodes is (neglecting all fringing-field effects)

$$C = C_0 + C_e \tag{21.53}$$

where

$$C_0 = \varepsilon_{11}\left(\frac{hL}{W}\right) \tag{21.54}$$

and

$$C_e = C_{11}d_{12}^2 \left(\frac{hL}{W}\right) \tag{21.55}$$

This completes the calculation of the equivalent capacitance in the electrical domain.

Returning once again to the flow chart, one can also build a lumped-element model directly, starting with the applied voltage. An element is needed to transform coordinates from the electrical domain to the mechanical domain. Using the $e \rightarrow V$ convention in both domains, we elect to use a transformer.

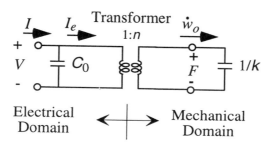

Figure 21.19.   Illustrating the use of a transformer to couple from the electrical domain to the mechanical domain as a first step in building a lumped mechanical model. The mechanical effort variable is an effective force F, and the mechanical displacement variable is the displacement of the tip of the tine w_o.

Figure 21.19 shows a quasi-static lumped-element model for the piezoelectric response. The capacitance C_0 due to the dielectric constant is shown in the electrical domain. The piezoelectric response is represented in the mechanical domain. A transformer with turns ratio n couples the electrical domain, in which the voltage V is the voltage applied to the electrode pair, to the mechanical domain, in which the tine tip displacement w_o is the displacement coordinate and F is the effective force causing this displacement.[6] We do not yet know how to specify the turns ratio; that will turn out to be the final quantity calculated. But we can identify that under quasi-static conditions, the tip displacement is proportional to the applied effective force with a factor $1/k$, where k is the effective mechanical spring constant. Hence, the quasi-static mechanical model contains a single capacitor with capacitance value $1/k$.

To calculate the spring constant, we must find the bending moment and resulting tip displacement, and relate it to the force F. We go back to the flow

---

[6]The origin of this effective force is the bending moment in the beam, as explained in the paragraphs to follow.

chart, and follow the "bending moment" path from the y-directed stress box. The moment is found from the integral

$$M = \int_{z_e-h/2}^{z_e+h/2} W \sigma_y z \, dz \qquad (21.56)$$

Carrying out the integration under the assumption that $\mathcal{E}_x$ does not vary with z yields

$$M = \sigma_y z_e h = C_{11} d_{12} z_e h V \qquad (21.57)$$

This moment is independent of position y along the length of the tine. Therefore, we have the ideal case of a beam with a constant bending moment. The result is a perfectly parabolic shape for the beam, $w(y)$:

$$w(y) = w_o \left(\frac{y}{L}\right)^2 \qquad (21.58)$$

where w_o is the tip displacement. To find the value of w_o, we go back to the relation between bending moment and radius of curvature ρ:

$$\frac{1}{\rho} = \frac{d^2 v}{dy^2} = -\frac{M}{C_{11} I} \qquad (21.59)$$

where I is the moment of inertia of the beam with respect to the neutral axis. The minus sign affects on the direction of bending relative to the coordinate axes. The magnitude of w_o is given by

$$w_o = \frac{ML^2}{2C_{11}I} = \left(\frac{d_{12} z_e L^2 h}{2I}\right) V \qquad (21.60)$$

This gives us one equation relating w_o to V. A second relation comes from the lumped model, in which the force F is related to the applied voltage by the transformer relation ($F = nV$):

$$w_o = \frac{F}{k} = \frac{nV}{k} \qquad (21.61)$$

Since these two expressions must be the same, we conclude that the spring constant k is given by

$$k = \frac{2nI}{d_{12} z_e L^2 h} \qquad (21.62)$$

The final step is to find the equivalent capacitance in the electrical domain for the mechanical capacitance $1/k$. To do this, we must use the admittance-transforming property of the transformer. If Y_m is an admittance in the mechanical domain, such that

$$\dot{w}_o = Y_m F \qquad (21.63)$$

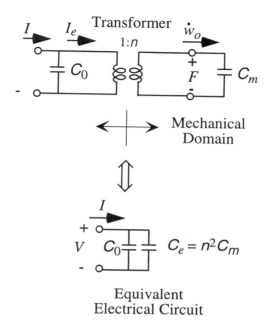

Figure 21.20. Equivalent circuit in the electrical domain for the combination of the direct electrical capacitance C_0 and the mechanical capacitance associated with the piezoelectric response C_u.

and we use the transformer relations

$$F = nV \quad \text{and} \quad \dot{w}_o = \frac{I_e}{n} \tag{21.64}$$

we obtain the equivalent admittance in the electrical domain

$$Y_e = n^2 Y_m \tag{21.65}$$

Therefore, referring to Fig. 21.20, the equivalent capacitance C_e in the electrical domain for a mechanical-domain capacitance of C_m is $n^2 C_m$. But Equation 21.55 already expresses C_e directly. Therefore, combining all quantities, including the substitution of $WH^3/12$ for I, we conclude that

$$n = \frac{C_{11} d_{12} H^3}{6 z_e L^2} \tag{21.66}$$

We note that the units of n are Coulombs/meter, which is what it should be to transform a displacement coordinate to a charge coordinate. This now completes the flow chart. We have identified every element of the quasi-static model.

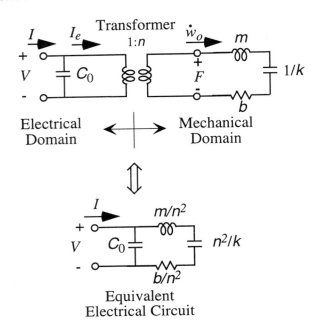

Figure 21.21.   Electrical equivalent circuit for the dynamic piezoelectric response, including the lumped-element mass and damping for one resonant mode of the structure.

It is now straightforward to create a dynamic model for the piezoelectric response by adding the appropriate mass and damping in the mechanical domain. If we consider only one mode of deformation, hence one mechanical resonance, we add a single mass and damper to the already-calculated capacitive stiffness, and obtain the equivalent lumped circuit of Fig. 21.21. Note that in the equivalent electrical circuit, each element has its admittance multiplied by n^2.

Every mechanical structure has multiple resonant modes, each one of which can be more or less tightly coupled to a given electrode structure. The equivalent circuit for a multi-mode piezoelectric device is shown in Fig. 21.22. Each mode has its own quasi-static stiffness (k or k'). And since each mode has its own resonant frequency and Q-factor, one can calculate equivalent masses and damping resistances for each mode, creating the multi-mode equivalent circuit.

The turns ratio for each mode is particularly important. It represents the *electrical-to-mechanical coupling* for each mode. The coupling depends in detail on the electrode geometry and the geometry and crystalline orientation of the mechanical structure. Consider, for example, the complete electrode sensing structure, with two pairs of symmetrically placed but oppositely connected electrodes (Fig. 21.15). The result of bending in the z-direction, which is mechanically equivalent to charging the mechanical capacitance for the sense

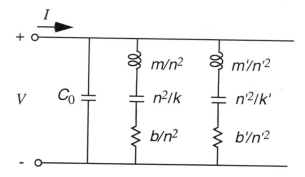

Figure 21.22. Electrical equivalent circuit for a piezoelectric device having two resonant modes.

mode, is to create voltages on the two electrode pairs that add together. However, if the same structure is deformed in the x-direction, corresponding to charging the mechanical capacitor for the drive mode, there is no output signal because the net turns ratio between the sense electrodes and the mechanical drive mode is zero. The effects of the two electrodes on the same face of the beam cancel each other out because they are oppositely connected. Of course, if the two electrodes are not exactly symmetrical, then there can be a second-order parasitic coupling from the drive mode to the sense mode. This is a major reason for separating the sensing tines from the drive tines, as discussed further below.

A more conventional way of expressing this coupling is with what is called the piezoelectric *electromechanical coupling coefficient*, conventionally denoted by the letter k. For clarity in distinguishing k from the mechanical spring constant and from the impedance-based definition of electromechanical coupling introduced in Section 6.6, we use the notation k_p. The k_p value for a particular mode is related to the size of the electrically equivalent capacitance C_e, and is defined as

$$k_p = \sqrt{\frac{C_e}{C_0}} = d12\sqrt{\frac{C_{11}}{\varepsilon_{11}}} \tag{21.67}$$

Thus we see that the electromechanical coupling is combination of the permittivity, the structural stiffness, and the strength of the piezoelectric coupling term d_{12}. The coupling can also be expressed in terms of the elements of the lumped electrical-mechanical model:

$$k_p = n\sqrt{\frac{1}{kC_0}} \tag{21.68}$$

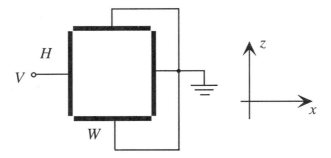

Figure 21.23. Illustrating one part of the superposition solution for the electric field distribution set up by the drive electrodes.

To calculate the corresponding quantities for the drive mode, we must take account of the more complex electrode geometry. Figure 21.23 shows an idealized drive electrode geometry, in which the electrode on each face occupies the entire face, but does not touch the electrode on the neighboring face. In practice, small gaps are required near the corners to prevent shorts. Also, it is necessary in general to take account of the anisotropy of the dielectric permittivity when finding electric fields. However, the dielectric anisotropy of quartz is small, so we shall assume that $\varepsilon_{33} = \varepsilon_{11}$. With these assumptions, the potential within the beam cross-section can be found from the Laplace equation. If we apply a voltage V to one of the electrodes and ground the other three, the Laplace equation has a well-known[7] Fourier-series solution for the potential within the beam:

$$\phi(x, z) = \sum_{l=1,odd} \frac{4V}{l\pi} \frac{\sinh[l\pi(W - x)/H]}{\sinh(l\pi W/H)} \sin\left(\frac{l\pi z}{H}\right) \qquad (21.69)$$

The complete solution, with four independently set potentials on each electrode, can be created by superposition of four such solutions, one arising from the potential on each electrode. The net result, assuming that the two x-face electrodes have voltage V on them while the two z-face electrodes are grounded, is

$$\phi(x, z) = \sum_{l=1,odd} \frac{4V}{l\pi} \left[\frac{\sinh[l\pi(W - x)/H]}{\sinh(l\pi W/H)} \sin\left(\frac{l\pi z}{H}\right) \right.$$

$$\left. + \frac{\sinh(l\pi x/H)}{\sinh(l\pi W/H)} \sin\left(\frac{l\pi z}{H}\right) \right] \qquad (21.70)$$

---

[7] See, for example, Haus and Melcher, p. 158, [129].

The x-directed electric field everywhere in the cross-section can be found from

$$\mathcal{E}_x(x, z) = -\frac{\partial \phi(x, z)}{\partial x} \tag{21.71}$$

The rest of the calculation proceeds exactly as for the sense mode, following the flow chart of Fig. 21.18. The capacitance is found from $\mathcal{D}_x$, which includes two components, one proportional to ε_{11}, the other to $C_{11}d_{12}^2$. However, when evaluating the two capacitances, one must integrate the value of $\mathcal{E}_x(0, z)$ over the electrode area, since $\mathcal{E}_x(0, z)$ is a function of z. Because the model of Fig. 21.23 and Eq. 21.70 assumes electrodes that extend right to the corners of the beam, there is a singularity in the electric field near the corners. As an approximate way of removing that singularity without having to solve the Laplace equation with a more complex set of boundary conditions, we choose to integrate over only the central 80% of the z-axis structural extent, ignoring the charge on strips at the top and bottom $H/10$ of the beam. The y-directed axial stress in the beam is equal to $C_{11}d_{12}\mathcal{E}_x$, and the bending moment is found by integrating the moment of this axial stress over the volume of the beam, again, ignoring the top and bottom 10% to avoid the field singularity. Unlike the sense-mode case, integrals over both x and z must be performed because σ_y depends on both x and z. Assuming that the electrodes run the full length of the tine in the y direction, the solution is, once again, a constant-moment case, with a parabolic deformation. From this, an effective spring constant can be found leading to an effective mechanical capacitance and a corresponding turns ratio for the electrical-to-mechanical transformer. The results of performing these calculations for the specific geometry of the QRS device can be found in Table 21.3 in Section 21.5.4.

Completion of the lumped-element model requires that we calculate the mass and damping for each mode of interest. We shall do this by estimating the resonant frequency using Rayleigh-Ritz methods, which will yield an equivalent lumped mass, given the spring constant. Then, using an estimate of the Q value of the mode, we can assign a value for the mechanical damping resistance.

We recall from Section 10.5 that we can estimate the resonance frequency from

$$\omega_o^2 = \frac{\mathcal{W}_e}{\displaystyle\int_{\text{volume}} \frac{1}{2}\rho_m w(y)^2 \, dxdydz} \tag{21.72}$$

where $\mathcal{W}_e$ is the elastic energy of bending for our assumed parabolic deformation $w(y)$, and ρ_m is the mass density of quartz. This calculation is readily performed, to yield a drive-mode resonant frequency estimate of

$$\omega_{o,drive} = \sqrt{\frac{5C_{11}W^2}{6\rho_m L^4}} \tag{21.73}$$

and a corresponding sense-mode resonant frequency estimate of

$$\omega_{o,sense} = \sqrt{\frac{5C_{11}H^2}{6\rho_m L^4}} \tag{21.74}$$

Thus, we see that by making W differ from H, the mode spacing can be controlled. The lumped mass for a mode is

$$m = \frac{k}{\omega_o^2} \tag{21.75}$$

and the resistance value b is obtained from

$$b = \frac{m\omega_o}{Q} \tag{21.76}$$

The only element not yet explored is the suspension between the drive and sense tines. That discussion is deferred until the actual device geometry is introduced.

21.5.3 QRS Specifications and Performance

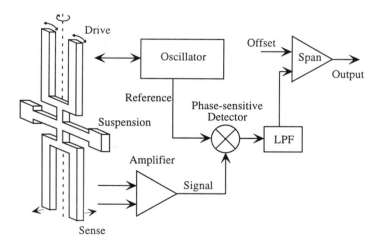

Figure 21.24.   Illustrating the basic system for the QRS rate sensor.

A schematic of the Model AQRS quartz angular rate sensor system is shown in Fig. 21.24. The quartz double-ended tuning fork with its supporting suspension and drive and sense electrodes is mounted as shown in Fig. 21.9, sealed, and mounted on a circuit board with the supporting electronics. The device has three leads: power, ground, and output. The details of the circuitry are not provided by Systron Donner, but all the components except the oscillator are familiar. In order to explore the oscillator, we will first develop a quantitative model for the device using the lumped-modeling methods of the previous section.

A variety of packages are used for the assembly, one of which is shown in Fig. 21.25. The tuning-fork axis is keyed to reference pins and fiducial marks on the package, so it can be oriented correctly when mounted.

Figure 21.25. One of the packages used for the QRS system sensor and system electronics of Fig. 21.24. (Source: Systron Donner; reprinted with permission.)

Table 21.2 lists the performance specifications the Model AQRS. We note that the output is an analog signal with a nominal value of 2.5 V in the absence of rotation. A positive rotation sense about the tuning-fork axis causes the output to increase, while a negative rotation causes it to decrease. The resolution threshold is stated as .004°/sec, but the output noise is specified as .025°/sec/$\sqrt{Hz}$ over a 100 Hz bandwidth, implying a 0.25°/sec RMS noise floor for the device over the full bandwidth, and suggesting that a narrower bandwidth is needed to resolve the stated resolution threshold. The intrinsic mechanical noise of the modes is very small because the Q-values for the modes of quartz are very large, typically greater than 5000. Therefore, the b values are quite small, along with their associated thermal noise contributions. And as discussed in Chapter 16, resolution thresholds are often set by $1/f$ noise and drifts. The resonant operation of these devices moves the signal frequency out of the $1/f$ noise region of typical amplifiers. Residual $1/f$ effects in the sense signal itself could arise from slow thermal fluctuations.

Table 21.2.   Specifications for a Systron Donner Model AQRS quartz angular rate sensor.

Parameter	Specification
Input voltage	5 V DC $\pm$ 5%
Input current	< 20 mA
Standard ranges	$\pm$ 64°/sec
Full range output	0.025 to 4.75 V DC
Scale factor calibration (at 22 °C)	$\pm$ 3% (including temp)
Offset bias	2.50 V DC
Bias variation over temperature	< 4.5°/sec
Short-term bias stability	< .05°/sec
(100 sec at const temp)	
Long-term bias stability	< 1.0°/sec
g-sensitivity	< .06°/sec/*g*
Bandwidth	> 50 Hz
Nonlinearity	< 0.05% of full range
Threshold/resolution	< .004°/sec
Output noise (DC to 100 Hz)	< 0.025°/sec/$\sqrt{\text{Hz}}$
Operating temperature	-40°C to 85°C
Storage temperature	-55°C to 100°C
Vibration (operating)	1.5 g_{RMS} 20 Hz to 2 kHz random
Vibration (survival)	10 g_{RMS} 20 Hz to 2 kHz random
Shock	200 g

Sensitivity to accelerations is always an issue for gyros. This design specifies a relatively low .06°/sec/*g* acceleration response, indicating that it would take a 10 *g* acceleration to get a response equal to 1% of full scale. This is quite acceptable for automotive use.

21.5.4   A Quantitative Device Model

Device dimensions for the tines and the support suspension are not published by Systron Donner, but plausible dimensions can be inferred from the device photograph, known standard wafer thicknesses for Z-cut α-quartz, and quantitative data on resonant frequencies in one of the related patents [130]. For purposes of modeling, we shall assume that the wafer thickness H is 0.5 mm, the tine length L is 6 mm, and the tine width W is 0.45 mm. Using Eqs. 21.73 and 21.73 and the data of Table 21.1, we calculate drive- and sense-mode resonant frequencies of 10.4 kHz and 11.6 kHz, respectively, which are very close the the resonant frequencies published in the patent.

Further, if we assume a sense-electrode configuration in which $z_e = H/3$ and $h = H/6$, so that the sense electrodes cover 5/6 of the x-directed face starting

Table 21.3.   Lumped-element model parameters for the the QRS device.

Quantity	Drive	Sense	Units
ω_o	6.54×10^4	7.27×10^4	rad/sec
$f_o = \omega_o/2\pi$	10.41	11.56	kHz
k_p	0.1077	0.1077	–
C_0	634	178	fF
C_e	7.34	2.06	fF
n	4.62×10^{-6}	4.19×10^{-6}	C/m
k	2.91×10^3	8.50×10^3	N/m
m	0.68×10^{-6}	1.61×10^{-6}	kg
b	0.89×10^{-6}	2.34×10^{-5}	(N-sec)/m
R	0.42	1.33	MΩ
L	3.18×10^4	9.18×10^4	H

at the top, and a large Q of 5000,[8] the calculated values of the various lumped element quantities for the sense mode are as shown in Table 21.3.

The corresponding drive-mode parameters are slightly more difficult to calculate because the Fourier series must be summed sufficiently to get conversion. The electrode model of Fig. 21.23, with vanishingly small gaps at the corners, has an elementary Fourier series, but it introduces a non-physical field singularity at the corners which skews the model. As a first-order estimate, the simple field solution was used, but $\mathcal{E}_x$ values for z-values within $H/10$ of either top or bottom face were ignored and treated as zero. The drive-mode values were then calculated, as shown in Table 21.3.

We note several interesting things from the table. First, the coupling constant k_p is small, of order 0.1. The ratios of $C_{e,d}$ to $C_{0,d}$ (drive mode) and of $C_{e,s}$ to $C_{0,s}$ (sense mode), which go as the square of k_p, are 1%. Therefore, unless we are at a frequency which is very close to one of the resonant frequencies of the structure, the equivalent electrical admittance is dominated by C_0. Second, the drive-mode and sense mode masses are of the same order as the total mass of an entire tine (3.6×10^{-6} kg), but are smaller, reflecting the fact that the mass at the free ends of the tine move more than the mass near the supports. We now use these lumped models to examine the quantitative behavior of the device.

21.5.5   The Drive Mode

The admittance of the equivalent circuit of Fig. 21.21, is

---

[8]We shall not be investigating mode damping here, so this assumed Q value is for purposes of modeling only. Q values in quartz resonators are typically very large, in some cases in the millions.

$$Y(j\omega) = j\omega C_0 + \frac{1}{j\omega L + R + 1/j\omega C_e} \qquad (21.77)$$

where L, R and C_e are the inductance, resistance, and capacitance correspond-ing to the scaled mechanical elements in the circuit. After some manipulation, this has the value

$$Y(j\omega) = j\omega C_0 \left[\frac{(1 + k_p^2) + j\omega RC_e - \omega^2 LC_e}{1 + j\omega RC_e - \omega^2 LC_e} \right] \qquad (21.78)$$

This admittance function gives the total current into the circuit for a given sinusoidal voltage. Virtually all of this current flows in C_0 unless we are at a frequency very close to the series RLC_e resonance. To get a clearer understanding of this, we plot the magnitude and angle of Y for the drive mode in the vicinity of the drive-mode resonance.

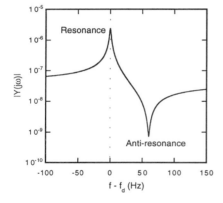 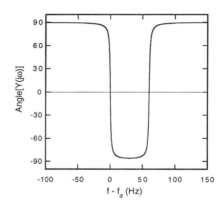

Figure 21.26. Admittance *vs.* frequency for the drive mode in the vicinity of the drive-mode resonance frequency, f_d.

Figure 21.26 shows the frequency dependence of the magnitude and angle of the admittance for the drive mode in the vicinity of the drive-mode resonant frequency f_d. Note the sharp peak in the admittance at f_d. Since the admittance measures current per volt, this peak indicates a large velocity in the tip at resonance, corresponding to a large current through the series-resonant RLC_e branch of the equivalent circuit. However, at a slightly higher frequency, there is a sharp dip in the admittance. This is due to the parallel resonance between C_0 and the rest of the circuit. Parallel resonances correspond to impedance maxima, hence, to minima in the admittance. In the piezoelectric literature, this is sometimes called the *anti-resonance*. At frequencies between the two resonances, the phase angle is -90 °. Exactly at the resonance and anti-resonance peaks, the phase is zero. In the case of the resonance, this means that the velocity of the tine is in phase with the drive voltage. At frequencies below the resonance and above the antiresonance, the phase is $+90°$, consistent with the admittance being dominated by C_0.

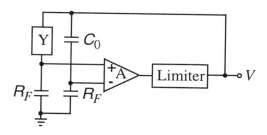

Figure 21.27.   A plausible oscillator circuit for the drive mode.

Systron Donner does not publish their drive circuit, but we can create a plausible elementary drive circuit, as shown in Figure 21.27. The resonator, represented by admittance Y, is placed in a bridge circuit. The right-hand branch does not have to match numerically the values of C_0 and R_F, but their ratio should match the ratio of C_0 to R_F. We assume numerical matching here to simplify the algebra. The differential voltage V_{in} applied to the input of the amplifier is

$$V_{in} = \left[\frac{Y}{Y + G_F} - \frac{j\omega C_0}{j\omega C_0 + G_F} \right] V \tag{21.79}$$

where G_F is the conductance $1/R_F$. This reduces to

$$V_{in} = \frac{G_F(Y - j\omega C_0)}{(Y + G_F)(j\omega C_0 + G_F)} V \tag{21.80}$$

Thus we see that the effect of the bridge circuit is to present to the amplifier an admittance from which the dominant C_0 component has been removed. The net result is, noting that in the denominator we can approximate Y by $j\omega C_0$,

$$V_{in} = \left[\frac{G_F}{(G_F + j\omega C_0)^2} \right] \left(\frac{1}{j\omega L + R + 1/j\omega C_e} \right) V \tag{21.81}$$

If we select G_F large compared to $j\omega C_0$, the prefactor is effectively a real number. This system is now equivalent to the oscillator circuit we analyzed in Section 15.5. The amplifier gain must be set sufficiently high to make the linearized loop unstable, and the limiter guarantees that the oscillation cannot grow in an unbounded manner. The result is a stable limit cycle.

A generic SIMULINK model for this system is shown in Fig. 21.28. The second ramp source is delayed slightly with respect to the first, and the two are subtracted in the adder. The net effect is a short triangular pulse which

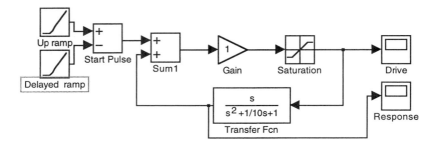

Figure 21.28.   A generic SIMULINK model for the drive oscillator of Fig. 21.27. A Q of 10 is used for this example, but a much higher Q consistent with quartz oscillators can be used.

moves the system away from the origin of its state space. Thereafter, the gain takes over, and a stable limit cycle is obtained. As expected, the waveforms labelled "drive" and "response" are in phase. The drive is a square wave and the response is a sine wave. The physical interpretation of the response is that since it derives from an admittance function, it is proportional to the velocity of the tine tip. The magnitude of the response is proportional to the Q of the resonator.

In practice, the drive-mode resonance is driven very hard to achieve a high velocity. One way to estimate the maximum velocity is to assume that the mode is driven so to achieve a certain maximum bending stress in the quartz beam, and that this maximum cannot approach the fracture strength. For purposes of estimating a transfer function, we examine the velocity for the case that the maximum stress in the quartz is 100 MPa.

The maximum stress in the drive mode at tip deflection x_o is

$$\sigma_{max} = \frac{C_{11} W x_o}{L^2} \tag{21.82}$$

If σ_{max} is 100 MPa, then the maximum drive-mode tip deflection is x_o = 92 μm. The corresponding maximum velocity is $\omega_d x_o$, equal to 6 m/sec.

21.5.6   Sense-Mode Displacement of the Drive Tines

When the gyro is rotated about the y-axis, the drive tines are driven in the sense mode. We can estimate the maximum displacement in the sense mode by noting that the magnitude of the lumped force exciting the sense mode is $2\Omega x_o \omega_d$. Since the drive mode and sense mode resonances are split by a large amount compared to the bandwidth of either resonance, we can ignore the velocity term in Eq. 21.12, to find that the maximum sense mode displacement z_o is

$$z_o \approx \frac{\Omega x_o}{\omega_s - \omega_d} \qquad (21.83)$$

If we assume a full-scale rotation rate of 64 °/sec, corresponding to 1.1 rad/sec, the maximum sense-mode deflection is

$$z_o = \frac{1.1 \times 92 \times 10^{-6}}{0.73 \times 10^4} = 14 \text{ nm} \qquad (21.84)$$

Thus, the estimated ratio of maximum sense-mode displacement to maximum drive-mode displacement *at full scale rotation* is only 1.5×10^{-4}, and at the noise threshold is more than two orders of magnitude smaller than this. We shall return to discuss the implications of this result after completing the transfer-function analysis.

21.5.7 Coupling to the Sense Tines

The suspension that couples the drive tines to the sense tines is a crucial element in the quartz gyro. Full details of its design are not published. However, the basic principles can be illustrated with the aid of Fig. 21.29, which is a somewhat simplified version of the suspension in the Gupta patent [130].

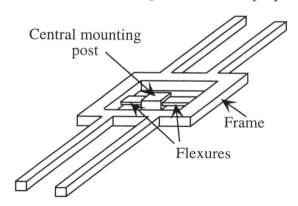

Figure 21.29. Illustrating one type of suspension and mounting for the quartz gyro. For purposes of making numerical estimates, the tines are assumed to be 1.5 mm apart (h_t), and the frame is a square with an inner dimension of 3 mm. The mounting post is assumed to be square, with dimension W_p of 1 mm on a side. Therefore, the length of each flexure L_f is 1 mm. Flexure thickness H_f is assumed to be 0.25 mm, and flexure width W_f 0.5 mm.

The drive and sense tines are both supported by a frame. The frame is, in turn, connected to a central mounting post by two relatively thin flexures that permit the frame to twist about the tine axis direction. When the drive tines are vibrated in the drive mode, the reaction forces and moments at the base of the two tines are, ideally, equal and opposite. Hence they sum to zero. Therefore, drive-mode vibration does not create any net force or torque on the frame. There is no direct mechanical coupling between the drive mode of the drive tines and the drive mode of the sense tines. In the absence of rotation, the sense tines remain stationary. When the device is rotated about the tine axis, exciting sense-mode motion in the drive tines, the reaction forces at the base of the tines do not cancel. Instead, they combine to exert a net torque on the frame. This torque is in phase with the tine displacement. Because the frame is not rigidly held, it can twist about the tine axis direction in response to this torque, imparting an inertial acceleration to the base of the sense tines which drives them into sense-mode motion.

This is not a resonant coupling, because the time-dependent torque driving the sense mode is at the drive-mode frequency. Nevertheless, the effect of the suspension coupling is to permit the sense-mode motion of the drive tines to be transferred to the sense-mode motion of the sense tines, where the electrode pattern can detect the resulting electric field.

To estimate the efficiency of this coupling, we note that even though quasi-static bending produces only a reaction moment, there is, because of the inertial mass of the tine, a dynamic net reaction force due to the sense-mode motion of each drive tine. The reaction force for each tine is $k_s z_o$, where k_s is the sense-mode spring constant. The two reaction forces are in opposite directions, creating a torque on the frame equal to $k_s z_o h_t$, where h_t is the spacing between tines. This torque drives the torsional motion of the frame.

To estimate the angular response of the frame, we note that the quasi-static bending of each the two flexures of length L_f when the frame is rotated is approximately equivalent to the bending of doubly clamped beam of length $2L_f$ with a point load at its center. The deflection w of such a beam with a point load F is

$$w = \frac{(2L_f)^3}{48EI_f} F_f \tag{21.85}$$

where E is the appropriate Young's modulus, I_f is the moment of inertia of the flexure, and F_f is the force loading each flexure. The actual load is a torque applied to the frame at the base of the drive tines. Therefore, the equivalent force applied to the end of the flexures must scaled down by the width of the frame compared to the spacing between tines. Thus,

$$F_f = \left(\frac{h_t}{W_p + 2L_f} \right) k_s z_o \tag{21.86}$$

The quasi-static angular displacement in response to this load is

$$\theta_{q-s} = \frac{w}{W_p + 2L_f} = \frac{(2L_f)^3 h_t k_s z_o}{48C_{11}I(W_p + 2L_f)^2} \tag{21.87}$$

where the value of C_{11} is being used as the Young's modulus. All quantities are now known (or, at least, estimated), so we can calculate the result. The angular deflection of the frame under quasi-static conditions is $\theta_{q-s} = 8.2 \times 10^{-8}$ rad.

The torque is not actually being applied quasi-statically, but rather at frequency ω_d. The torsional resonance frequency for the frame plus flexures about the tine axis is typically lower than either the drive-mode or sense-mode resonance frequency. A value quoted in the previously cited patent is 6.8 kHz. The drive-mode frequency of 10.4 kHz is well beyond the resonant peak of the torsional response. As a result, the torsional displacement resulting from the torque is scaled down from the quasi-static torsional response by a factor of roughly $(6.8/10.4)^2$, or 0.42. Thus, the dynamic angular displacement θ_d is 3.4×10^{-8} radians, and the corresponding angular acceleration is ω_d^2 times this quantity. Hence the inertial acceleration at the base of each sense tine is

$$a_s = \frac{\omega_d^2 h_t \theta_d}{2} \tag{21.88}$$

This has magnitude 0.11 m/sec². Using this value as the amplitude of a sinusoidal-steady-state excitation of the sense-tine motion in the sense-mode direction at frequency ω_d, we conclude that the maximum displacement of the sense tine tip is

$$z_{max} = \frac{a_s}{\omega_s^2 - \omega_d^2} = 0.11 \text{ nm} \tag{21.89}$$

The maximum stress in a sense tine is

$$\sigma_{max} = C_{11} H z_{max} / L^2 = 134 \text{ Pa} \tag{21.90}$$

Therefore, the x-directed displacement $\mathcal{D}_x$ in the sense tines has a magnitude ranging from zero at the neutral axis to a maximum (at the top and bottom of the beam) of $\sigma_{max} d_{12}$, or 3.2×10^{-10} Coul/m². The average $\mathcal{D}_x$ at the sense electrode is about half this value, so the total charge on the C_0 capacitance is about 3.8×10^{-16} Coulombs (using 5/6 of the face area for the electrode). Since the C_0 capacitor has a value of 178 fF, we conclude that the full-scale output voltage, prior to amplification, is a sinusoid at frequency ω_d of amplitude 2.1 mV. This is quite a healthy signal. The resulting nominal scale factor for this device, based on the many assumptions used here, is 3.28×10^{-5} (V-sec)/°.

The phase of this sinusoid is the same as the original drive voltage. The reasoning is as follows: The velocity of the drive-mode motion is in phase with the drive voltage. The sense-mode displacement and corresponding reaction force is shifted in phase by 90°, which is recovered because of the 90° phase shift in the suspension response. The sense-tine response is in phase with the torsional acceleration which is in phase with the torsional displacement.

21.5.8   Noise and Accuracy Considerations

We can now examine the noise specification, based on the assumptions used to estimate the scale factor. The output noise in a 100 Hz bandwidth is 0.25°/sec RMS, which translates to an equivalent 8.2 μV RMS noise source at the input of the sense amplifier. The sense signal corresponding to the stated sensitivity threshold of .004°/sec is 0.13 μV. These are reasonable noise levels for commercial-grade devices.

The sense signal must be measured in the presence of whatever interferences are present from drifts and other effects. For example, the stated short-term bias stability of 0.05°/sec is an order of magnitude larger than the sensitivity threshold, and the long-term bias stability at the output is equivalent to 1.0°/sec, which is significantly larger than the random noise. Therefore, it is fair to state that this is a device in which performance limits are dominated by stability and drifts, rather than by random noise.

Further, we recall that our estimated ratio of sense-mode motion of the drive tines relative to the drive-mode motion was on the order of 1.5×10^{-4} at full scale rotation, and two orders of magnitude smaller when measuring at the sensitivity limit. This means that the mechanical structure must be exceptionally well balanced so that the drive mode is mechanically pure. Very small amounts of mass imbalance on the drive tines could result in imperfect cancellation of the reaction forces at the base of the tines. The net result could be a y-axis torque on the frame at the drive-mode frequency. Because this torque has the same frequency as the desired sense-mode signal, it is a source of direct interference with the measurement. Elimination of this parasitic signal is perhaps the most difficult aspect of gyro system design. Systron Donner does not publish its methods of achieving optimized mode purity. One plausible method is to laser trim the shape of gold deposits on the drive tines while measuring the output signal at zero rotation, so as to achieve the best combination of mass balance, moment-balance and reaction-force cancellation for the drive tines when excited in the drive mode.

21.5.9   Closing Comments

This example of a quartz rate gyro illustrates the many issues that must be addressed in successful gyro design. The MEMS literature is rich in silicon gyro designs, most using capacitive actuation of the drive mode and using either capacitive position-sensing of motion in the sense mode [125]. Some of them are beginning to appear in the market. It is suggested that the reader examine one or more of these examples, and go through the same kind of analysis that was done here for the quartz gyro, drawing on the discussions of capacitive actuation and sensing of the previous two chapters. It will be discovered that the principles of these various devices are the same, but the detailed implementations differ in many ways, both great and small.

Related Reading

D. S. Ballantine, R. M. White, S. J. Martin, A. J. Ricco, E. T. Zellers, G. C. Frye, and H. Wohltjen, *Acoustic Wave Sensors: Theory, Design, and Physico-Chemical Applications*, Boston: Academic Press, 1997.

T. Ikeda, *Fundamentals of Piezoelectricity*, Oxford: Oxford University Press, 1990.

W. P. Mason, *Piezoelectric Crystals and Their Application to Ultrasonics*, New York: Van Nostrand, 1950.

A. Nathan and H. Baltes, *Microtransducer CAD: Physical and Computational Aspects*, Vienna: Springer, 1999.

J. F. Nye, *Physical Properties of Crystals*, Oxford: Oxford University Press, 1960.

N. Yazdi, F. Ayazi, and K. Najafi, Micromachined inertial sensors, *Proc. IEEE*, Vol. 86, pp. 1640-1659, 1998.

Problems

21.1 Consider the sense mode governing equation (Eq. 21.12) with $\omega_x = \omega_y = 2\pi \times 10^4$ rad/sec, a mass of 10^{-6} kg, and $Q_y = 500$. Assume $x = x_o \cos \omega_x t$ with $x_o = 100$ μm. Find the response of the gyro to a rotation pulse that might occur when turning a corner in a car. Specifically, find the response to $\Omega(t)$, where

$$\Omega(t) = \frac{\pi \omega_r}{4} \sin \omega_r t$$

for $0 \leq \omega_r t \leq \pi/2$, and zero thereafter. What happens to the response when ω_r exceeds δ_y?

21.2 Design a silicon cantilever with a 2 μm thick ZnO film on it and electrodes above and below the ZnO film such that an applied voltage can result in the free end of the cantilever being tipped by 30°. Such a device could be used to manipulate a micromirror attached to the free end of the cantilever. Specify the length and thickness of the cantilever, and the voltage required to achieve the bending. (This is an underconstrained problem. There are many possible solutions. Try to select one that can be readily fabricated.)

21.3 The drive-mode electrode configuration differs from that in Fig. 21.13 in that the electrodes occupy only about 80% of the sidewalls and about 50% of the top and bottom. Using finite-difference methods, determine a solution to the Laplace equation for this geometry. To simplify the calculation, assume that all of space is filled with a uniform dielectric with dielectric constant equal to that of quartz. Based on this result, calculate the bending moment and compare it to the result using the analytic solution in the text together with the assumption that the fields in outer 20% of the sidewalls were ignored.

21.4 Set up the oscillator model of Fig. 21.28 and compare the drive and response waveforms. How purely sinusoidal is the motion?

21.5 A Gyro Treasure Hunt: go to the literature or to a commercial web site and locate a rate gyroscope design. Determine the two modes of motion and the corresponding elements of a lumped-parameter model for the device.

Chapter 22

MICROSYSTEMS FOR DNA AMPLIFICATION

Naturae enim non imperatur, nisi parendo.
[Tr: Nature cannot be ordered about, except by obeying her.]

—Francis Bacon

22.1   Introduction

Biological and medical applications of MEMS technology offer opportunities for product and system development that stagger the imagination. Our previous case studies focused on mechanical sensors and displays, with application to automotive, consumer and industrial markets and, to a limited degree, to health and medicine (for example, blood-pressure measurement). A crude estimate of the global market potential for micromechanical devices is that each human on earth might use one of each type at a time. In contrast, each human on earth has on the order of 50,000 genes that might play a critical health-related role. Microsystems that can perform genetic analysis, therefore, have a market potential that could dwarf that of mechanical devices.

Biological and medical MEMS, abbreviated *bio-MEMS*, is a blossoming field. It includes devices for manipulating tissue fragments, cells, and individual biomolecules. A number of companies have entered the bio-MEMS market with "lab-on-a-chip" techniques for drug screening and various types of genetic analysis [131, 132]. The field is simply too vast to cover in a single chapter. We have chosen a single example, DNA amplification, to illustrate some of the potential benefits and some of the challenges of bio-MEMS. This example has an additional virtue: it involves a *micro-chemical reactor*, another area of MEMS that is beginning to exhibit rapid growth.

After a brief description of the essential molecular biology, we will examine two very different approaches for building DNA amplification systems using microfabrication. Neither of these is a commercial product at present, so we will compare some of the characteristics of the two microsystems to the specifications of a commercial macroscopic DNA amplification system. Finally, we shall look closely at some of the thermal-control issues that affect the performance of these micro-reactor examples.[1]

22.2   Polymerase Chain Reaction (PCR)

The discovery of the double-helix nature of deoxyribonucleic acid (DNA), the genetic-code-carrying molecule of every living organism, was one of the most dramatic stories of the mid-twentieth century [133]. Since then, molecular biologists have made immense strides not only in understanding and applying genetic knowledge, but also in working with the molecular building blocks, including methods for synthesizing short molecules that mimic sections of DNA molecules. *Polymerase chain reaction* (PCR) is a particular reaction sequence that starts with an original DNA molecule and creates an exponentially growing population of copies of fragments of that molecule. To understand how it works and to begin to develop the specifications needed for a PCR system, we must examine the fundamentals of DNA replication.

22.2.1   Elements of PCR

DNA consists of a helically wound double strand of *nucleotides*. A nucleotide consists of a sugar linkage, a phosphate linking group at one end, and an amino acid group called a *base* attached to the sugar. A cartoon representation is shown in Fig. 22.1. The nucleotide is not symmetric end-to-end. In a strand of DNA, the end of the nucleotide labeled 5' is attached to the 3' end of the next nucleotide.

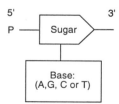

Figure 22.1.   A cartoon representation of a nucleotide, consisting of a sugar linkage, a phosphate end-group (P) which serves to link to the next nucleotide, and an amino-acid group, the base. (After [131].)

In DNA molecules, the base is one of four *nucleic acids*: adenine (A), guanine (G), cytosine (C), or thymine (T). Figure 22.2 shows at the top an

---

[1]Mathew Varghese of MIT made significant contributions to the preparation of this chapter.

example of a single DNA strand. The exact sequence of bases along the chain is what carries the genetic information.

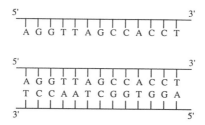

Figure 22.2. Single- and double-stranded DNA fragments.

The two strands that combine to form the double-helix molecule have complementary pairs of bases bonded to each other, as shown at the bottom of Fig. 22.2. A is always bonded to T, and G is always bonded to C. The helical twist of the double strand is not shown here.

To copy a DNA molecule using PCR, one first separates the two bound strands into individual strands by raising the temperature to ~95 °C. This step is called *denaturing*. Then one cools the separated strands to ~65 °C in the presence of a small molecule, called the *primer*, which is a synthetically produced single-strand molecule with a specific target sequence of bases. This primer can attach only to a point in the original DNA strand that has the correct complementary base sequence. This step is called *annealing*.

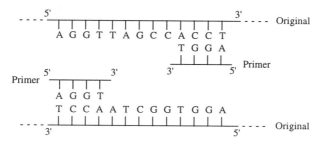

Figure 22.3. Annealing: the primer can attach to single-stranded DNA wherever the target sequence is complementary to the primer sequence.

Figure 22.3 illustrates this first attachment of a primer using a primer with the 5'-3' sequence AGGT. Note that the primer can attach to either strand wherever the sequence complements that of the primer. The example in this section shows the effect of a single primer which, for the particular DNA strand shown, has two potential attachment sites. In general, though, two primers can be used with different target sequences. It is the DNA fragment between the two target sequences that is amplified by PCR. When using a single primer, as in the example here, the two target sequences are the sequence of the primer itself and the complement of that sequence.

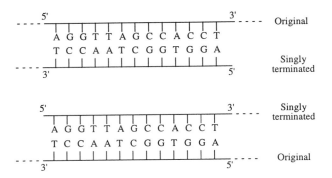

Figure 22.4. After the first extension reaction, the original DNA strands are copied as singly-terminated strands, each one starting from the point of attachment and extending from the 3' end.

Once the primer is attached, the sample is heated to ~72 °C in the presence of reagents that supply nucleotides and a polymerase enzyme that promotes the *extension* of nucleotide chains. The enzyme functions in a particular direction, attaching new nucleotides that complement the main chain to the 3' end of the attached partial chain. This leads to the situation pictured in Fig. 22.4. The copies start at a particular point on the DNA chain, but can extend without limit. Hence, they we call them *singly-terminated* chains.

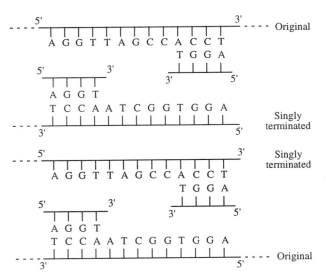

Figure 22.5.   After the second anneal, primer can attach to all four strands.

If the denature-anneal cycle is repeated, the situation in Fig. 22.5 is reached, and following a second extension cycle, the situation in Fig. 22.6 is reached. Here, the original DNA fragments are once again copied to create singly-terminated chains, and each singly-terminated chain is copied to create a doubly terminated chain. Now, the basis for a chain reaction is established.

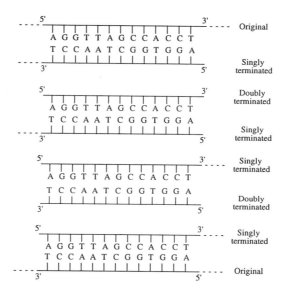

Figure 22.6. After the second extension cycle, there are original, singly-terminated, and doubly-terminated strands.

If the denature-anneal-extension sequence is repeated again, each original strand produces a singly-terminated strand, each singly-terminated strand produces a doubly-terminated strand, and each doubly-terminated strand *also* produces a doubly terminated strand. Thus if we denote by $N_S[n]$ the number of singly-terminated strands after the nth cycle, we find that

$$N_S[n] = nN_O \qquad (22.1)$$

where n is the number of complete cycles, and N_O is the number of original DNA strands. For a single original molecule of DNA, N_O would be 2. The number of singly-terminated strands grows linearly with the number of cycles.

On the other hand, the number of doubly-terminated strands produced in the nth cycle is

$$(N_D[n] - N_D[n-1]) = N_D[n-1] + N_S[n-1] \qquad (22.2)$$

which can be transposed to

$$N_D[n] = 2N_D[n-1] + nN_O \qquad (22.3)$$

Thus, the number of doubly-terminated strands more than doubles each cycle. It is this exponential growth in the number of doubly-terminated strands that is at the heart of PCR amplification. After 20 cycles, assuming perfect fidelity at each copying step, a single original DNA molecule produces 40 singly-terminated strands and 2^{19}, or 542,000 doubly-terminated strands. If the mixture that results from PCR amplification is analyzed to determine the base-pair sequence of the strands, the results will be dominated by the amplified doubly-terminated strands.

22.2.2   Specifications for a PCR System

The specifications for a PCR system are determined by the thermal cycle requirements. Denaturing and annealing each require only a few seconds once the sample has reached the correct temperature. The extension time depends on the specific polymerase and on the length of the DNA segment being amplified. If Taq Polymerase is used, with an extension rate of 50 base-pairs per second (bp/s), and the segment to be amplified is 500 base pairs (bp), then 10 seconds of extension time is required. Therefore, in aggregate, the total reaction time required for a PCR cycle is about 15 seconds.

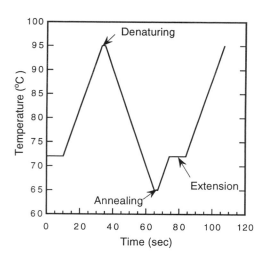

Figure 22.7.   PCR temperature cycle of a commercial PCR system.

In a macro-scale instrument, which is a bench-top instrument that holds a standard 96-well plate used in molecular biology, the actual cycle time is dominated by the temperature ramps, up and down. Figure 22.7 shows the

Table 22.1. Selected specifications for the GeneAmp 9700 PCR system.

Property	Specification
Temperature range	5-100 °C
Set-point accuracy	±0.25°C
Static temperature uniformity	±0.5°C
Heating/cooling rate	~ 1°C/sec
Sample volume	0.2 ml
Number of samples	96
Cycle time	65 sec (calculated)
Power required	1800 W

thermal cycle for a commercial macro-scale PCR machine,[2] using the published specification on heating and cooling rate of 1 °C/sec (see Table 22.1). A total of 65 seconds is required for the various temperature ramps during one cycle, while only 15 seconds is required for performing the reactions. This fact alone is enough to motivate the search for a microfabricated implementation of PCR, because thermal masses and response times can be made smaller than for corresponding macro-scale devices, with the result that temperature-ramp times can be reduced. A reduction in cycle time translates directly into instrument throughput.

22.3 Microsystem Approaches to PCR

There are two fundamentally different approaches to creating microfabricated PCR systems. Northrup [134] has reported a bulk-micromachined silicon PCR chamber. A variant of this design, with improved thermal isolation of the chamber from the substrate, has been reported by Daniel *et al.* [135]. Both of these devices are basically miniaturized versions of macro-scale PCR reactors. Miniaturization achieves reduced cycle time, as expected. A radically different continuous-flow design has been reported by Kopp *et al.* [136], in which a sample flows from one fixed-temperature zone to the next, executing the PCR cycle along the way. Because these two approaches are so different, we can gain valuable insights into microfabricated reactor systems by comparing them.

22.3.1 Batch System

Northrup's miniaturized batch PCR system [134] is illustrated in Fig. 22.8. The chamber is etched from a silicon wafer using anisotropic bulk microma-

---

[2]GeneAmp PCR System 9700, manufactured by Perkin Elmer Biosystems, Foster City, CA.

chining. The bottom of the chamber is a low-stress silicon-nitride membrane. The top is a cover glass attached with silicone rubber. Polyethylene tubing is embedded into the silicone rubber seal to provide access for sample and reagent insertion and removal. A resistive heater, formed from a 0.25 μm layer of polysilicon, is deposited onto the outside of the nitride membrane. The layout of this resistor is not reported, but it is presumably a long narrow line that meanders across the membrane so as to provide reasonably uniform heating. Aluminum bond pads for the heater are located on the bulk silicon portion of the device.

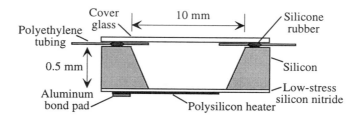

Figure 22.8.   The miniaturized PCR chamber reported by Northrup [134].

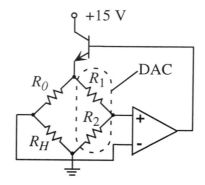

Figure 22.9.   Control circuit for Northrup's PCR chamber [134].

The control system is shown in Figure 22.9. R_H is the polysilicon heater resistor, which has a positive temperature coefficient reported as 1.22×10^{-3} Kelvin^{-1}. The resistor labeled R_0 is a fixed resistor, with a small self-heating thermal coefficient. The pair of resistors shown as R_1 and R_2 can be thought of as a programmable voltage divider, made from a digital-to-analog converter (DAC) driven with the voltage applied at the top of the bridge. As the ratio of R_1 to R_2 is changed, the balance point of the bridge is modified. At balance, there must be just enough current through the R_0-R_H arm of the bridge to raise the temperature of R_H so that the bridge becomes balanced.[3] Using a 15 V supply, the circuit draws 150 mA of current at a maximum heating rate of greater than 35 °/sec for a chamber with a volume of 25 μl. Cooling is passive. That is, the transistor turns off when the bridge balance point is set below the balance point determined by R_H, dropping the resistor current to zero. The authors claim a cooling rate comparable to the achieved heating rate. Numerical values for the various components are not provided by Northrup.

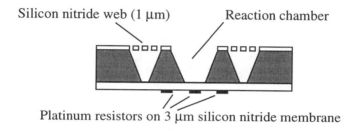

Figure 22.10. Cross-section of the Daniel design of a bulk-micromachined PCR reactor [135].

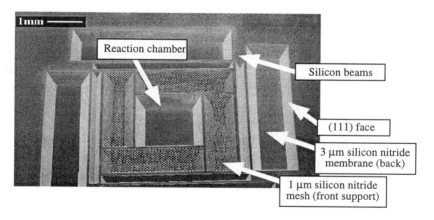

Figure 22.11. Photograph of the Daniel PCR chamber viewed from the top [135]. KOH is used to etch pits in the silicon beneath the mesh, and to etch additional pits outside the mesh region. A network of silicon beams supports the frame of the reaction chamber, providing both mechanical support and a controlled thermal resistance between the chamber and the substrate. (Source: David Moore, Cambridge University; reprinted with permission.)

One problem with the Northrup design is that the walls of the chamber are formed from the silicon wafer itself, which has high thermal conductivity. The heat-flow path from the heater to the substrate passes through the fluid volume, but the side walls of the chamber could be significantly cooler than the nitride

---

[3]In the original paper [134], the signs on the op-amp inputs are accidentally reversed.

membrane. Daniel *et al.* [135] isolate the PCR chamber from the rest of the silicon wafer using a silicon nitride membrane on the back side of the wafer and a thin mesh of silicon nitride on the front side (see Fig. 22.10). The mesh features are oriented along <100> directions, which results in complete under-etching of the mesh by KOH, creating pits in the substrate adjacent to the chamber. These pits improve the thermal isolation of the chamber. The mesh, whose surface is made hydrophobic, also serves to prevent aqueous solutions from entering the pits when the reaction chamber is being filled by pipette. Platinum resistors on the back side are used for heating the silicon beams that form the walls of the reaction chamber and for temperature sensing. Two sense resistors are used. One is placed on the silicon beams to provide the rapid response needed for feedback control of the beam temperature. The second is placed on the backside nitride membrane to measure the temperature of the liquid inside the reaction chamber. A photo of the device is shown in Fig. 22.11

Following fabrication of the microstructure, a 200 nm coating of silicon dioxide is deposited on the walls of the chamber by PECVD, followed by a surface treatment such as bovine serum albumin (BSA) for chemical compatibility with the PCR reagents. After filling the chamber, it is covered with an oil drop to prevent evaporation. Chamber volume is 2 μl. The authors report heating and cooling rates in the 60-90 °C/sec range, and a maximum static heater power of 1.9 W at the highest operating temperature. The total heating and cooling time required for one PCR cycle is about 1.5 seconds, which is now noticeably shorter than the required reaction times at the various temperatures. Further analysis of the thermal performance of this device is presented in Section 22.4.

22.3.2   PCR Flow System

A radically different approach to microfabricated PCR systems has been reported by Kopp *et al.* [136]. Instead of building a fixed-volume chamber whose temperature is cycled, Kopp constructs a continuous flow path etched in glass and clamps the glass to a support with three copper heat sinks, each one at one of the PCR temperatures. Figure 22.12 illustrates the principle for a PCR process with denaturing at 95 °C, annealing at 60 °C, and extension at 77 °C. The length of time in each temperature zone is controlled by the channel path length. The figure is only a schematic illustration. The actual device is shown in Fig. 22.13. Extra path length is provided during the first denaturing step to assure that the initial sample of double-stranded DNA is fully denatured. The path then flows to the annealing zone and then to the extension zone. Extra path length is also provided in the extension zone because of the longer reaction time needed at that temperature. In the device photograph, the extension zone appears opaque because there are so many path loops.

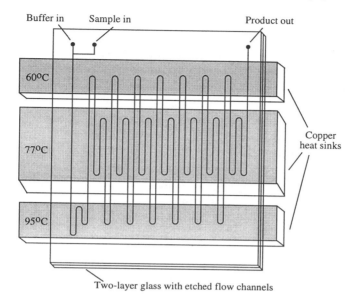

Buffer in   Sample in   Product out

60°C

77°C   Copper
                                             heat sinks

95°C

Two-layer glass with etched flow channels

Figure 22.12.   A continuous-flow PCR system [136].   A two-layer glass sample with flow channels etched into it is clamped to a support containing three copper heat sinks, each one controlled to a fixed temperature.   As fluid flows through the channel, it encounters a typical PCR temperature cycle.

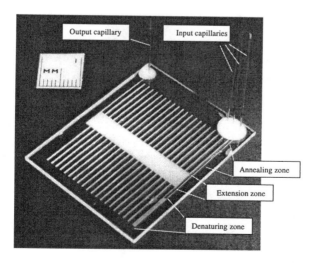

Output capillary   Input capillaries

Annealing zone

Extension zone

Denaturing zone

Figure 22.13.   Photo of the continuous-flow PCR cell of Kopp *et al.*. (Source: Andreas Manz, Imperial College; reprinted with permission.)

Fabrication of this device is straightforward.   Flow channels 40 μm deep and 90 μm wide are etched into 0.55 mm thick sheets of Corning 0211 glass using conventional lithography and wet etching. A second glass sheet has 400 μm holes drilled in it to provide flow-channel access. After bonding the cover sheet to the sheet containing the channels, standard 375 μm diameter capillaries (inside diameter 100 μm) are glued into the holes with epoxy, providing low-dead-volume fluid entry and exit ports. Flow of buffer and sample is controlled with syringe pumps. The copper blocks are heated with 5 W cartridge heaters, and block temperature is monitored with platinum thin-film resistance thermometers mounted on the block surface.

Unlike the batch reactor, whose throughput is determined by the sample volume and cycle time, the throughput of the continuous flow reactor depends on the flow rate. Increasing the flow rate, however, reduces the time spent in each zone. Thus each device must be designed for a specific ratio of dwell times in each of the three zones, and the overall flow rate must then be adjusted to fix the actual time. The throughput, measured in volume of sample carried through the complete PCR cycle, is proportional to flow rate. However, it is important that the residence time in each zone be long compared to the time required to bring the fluid to temperature as it enters each zone. This issue is examined in Section 22.5. Finally, in this system, it is possible to pipeline multiple samples by inserting only buffer between sample injections. Thus a single flow channel can process a sequence of PCR samples.

Both systems depend on heat transfer to and from the walls of a chamber in order to control the temperature of the sample. As we shall see below, both devices are amenable to lumped-element modeling of their thermal circuits.

22.4   Thermal Model of the Batch Reactor

For the Daniel batch device, the silicon sidewalls are heated, but not the top or bottom. Heat is conducted from these sidewalls directly into the fluid sample. Heat also flows through the nitride that forms the bottom of the chamber and the oil that caps the sample, but because the silicon nitride is thin and is not a good thermal conductor, the direct heat flow from the sidewalls is the dominant heat-transfer method. Therefore, to first order, we can consider the heat transfer in and out of the batch chamber as a two-dimensional heat-flow problem. We choose to model the chamber as perfectly rectangular in cross section, ignoring the sloping of the sidewalls. With this assumption, we can use the methods of Chapter 12 to create a lumped thermal model for the fluid in the chamber.

We use the eigenfunction expansion of the solution to the two-dimensional heat-flow equation and concentrate on the lowest eigenmode. It has the functional form

$$T(x,t) = T_0 \cos\left(\frac{\pi x}{L}\right) \cos\left(\frac{\pi y}{L}\right) e^{-t/\tau_c} \qquad (22.4)$$

where T_0 is a constant, L is the lateral dimension of the chamber, and where the fluid thermal time constant τ_c is given by

$$\tau_c = \frac{\rho_m \tilde{C}_m L^2}{2\kappa\pi^2} \tag{22.5}$$

where ρ_m is the mass density of water (1000 kg/m^3), $\tilde{C}_m$ is the specific heat of water per unit mass (4182 J/(kg-K)), and κ is the thermal conductivity of water (0.6 W/(K-m)) [52]. This time constant can be thought of as being formed from the product of a heat capacity and a thermal resistance. The heat capacity is found by weighting the distributed heat capacity over the spatial eigenmode, with the result

$$C_c = \frac{4}{\pi^2} L^2 H \rho_m \tilde{C}_m \tag{22.6}$$

where H is the height of the chamber. Therefore, since $\tau_c = R_c C_c$, the thermal resistance associated with the fluid in the chamber is

$$R_c = \frac{1}{8\kappa H} \tag{22.7}$$

The Daniel chamber is supported by four beams. We need to model both the thermal resistance and the heat capacity of these beams. Figure 22.14 shows two alternate lumped models we could use.

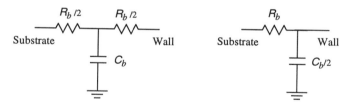

Figure 22.14. Two alternate methods for modeling the beams of the Daniel PCR chamber.

The model on the left of Fig. 22.14 uses the full heat capacity of the beam in between two thermal resistors of value $R_b/2$, where R_b is the thermal resistance of the beam. We see that in this model, the full thermal resistance between the wall and substrate is correctly modeled, and that in steady state, the point at which the capacitor is attached will come to a temperature halfway between the wall temperature and the substrate temperature (which we take as thermal ground). Therefore, the total energy stored in the capacitor is $T_w C_b/2$, where T_w is the wall temperature. The model on the right places the capacitor at the wall. In order to get the correct total energy storage in steady state, the capacitor value is reduced to $C_b/2$. We prefer the right-hand model when connecting the beams to the chamber. And, of course, there are four beams, so we must include four times the heat capacity of one beam in C_b and 1/4 the thermal resistance of one beam in R_b.

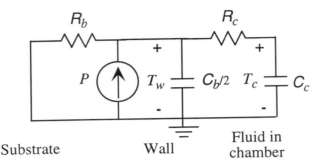

Figure 22.15. Combination of the chamber thermal model with the beam thermal model

Figure 22.15 shows the complete lumped-element model of the batch PCR reactor. Numerical values for the various parameters based on the data of Daniel are given in Table 22.2. The transfer functions relating the Joule heat power P to the wall and fluid temperatures are

$$T_w = \frac{R_b(1 + sR_fC_c)}{\frac{R_cR_bC_cC_b}{2}s^2 + \left[\frac{R_bC_b}{2} + (R_c + R_b)C_c\right]s + 1} P \qquad (22.8)$$

and

$$T_c = \frac{1}{1 + \tau_c s}T_w \qquad (22.9)$$

We note that the relationship between the fluid center temperature T_c and the wall temperature T_w is that of a single-pole response. Therefore, if we create a feedback loop to control the wall temperature, we expect the fluid temperature to show a time delay in its response of about $3\tau_c$. What would happen if we use the temperature sensor on the face of the chamber as the source of our feedback signal? Would the situation improve, or get worse? We examine this in the next section.

22.4.1   Control Circuit and Transient Behavior

Figure 22.16 shows a SIMULINK model with a proportional controller used to control either the wall temperature or the fluid temperature (as sensed from the sense resistor on the nitride window), with an assumed reference temperature of zero. The output of the op-amp is used to heat a 100 Ω resistor, creating Joule heat P. This Joule heat is the input to the wall-temperature transfer function

Table 22.2. Lumped-element model parameters for batch PCR reactor.

Symbol	Parameter	Value
L	Chamber edge length	2.2 mm
H	Chamber height	400 μm
τ_c	Thermal time constant of fluid	1.77 sec
C_c	Heat capacity of fluid	3.39 mJ/K
R_c	Thermal resistance of fluid	520 K/W
C_b	Effective heat capacity of 4 beams	1.23 mJ/K
R_b	Effective thermal resistance of 4 beams	38 K/W
R_h	Resistance of heater	100 Ω
K	Variable amplifier gain	100
As_o	Op-amp gain-bandwidth product	$2\pi \times 10^5$ rad/sec

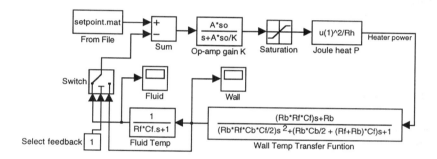

Figure 22.16. SIMULINK model of a control circuit for the PCR batch reactor. There are two feedback options: control from the wall temperature sensor, or control from the temperature sensor at the center of the chamber surface. The reference temperature in this model is zero degrees.

(Eq. 22.8), and the wall temperature is the input to the wall-temperature-to-fluid-temperature transfer function (Eq. 22.9). The loop gain is set by the factor K in the op-amp block. In the simulations shown below, this gain is set at 100.

Figure 22.17 shows the time evolution of the temperature for a typical PCR cycle using two different feedback sources. On the left, the temperature of the wall is used for feedback, and the wall temperature tracks the set-point temperature so closely that it is not possible to see any difference on this graph. The fluid temperature, however, lags the wall temperature with characteristic exponential responses. Note that the wall-to-fluid delay is the dominant speed-limiting factor in this system!

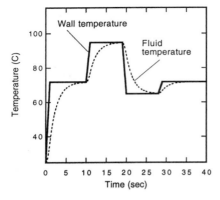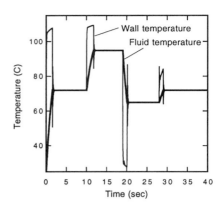

Figure 22.17.   Temperature behavior of the system of Fig. 22.16 with the gain K set to 100. On the left is the effect of feedback from the wall temperature. On the right is the effect of feedback from the fluid temperature as measured from the sensor on the nitride. A reference temperature value of 25°C has been added to all waveforms so that the plotted temperatures are comparable to actual PCR cycle values.

On the right side of Fig. 22.17 is a simple attempt to improve the response time of the fluid by using the sensor on the nitride layer as the source of the feedback signal. And, indeed, the fluid response now matches the set point almost perfectly, but at a serious cost. The wall temperature now severely overshoots or undershoots the desired set point in an attempt by the feedback loop to compensate for the time delay between the wall temperature and the fluid temperature. In fact, the temperature exceeds the boiling point during the overshoot, an absolutely unacceptable thermal behavior for a PCR system.

Our conclusion is that it is necessary either to develop a better controller (we used only proportional control), so that the wall overshoot is eliminated, or to accept the wall-to-fluid time delay and control the wall temperature. In the latter case, there is no danger of overshoot, but the system is noticeably slower.

One way to reduce the wall-to-fluid response time is to make the chamber smaller, sacrificing volume throughput in exchange for cycle time. Optimization of this tradeoff is the subject of Problem 22.1.

If some amount of overshoot can be tolerated, then it is possible to design a better controller that uses sense signals from both the wall and the fluid in order to speed up the fluid response while limiting the wall temperature. Exploration of this is the subject of Problem 22.2.

Alternatively, one could mount an additional heater on the nitride. When using both wall and nitride heaters, one would need to control both wall and nitride temperatures so as to prevent overshoot of either. The benefit of this approach is that the longest dimension for heat flow becomes the thickness rather than half the edge length, so the response would speed up. This possibility is explored in Problems 22.3.

22.5 Thermal Model of the Continuous Flow Reactor

The continuous flow PCR system of Koop *et al.* presents a completely different set of thermal modeling challenges. In this case, the temperature zones are fixed, but the fluid moves through them. Therefore, while most of the heat transfer is by conduction, there is some heat transfer due to convection.

As the fluid moves from one zone to another, it experiences a changing wall temperature. Complete modeling of the time dependence of the fluid temperature must address a coupled mass-flow and heat-transfer problem. While solving such a problem is well within the capabilities of modern numerical simulations, our goal here is to gain analytical insight by creation of lumped models. We do this in several steps.

First, we consider the wall-to-fluid time constant, just as with the batch PCR reactor. Specifically, we begin by dealing with a section of fluid channel located in a region of uniform wall temperature T_w and ask how to model the heat transfer between the wall and the fluid center. This is exactly the same problem we solved in the last section, but with different dimensions. We model the channel has having a rectangular cross-section with dimensions 40 μm $\times$ 90 μm. The thermal time constant τ_c is, in this case, about 1 msec. This time constant can be further resolved into a fluid heat capacitance per unit channel length C'_c divided by a conductance per unit channel length G'_c, where

$$G'_c = 8\kappa = 4.8 \text{ Watts/(Kelvin} - \text{meter)} \tag{22.10}$$

The total thermal resistance R_c between the wall and the center of a channel segment of length L_c is

$$R_c = \frac{1}{G'_c L_c} \tag{22.11}$$

in agreement with Eq. 22.7.

The fluid, of course, is flowing. Therefore, based on the time-constant calculation above, if a segment of fluid at temperature T_1 enters a region with uniform wall temperature T_2, moving with average fluid velocity v_f, then because it takes a time of order τ_c for the center of the fluid to reach temperature T_2, we can estimate that there is a characteristic distance equal to $v_f \tau_c$ that the fluid must flow before it reaches temperature T_2. (More precisely, it would be three times this characteristic distance, just as we typically assume that it takes three time constants for effective completion of an exponential decay.) Volume flow rates Q_f in the Koop *et al.* device range from 6 to 72 nanoliter/sec. For the given cross-section of the channel, the maximum flow velocity in this range is 0.02 m/sec. The characteristic distance corresponding to one time constant is 20 μm. Under worst-case circumstances, therefore, a distance of $\sim$60 μm is needed for full equilibration of a fluid sample entering a temperature zone. Since the zone dimensions are much greater than this, on the order of tens of millimeters, it is safe to assume that the fluid sample is truly isothermal for most of its travel through a zone.

In practice, the distance required for equilibration will be less than this estimate because some amount of equilibration takes place while the fluid travels through the transition zone between one temperature-controlled zone and another. To first order, there is a uniform temperature gradient supporting heat conduction through the glass from one temperature-controlled zone to the next. In the absence of fluid motion, the fluid would, because of heat conduction in the fluid in the axial channel direction, have the same temperature profile as the glass throughout the transition zone, and would already be exactly at temperature T_2 when reaching the boundary of the T_2 zone. The only thing upsetting this situation is the fluid flow itself.

In order to understand the thermal effect of the fluid flow, we can compare the heat flows due to conduction and to convection down the fluid channel. The conductive heat flow down the channel P_{cond} in the transition zone, assuming that the ends of the channel are at T_1 and T_2 is

$$P_{cond} = \frac{\kappa A_c}{L_c}(T_1 - T_2) \tag{22.12}$$

where A_c is the channel cross-sectional area and L_c is the length of the transition zone (assumed to be 1 mm for purposes of making this estimate; actual zone lengths might be several mm). The convective heat flow into the channel from temperature T_1 is

$$P_{conv,T_1} = \lambda T_1 \tag{22.13}$$

where

$$\lambda = Q_f \tilde{C}_m \rho_m \tag{22.14}$$

where Q_f is the volume flow rate of the fluid. The corresponding convective heat flow out of the channel from temperature T_2 is

$$P_{conv,T_2} = \lambda T_2 \tag{22.15}$$

The difference between these two convective flows represents heat flow that must spread out into the glass and be conducted to the T_2 zone through the larger area of the glass cross section. Convective heat transfer becomes potentially important when the difference in convective flows is comparable to the conductive flow. That occurs when

$$\frac{\kappa A_c}{L_c}(T_1 - T_2) \approx Q_f \tilde{C}_m \rho_m (T_1 - T_2) \tag{22.16}$$

Thus, the flow rate above which convective effects will noticeably perturb the temperature distribution is

$$Q_f = \frac{\kappa A_c}{L_c \tilde{C}_m \rho_m} \qquad (22.17)$$

Substituting values for the various quantities, this flow rate turns out to be 5×10^{-13} m^3/sec, or 0.5 nl/sec. This is far less than the range of flow rates actually used. Therefore, convection of heat into the transition zones will affect the temperature distribution.

What does all this mean? On the one hand, if we assume that the convection dominates so that fluid entering the T_2 zone is still very close to T_1 in temperature, then a length of 60 μm is required to equilibrate the fluid. On the other hand, if the convected heat can spread easily into the glass without an excessive perturbation of the fluid temperature profile through the transition region, then the fluid enters the T_2 zone nearly at temperature T_2. To go further, we need a more elaborate model, as shown in Fig. 22.18.

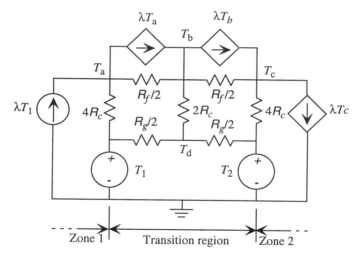

Figure 22.18. A lumped equivalent circuit for a transition zone between two temperature-controlled zone. The two temperature zones have temperatures T_1 and T_2. The current sources represent convection of heat by the flowing fluid. Entering from the left is a constant heat flux proportional to T_1, while the other three convective terms are represented as dependent current sources, since the amount of convected heat at the intermediate nodes depends on the temperatures at these nodes. R_g is the lumped thermal resistance across the transition zone for a portion of the glass surrounding the channel, R_f is the axial thermal resistance of the fluid in the channel, and R_C is the total wall-to-center thermal resistance of the channel segment in the transition zone. The various resistors have been divided up so that T_b represents the fluid temperature halfway through the transition zone and T_d represents the wall temperature halfway through the transition zone.

The circuit of Fig. 22.18 looks rather complicated, but it is a straightforward extension of the ideas already presented when examining the batch reactor. First, the continuous-flow system operates in a thermal steady state. Therefore, all of the heat capacitances disappear from the model. Second, one of the goals in constructing this model is to estimate the difference between the fluid temperature and the wall temperature as the fluid crosses a transition zone. The dominant heat-flow path between zones is through the glass. However, only the glass near the channel wall interacts thermally with the channel. As a rough estimate, therefore, we compute the axial thermal resistance for a volume of glass surrounding the channel having dimensions 5 times as large as the channel, and assign it the value R_g. This segment of glass has a cross-section equal to 24 times the cross section of the channel, and a length equal to the length of the transition zone, assumed here for estimation purposes to be 1 mm (in practice, it might be a few mm). The thermal resistance of this segment, using a thermal conductivity of glass equal to 1 W/(K-m) [52], is 1.2×10^4 K/W. The corresponding axial thermal resistance of the fluid in the channel is 4.63×10^5 K/W. The total wall-to-fluid thermal resistance R_c for a length of channel equal to 1 mm is 208 K/W. The individual resistors are broken up to create a zeroth-order finite-difference approximation containing two nodes (T_a and T_c) at the boundaries of the temperature zones driven by thermal voltage sources, and two nodes halfway between these points, one in the fluid (T_b) and one at the wall (T_d). The R_c resistor is distributed so that the total *conductance* between wall and fluid is R_c, and for each half section between a temperature zone and the midpoint, half that conductance is placed at at the temperature zone end and half at the midpoint. Hence the resistor $2R_c$ connected to the midpoint has twice the conductance of the two resistors at the zone edges.

If we assume an extreme case, with $T_1 = 95°C$ and $T_2 = 65°C$, then with no flow, T_d and T_b both equal 80°C, the halfway temperature. However, with a flow rate of 72 nl/sec, corresponding to a value of λ of 3×10^{-4} W/K, the midpoint temperatures go up noticeably: the fluid temperature goes up to 87.4°C and the wall temperature goes up to 86.5°C. Note that the difference between the fluid and wall temperatures is small even though the midpoint temperature has been increased by the convected heat. Further, in this very crude model, the temperature T_c corresponding to the fluid temperature entering the 65°C zone is 69.5°C. In other words, the high temperature T_1 zone, thanks to the convection by the fluid, actually extends a little into the transition zone and the fluid reaching the T_2 zone is slightly hotter than T_2. Of course, this temperature difference completely disappears within 60 μm of distance into the T_2 zone.

The above results are rather sensitive to the assumptions about R_g, a fact that underscores how crude the model of Fig. 22.18 really is. Therefore, the model should be viewed as providing a qualitative insight into the effects of convection rather than a quantitative result. In order to do a better job, we would have to solve a three dimensional heat-transfer problem with convection, a perfectly straightforward thing do to using finite-difference methods, but not an appropriate problem for the level of this text.

22.6   A Comparison

The two approaches to PCR are very different. The continuous-flow system has the virtue of providing very good temperature control, no risk of overshoot, and it requires a rather simple control system (three fixed temperature zones). However, it is not as flexible as the batch system because the relative times spent in each zone must be designed into the chip. For a different ratio of times in the various zones, a differently designed chip is required. Finally, the continuous-flow system has a limited throughput, 72 nl/sec in the example analyzed here. Scaling to higher throughput requires larger channels or multiple parallel channels.

The batch PCR system requires a more sophisticated control system and, because of the relatively larger chamber size, has heat-transfer time delays that are substantially larger than for the continuous-flow system. There is flexibility in programming the cycle, but all cycles are limited by the thermal time constant of the chamber. Interestingly enough, because of the cycle-time penalty associated with this time constant, the two systems have about the same throughput. The batch reactor, with a chamber edge length of 2.2 mm and a height of 0.4 mm, has a volume of 1.9 μliter. The minimum cycle time that could be achieved assuming control of the wall temperature (to eliminate unacceptable overshoot) is 15 seconds for the basic cycle, about 2 seconds for ramping between temperatures, and three time constants per temperature hold, or about an additional 16 seconds in thermal equilibration time. The total cycle time is therefore 33 seconds. Even without assuming any overhead for filling and emptying the chamber, the average throughput of sample for the batch reactor is about 58 nl/sec, actually slightly less than what has been reported for the continuous flow system.

Of course we analyzed only the thermal behavior. One must also be concerned with the chemical behavior of the system, including the ease of cleaning so as to prevent cross-contamination of samples. The continuous flow system appears to be beautifully adapted to performing PCR on a sequence of samples that are introduced into the sample stream, passed through the PCR cycle, and then separated after they exit. However, if residue on the walls from one sample contaminates the next sample, unacceptable cross-talk could occur. Evaluation of these very important details lies beyond the scope of what we can cover here.

As always, it is the marketplace that will determine which approach will dominate. Based on the thermal analyses presented here, the continuous-flow system appears to have some significant advantages for miniaturizing PCR systems. If it is successful, it will be yet another demonstration of a paradigm shift that occurs through incisive use of the tools of microfabrication.

Related Reading

J. H. Daniel, S. Iqbal, R. B. Millington, D. F. Moore, C. R. Lowe, D. L. Leslie, M. A. Lee, and M. J. Pearce, Silicon microchambers for DNA amplification, *Sensors and Actuators A*, Vol. A71, pp. 81-88, 1998.

E. A. Ehrlich, *PCR Technology: Principles and Applications for DNA Amplification*, Oxford, UK: Oxford University Press, 1992.

M.U. Kopp, A. J. de Mello, and A. Manz, Chemical amplification: continuous-flow PCR on a chip, *Science*, Vol. 280, pp. 1046-1048, 1998.

C. H. Mastrangelo, M. A. Burns, and D. T. Burke, Microfabricated devices for genetic diagnostics, *Proc. IEEE*, Vol. 86, pp. 1769-1787, 1998.

M. J. McPhearson, P. Quirke, and G. R. Taylor, *PCR: Clinical Diagnostics and Research*, New York: Springer-Verlag, 1992.

K. B. Mullis, F. Ferre, and R. A. Gibbs, *The Polymerase Chain Reaction*, Boston, MA: Birkhauser, 1994.

M. A. Northrup, M. T. Ching, R. M. White, and R. T. Watson, DNA amplification with a microfabricated reaction chamber, *Proc. Int'l. Conf. Solid-State Sensors and Actuators (TRANSDUCERS '93)*, Yokohama, June, 1993, pp. 924-926.

Problems

22.1 Determine whether the reduction in cycle time that results from a decrease in the chamber volume for a batch PCR chamber results in a net increase in throughput. Is the tradeoff monotonic, or is there a clear optimum chamber size?

22.2 Assume that a 2°C overshoot in the wall temperature of the batch PCR reactor could be tolerated. Can you develop a feedback system that uses feedback from both the wall temperature and the chamber temperature to improve the response time? (This will require building the SIMULINK model of Fig. 22.16, and modifying it so that the control law involves feedback from both sensors.)

22.3 Assume that a heater is mounted on the nitride surface of the batch PCR chamber so that both the silicon wall and nitride temperatures can be controlled. Design a simple feedback system that controls both temperatures, and determine how the fluid-temperature response time compares with what is achieved in the text using only wall-temperature control.

22.4 A PCR Treasure Hunt: go to the literature or to a commercial web site and locate a microfabricated PCR system that differs from those presented in this chapter. Compare the thermal performance and throughput of that system to the batch and continuous-flow systems of this chapter.

Chapter 23

A MICROBRIDGE COMBUSTIBLE-GAS SENSOR

*It goes so heavily with my disposition that this goodly
frame, the earth, seems to me a sterile promontory;
this most excellent canopy, the air, look you, this
brave o'erhanging firmament, this majestical roof
fretted with golden fire, why, it appears no other thing
to me but a foul and pestilent congregation of vapours.*

—William Shakespeare in *Hamlet*

23.1   Overview

This Case Study concerns a class of catalytic detectors for Hamlet's "pestilent congregation of vapours," namely, potentially dangerous combustible gases.[1] The principle of operation is that a suitable catalyst, when heated to an appropriate temperature, promotes the oxidation of combustible gases. The additional heat released by the oxidation reaction can be detected. The oldest version of this type of sensor is the Pellistor, which is basically a heater resistor embedded in a sintered ceramic pellet on which catalytic metal (platinum) is deposited.

Several microfabricated versions of this type of sensor have been reported. This Case Study is based on the microbridge-filament approach developed by Manginell *et al.* at the Sandia National Laboratories [137, 138, 139]. It is selected, in part, because the thesis of Dr. Manginell[2] has a very complete discussion not only of the entire history of this type of device, but very careful modeling at each step, including procedures to determine needed constitutive properties.

---

[1]This Case Study is based in part on materials and commentary provided by Dr. Ronald Manginell of the Sandia National Laboratory.
[2]Available over the web at www.mdl.sandia.gov/micromachine/biblog_sensors.html.

The subject of this Case Study is not a commercial product. Hence, some of the issues associated with manufacturability, yield, final device packaging, and device repeatability, calibration, sensitivity, and accuracy, are treated at a rather pragmatic level: the goal has been to obtain devices that work sufficiently well and repeatably to permit quantitative assessment of performance, and comparison with models.

23.2   System-Level Issues

The approach being investigated is a microfabricated filament onto which the catalyst is deposited. The filament is Joule-heated to its operating temperature, and resistance changes produced by the heat of reaction are then detected.

There are several architectural issues to address: First, does one build separate heater and detection resistors, or use the same resistor for both purposes? The advantage of separate resistors is that the sensing resistor can be accurately calibrated, and its value can be measured at low currents without additional self-heating effects. The disadvantage, however, is that a larger structure is needed, such as a platform on which to place the two resistors. Such a structure will have a larger heat capacity and hence a slower response time. Further, the two-resistor approach can be expected to require a more complex and costly fabrication sequence.

A second issue is the measurement method. When measuring the effect of the heat of reaction, does one use a Wheatstone bridge circuit or equivalent to measure a resistance change, or does one use a feedback circuit to maintain the temperature and measure the decrease in power required when combustion is present?

The primary goal of this particular development effort was to achieve a low thermal heat capacity on the theory that this leads to both a fast response time and a large sensitivity. Hence, the single-resistor approach was selected, fabricated as a self-supporting filament with no additional structural elements. This choice poses a daunting modeling and design problem which is what makes it a good Case-Study example.

We have already studied self-heating in resistors. We have learned that resistors with positive temperature coefficients should be driven from voltage sources to avoid the fusing instability. A voltage applied to such a resistor produces Joule heating throughout its length, but because our filament is suspended, heat loss occurs primarily by conduction to the ends of the filament. This leads to a roughly parabolic temperature profile along the filament length, with the maximum temperature at the center. Given the positive temperature coefficient of resistance, this means that the resistivity in the central portion of the filament is higher than at the ends. Another complication results from the fact that both the resistivity and thermal conductivity of the filament material will be temperature dependent. The filament is to be coated with a catalyst, which further modifies the heat capacity and thermal conductivity of those portions of the filament that are coated. Finally, the resistance change measured at the terminals depends on whether the oxidation reaction is reaction-rate controlled, which is strongly temperature dependent, or mass-flow controlled (i.e., limited by the rate of diffusion of reaction products to the catalyst surface). In the former case, it is possible to get hysteresis, as follows: the filament is Joule heated to an "ignition" temperature, a temperature sufficient to increase the rate of oxidation such that the heat necessary to sustain the filament temperature is supplied by the heat of reaction. Once the reaction is started, it is possible to reduce the Joule-heating power but sustain the filament at an elevated temperature. Catalytic chemical reactors use this principle in chemical manufacture. However, when sensing combustible gases, this ignition regime must be avoided. Because the heat of reaction and the increasing reaction rate with increasing temperature can create a runaway situation, it is desirable to operate the sensing device in the constant-temperature mode. As long as the heat of reaction provides only part of the heat needed to sustain the filament temperature, a system that controls the filament temperature will be stable against runaway.

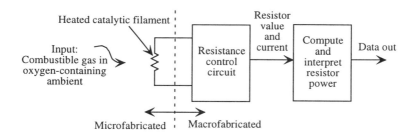

Figure 23.1. System architecture for the combustible-gas sensor.

The system architecture for the combustible-gas sensor is shown in Fig. 23.1. As will be demonstrated, it is difficult to control the *temperature* of the filament directly. Instead, one can control its *resistance*, and measure the power needed to maintain that resistance value. Interpretation of the result then requires careful modeling of the relation between heat entering the resistor from the heat of reaction and total device resistance as measured at its terminals. The model plays an integral role in interpreting device behavior.

System partitioning between the microfabricated and macrofabricated parts is straightforward. There are no performance issues that require a monolithic approach. Indeed, there is an advantage in a hybrid approach since the sensor element itself is to be heated to 500°C. With an integrated sensor, one might be concerned about heating of the electronic components.

The packaging selected is based on the need for insertion of the sensor into a gas flow cell. A standard integrated-circuit ceramic dual-in-line package was used, without a cover.[3]

23.3   First-Order Device and System Models
23.3.1   Filament Characteristics

To create a first-order model of the device and its associated circuit, it is usefi.l to ignore as many subtleties as possible. Specifically, we shall assume that *for purposes of estimating the temperature distribution in the filament,* all constitutive properties (resistivity, heat capacity, thermal conductivity) are temperature-independent. Therefore, the Joule heating is uniform along the filament length. However, once we estimate the temperature profile, we will use the temperature coefficient of resistance together with this profile to calculate a total filament resistance.

We model the temperature distribution using a one-dimensional model along the filament length, assuming perfect heat sinking at the ends, and assuming that heat conduction down the length of the filament is the only heat-loss mechanism. In steady state, the temperature distribution $T(x)$ of a uniformly Joule-heated filament is given by the solution to the steady-state heat-flow equation with (by assumption) constant coefficients:

$$\frac{\partial^2 T(x)}{\partial x^2} = -\frac{I^2 \rho_e}{W^2 H^2 \kappa} \tag{23.1}$$

where I is the current in the filament of cross-sectional area WH, ρ_e is the electrical resistivity of the filament, and κ is the thermal conductivity. The solution for a filament that extends from $x = -L/2$ to $x = L/2$ is

$$T(x) = T_0 + T_m \left[1 - \left(\frac{x}{L/2} \right)^2 \right] \tag{23.2}$$

where T_0 is the substrate temperature (which we will take as zero hereafter), and where the maximum temperature T_m is given by

$$T_m = \left(\frac{I^2 \rho_e L}{WH} \right) \left(\frac{L}{8WH\kappa} \right) \tag{23.3}$$

The first factor is the total Joule heat dissipated in the filament (for constant ρ_e), and has the unit of Watts. The second factor is the effective thermal

---

[3] An alternate is to use a cover with a few small holes in it to provide gas access. Such an approach has been successfully used with a palladium-nickel based hydrogen sensor developed at Sandia National Laboratories [140]. However, the cover introduces a potential delay in the response. For research purposes, therefore, no lid was used.

resistance (for constant κ), having the units of Kelvins/Watt. It is evident that to get a large temperature rise, it is desired to have the filament long, and with a small cross section. Such structures, however, are very prone to out-of-plane bending, stiction, and buckling, and since the filament will be heated, attention to thermally-induced stresses is required. It appears that to get a large temperature rise, one seeks a high resistivity. However, that is deceptive. All metals, and silicon at temperatures well below that at which its conductivity is dominated by intrinsic carriers, have positive temperature coefficients. Therefore, for stability, this filament should be driven from a voltage source. Thus, the total Joule heat goes as $V^2WH/\rho_e L$, which argues for a low resistivity to obtain a high T_m.

Using this temperature profile, we can estimate the total resistance by integrating the now temperature-dependent resistivity along the length of the filament

$$R = \frac{1}{WH} \int_{-L/2}^{L/2} \rho_e(1 + \alpha_R T)dx \qquad (23.4)$$

where α_R is the temperature coefficient of resistance for the filament material, assumed to be a constant over the temperature range of interest.[4] The result is

$$R = \frac{\rho_e L}{WH}(1 + \frac{2}{3}\alpha_R T_m) \qquad (23.5)$$

To estimate the thermal response time, we will use the decay constant for the lowest eigenmode of the system. That is, we will assume that a temperature profile has been established, and find the decay constant when the Joule heating is turned off. Of course, the steady-state parabolic temperature profile is a superposition of many eigenmodes, each having its own decay constant. However, to get a rough estimate, we can use the lowest eigenmode and solve the time-dependent heat-flow equation with no Joule heating, but with a trial function of the form

$$T(x) = T_m \cos\left(\frac{\pi x}{L}\right) e^{-t/\tau_{th}} \qquad (23.6)$$

where τ is the time constant governing the thermal decay. The equation that must be solved is

$$\frac{\partial^2 T}{\partial x^2} = \frac{\tilde{C}}{\kappa}\frac{\partial T}{\partial t} \qquad (23.7)$$

where $\tilde{C}$ is the heat capacity per unit volume. The result is

---

[4]Note that if this factor were included into the ρ_e term in the heat-flow equation, that equation would become nonlinear and require a much more elaborate solution.

$$\tau_{th} = \frac{\tilde{C}L^2}{\pi^2 \kappa} \tag{23.8}$$

We now have all the pieces to create a very rough first-order dynamic model of the filament. If we treat this time constant has the product of a thermal resistance and a lumped capacitance, the two appropriate quantities are:

$$R_{th} = \frac{L}{8WH\kappa} \tag{23.9}$$

$$C_{th} = \frac{8\tilde{C}WHL}{\pi^2} \tag{23.10}$$

The units of R_{th} and C_{th} are Kelvins/Watt and Joules/Kelvin, respectively, giving a product in seconds.

We see that the first-order lumped thermal capacitance is approximately equal to the heat capacity of the entire filament. We can now write a first-order state-equation for the maximum temperature:

$$\frac{dT_m}{dt} = -\frac{T_m}{R_{th}C_{th}} + \frac{V^2}{C_{th}R(T_m)} + \frac{P_c}{C_{th}} \tag{23.11}$$

where the Joule heating term has now been written has $V^2/R(T)$, in anticipation of requiring voltage control, and a new power term P_c/C_{th} has been added, in anticipation of the additional heat input to the filament from the heat of reaction. With this dynamic model, we can now investigate the resistance-control circuit. Since there is a direct link between T_m and the total resistance, we expect that, at least to first order, control of resistance will serve to control T_m.

23.3.2   Resistance-Control System

The actual circuit used by Manginell et. al. for control of the filament resistance has several subtleties. Rather than address those at the start, we instead build a dynamic model in SIMULINK that incorporates the various functions required for a resistance-control scheme.

Figure 23.2 shows the basic resistance-control system in block-diagram form.[5] We see that this is a canonical feedback system, with a resistance set point, a controller, and a plant (the filament) driven by two inputs, the voltage from the controller and the combustion power P_c. We recognize that P_c functions as a "disturbance." This is a good example of the use of a feedback system to *measure* the value of the disturbance. Panel meters are shown connected to the various variables for convenience in visualization.

---

[5]Fig. 23.2 uses SIMULINK subsystems that can read parameter values from the MATLAB workspace. The various subsystems are expanded in the following figures.

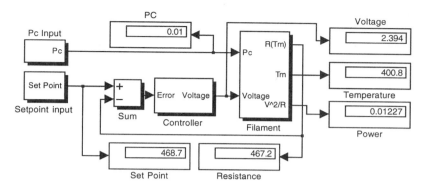

Figure 23.2. SIMULINK block diagram of a resistance-control feedback system for the combustible-gas sensor.

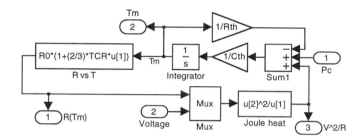

Figure 23.3. Expanded view of the filament subsystem. Temperature is the only internal state variable. The temperature dependence of the resistance and the Joule heating are calculated with SIMULINK function blocks.

An expanded view of the filament subsystem is shown in Fig. 23.3. The integrator and associated inputs implement Eq. 23.11. The temperature dependence of the resistance is calculated from the temperature, and the resistance is used to calculate the Joule heating.

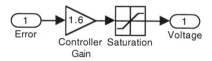

Figure 23.4. The controller subsystem. The saturation level is set at 15 V to represent the saturation of practical amplifiers. Without this limit, the model is susceptible to numerical instabilities in response to step changes.

An expanded view of the controller subsystem is shown in Fig. 23.4. A linear gain block, set to 1.6 to model the gain to be used in the actual circuit, is followed by a saturator with its output set to limit at 15 V. This prevents unrealistically large voltages in response to step changes; such voltage spikes can produce numerical instabilities in the model.

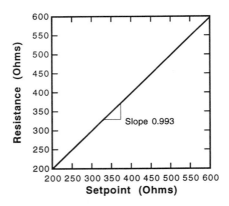

Figure 23.5.   Steady state resistance *vs.* set-point resistance with P_c equal to zero.

To use this model, parameter values are needed. These are based on the published devices. The device dimensions used were $L = 200 \ \mu$m, $W = 10 \ \mu$m, and $H = 2 \ \mu$m. The filament was assumed to be extremely heavily doped, with a resistivity at room temperature of 0.002 Ω-cm, leading to a room temperature resistance of 200 Ω. A TCR of 5×10^{-3} Kelvin^{-1} and a thermal conductivity of 0.7 Watts/Kelvin-centimeter were used.[6] The effective thermal resistance R_{th} is then 1.8×10^4 Kelvin/Watt. The thermal capacitance value used in the simulations shown below was $1/R_{th}$, so that the time-axis of the simulations is measured in thermal time constants. However, the physical value of C_{th} based on a room-temperature specific heat of silicon of 0.77 J/g-K and the device dimensions is 5.7×10^{-9}J/K. Thus the thermal time constant is about 0.1 msec, which is (as originally desired) extremely short.

The steady-state resistance *vs.* set-point resistance is plotted in Fig. 23.5. The relationship is linear, with a slope of 0.993. From this slope, we deduce that the effective quasi-static loop gain is 14.2.

Steady-state responses of the other variables are shown in Fig. 23.6. Note that the resistor power and temperature *vs.* set-point resistance are very linear, but the voltage driving the resistor is nonlinear. The values of the design parameters work well with voltages that are readily obtained using standard circuit elements and conventional power supplies.

---

[6]These are plausible, but not exact, values. A TCR of 1.1-1.2 $\times$ 10^{-3} Kelvin^{-1} would be more accurate. The effects of changing the TCR to this value are explored in Problem 23.2.

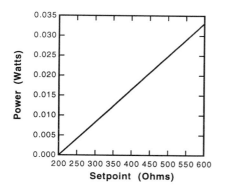 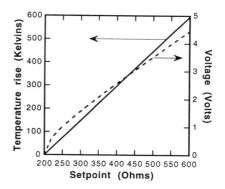

Figure 23.6.   Resistor power, temperature, and voltage *vs.* set-point resistance with P_c set to zero.

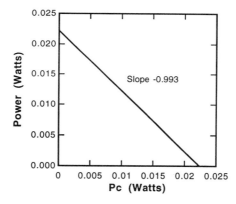

Figure 23.7.   The steady-state relation between Joule-heat power and heat-of-reaction power P_c at a set-point resistance of 468.7 Ω, corresponding to an operating temperature of 400° above room temperature. The system cannot keep the resistance value controlled at the set point for P_c values that exceed 0.022.

The effect of the heat of reaction when combustible gases are present is shown in Fig.   23.7. As expected, the Joule heat decreases linearly with increasing heat of reaction. The slope is the same 0.993 obtained for the slope of the resistance *vs.* set-point graph. It is seen that this sensor can only function as a sensor (as opposed to a combustion catalyst) when P_c is less than the steady-state power required at the set-point temperature.

A transient response[7] to a square-pulse P_c waveform is shown in Fig. 23.8. The amplitude is 0.01, about 50% of the steady-state power needed to maintain

---

[7]In calculating these transients, it was necessary to reduce the relative tolerance in SIMULINK . A value of $1e^{-6}$ was used. If the default value of $1e^{-3}$ is used, numerical instabilities can result from step changes.

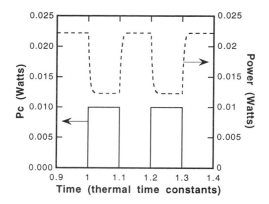

Figure 23.8. Transient response of the closed-loop system to pulses of combustible gas at a set-point resistance of 468.7 Ω. Note how rapidly the closed-loop system responds to the square pulse of P_c.

the 400° set point. Note that the time axis is measured in thermal time constants. Under open loop conditions, the decay time would be one time constant. However, as expected for a closed-loop system, the response of the closed-loop system is much faster. We also note that there is almost a one-to-one match between the increase in P_c and the corresponding decrease in Joule-heat power, consistent with Fig. 23.7.

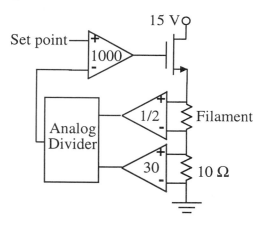

Figure 23.9. A simplified version of the Manginell [141] circuit for the combustible-gas sensor.

A simplified version of the published Manginell circuit [141] is shown in Fig. 23.9. The drive for the filament comes from the source of an FET. In series with the filament is a 10 Ω current-sense resistor. Two differential amplifiers measure the voltage across the filament and 10 times the current through it. The outputs are divided in an Analog Devices AD534 analog divider. The result is a voltage proportional to the filament resistor. A set-point voltage is then applied as shown. The voltage at the source of the FET is approximately one threshold voltage below the output of the $\times 1000$ amplifier. The combined effect of the three amplifiers and the analog division is a factor of 1.6, which is what was used in the resistance-control system. The offset due to the FET threshold voltage was ignored in the block diagram version since, in the closed-loop configuration, it has a negligible impact on the actual voltage across the filament resistor. What is notably absent in this circuit is an output for the Joule heating power in the resistor. It could be obtained from the product of the voltage-sense and current-sense op-amps, but in the published version, this output was computed off-line from the current-sense signal and the output of the analog divider.

We now have good insight into how this device works, and can consider how it might be built.

23.4   A Practical Device and Fabrication Process

There are two issues in the fabrication process. The first is to build the filament itself. The second is to create a robust contact-pad metallurgy to support connection to the outside world when performing measurements of constitutive properties at high temperature. These two process steps interact badly, because release etches tend to attack metallization on bond pads. The solution used here was to make large bond pads, cover them with a multi-layer metallization that could withstand the elevated temperature of operation, and live with the damage at the edges of these bond pads that resulted from the release etch.

23.4.1   Creating the Filament

A filament can be made with standard surface micromachining. It is desirable to have a low-resistivity material for the filament, which argues for either a metal or heavily-doped polysilicon. Polysilicon was chosen, in part, because of well-established fabrication procedures, and, in part, to avoid thermal-expansion effects.

The thermal expansion coefficient of a silicon substrate is 2.5×10^{-6} K^{-1}. If a metal, such as tungsten, is used for the filament, with a thermal expansion coefficient of 4.5×10^{-6} K^{-1}, then the compressive axial strain in the filament when operating 500° above room temperature is on the order of $\epsilon = 10^{-3}$. Of course, because there is a temperature distribution in the beam, the actual axial strain would be less than this amount, but only by a factor of order 3. And the axial strain at which the filament would buckle[8] for a length of 200 μm and a

---

[8]see Section 9.6.3

thickness of 2 μm is 3×10^{-4}, which is indeed a factor of three less than the strain estimate above. Thus a tungsten filament would be right at the buckling threshold at the maximum operating temperature.

The preferred catalyst to be deposited on the filament is platinum. The procedure for depositing this catalyst takes advantage of the fact that the filament is to be heated. An organometallic, platinum acetylacetonate (PtAcAc), will decompose at temperatures above 350 °C. By operating the filament at these temperatures in a gas cell to which PtAcAc vapor is supplied, *selective chemical vapor deposition* can be achieved. This procedure is documented in the references.

Because platinum reacts with silicon at elevated temperatures to form a silicide, and because the presence of conducting platinum would alter the electrical power dissipation pattern, it was decided to insulate the filament from the catalyst with a thin silicon nitride film. The process is discussed below.

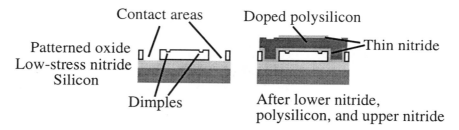

Figure 23.10.   The fabrication process for a nitride-coated polysilicon filament. The contact regions are very distorted in scale; the actual contact regions are 500 μm wide, while the filament length is only 200 μm.

The first steps of the process are illustrated schematically in Fig. 23.10. A 0.8 μm layer of silicon-rich nitride is deposited onto a silicon wafer. This serves to insulate the filament from the substrate, and also provides a low-stiction surface for the filament. A 2 μm silicon dioxide layer is then deposited by the pyrolytic oxidation of tetraethylorthosilane (TEOS). An anneal at 850° is performed to densify the oxide, after which two patterning steps are performed: a shallow etch to create dimples that help prevent stiction, and a deep etch to create the contact areas. The horizontal features in the figure are not at all to scale. The actual width of the contact regions is 500 μm, while the beam length is 200 μm. A 0.25 μm layer of LPCVD nitride is then deposited. Following this step, a 2 μm layer of doped polysilicon is deposited. The Sandia group used two different recipes for obtaining the doped poly. The first was to add 1% phosphine (PH_3) to the CVD poly reactor, creating "in-situ-doped" poly. The second method was to create a three-layer structure, with in-situ-doped poly on the bottom and top and undoped poly in the middle. A 0.3 μm layer of

TEOS is then deposited (not shown in the figure). The entire structure is then annealed at 1100°C in nitrogen to harden the oxide, relieve the stress in the polysilicon, and homogenize the dopant distribution by diffusion. The upper TEOS is patterned to serve as a hard mask for plasma etching of the polysilicon, after which the upper TEOS is removed. A final 0.25 μm layer of nitride is then deposited. This final nitride coats the tops and sidewalls of the filament. This layer is patterned to the shape of the beam, which opens up access to the sacrificial oxide.

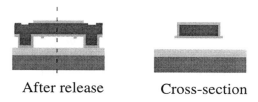

After release   Cross-section

Figure 23.11.   Transverse view and cross-section of filament after release.

Figure 23.11 shows the structure after release. Release is performed with concentrated HF, with HCl added which improves the etch selectivity to nitride. The oxide is completely removed beneath the filament, after which the etch is terminated. Some oxide remains around the borders of the large contact regions (not shown in the figure). Surface tension pulls the beams down as the release etchant is withdrawn, but beams 200 μm or shorter had sufficient stiffness to lift free without stiction problems. The release etch could be performed at the wafer level or the die level. In practice, it was usually performed after die separation.

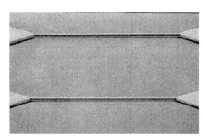

Figure 23.12.   A pair of filaments. The bond-pad regions are outside the field of view. The tapered supports are an extension of the bond-pad region, which are visible in Fig. 23.13. (Source: Sandia National Laboratory; reprinted with permission.)

23.4.2   High-Temperature Bond Pads

Even though the middle of the filament gets hot during operation, the bond pads located in the large contact areas do not get hot. Thus it is not necessary to use a high-temperature bond-pad metallurgy for purposes of testing the devices. However, in order to model the devices well, it was necessary to measure the resistivity and thermal diffusivity as a function of temperature. This required placing the entire device in an oven, and this, in turn, required high-temperature contact metallurgy.

The principle of high-temperature contacts is to make a transition between the silicon and an upper metal layer that can be wire bonded, such as gold. However, gold readily diffuses into silicon. Hence a barrier material, such as titanium nitride (TiN) is used. But adhesion between TiN and other materials is not good, so intermediate adhesion layers are used. The metal stack used by the Sandia group was Ti (to provide adhesion to the silicon), TiN (the diffusion barrier), a second Ti layer (to promote adhesion between the TiN and the gold), and a gold upper layer. The lower Ti and TiN layers were sputtered and pattered so as to cover the polysilicon bond pads. Then the upper Ti layer and the gold bond pad were patterned with liftoff.

The HF release etch tends to attack the titanium adhesion layers, undercutting the bond pads. In this case, the workers elected to use very large bond pads, and live with the undercut.

23.4.3   Catalyst Coating

Once the die are released, with or without the high-temperature metallurgy, the catalyst coating can take place by using the resistance-control circuit to heat the filament to the neighborhood of 350°C in the presence of PtAcAc. Figure 23.13 shows an example of a pair of filaments, one of which was coated with this selective CVD process.

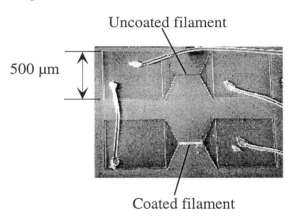

Figure 23.13. A pair of wire-bonded filaments, in this case, without the high-temperature metallurgy. The lower of the two filaments has been platinum coated using selective CVD. (Source: Sandia National Laboratory; reprinted with permission.)

23.5   Sensor Performance

23.5.1   Demonstration of Hydrogen Detection

Operation of the device as a sensor is demonstrated in Fig. 23.14. A typical device is operated with varying percentages of hydrogen in 20% oxygen, with the balance being nitrogen [137]. The "Sensor Response" is not identified with respect to the detection circuit discussed here, but its behavior is consistent with the voltage that measures resistor current. Clearly, hydrogen can be detected.

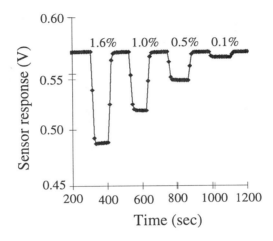

Figure 23.14.   Sensor response of a Pt-coated filament to hydrogen in synthetic air. The sensor is operated with 70 mW static power. (Source: Sandia National Laboratory; reprinted with permission.)

The authors report a detection limit of 100 ppm hydrogen in synthetic air at the stated static power level of 70 mW. However, because the filament comes to temperature very quickly, the device can be operated in pulsed mode, greatly reducing the average power. The device speed appears to be much slower than what we estimated from the thermal model. This is due to the finite time required to establish new gas concentrations in the flow system, and is not related to the intrinsic speed of the sensor. Fill time for the 0.5 liter test chamber was about 30 seconds, which agrees with the apparent settling time of the sensor response.

The sensitivity of the device depends strongly on the morphology of the platinum catalyst. This is one problem with this device-manufacture method. Manginell *et al.* devoted a great deal of effort to determining a procedure for depositing the platinum catalyst so that it had a grainy morphology, with a large surface area for catalytic activity. This included the use of pulsed heating sequences during the catalyst-deposition steps. The preferred morphology is difficult to reproduce, and also difficult to keep completely clean during operation. Therefore, calibration of every device is required, and periodic re-calibration may be required for operation over long times. Such difficulties are common with chemical sensors that rely on direct surface interaction between a chemically active surface and a target species.

23.5.2   Mass-Transport-Limited Operation

The sensitivity of the device, that is, the overall scale factor between the concentration of combustible gas and the change in sensor power, depends on several factors: the composition of the combustible gas which determines the heat of reaction, the composition of the background gas (presumably air), and the diffusivity of the combustible gas species through the air. Assuming that the combustion reaction forces the combustible-gas concentration to near zero at the surface of the filament, there is a depletion zone near the filament through which the combustible species must diffuse. The exact geometry of this depletion zone depends on the gas flow around the sensor, but a rough estimate can be made based on a cylindrical depletion model. If we assume that the combustible-gas concentration $[C_g]$ is a constant $[C_{g,0}]$ at a radius r_o from the filament surface, and decreases to zero at the "radius" of the filament a, then the steady-state gas concentration around the hot region of the filament is found from the logarithmic solution of the Laplace equation in cylindrical coordinates:

$$[C_g(r)] = [C_{g,0}] \ln \left(\frac{r}{r_o} \right) \tag{23.12}$$

The radial diffusive flux toward the filament $\mathbf{J}_g$ is

$$\mathbf{J}_g(r) = -\frac{D[C_{g,0}]}{r} \mathbf{1}_r \tag{23.13}$$

where D is the diffusivity of the combustible-gas species and $\mathbf{1}_r$ is the radial unit vector. The total rate of arrival of combustible gas at the sensor surface, and hence the heat of reaction P_c (assuming an ample supply of oxygen at the catalyst surface, and assuming that the reaction-rate is *not* the rate-limiting factor) is given by the integral of the normal component the flux over the surface area of the hot portion of the filament. Thus we can estimate P_c as

$$P_c = 2\pi a L_h \mathbf{J}_g(a)| = 2\pi D[C_{g,0}] L_h \Delta H \tag{23.14}$$

where L_h is the length of the heated portion of the filament (clearly an approximation, since the temperature of the filament is not uniform), ΔH is the heat of reaction, and the filament "radius" a cancels out. The units of ΔH must conform to the units of $[C_{g,0}]$. While very crude, this model argues for a reasonable proportionality between P_c and $C_{g,0}$, and this appears to be supported by the data of Fig. 23.14.

23.5.3   Reaction-Rate-Limited Operation

If the rate of arrival of combustible gas exceeds the achievable reaction rate at the surface of the catalyst, then the concentration will build up near the catalyst surface, reducing the concentration gradient and hence the diffusive flux to the filament. This build-up continues until the steady-state rate of arrival is reduced to match the rate of reaction. Because of the reduction in total reaction rate, this results in a smaller value of P_c than what was calculated in Eq. 23.14.

However, when reaction-rate-limited operation occurs, the intrinsic temperature dependence of the reaction rate, if strong enough, can lead to a runaway ignition situation. The hotter the filament, the greater the rate of release of heat for a given supply of reactants. We can model this with a simple linearized temperature-dependent P_c.

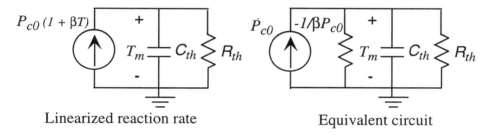

Linearized reaction rate   Equivalent circuit

Figure 23.15.   An increase in P_c with temperature adds a negative resistance to the thermal model, and can lead to runaway.

Figure 23.15 shows the thermal portion of the lumped filament model (no Joule heating), but now with a positive temperature coefficient term added to the P_c source. This only makes sense if the device is operating in the reaction-rate-limited regime, meaning that there is an ample supply of reactants to support an increased reaction rate at elevated temperature. If we expand the expression for P_c, we get two terms:

$$P_c = P_{c0} + \beta P_{c0} T \tag{23.15}$$

The first term represents an independent current source in the original thermal model. The second term, however, represents a *negative thermal resistance* of value $-1/\beta P_{c0}$. It is negative because it forces positive current out of the positive polarity terminal (a normal positive Ohm's Law resistor has positive current entering the positive polarity terminal). This leads to the equivalent circuit on the right of Fig. 23.15.

The net resistance in the thermal circuit is the parallel combination of the two resistors

$$R_{net} = \frac{R_{th}}{1 - \beta P_{c0} R_{th}} \qquad (23.16)$$

As long as this net resistance is positive, the system is stable, albeit elevated in temperature because of the increase in net thermal resistance. However, when $\beta P_{c0} R_{th}$ approaches unity, runaway occurs, and the temperature grows without limit or, more precisely, it grows until mass transport limits the rate of combustion.[9] Thus we see that in the reaction-rate-limited regime, the positive temperature dependence of the reaction rate creates positive feedback. It tends to drive the filament toward the mass-transport-limited regime by depleting the local supply of combustible gas with an ever increasing reaction rate driven by an ever increasing temperature. Once the mass-transport limit is reached, operation is once again stable.

In the reaction-rate-limited operating regime, the device also exhibits hysteresis with respect to the Joule heating. Starting from the substrate temperature, the filament is gradually Joule heated to the instability point, at which point the temperature jumps until the mass-transport limit is reached. The Joule heat can then be reduced, and while the filament temperature reduces, it remains higher than it was without the reaction. As the Joule heat is reduced toward zero, either the temperature drops back through the instability point, quenching the reaction rate, or, in an extreme case, the Joule heat can be turned all the way to zero and the reaction continues. In this latter case, the only way to turn off the reaction is to reduce the supply of reactants (or call the fire department).

23.6   Advanced Modeling

In addition to the complexities of modeling mass transport and reaction kinetics on the heated catalyst, the actual device is intrinsically much more complex than the simple model we developed earlier in this chapter. In particular, because the resistivity and thermal conductivity of polysilicon are functions of temperature, our model using a parabolic temperature profile and constant constitutive properties is incorrect. A more correct model can be created using the finite-difference approach to modeling dissipative systems presented in Chapter 12. In this case, though, it is necessary to create a finite-difference mesh for both the thermal and electrical portions of the system.

In the circuit model of Fig. 23.16, the lower portion is the thermal model of a segment of the filament of length h, and the upper resistor is the temperature-dependent resistance of a similar length of filament. The thermal portion of the

---

[9]Or, if there is a large enough supply of reactants, until the filament reaches the melting temperature and self-destructs.

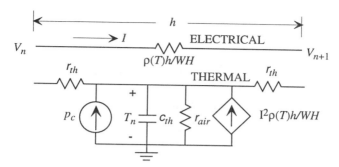

Figure 23.16. Finite-difference-model for an interior segment of the filament, coupling the electrical domain and the thermal domain.

model has a thermal capacitance c_{th} equal to the heat capacity of a segment of length h, and three thermal resistors. Two of them, labeled r_{th}, represent the filament-conduction thermal resistance of a segment of length $h/2$, while the resistor labeled r_{air} models the thermal conductivity from the filament to the substrate through the small 2 μm gap. In practice, this gap conduction accounts for about 15% of the heat loss from the filament. There are two current sources providing power to the thermal filament model, a dependent source representing the Joule heat and an independent source representing the local heat of reaction p_c. One could also build temperature-dependent reaction kinetics into p_c if one wished to model reaction-rate-limited operation. The Joule heat is written in terms of the electrical current I, but in practice, this current is obtained from the difference in voltages on adjacent electrical nodes, labeled V_n and V_{n+1}. Hence a system of equations can be constructed by connecting multiple copies of this circuit model together, end to end.

The modeling of the termination at the substrate is important. It is necessary to establish the potential on the electrical resistor with respect to a reference ground. For example, if inserted into the circuit of Fig. 23.9, such a reference potential would be created. On the thermal side, it is necessary to apply thermal boundary conditions at the ends. In our elementary model, we assumed that the ends of the filament were at the substrate temperature. In practice, however, there is a thermal spreading resistance accounting for the heat flux from the ends of the filaments. A lumped thermal resistor of value $1/2\pi\kappa a$, where a is an estimated filament radius, can be used to model the thermal resistance of the filament-substrate contact.

In order to use this model, it is necessary to have measured values of the electrical resistance, heat capacity, and thermal conductivity of polysilicon, and of the thermal conductivity of the air in the gap. Manginell conducted experiments using the filament to obtain these quantities. Basically, the TCR of the polysilicon is nearly constant, the thermal conductivity decreases nearly linearly with temperature, and the heat capacity is relatively constant. The thermal conduction in the air gap is estimated from published data. Details are in Manginell [141].

With appropriate data tables or polynomial functions representing the temperature dependencies in hand, one can construct increasingly sophisticated models of the filament to replace the SIMULINK "filament" block in Fig. 23.3. A practical procedure is to construct a subsystem that implements Fig. 23.16, and link multiple copies of that system together in SIMULINK. Alternatively, one can insert multiple copies of the circuit model directly into a circuit modeling program such as SPICE. The resistance-control circuit, including the FET and the analog divider, can also be inserted into SPICE, permitting a complete modeling of dynamic system behavior in response to a given P_c waveform using an equivalent-circuit model. In order to use SPICE, however, an additional current-controlled voltage-source element must be added to the electrical circuit to "measure" the current in the electrical resistor in order to compute the Joule heating in the thermal domain. Detailed SPICE modeling of this device is reported in [139].

23.7   Epilogue

This concludes our discussion of Case Studies. Those who have reached this point should have a good appreciation of how device performance is linked to the details of device design and fabrication, and how quantitative models, even quite approximate ones, yield valuable insight into device behavior. It should not be forgotten, though, that while the models provide the tools with which to think, they do not replace the intellectual creativity and imagination that are the cornerstones of good design.

Related Reading

R. P. Manginell, *A Polysilicon Microbridge Gas Sensor*, Ph.D. Thesis, University of New Mexico, 1997. Available on the web at www.mdl.sandia.gov/micromachine/biblog_sensors.html.

Problem

23.1 The thermal model of the filament developed in this chapter assumed that the only heat-loss path was conduction down the filament. In practice, about 15% of the total heat loss is through the air gap. How does the steady-state parabolic heat-flow solution change if one adds a heat conduction path through the air gap? (You can explore this either with analytical methods or with a modification of the finite-difference solution of Section 12.4.1.)

23.2 Build the SIMULINK model of Fig.s 23.2 through 23.4, and repeat the calculations of Figs. 23.5 through 23.7 using a TCR value of 1.2×10^{-3} Kelvin^{-1}. What are the effects of the change of TCR value from that used in the calculations in the text?

23.3 Set up the governing equations for a single section of the finite difference model of Fig. 23.16, without r_{air}, but including a linear TCR of 1.2×10^{-3} Kelvin^{-1} for the electrical resistivity, and an expression for the thermal conductivity of $0.7 - 8.75 \times 10^{-4}T$ W/(K-cm), where T is the temperature rise of the filament. Then create a model for one-half of the filament using four full sections and one half-section (representing the center of the filament). Identify the electrical and thermal boundary conditions that are required. Find the steady-state temperature distribution at various applied voltages. Test the relative effect of the TCR compared to the temperature dependence of the thermal conductivity. What does each do to the temperature distribution?

Appendix A
Glossary of Notation

This Appendix lists the symbols used in the book, a brief definition of each symbol's meaning, typical units, and a page reference to where that symbol is first introduced and defined. Values of fundamental constants and selected conversion factors for units are also included here. Some symbols have different meanings in well-separated parts of the text, matching well-established notation conventions.

Symbol	Description	Units	Page
(100), $\{100\}$	Crystal-plane notation		30
$[100]$, $<100>$	Crystal-direction notation		30
α	Damping constant	$\sec^{-1}$	155
α_R	Temperature coefficient of resistance	Kelvin^{-1}	279
α_S	Seebeck coefficient	Volts/Kelvin	293
α_T	Coefficient of thermal expansion	Kelvin^{-1}	193
γ	Shear strain		186
Γ	Surface tension	Newtons/meter	320
$\delta(t)$, $\delta(x)$	Unit impulse	$\sec^{-1}$, meters^{-1}	157, 232
δ_y	Bandwidth of resonance	radians/sec	565
δ_ω, Δ_ω	Frequency difference	radians/sec	566, 566
ϵ	Axial strain		186
ε	Feedback system error		398
ε	Dielectric permittivity	Farads/meter	111
ε_0	Permittivity of free space 8.854×10^{-12}	Farads/meter	658
ζ	Normalized displacement		136
η	Viscosity	Pascal-sec	318
η^*	Kinematic viscosity	$\text{meter}^3/\text{sec}$	318
θ	Phase angle	radians	159
κ	Thermal conductivity	Watts/(Kelvin-meter)	274

Symbol	Description	Units	Page
λ	Eigenvalue		161
λ	Convection coefficient	Watts/Kelvin	622
λ	Wavelength	meters	545
λ_m	Mean free path	meters	558
λ_n, λ_p	Channel-length modulation factor	Volts^{-1}	370
μ	Magnetic permeability	Henrys/meter	121
μ_{ep}	Electrophoretic mobility	m^2/(Volt-sec)	344
μ_f	Coefficient of friction		286
μ_n, μ_p	Electron and hole mobility	m^2/(Volt-sec)	268
μ_0	Permeability of free space $4\pi \times 10^{-7}$	Henrys/meter	660
μstrain	Microstrain (unit for strain)	10^{-6}	186
ν	Poisson ratio		187
Π	Peltier coefficient	Watts/Amp = Volts	293
Π, Π_{ijkl}	Piezoresistive tensor	Ohm-meters/Pascal	472
π_{IJ}	Piezoresistive coefficients	Pascals^{-1}	472
π_t	Transverse piezoresistive coefficient	Pascals^{-1}	473
π_l	Longitudinal piezoresistive coefficient	Pascals^{-1}	473
ρ	Radius of curvature	meters	212
ρ_e	Electrical resistivity	Ohm-meters	268
ρ_m	Mass density	kg/m^3	196
σ	Normal stress	Pascals	184
σ_d	Squeeze number		337
σ_e	Electrical conductivity	Siemens/meter	268
σ_o	Residual stress of thin film	Pascals	196
σ_{SB}	Stefan-Boltzmann constant 5.67×10^{-8}	Watts/(m^{2}K^4)	274
σ_w	Wall charge density	Coulombs/meter^2	342
τ_{DG}	Deal-Grove initial thickness	μm	35
τ_m	Minority carrier lifetime	sec	356
τ_{xz}	Shear stress on x face in z-direction	Pascals	184
ϕ	Electrostatic potential	Volts	268
ϕ	Magnetic flux	Webers	121
φ	Impedance transformation factor	Newtons/Volt	144
ϕ_1, ϕ_2	Clock signals	Volts	393
ϕ_B	Built-in diode potential	Volts	358
ϕ_w	Wall potential	Volts	342
$\Psi(x,t)$	Wave amplitude		544
$\psi_n(x)$	Eigenfunction		232
Ω, Ω	Rotation rate	radians/sec	562
ω_c	Damping cutoff frequency	radians/sec	337
ω_d	Damped resonant frequency	radians/sec	156
ω_o	Undamped resonant frequency	radians/sec	155
a	Acceleration	meters/sec^2	500
A	Area	meters^2	111
$\mathbf{A}$	State-equation system matrix		151
A_{DG}	Deal-Grove oxidation rate constant	μm	35
A_0	Op-amp open-loop DC gain		382

Symbol	Description	Units	Page
b	Damping constant	(Newton-sec)/m	25
$\mathbf{B}$	State-equation input matrix		151
$\boldsymbol{B}, \mathcal{B}$	Magnetic flux density	Webers/meter^2	660
B_{DG}	Deal-Grove oxidation rate constant	μm/hr	35
C	Heat capacity	Joules/Kelvin	273
$\tilde{C}$	Heat capacity per unit volume	J/(K-m^3)	273
C	Capacitance of a capacitor	Farads	111
$\mathbf{C}$	State-equation output matrix		151
c	Speed of sound	meters/sec	326
c	Phase velocity	meters/sec	544
$[C_g]$	Concentration of reactant	various	644
C_i	Concentration	meters^{-3}	339
c_i	Coefficient in variational trial solution		244
$\tilde{C}_m$	Heat capacity per unit mass	J/(kg-K)	273
C_b, C_s, C_r	Coefficients in load-deflection model	varies	254
C_{IJ}	Stiffness coefficients	Pascals	192
$\hat{C}_{ov}$	Overlap capacitance per unit length	F/cm	371
$\hat{C}_{ox}$	Oxide capacitance per unit area	F/cm^2	368
d	Beam deflection	meters	546
D, D_0	Diffusion constant	meters^2/sec	41
D	Flexural rigidity of plate	meters^4	221
$\mathbf{D}$	State-equation feedthrough matrix		151
$\boldsymbol{D}, \mathcal{D}$	Electric displacement field	Coulombs/meter^2	659
D_h	Hydraulic diameter	meters	331
d_{iJ}	Piezoelectric coefficient	Newton/Coulomb	572
d_t	Tunneling gap	nm	526
D/Dt	Material derivative	sec^{-1}	323
dyne/cm^2	Unit of stress	0.1 Pascal	185
E	Young's modulus	Pascals	187
$\tilde{E}$	Biaxial modulus	Pascals	225
$\boldsymbol{\mathcal{E}}, \mathcal{E}$	Electric field	Volts/meter	128, 658
e	Emissivity		274
E_A	Activation energy	Joules or eV	41
E_G	Semiconductor energy gap	electron-Volts	354
e_{iJ}	Piezoelectric coefficient	Coulombs/meter^2	573
$e(t)$	Generalized effort variable		105
eV	Electron-Volt: 1.6×10^{-19}	Joules	354
F	Force	Newtons	25
F'	Force per unit width	Newtons/meter	208
F_{MM}	Magnetomotive force	Amperes	121
$F(s)$	Laplace transform of a function $f(t)$		152
$f_s(\nu)$	Poisson-ratio load-deflection factor		259
$f(t)$	Generalized flow variable		105
$F(\omega)$	Fourier transform of function $f(t)$		157

Symbol	Description	Units	Page
g, G	Gap (between electrodes)	meters	111, 502
g	Acceleration of gravity: 9.8	meters/sec^2	324
G	Shear modulus	Pascals	188
G	Carrier generation rate	cm^{-3}sec^{-1}	356
G	Conductance	Siemens	303
g_m	Transconductance	Siemens	371
H, h	Height or thickness of structural element	meters	201, 225
$\mathcal{H}, \mathcal{H}$	Vector magnetic field and magnitude	Amperes/meter	121, 659
ΔH	Heat of reaction	various	644
$H(s), \mathbf{H}(s)$	System function		154
$h(t)$	Impulse response		157
I	Moment of inertia	meters^4	213
I, i	Current	Amperes	
I_d	Diffracted intensity		547
I_0	Diode reverse saturation current	Amperes	359
$\mathbf{I}$	Identity matrix		153
I_Q	Heat current	Watts	280
j	Square-root of -1		155
$J, \mathbf{J}_e$	Current density	Amperes/meter^2	659
$\mathbf{J}_Q$	Heat flux	Watts/meter^2	273
k	Spring constant for a force load	Newtons/meter	25
k'_{eff}	Spring constant for a pressure load	Pa/m	542
k_B	Boltzmann's Constant: 1.38×10^{-23}	Joules/Kelvin	41
k_e	Electromechanical coupling constant	Newtons/Volt	145
k_n, k_o	Wavenumber	meters^{-1}	236, 230
K_n	Knudsen number		333
K_P	Pressure sensor sensitivity	mV/(V-kPa)	488
k_p	Piezoelectric coupling coefficient		589
L	Length of beam or structure	meters	201
L	Inductance of an inductor	Henrys	123
L_D	Debye length	meters	340
L_m	Magnetic circuit path length	meters	120
L_n	Minority carrier diffusion length	cm	357
M	Mechanical moment	Newton-meters	208
M	Mach number		326
M'	Moment per unit width	Newtons	221
$\mathcal{M}, \mathcal{M}$	Magnetization	Amperes/meter	660
m	Modulation index		431
m	Mass	kg	
N	Tension in beam	Newtons	229
n, n_0	Electron carrier concentration	cm^{-3}	354
n'	Excess electron concentration	cm^{-3}	356
n_i	Intrinsic carrier concentration	cm^{-3}	354
N'	Tension per unit width	Newtons/meter	236
n	Turns ratio (transformer); gyrator parameter		119
$N(x, t)$	General dopant concentration	cm^{-3}	41
$N_I(x)$	Implanted dopant concentration	cm^{-3}	40

Symbol	Description	Units	Page
$N_{I,p}$	Peak implanted dopant concentration	cm^{-3}	40
P	Pressure	Pascals	208
$\mathcal{P}, \mathscr{P}$	Dielectric polarization	Coulombs/meter^2	659
$\mathcal{P}$	Power	Watts	104
$\tilde{\mathcal{P}}$	Power density	W/m^3	273
p, p_0	Hole carrier concentration	cm^{-3}	354
p	Coverage parameter		553
p'	Excess hole concentration	cm^{-3}	356
P_c	Heat released by combustion	Watts	634
$p(t)$	Generalized momentum variable		105
Pa	Pascal: unit of stress	1 Newton/meter^2	185
psi	Pound per square inch	69 kPa	185
Q	Charge (on capacitor)	Coulombs	111
Q	Quality factor of resonance		156
Q	Volumetric flow rate	meter^3/second	329
Q	Heat energy	Joules	108
$\tilde{Q}$	Heat energy per unit volume	J/m^3	273
Q_I	Total ion dose	cm^{-2}	38
q	Distributed transverse load on a beam	Newtons/meter	207
q_e	Electronic charge: 1.6×10^{-19}	Coulombs	268
$q(t)$	Generalized displacement variable		105
R	Universal gas constant: 8.3×10^3	Joules/(kg-Kelvin)	319
R	Resistance of a resistor	Ohms	110
$\mathcal{R}$	Reluctance of magnetic circuit	Amperes/Weber	121
Re	Reynolds number		325
r_o	Incremental output resistance	Ohms	372
$R_p, \Delta R_p$	Projected range and its deviation	μm	40
s_1, s_2	Poles of a transfer function	radians/sec	116
S	Entropy	Joules/Kelvin	276
$\tilde{S}(\mathbf{r}, t)$	Source term per unit volume	various	300
$S_n(f)$	Spectral density function		435
S_{IJ}	Compliance coefficients	Pascals^{-1}	192
S/N	Signal-to-noise ratio	decibels	435
t	Time	sec, min, hr	
T	Temperature	Kelvins or °C	
T_{EM}, T_{ME}	Transduction parameters	Newtons/Ampere	143
$\hat{u}$	Trial displacement function	meters	244
$\mathbf{u}$	Vector of system inputs		151
$U, \mathbf{U}$	Velocity	meters/sec	143
$\bar{U}$	Average velocity	meters/sec	329
$d\mathcal{U}$	Differential internal energy	Joules	275
V, v	Voltage; potential difference	Volts	
$\bar{v}, \overline{v^2}$	Mean and mean-square voltage	Volts	434
v	Velocity	meters/sec	113
V	Volume	meters^3	
$\mathcal{V}$	Volume (when needed for clarity)	meters^3	273
$\hat{v}$	Trial displacement function	meters	244

Symbol	Description	Units	Page
$\mathbf{v}_i$	ith eigenvector		161
$\mathbf{V}$	Matrix of eigenvectors		163
V_{PI}	Pull-in voltage	Volts	135
V_{T_n}, V_{T_p}	MOSFET threshold voltages	Volts	369
W	Width of beam or structure	meters	201
$\mathcal{W}, \mathcal{W}^*$	Stored energy and co-energy	Joules (J)	110
$\tilde{\mathcal{W}}$	Stored energy density	Joules/meter^3	240
w	Displacement of beam or plate	meters	213
$\hat{w}$	Trial variational displacement solution	meters	244
$\dot{x}$	Velocity	meters/sec	25
$\mathbf{x}$	Vector of state variables		151
x_1, x_2	General state variables		118
x_f	Final oxide thickness (Deal-Grove)	μm	35
x_i	Initial oxide thickness (Deal-Grove)	μm	35
X_J	Width of space-charge layer	cm	358
$\mathbf{X}_{\mathrm{zir}}(s)$	Zero-input response (Laplace transform)		153
$\mathbf{X}_{\mathrm{zsr}}(s)$	Zero-state response (Laplace transform)		153
$Y, Y(s)$	Complex admittance	Siemens	586
$Z, Z(s)$	Complex impedance	Ohms	116
z	Displacement of mechanical element	meters	130
z_n	Charge number of charge carrier		291
Z_{EB}, Z_{MO}, Z_{MS}	Impedance parameters	Ohms	143

Appendix B
Electromagnetic Fields

B.1 Introduction

This Appendix contains a very brief review of electromagnetic fields at the freshman physics level. The emphasis is on the kinds of structures that appear in MEMS devices, such as parallel-plate capacitors and simple coils. For a more complete discussion, the reader is referred to standard undergraduate texts.

B.2 Quasistatic Fields

When device dimensions are much less than the wavelength of electromagnetic radiation at a particular frequency, then the response of the system at that frequency can be considered *quasistatic* in that the emission, transmission, or absorption of electromagnetic radiation can be ignored.[1] Most MEMS devices meet this criterion. Even those devices designed for use in optical systems behave quasistatically with respect to electric or magnetic forces used to actuate them so as to control position.

B.3 Elementary Laws

There are four elementary laws of electromagnetism: Gauss's Law, Ampere's Law, Faraday's Law, and the Law of Magnetic Flux Continuity. These laws are expressed in terms of six vector fields: the *electric field* $\mathcal{E}$, the *electric displacement* $\mathcal{D}$, the *dielectric polarization* $\mathcal{P}$, the *magnetic field* $\mathcal{H}$, the *magnetic flux density* $\mathcal{B}$, and the *magnetization* $\mathcal{M}$. These are defined below.

---

[1] This is a somewhat simplified statement of the criterion. See pp. 71-76 of [129] for the details.

B.4   Electroquasistatic Systems

An *electroquasistatic system* is a quasistatic system in which the effect of Faraday's Law of Induction on the electric fields can be neglected. The capacitor-like structures that we shall encounter, such as the parallel-plate capacitor of Fig. B.1, can be considered as electroquasistatic.

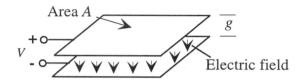

Figure B.1.   A parallel-plate capacitor in vacuum

The electric field is set up by charges. We can separate materials into conductors and insulators. In MEMS, we rarely need to consider the details of charge distributions inside materials. Therefore, for the purpose of this discussion, charge will exist as a thin sheet only at the surface of a conductor, and the electric field near the surface of a conductor is always perpendicular to that surface. Thus, if we consider a pair of parallel plates as in Fig. B.1 with a voltage V applied between them, we can reason that except near the edges of the plates, where there can be so-called *fringing fields*, the electric field is uniform between the plates with a direction perpendicular to the plate surfaces.

The magnitude of the electric field is determined by the applied voltage. The relation between field and voltage differences is expressed as a line integral of the electric field:

$$V(x_1) - V(x_2) = -\int_{x_2}^{x_1} \mathcal{E} \cdot d\mathbf{s} \tag{B.1}$$

Thus, in the parallel-plate example, $V = \mathcal{E}g$, where $\mathcal{E}$ is the magnitude of $\mathcal{E}$. To set up this electric field, the more positive plate has positive charge at its lower surface, and the more negative plate has negative charge at its upper surface. The relation between the charge density at a surface of a conductor and the electric field adjacent to the conductor is given by Gauss's Law which, in integral form, is

$$\int_{\text{closed surface}} \mathcal{D} \cdot d\mathbf{A} = Q \tag{B.2}$$

where Q is the total charge inside the closed surface, and where in vacuum $\mathcal{D} = \varepsilon_0 \mathcal{E}$. The quantity ε_0 is called the *permittivity of free space*. It is a fundamental constant with value 8.85×10^{-12} Farads/meter. Using Gauss's Law applied to a disk-shaped element which intersects one plate, we obtain the result that the surface charge density (in Coulombs/meter^2) on the surface of the positive electrode has magnitude $\varepsilon_0 \mathcal{E}$. The surface charge density on the other electrode is equal in magnitude, but negative.

When the space between the plates is filled with an insulating material other than vacuum (that is, with a *dielectric* medium), the bound charges inside the dielectric can displace slightly under the influence of the applied electric field, setting up a polarization $\mathcal{P}$. The physical interpretation of $\mathcal{P}$ is the average dipole moment per unit volume within the material. The polarization adds to the electric displacement:

$$\mathcal{D} = \varepsilon_0 \mathcal{E} + \mathcal{P} \tag{B.3}$$

In a *linear dielectric medium*, the polarization and electric field are proportional, permitting the more compact expression

$$\mathcal{D} = \varepsilon \mathcal{E} \tag{B.4}$$

where ε is the permittivity of the dielectric medium. The permittivity includes both the permittivity of free space and an added term (called the dielectric susceptibility) representing the polarization. The *dielectric constant* of a medium is the ratio of ε to ε_0, and is greater than unity in quasistatic systems. If we apply Gauss's Law to a capacitor that contains a dielectric medium, we find that the charge density required on each plate to set up the same electric field V/g has increased to $\varepsilon \mathcal{E}$, or $\varepsilon V/g$. This result is the starting point for the discussion of capacitors in Section 5.3.4.2.

B.5   Magnetoquasistatic Systems

A *magnetoquasistatic system* is a quasistatic system in which the effect time dependence of the electric displacement $\mathcal{D}$ on the magnetic field can be neglected. With this limitation, Ampere's Law in integral form becomes

$$\int_{\text{closed path}} \mathcal{H} \cdot d\mathbf{s} = \int_S \mathbf{J} \cdot d\mathbf{A} \tag{B.5}$$

where S is any surface bounded by the closed path and J is the current density through that surface. We illustrate Ampere's Law with a coil of N turns carrying current I (Fig. B.2).

The loops of magnetic field lines provide a pictorial representation for visualizing the field. The field strength is stronger where the lines are closer together, weaker where they are farther apart. If we apply Ampere's Law to this structure using the closed path suggested in Fig. B.2, we find that

$$\int_{\text{closed path}} \mathcal{H} \cdot d\mathbf{s} = NI \tag{B.6}$$

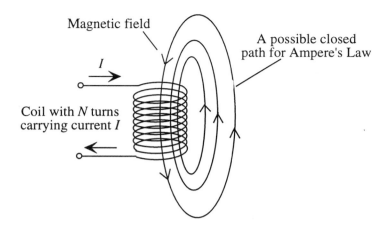

Figure B.2. An N-turn coil carrying current I sets up a magnetic field along the coil axis within the coil, and fringing fields that are oppositely directed outside the coil.

since the closed path is cut N times by a current density whose areal integral across the cross-section of each turn of coil wire is I. When making lumped-element models for magnetic systems, this line integral of $\mathcal{H}$ around a suitable closed path is called the *magnetomotive force* for that path, abbreviated MMF.

If we further assume that the coil is very long compared to its diameter, then the magnetic field strength inside the coil is much greater than that outside the coil, and we can approximate Ampere's Law as

$$\mathcal{H}_a \approx \frac{NI}{L} \qquad (\text{B.7})$$

where $\mathcal{H}_a$ is the axial field inside the coil, and L is the length of the coil.

In vacuum, the magnetic flux density is related to the magnetic field by $\mathcal{B} = \mu_0 \mathcal{H}$, where μ_0 is called the *permeability of free space*. It is a fundamental constant with value $4\pi \times 10^{-7}$ Henrys/meter. When a magnetic field is set up in a material that contains magnetic dipoles, these dipoles can align, setting up an internal magnetization $\mathcal{M}$. The physical interpretation of the magnetization is the average magnetic dipole moment per unit volume. When the effect of the magnetization is taken into account, the magnetic flux density becomes

$$\mathcal{B} = \mu_0(\mathcal{H} + \mathcal{M}) \qquad (\text{B.8})$$

Magnetization effects are most important in *ferromagnetic* materials, such as iron, nickel, and nickel-iron alloys, in which the magnetization can be very large in response to a small applied magnetic field. However, the magnetization of a ferromagnetic material cannot increase without limit; it saturates at some maximum value at high fields. Therefore, the relation between magnetic field and magnetization in ferromagnetics is nonlinear, and may also exhibit hysteresis. For small enough magnetic fields, though, we can approximate the relationship between magnetization and magnetic field as strictly linear. When we can do so, the magnetic flux density becomes

$$\mathcal{B} = \mu \mathcal{H} \tag{B.9}$$

where μ is the *permeability* of the material.

There is an equivalent to Gauss's Law for the magnetic case, but with the restriction that the "charge" for magnetic systems must always be zero. Thus, the Law of Continuity of Magnetic Flux states that the surface integral of the magnetic flux density through any closed surface must be zero. This means that the flux component normal to the surface of a material must be continuous.

We now consider Faraday's Law of Induction. An integral expression of this law is

$$\int_{\text{closed path}} \mathcal{E} \cdot d\mathbf{s} = - \int_S \frac{\partial}{\partial t} \mathcal{B} \cdot d\mathbf{A} \tag{B.10}$$

The surface integral of flux density is called the *flux* through that surface. The Law of Induction states that if the net flux through a surface S is changing with time, then there is an electric field set up along the bounding line of that surface, and the line-integral of that electric field is determined by the time-variation of the flux. This line integral of electric field is called the *electromotive force*, abbreviated EMF.

We have now reached the starting point for the discussion of inductors in Section 5.6.2.

Related Reading

H. A. Haus and J. R. Melcher, *Electromagnetic Fields and Energy*, Englewood Cliffs, NJ: Prentice Hall, 1989.

Appendix C
Elastic Constants in Cubic Material

In Chapter 8 it was demonstrated that the stiffness matrix in cubic material, when expressed in the cubic coordinate system, has three independent non-zero components, C_{11}, C_{12}, and C_{44}. The compliance matrix is the inverse of the stiffness matrix. It, too, has three independent components, S_{11}, S_{12}, and S_{44}.

When seeking the elastic behavior in directions or orientations other than the three principal cubic axes, tensor transformations are required [53]. Brantley [101] has applied these transformations for cubic materials, and has published tables of elastic constants for several cubic semiconductors. His formulae for obtaining Young's modulus, the Poisson ratio, and the biaxial modulus are presented here.

For an arbitrary crystallographic direction with directions cosines (l_1, l_2, l_3) with respect to the cubic axes, the Young's modulus is given by

$$\frac{1}{E} = S_{11} - 2(S_{11} - S_{12} - \frac{1}{2}S_{44})(l_1^2 l_2^2 + l_2^2 l_3^2 + l_1^2 l_3^2) \qquad \text{(C.1)}$$

The Poisson ratio in an isotropic material is given by $-S_{12}/S_{11}$. In a cubic material, in a coordinate system with two orthogonal directions with direction cosines (l_1, l_2, l_3) and (m_1, m_2, m_3) with respect to the cubic axes, the Poisson ratio is

$$\nu = -E \left[S_{12} + (S_{11} - S_{12} - \frac{1}{2}S_{44})(l_1^2 m_1^2 + l_2^2 m_2^2 + l_3^2 m_3^2) \right] \qquad \text{(C.2)}$$

Silicon wafers typically have {100} surface normals. Assuming that the x-axis is normal to the plane, then for in-plane directions between [010] and [001], the expressions for Young's modulus and Poisson ratio are

$$\frac{1}{E_{(100)}} = S_{11} - 2S\, l_2^2 l_3^2 \qquad (C.3)$$

and

$$\nu_{(100)} = -E_{(100)}\, (S_{12} - 2S\, l_2 l_3 m_2 m_3) \qquad (C.4)$$

where $S = (S_{11} - S_{12} - S_{44}/2)$.

The biaxial modulus in a (100) plane is independent of orientation, and has the value

$$\left(\frac{E}{1-\nu}\right)_{(100)} = \frac{1}{S_{11} + S_{12}} \qquad (C.5)$$

Finally, one can compute various types of isotropic averages over the elastic constants in order to estimate the Young's modulus for a random polycrystalline material. Here, the averaging method has some impact on the results. If we perform the angular average of the expression for $1/E$, we obtain

$$\left.\frac{1}{E}\right|_{avg} \approx 0.6 S_{11} + 0.4 S_{12} + .25 S_{44} \qquad (C.6)$$

Angular averaging of the Poisson ratio is more complicated, as the relative directions of the two axes enters. However, a good first estimate of the Poisson ratio of most amorphous or polycrystalline inorganic materials is a value of 0.25.

References

[1] Systems Planning Corporation, "MicroElectroMechanical Systems (MEMS), An SPC Market Study," January, 1999.

[2] K. Petersen, "Bringing MEMS to market," in *Technical Digest, Solid-State Sensors and Actuator Workshop*, (Hilton Head, SC), pp. 60–64, June 2000.

[3] S. A. Campbell, *The Science and Engineering of Microelectronic Fabrication*. New York: Oxford University Press, 1996.

[4] R. A. Colclaser, *Microelectronics: Processing and Device Design*. New York: Wiley, 1980.

[5] Fairchild Corporation, *Semiconductor & Integrated Circuit Fabrication Techniques*. Reston, VA: Reston, 1979.

[6] S. M. Sze, *Semiconductor Devices: Physics and Technology*. New York: Wiley, 1985.

[7] S. Wolf and R. N. Tauber, *Silicon Processing for the VLSI Era*, vol. 1: Process Technology. Sunset Beach, CA, USA: Lattice Press, second ed., 2000.

[8] M. Madou, *Fundamentals of Microfabrication*. New York: CRC Press, 1997.

[9] G. T. A. Kovacs, *Micromachined Transducers Sourcebook*. New York: WCB McGraw-Hill, 1998.

[10] W. P. Mason, *Piezoelectric Crystals and Their Application to Ultrasonics*. New York: Van Nostrand, 1950.

[11] T. Ikeda, *Fundamentals of Piezoelectricity*. New York: Oxford University Press, 1990.

[12] A. B. Frazier and M. G. Allen, "Metallic microstructures fabricated using photosensitive polyimide electroplating molds," *J. Microelectromech. Syst.*, vol. 2, pp. 87–94, 1993.

[13] H. Guckel, "High-aspect-ratio micromachining via deep X-ray lithography," *Proc. IEEE*, vol. 86, pp. 1586–1593, 1998.

[14] M. A. Schmidt, "Wafer-to-wafer bonding for microstructure formation," *Proc. IEEE*, vol. 86, pp. 1575–1585, 1998.

[15] L. Parameswaran, C. H. Hsu, and M. A. Schmidt, "IC process compatibility of sealed cavity sensors," in *Proc. 1997 International Conference on Solid-State Sensors and Actuators (TRANDSUCERS '97)*, (Chicago), pp. 625–628, June 1997.

[16] Y. Xia and G. M. Whitesides, "Soft lithography," *Angew. Chem. Int. Ed.*, vol. 37, pp. 550–575, 1998.

[17] W. Kern and C. A. Deckert, "Chemical etching," in *Thin Film Processes* (J. L. Vossen and W. Kern, eds.), pp. 401–496, New York: Academic Press, 1978.

[18] K. R. Williams and R. S. Muller, "Etch rates for micromachining processing," *J. Microelectromechanical Systems*, vol. 5, pp. 256–269, 1996. The tables of data from this paper are also available at the web site of the Berkeley Sensors and Actuators Center, http://www-bsac.EECS.Berkeley.EDU/db/.

[19] G. T. A. Kovacs, N. I. Maluf, and K. E. Petersen, "Bulk micromachining of silicon," *Proc. IEEE*, vol. 86, pp. 1536–1551, 1998.

[20] J. M. Bustillo, R. T. Howe, and R. S. Muller, "Surface micromachining for microelectromechanical systems," *Proc. IEEE*, vol. 86, pp. 1552–1574, 1998.

[21] M. A. Schmidt, R. T. Howe, S. D. Senturia, and J. H. Haritonidis, "Design and calibration of a microfabricated floating-element shear-stress sensor," *IEEE Trans. Elec. Dev.*, vol. 35, pp. 750–757, 1988.

[22] P. B. Chu, J. T. Chen, R. Yeh, G. Lin, J. C. P. Huang, B. A. Warneke, and K. S. J. Pister, "Controlled pulse-etching with xenon diflouride," in *Proc. 1997 Int'l. Conf. Solid-State Sensors and Actuators (TRANSDUCERS '97)*, (Chicago), pp. 665–668, June 1997.

[23] P. F. Van Kessel, L. J. Hornbeck, R. E. Meier, and M. R. Douglass, "A MEMS-based projection display," *Proc. IEEE*, vol. 98, pp. 1687–1704, 1998.

[24] C. Hedlund, U. Lindberg, U. Bucht, and J. Söderkvist, "Anisotropic etching of Z-cut quartz," *J. Micromechanics and Microeng.*, vol. 3, pp. 65–73, 1993.

[25] H. Seidel, L. Csepregi, A. Heuberger, and H. Baumgärtel, "Anisotropic etching of crystalline silicon in alkaline solutions I: Orientation dependence and behavior of passivation layers," *J. Electrochem. Soc.*, vol. 137, pp. 3612–3626, 1990.

[26] H. Seidel, L. Csepregi, A. Heuberger, and H. Baumgärtel, "Anisotropic etching of crystalline silicon in alkaline solutions II: Influence of dopants," *J. Electrochem. Soc.*, vol. 137, pp. 3626–3632, 1990.

[27] H. D. Goldberg, K. S. Breuer, and M. A. Schmidt, "A silicon wafer bonding technology for microfabricated shear-stress sensors with backside contacts," in *Technical Digest, IEEE Solid-State Sensor and Actuator Workshop*, (Hilton Head, SC), pp. 111–115, June 1994.

[28] F. Lärmer and P. Schilp, "Method of anisotropically etching silicon," German Patent DE 4241045, 1994.

[29] L. G. Fréchette, S. A. Jacobson, K. S. Breuer, F. F. Ehrich, R. Ghodssi, R. Khanna, C. W. Wong, X. Zhang, M. A. Schmidt, and A. H. Epstein, "Demonstration of a microfabricated high-speed turbine supported on gas bearings," in *Technical Digest, IEEE Solid-State Sensor and Actuator Workshop*, (Hilton Head, SC), pp. 43–47, June 2000.

[30] C. A. Spindt, I. Brodie, L. Humphrey, and E. R. Westerberg, "Physical properties of thin-film field emission cathodes with molybdenum cones," *J. Appl. Phys.*, vol. 47, p. 5248, 1976.

[31] H. Lorenz, M. Despont, N. Fahrni, N. LaBianca, P. Renaud, and P. Vettiger, "SU-8: A low-cost negative resist for MEMS," *J. Micromech. Microeng.*, vol. 7, pp. 121–124, 1997.

[32] M. A. Schmidt, "Microelectronic fabrication," Massachusetts Institute of Technology, 1999. Chapter 3 of MIT Course Notes for 6.77S, Microsystems: Mechanical, Chemical, Optical.

[33] C. H. Hsu, *Silicon Microaccelerometer Fabrication Technologies*. PhD thesis, Massachusetts Institute of Technology, September, 1977.

[34] N. Hogan, "Integrated modeling of physical system dynamics," Massachusetts Institute of Technology, 1994. Class Notes for Subject 2.141, "Modeling and Simulation of Dynamic Systems.".

[35] H. A. C. Tilmans, "Equivalent circuit representation of electromechanical transducers I: Lumped-parameter systems," *J. Micromech. Microeng.*, vol. 6, pp. 157–176, 1996. Erratum in *J. Micromech. Microeng.* vol. 6., p. 359, 1996.

[36] H. A. C. Tilmans, "Equivalent circuit representation of electromechanical transducers II: Distributed-parameter systems," *J. Micromech. Microeng.*, vol. 7, pp. 285–309, 1997.

[37] L. Beranek, *Acoustics*. New York: McGraw-Hill, 1954.

[38] B. F. Romanowicz, *Methodology for the Modeling and Simulation of Microsystems*. Boston: Kluwer Academic Press, 1998.

[39] S. D. Senturia and B. D. Wedlock, *Electronic Circuits and Applications*. New York: Wiley, 1975. Now available from Krieger reprints.

[40] C. A. Desoer and E. S. Kuh, *Basic Circuit Theory*. New York: McGraw-Hill, 1969.

[41] F. V. Hunt, *Electroacoustics: The Analysis of Transduction, and Its Historical Background*. New York: American Institute of Physics, 1982.

[42] M. Rossi, *Acoustics and Electroacoustics*. Norwood, MA: Artech House, 1988. Translation by Patrick Rupert Windsor Roe.

[43] W. M. Siebert, *Circuit, Signals, and Systems*. Cambridge, MA: MIT Press, 1986.

[44] R. N. Bracewell, *The Fourier Transform and its Applications*. New York: McGraw-Hill, 1978.

[45] S. H. Strogatz, *Nonlinear Dynamics and Chaos*. Reading, MA: Addison-Wesley, 1994.

[46] L. M. Castañer and S. D. Senturia, "Speed-energy optimization of electrostatic actuators based on pull-in," *J. Microelectromechanical Systems*, vol. 8, pp. 290–298, 1999.

[47] A. H. Nayfeh and D. T. Mook, *Nonlinear Oscillations*. New York: Wiley, 1979.

[48] S. P. Timoshenko and J. N. Goodier, *Theory of Elasticity*. New York: McGraw-Hill, third ed., 1970. Reissued in 1987.

[49] O. Sigmund, "Topology optimization: a tool for the tailoring of structures and materials," *Phil. Trans. Royal Soc. London, Series A*, vol. 358, pp. 211–227, Jan 15, 2000.

[50] J. M. Gere and S. P. Timoshenko, *Mechanics of Materials*. Monterey, CA: Brooks/Cole Engineering Division, second ed., 1984.

[51] W. M. Lai, D. Rubin, and E. Krempl, *Introduction to Continuum Mechanics*. Oxford: Pergamon Press, third ed., 1993.

[52] D. R. Lide, ed., *Handbook of Chemistry and Physics*. Boca Raton, FL: CRC Press, 78th ed., 1997.

[53] J. F. Nye, *Physical Properties of Crystals*. London: Oxford University Press, 1957.

[54] R. R. Tummala and E. J. Rymaszewski, eds., *Microelectronics Packaging Handbook*. New York: Van Nostrand Reinhold, 1989.

[55] American Institute of Physics, *American Institute of Physics Handbook*. New York: McGraw-Hill, 1972.

[56] C. H. Mastrangelo, Y.-C. Tai, and R. S. Muller, "Thermophysical properties of low-residual-stress, silicon-rich LPCVD silicon nitride films," *Sensors & Actuators*, vol. A23, pp. 856–880, 1990.

[57] S. P. Timoshenko and S. Woinowsky-Krieger, *Theory of Elasticity*. New York: McGraw-Hill, second ed., 1959. Reissued in 1987.

[58] K. Bathe, *Finite Element Procedures in Engineering Analysis*. Englewood Cliffs, NJ: Prentice Hall, 1982.

[59] P. Lin and S. D. Senturia, "The in-situ measurement of biaxial modulus and residual stress of multi-layer polymeric thin films," in *Thin Films: Stresses and Mechanical Properties II* (M. F. Doerner, W. Oliver, G. M. Pharr, and F. R. Brotzen, eds.), vol. 188, (San Francisco, CA), pp. 41–46, Materials Resarch Society Symposium Proceedings, April 16-19 1990.

[60] J. Y. Pan, F. Maseeh, P. Lin, and S. D. Senturia, "Verification of FEM analysis of load-deflection methods for measuring mechanical properties of thin films," in *Technical Digest: Solid-State Sensor and Actuator Workshop*, (Hilton Head, SC), pp. 55–61, June 1990.

[61] W. M. Deen, *Analysis of Transport Phenomena*. New York: Oxford University Press, 1998.

[62] F. P. Incropera and D. P. DeWitt, *Fundamentals of Heat and Mass Transfer*. New York: Wiley, fourth ed., 1996.

[63] R. B. Lindsay, *Introduction to Physical Statistics*. New York: Wiley, 1941.

[64] L. D. Landau and E. M. Lifshitz, *Stastical Physics*. Reading, MA: Addison-Wesley, 1958.

[65] A. D. Khazan, *Transducers and Their Elements*. Englewood Cliffs, NJ: PTR Prentice Hall, 1994.

[66] A. J. Kinlock, *Adhesion and Adhesives: Science and Technology*. London: Chapman Hall, 1987.

[67] A. van der Ziel, *Solid State Physical Electronics*. Englewood Cliffs, NJ: Prentice-Hall, third ed., 1976.

[68] A. H. Epstein and S. D. Senturia, "Macro power from micro macinery," *Science*, vol. 276, p. 1211, 1997.

[69] A. London, *Development and Test of a Microfabricated Bipropellant Rocket Engine*. PhD thesis, Massachusetts Institute of Technology, June 2000.

[70] J. A. Fay, *Introduction to Fluid Mechanics*. Cambridge, MA: MIT Press, 1994.

[71] A. M. Kuethe and C. Chow, *Foundations of Aerodynamics*. New York: Wiley, fifth ed., 1998.

[72] W. S. Griffin, H. H. Richardson, and S. Yamanami, "A study of fluid squeeze-film damping," *J. Basic Engineering*, pp. 451–456, June,1966.

[73] J. J. Blech, "On isothermal squeezed films," *J. Lubrication Technology*, vol. 105, pp. 615–620, 1983.

[74] A. Burgdorfer, "The influence of the molecular mean free path on the performance of hydrodybnamic gas lubricated bearings," *J. Basic Engineering*, pp. 94–100, March, 1959.

[75] A. Beskok and G. E. M. Karniadakis, "A model for rarefied internal gas flows," *J. Fluid Mechanics*, vol. X, pp. 1–37, 1996.

[76] T. Veijola, H. Kuisma, and J. Lahdenperä, "Model for gas film damping in a silicon accelerometer," in *Proc. Int'l. Conf. Solid-State Sensors and Actuators (TRANSDUCERS '97)*, (Chicago), pp. 1097–1100, June 1997.

[77] H. Seidel, H. Riedel, R. Kolbeck, G. Mück, W. Kupke, and M. Königer, "Capacitive silicon accelerometer with highly symmetrical design," *Sensors and Actuators*, vol. A21-A23, pp. 312–315, 1990.

[78] M. Andrews, I. Harris, and G. Turner, "A comparison of squeeze-film theory with measurements on a microstructure," *Sensors and Actuators*, vol. A36, pp. 79–87, 1993.

[79] T. Veijola, H. Kuisma, J. Lahdenperä, and T. Ryhänen, "Equivalent-circuit model of the squeezed gas film in a silicon accelerometer," *Sensors and Actuators*, vol. A48, pp. 239–248, 1995.

[80] Y.-J. Yang, M.-A. Grétillat, and S. D. Senturia, "Effects of air damping on the dynamics of nonuniform deformation of microstructures," in *Proc. Int'l. Conf. Solid-State Sensors and Actuators (TRANSDUCERS '97)*, (Chicago), pp. 1093–1096, June 1997.

[81] A. J. Grodzinsky and M. L. Yarmush, "Electrokinetic separations," in *Biotechnology* (H.-J. Rehm and G. Reed, eds.), vol. 3, Bioprocessing, pp. 680–693, Weinheim: VCH Ferlagsgesellschaft mbH, second ed., 1993.

[82] D. J. Harrison, K. Fluri, K. Seiler, Z. Fan, C. S. Effenhauser, and A. Manz, "Micromachining a miniaturized capillary electropohoresis-based chemical analysis system on a chip," *Science*, vol. 261, pp. 895–897, 1993.

[83] R. T. Howe and C. G. Sodini, *Microelectronics: An Integrated Approach*. Upper Saddle River, NJ: Prentice Hall, 1997.

[84] S. M. Sze, *Physics of Semiconductor Devices*. New York: Wiley, second ed., 1981.

[85] S. D. Senturia, N. F. Sheppard, Jr., S. L. Garverick, H. L. Lee, and D. R. Day, "Microdielectrometry," *Sensors and Actuators*, vol. 2, pp. 263–274, 1982.

[86] M. S. Adler, S. D. Senturia, and C. R. Hewes, "Sensitivity of marginal oscillator spectrometers," *Rev. Sci. Instr.*, vol. 42, pp. 704–712, 1971.

[87] A. van der Ziel, *Noise: Sources, Characterization, Measurement*. Englewood Cliffs, NJ: Prentice-Hall, 1970.

[88] P. R. Gray and R. G. Meyer, *Analysis and Design of Analog Integrated Circuits*. New York: Wiley, second ed., 1984.

[89] D. Monk. Personal communication.

[90] L. Ristic, ed., *Sensor Technology and Devices*. Boston: Artech House, 1994.

[91] N. Maluf, *An Introduction to Microelectromechanical Systems Engineering.* Boston: Artech House, 2000.

[92] Motorola, *Sensor Device Data/Handbook.* Phoenix, AZ: Motorola, Inc., fourth ed., 1998.

[93] D. J. Monk and M. K. Shah, "Packaging and testing considerations for commercialization of bulk micromachined piezoresistive pressure sensors," in *Proc. Commercialization of Microsystems '96* (S. Walsh, ed.), (Kona HA), pp. 136–149, October 1996.

[94] J. L. McKechnie, ed., *Webster's New Universal Unabridged Dictionary.* Dorset & Baber, second ed., 1984.

[95] K. Matsuda, K. Suzuki, K. Yamamura, and Y. Kanda, "Nonlinear piezoresistance effects in silicon," *J. Applied Physics,* vol. 73, pp. 1838–1847, 1993.

[96] W. P. Mason and R. N. Thurston, "Use of piezoresistive materials in the measurement of displacement, force, and torque," *J. Acoustical Soc. Amer.,* vol. 29, pp. 1096–1101, 1957.

[97] O. N. Tufte and D. Long, "Recent developments in semiconductor piezoresistive devices," *Solid-State Electronics,* vol. 6, pp. 323–338, 1963.

[98] C. S. Smith, "Piezoresistance effect in germanium and silicon," *Physical Review,* vol. 94, pp. 42–29, 1954.

[99] K. Petersen, P. Barth, J. Poydock, J. Brown, J. Mallon, Jr., and J. Bryzek, "Silicon fusion bonding for pressure sensors," in *Technical Digest, Solid-State Sensors and Actuator Workshop,* (Hilton Head, SC), pp. 144–147, June 1988. Results on maximum stress location were shown only in the oral presentation.

[100] P. J. French and A. G. R. Evans, "Polysilicon strain sensors using shear piezoresistance," *Sensors and Actuators,* vol. 15, pp. 257–272, 1988.

[101] W. A. Brantley, "Calculated elastic constants for stress problems associated with semiconductor devices," *J. Appl. Phys.,* vol. 44, pp. 534–535, 1973.

[102] G. Bitko, A. McNeil, and R. Frank, "Improving the MEMS pressure sensor," *Sensors Magazine,* pp. 62–67, July 2000.

[103] S. K. Clark and K. D. Wise, "Pressure sensitivity in anisotropically etched thin-diaphragm pressure sensors," *IEEE Trans. Electron Devices,* vol. ED-26, pp. 1887–1896, 1979.

[104] J. E. Gragg, W. E.McCulley, W. B. Newton, and C. E. Derrington, "Compensation and calibration of a monolithic four terminal silicon pressure transducer," in *Tech. Digest, Solid-State Sensors and Actuators Workshop*, (Hilton Head, SC), pp. 21–27, June 1984.

[105] NovaSensor, *Silicon Sensors and Microstructures*. Fremont, CA: NovaSensor, 1990.

[106] Y. Kanda, "Piezoresistance effect of silicon," *Sensors and Actuators*, vol. A28, pp. 83–91, 1991.

[107] Microcosm Technologies, "Damping analysis of the ADI XL76," March 2000. Unpublished internal memorandum.

[108] Analog Devices, "iMEMS II short-course notes," April 1994.

[109] C. M. Roberts, Jr., L. H. Long, and P. A. Ruggerio, "Method for separating circuit dies from a wafer." U. S. Patent 5,362,681, Nov. 8 1994.

[110] S. F. Bart and H. R. Samuels, "Design techniques for minimizing manufacturing variations in surface micromachined accelerometers," in *Vol. 59: Microelectromechanical Systems (Mems)*, no. G01036 in DSC, pp. 427–433, ASME International, 1996.

[111] C.-H. Liu, A. M. Barzilai, J. K. Reynolds, A. Partridge, T. W. Kenny, J. D. Grade, and H. K. Rockstad, "Characterization of a high-sensitivity micromachined tunneling accelerometer with micro-g resolution," *J. Microelectromechanical Systems*, vol. 7, pp. 235–244, 1998.

[112] C.-H. Liu, H. K. Rockstad, and T. W. Kenny, "Robust controller design via μ-synthesis for high-performance micromachined tunneling accelerometers," in *Proc. American Control Conference*, (San Diego, CA), pp. 247–252, June 1999.

[113] C.-H. Liu and T. W. Kenny, "A high-precision wide-bandwidth micromachined tunneling accelerometer," *J. Microelectromechanical Systems*, vol. 9, September 2000. In press.

[114] S. J. Walker and D. J. Nagel, "Optics and MEMS," Tech. Rep. NRL/MR/6336–99-7975, Naval Res. Lab., Washington DC, 1999. PDF version available at http://mstd.nrl.navy.mil/6330/6336/moems.htm.

[115] O. Solgaard, F. S. A. Sandejas, and D. M. Bloom, "Deformable grating optical modulator," *Optics Letters*, vol. 17, pp. 688–690, 1992.

[116] L. J. Hornbeck, "Deformable-mirror spatial light modulators," *SPIE Critical Reviews Series*, vol. 1150, pp. 86–102, 1989.

[117] P. M. Osterberg and S. D. Senturia, "M-Test: A test chip for MEMS material property measurement using electrostatically actuated test structures," *J. Microelectromechanical Systems*, vol. 6, pp. 107–118, 1997.

[118] E. S. Hung and S. D. Senturia, "Extending the travel range of analog-tuned electrostatic actuators," *J. Microelectromechanical Systems*, vol. 8, pp. 497–505, 1999.

[119] R. K. Gupta, "Material property measurement of micromechanical polysilicon beams," in *Proc. SPIE Conf. on Microlithography and Metrology in Micromachining II*, (Austin, TX), pp. 39–45, October 1996.

[120] E. R. Deutsch, "Development of calibration standards for accurate measurement of geometry in microelectromechanical systems," Master's thesis, Massachusetts Institute of Technology, 1998.

[121] A. P. Payne, B. Staker, C. S. Gudeman, M. Daneman, and D. E. Peter, "Resonance measurements of stresses in Al/Si_3N_4 micro-ribbons," in *MEMS Reliability for Critical and Space Applications, SPIE Proceedings*, vol. 2880, pp. 90–100, 1999.

[122] D. Amm. Personal communication.

[123] C. S. Gudeman, B. Staker, and M. Daneman, "Squeeze film damping of doubly supported ribbons in noble gas atmospheres," in *Tech. Digest, Solid-State Sensors and Actuators Workshop*, (Hilton Head, SC), pp. 288–291, June 1998.

[124] H. Goldstein, *Classical Mechanics*. Reading, MA: Addison-Wesley, 1950.

[125] N. Yazdi, F. Ayazi, and K. Najafi, "Micromachined inertial sensors," *Proc. IEEE*, vol. 86, pp. 1640–1659, 1998.

[126] D. S. Ballantine, R. M. White, S. J. Martin, A. J. Ricco, E. T. Zellers, G. C. Frye, and H. Wohltjen, *Acoustic Wave Sensors: Theory, Design, and Physico-Chemical Applications*. Boston: Academic Press, 1997.

[127] A. Nathan and H. Baltes, *Microtransducer CAD: Physical and Computational Aspects*. Vienna: Springer, 1999.

[128] M. Tabib-Azar, *Microactuators: Electrical, Magnetic, Thermal, Optical, Mechanical, Chemical and Smart Structures*. Boston: Kluwer Academic Press, 1998.

[129] H. A. Haus and J. R. Melcher, *Electromagnetic Fields and Energy*. Englewood Cliffs, NJ: Prentice Hall, 1989.

[130] P. K. Gupta and C. E. Jenson, "Rotation rate sensor with center mounted tuning fork." U. S. Patent 5,396,114, March 7 1995.

[131] C. H. Mastrangelo, M. A. Burns, and D. T. Burke, "Microfabricated devices for genetic diagnostics," *Proc. IEEE*, vol. 86, pp. 1769–1787, 1998.

[132] S. F. Brown, "Good-bye, test tubes. hello, labs-on-a-chip," *Fortune Magazine*, pp. 282[C]–282[J], October 11 1999.

[133] J. D. Watson, *The Double Helix*. New York: Atheneum, 1968.

[134] M. A. Northrup, M. T. Ching, R. M. White, and R. T. Watson, "DNA amplification with a microfabricated reaction chamber," in *Proc. 1997 International Conference on Solid-State Sensors and Actuators (TRANS-DUCERS '93)*, (Yokohama), pp. 924–926, June 1993.

[135] J. H. Daniel, S. Iqbal, R. B. Millington, D. F. Moore, C. R. Lowe, D. L. Leslie, M. A. Lee, and M. J. Pearce, "Silicon microchambers for DNA amplification," *Sensors and Actuators*, vol. A71, pp. 81–88, 1998.

[136] M. U. Kopp, A.J. de Mello, and A. Manz, "Chemical amplification: continuous-flow PCR on a chip," *Science*, vol. 280, pp. 1046–1048, 1998.

[137] R. P. Manginell, J. H. Smith, A. J. Ricco, D. J. Moreno, R. C. Hughes, R. J. Huber, and S. D. Senturia, "Selective, pulsed CVD of platinum on microfilament gas sensors," in *Tech. Digest, Solid-State Sensors and Actuators Workshop*, (Hilton Head, SC), pp. 23–27, June 1996.

[138] R. P. Manginell, J. H. Smith, A. J. Ricco, R. C. Hughes, D. J. Moreno, and R. J. Huber, "In-situ monitoring of micro-chemical vapor deposition (μ-CVD): Experimental results and SPICE modeling," in *Tech. Digest, Solid-State Sensors and Actuators Workshop*, (Hilton Head, sC), pp. 371–374, June 1998.

[139] R. P. Manginell, J. H. Smith, A. J. Ricco, R. C. Hughes, D. J. Moreno, and R. J. Huber, "Electro-thermal modeling of a microbridge gas sensor," in *SPIE Symposium on Micromaching and Microfabrication*, (Austin, TX), pp. 360–371, Sept. 1997.

[140] J. L. Rodriguez, R. C. Hughes, W. T. Corbett, and P. J. McWhorter, "Robust, wide range hydrogen sensor," in *IEEE International Electron Devices Meeting Technical Digest*, (San Francisco), pp. 521–524, December 1992.

[141] R. P. Manginell, *A Polysilicon Microbridge Gas Sensor*. PhD thesis, University of New Mexico, 1997. PDF version vailable at www.mdl.sandia.gov/micromachine/biblog_sensors.html.

Index